Biological Management of Diseases of Crops

Progress in Biological Control

Volume 16

Published:

Volume 7
Jonathan G. Lundgren:
Relationships of Natural Enemies and Non-Prey Foods. 2008
ISBN: 978-1-4020-9234-3

Volume 8
S.S. Gnanamanickam:
Biological Control of Rice Diseases
ISBN: 978-90-481-2464-0

Volume 9
F.L. Cônsoli, J.R.P. Parra and R.A. Zucchi (eds.):
Egg Parasitoids in Agroecosystems with Emphasis on *Trichogramma*
ISBN: 978-1-4020-9109-4

Volume 10
W.J. Ravensberg:
A Roadmap to the Successful Development and Commercialization of Microbial Pest Control Products for Control of Arthropods
ISBN: 978-94-007-0436-7

Volume 11
K. Davies and Y. Spiegel (eds.):
Biological Control of Plant-Parasitic Nematodes. 2011
ISBN: 978-1-4020-9647-1

Volume 12
J.M. Mérillon and K.G. Ramawat (eds.):
Plant Defence: Biological Control
ISBN: 978-94-007-1932-3

Volume 13
H. Roy, P. De Clercq, L.J. Lawson Handley, J.J. Sloggett, R.L. Poland and E. Wajnberg (eds.):
Invasive Alien Arthropod Predators and Parasitoids: An Ecological Approach
ISBN: 978-94-007-2708-3

Volume 14
G. Smagghe and I. Diaz (eds.):
Arthropod-Plant Interactions: Novel Insights and Approaches for IPM
ISBN: 978-94-007-3872-0

Volume 15
P. Narayanasamy:
Biological Management of Diseases of Crops: Volume 1: Characteristics of Biological Control Agents
ISBN: 978-94-007-6379-1

For further volumes:
http://www.springer.com/series/6417

Biological Management of Diseases of Crops

Volume 2: Integration of Biological Control Strategies with Crop Disease Management Systems

P. Narayanasamy

Former Professor and Head, Department of Plant Pathology
Tamil Nadu Agricultural University, Coimbatore, India

Springer

P. Narayanasamy
Grand Parade Apartments
32 D Thilagar Street, R. S. Puram
Coimbatore 641 002
India
pnsamyraj@gmail.com

ISBN 978-94-007-6376-0 ISBN 978-94-007-6377-7 (eBook)
DOI 10.1007/978-94-007-6377-7
Springer Dordrecht Heidelberg New York London

Library of Congress Control Number: 2013932136

Printed on acid-free paper

Springer is part of Springer Science+Business Media (www.springer.com)

Dedicated to the
Memory of My Parents for
Their Love and Affection

Progress in Biological Control

Series Preface

Biological control of pests, weeds, and plant and animal diseases utilising their natural antagonists is a well-established and rapidly evolving field of science. Despite its stunning successes world-wide and a steadily growing number of applications, biological control has remained grossly underexploited. Its untapped potential, however, represents the best hope to providing lasting, environmentally sound, and socially acceptable pest management. Such techniques are urgently needed for the control of an increasing number of problem pests affecting agriculture and forestry, and to suppress invasive organisms which threaten natural habitats and global biodiversity.

Based on the positive features of biological control, such as its target specificity and the lack of negative impacts on humans, it is the prime candidate in the search for reducing dependency on chemical pesticides. Replacement of chemical control by biological control – even partially as in many IPM programs – has important positive but so far neglected socio-economic, humanitarian, environmental and ethical implications. Change from chemical to biological control substantially contributes to the conservation of natural resources, and results in a considerable reduction of environmental pollution. It eliminates human exposure to toxic pesticides, improves sustainability of production systems, and enhances biodiversity. Public demand for finding solutions based on biological control is the main driving force in the increasing utilisation of natural enemies for controlling noxious organisms.

This book series is intended to accelerate these developments through exploring the progress made within the various aspects of biological control, and via documenting these advances to the benefit of fellow scientists, students, public officials, policy-makers, and the public at large. Each of the books in this series is expected to provide a comprehensive, authoritative synthesis of the topic, likely to stand the test of time.

Heikki M.T. Hokkanen, Series Editor

Preface

Various crop plants have been domesticated, after careful selection from innumerable wild plant species over several millennia, because of their potential for higher yield and better quality of the produce. Crop production systems have been developed primarily to fulfill philosophic and economic objectives of feeding humans and animals and providing better livelihood for the growers. Microbial plant pathogens continue to be a scourge of mankind from the prehistoric period, as the causative agents of numerous devastating diseases of plants that provide food, feed, fiber and all other materials essentially required for man and animals. Continuous and sustained efforts have been made to minimize the quantitative and qualitative losses of crops due to diseases incited by the microbial plant pathogens – oomycetes, fungi, bacteria, phytoplasmas, viruses and viroids – in different ecosystems. Managing crop diseases through development of cultivars resistant to diseases has been successful only to a limited extent, because of the unavailability of dependable sources of resistance genes for incorporation into susceptible cultivars. Application of chemicals is being practiced for several centuries and selective chemicals with systemic action could provide protection against microbial pathogens for short periods only. Development of resistance in plant pathogens to chemicals, accumulation of chemical residues in grains and food materials and environmental pollution due to indiscriminate use of chemicals, gave negative signals for their continued use for crop protection. Biological management of crop diseases has emerged as an attractive, alternative approach for minimizing the incidence and severity of diseases of crops caused by microbial pathogens.

Biological management of crop diseases involves the utilization of biotic and/or abiotic agents that act through one or more mechanisms to reduce the infection potential of microbial pathogens directly and/or indirectly by activating the host defense systems to reduce the disease incidence and/intensity. The biotic agents include oomycetes, fungi, bacteria and viruses that suppress the development of crop diseases caused by microbial pathogens in various crops. The abiotic agents such as solar energy, heat, ultraviolet light, organic amendments, organic and inorganic compounds and naturally-derived substances of plant and animal origin also possess the ability to restrict the development of crop pathogens through direct and/indirect effects, as the biotic agents. Although innumerable biotic and abiotic agents have been demonstrated to have high level of biocontrol activity in in vitro assays, very few have been found to have the expected level of biosuppressive activity under field conditions where they have to compete with the pathogens for the available nutrients and niche for survival. In the recent years, due to growing concern and awareness for protecting the environment and the need for providing chemical-free food to the consumers, several formulations based on microbial antagonists and resistance inducers (plant activators) have been commercialized. Consistency in their performance under various agroclimatic zones is one of the major requirements to make the production of biocontrol products as a viable economic industrial venture. The advantages of biocontrol strategies and their limitations are discussed in detail in two volumes of this treatise. The investigations to study the nature and characteristics of the biological control agents are presented in the first volume. The possibilities of integrating different biological strategies with crop disease management systems are highlighted in the second volume.

The information presented in this book represents extensive literature search (over 2,500 citations) and it is aimed to provide a comprehensive knowledge to the upper level undergraduate and graduate students, researchers and teachers associated with teaching courses as a component of biological management of diseases of crops in the Departments of Plant Pathology, Plant Protection, Microbiology, Molecular Biology, Botany, Ecology, Agriculture and Horticulture. Provision of several protocols appended at the end of relevant chapters to assist the researchers in planning their experiments is a unique feature of this book.

Coimbatore, India P. Narayanasamy

Acknowledgement

The author expresses his appreciation to his colleagues, particularly Dr. T. Ganapathy, Professor, and students of the Department of Plant Pathology, Tamil Nadu Agricultural University, Coimbatore, India, for their suggestions and help in different ways to make the book more comprehensive and useful. Secretarial assistance of Ms. K. Mangayarkarasi reduced part of the strain in the preparation of the typescript of the book. I profusely thank my wife Mrs. N. Rajakumari for the constant and patient tolerance of my inadequacies, allowing me to bestow my undivided attention from initiation to completion of this book to my satisfaction. Affectionate support and encouragement provided by my family members Mr. N. Kumar Perumal, Mrs. Nirmala Suresh, Mr. T. R. Suresh and Mr. S. Varun Karthik placed me always in comfortable zone to prepare the book with required information for the intended audience. Permission to use the figure(s) granted by various publishers/copyright holders is gratefully acknowledged in respective pages where the figures appear.

Contents

Table of Contents for Volume 1

Chapter 1
Introduction

Biological management of diseases of crops has the foundation on the research efforts to identify the various ecofriendly approaches (i) to mitigate the ill effects of infection by microbial plant pathogens – oomycetes, fungi, bacteria, phytoplasmas, viruses and viroids, by using antagonistic microorganisms, naturally-derived materials of plant and animal origin and inorganic and organic compounds, (ii) to reduce or replace the use of synthetic chemicals and (iii) to integrate the compatible and synergistic strategies for enhancing the effectiveness of disease suppression. These approaches are expected to protect the environment and to create the possibilities of increasing the yields of crops and supplying chemical residue-free farm produce to the consumer.

1.1 Concepts and Aims of Biological Management of Crop Diseases

Crops in different ecosystems have been under biotic and abiotic stresses that constitute a major constraint in realizing the full yield potential of cultivars of various crops. The extent of loss induced, may vary depending on the nature of the cause and availability of environments favorable for the development and persistence of disease-inducing agent(s). Among the biotic causes, microbial plant pathogens play a predominant role in adversely affecting the crop production systems, leading, to huge quantitative and qualitative losses in attainable yield levels. In order to minimize the incidence and severity of diseases of crops, the effectiveness of various short- and long-term strategies has been assessed in different geographical locations. No single disease management strategy or combination of strategies could be advocated for successful cultivation of crops in all areas with desirable profit margin to the growers. However, with growing concern for environmental pollution and presence of chemical residues in grains, vegetables, fruits and other food materials, intensive efforts have been made to restrict the use of plant protection chemicals which were earlier considered as indispensable requirement for containing the plant diseases.

P. Narayanasamy, *Biological Management of Diseases of Crops*,
Progress in Biological Control 16, DOI 10.1007/978-94-007-6377-7_1,

Crop management methods have been developed primarily to maintain soil fertility at high levels and to enhance the crop yields by manipulating application of fertilizers, tillage and irrigation practices and maintaining appreciable levels of organic matter in the soil. Crop management practices have significant influence on the incidence and severity of diseases caused by microbial pathogens. It is essential that disease management strategies are compatible with the current cultural practices adopted in a given geographical location to have the acceptance of the farmers. Biological management tactics applicable for crop disease management seek to be a component of the synergistic orchestrated symphony, the benefit of which will reach both the producer and consumer. Concepts of biological management of diseases have evolved from time to time, based on the information and techniques to assess the nature of interactions between the pathogens and other organisms and the plants. The term 'biological control' or 'biocontrol' is applied in a narrow sense to indicate the control of "one organism by another organism" (Beirner 1967). This term has been used also in a wider sense to indicate the "use of natural or modified organisms, genes or gene product to reduce the effects of undesirable organisms (pathogens) and to favor desirable organisms such as crops, trees, animals, beneficial insects and microorganisms". This broad definition was proposed in the Report of the Research Briefing Panel on Biocontrol in Managed Ecosystems (Cook 1987). It is well known that no organism can exist independently and interdependence is the basic factor that decides the survival and persistence of life forms through subsequent generations. Restrictive application of the term 'biological control' to one organism controlling another organism appears to be unrealistic. The development of a microorganism controlling another depends on several factors that favor its antagonistic activity. Further, some microorganisms and naturally-derived organic or inorganic compounds act on the host plants in the same manner by inhibiting pathogen development directly or enhancing the level of resistance in host plants against the target pathogens indirectly. Both biotic and abiotic agents that have been commercialized, are applied to crops against several diseases and both types have direct and indirect effects on the pathogen and host plants. Hence, biological management is defined as the utilization of biotic and abiotic agents that act through one or more mechanisms to reduce the potential of the pathogen directly or indirectly by activating host defense systems to reduce the disease incidence and/intensity. Several investigations have shown that combination of biotic and abiotic agents results in synergism, improving the effectiveness of disease control.

Green revolution leading to higher crop yields achieved in several countries including developing countries was possible by applying different crop management strategies such as use of high yielding varieties, high doses of fertilizers and frequent and indiscriminate use of pesticides. Adoption of these strategies was also responsible for emergence of new problems that reduced yields in due course. As regards crop disease management, development of resistance in pathogens to fungicides and bactericides became a serious constraint for reaching the intended targets of high yield. The need for finding out effective alternative strategies for minimizing the incidence and intensity of crop diseases was fully realized. In addition, the possibility of restricting the use or replacing the synthetic chemicals that were apparently found to be highly effective against some pathogens had to be explored. The studies on the compatibility of biotic agents with fungicides or bactericides revealed that it would be possible to select strains of biocontrol agents (BCAs) showing tolerance to synthetic chemicals and they could be applied in combination with the chemicals resulting in restricted use of the chemicals. Thus the aims of the biological management of crop diseases are to (i) select the most effective biotic agents; (ii) identify the abiotic agents that can act individually or in combination with biotic agents; (iii) assess their effects on growth promotion in treated plants; and (iv) examine the possibility of reducing the use or replacing the chemicals without compromising the effectiveness of disease control.

1.2 Landmarks in the Development of Biological Disease Management Systems

Occurrence of diseases of crops of economic importance is known even in prehistoric times as indicated by examination of fossils about 25,000 years old (Klausner 1987; Chu et al. 1989). Although concepts about the nature of the causes of plant diseases underwent changes due to the availability of information, efforts to contain the diseases were made constantly to reduce the losses. Several ancient conventional practices were adopted to preserve soil fertility. Crop rotation, application of green manures, tillage and irrigation methods had been followed with a view to reducing disease incidence by increasing the period of absence of susceptible crops, encouraging the microbial activities and altering soil moisture conditions to make them unfavorable for pathogen development (Grigg 1974; Katan 2010). Injuries to branches and stems of plants due to pruning were protected with a mixture of cow dung and urine to prevent infection of apple canker (Austen 1657), indicating that the points of pathogen entry had to be sealed with protectants.

Antagonism of one microorganism by another microorganism was observed by many earlier researchers. But the practical utility of this phenomenon became a reality, only after the discovery of Alexander Fleming (1929). Production of penicillin and its application in medicine provided great impetus to initiation of numerous studies on antagonists of plant pathogens. Investigations to determine the usefulness of application of antagonistic fungi for the suppression of damping-off of pine seedlings (Hartley 1921) and potato scab disease (Millard and Taylor 1927) appear to be the pioneering attempts that laid the foundation for the development of biological management as a feasible strategy. Mycoparasitism of the pathogen *Rhizoctonia solani* by *Trichoderma (Gliocladium) virens* was first reported by Weindling and Fawcett (1936). The evidence for the existence of soil suppressive to soilborne pathogens was obtained by Henry (1931) and the transferability of microbiota responsible for soil suppressiveness to conducive soil was demonstrated by Shipton et al. (1973). The role of ectomycorrhizal fungi in increasing the resistance of pine plants against *Phytophthora cinnamomi* was investigated by Marx (1969). The possibility of commercial exploitation of cross-protection phenomenon for reducing the incidence of *Citrus tristeza virus* (CTV) was demonstrated in Brazil by pre-immunization of susceptible plants with attenuated strain of CTV (Costa and Miller 1980).

Antagonistic microorganisms could compete with and inhibit the development of the microbial plant pathogens present in the soil by producing antibiotics. The plant growth-promoting rhizobacteria (PGPRs), *Pseudomonas* spp. and *Bacillus* spp. are excellent antibiotics producers. Hence, they may have the potential for suppressing the development of crop diseases caused by microbial pathogens (Cook and Rovira 1976). Fluorescent *Pseudomonas* spp. are present abundantly in the rhizosphere and their ability to protect plants against microbial plant pathogens and to promote growth of plants has been demonstrated. *Bacillus* spp. produce endospores that are resistant to adverse environmental conditions, facilitating their survival for long periods. These bacteria produce different kinds of antibiotics, siderophores and growth-promoting compounds that could have a major role in disease-suppressive and growth-promotion effects on plants treated with them. Many PGPRs are able to transcend the endodermis barrier, cross the root cortex to the vascular system and subsequently thrive as endophytes in stem, leaves, tubers and other organs (Bell et al. 1995; Hallman et al. 1997).

1.3 Biological Disease Management Strategies

Biological management of any specific disease has to begin by making maximal use of indigenous or constitutive biological control principles. Cultural practices, meant for preserving the soil fertility and enhancing yields, have significant influence on the incidence and severity of many diseases. Crop rotation, crop sanitation, and elimination of infected plants and debris reduce the chances of pathogen inoculum buildup. On the other hand, application of fertilizers, tillage and irrigation practices alter the soil microbial community structure and resistance/susceptibility levels of plants to infection by microbial pathogens (Chap. 2). Physical agents such as heat, solar energy and irradiation have been applied either to reduce the population of pathogens or weaken their pathogenic potential (virulence). Hot air, hot water and aerated steam are employed to obtain disease-free seeds and planting materials. Soil solarization using transparent polyethylene sheets has effectively suppressed the development of many soilborne diseases. In addition, solarization reduces the weed population significantly and consequently the crop plant could obtain adequate nutrition resulting in higher yields. Hot water and hot air application is effective for the control of postharvest diseases of fruits and vegetables. Thermotherapy has been an effective treatment for elimination of viruses from infected plant materials, when coupled with tissue culture methods. UV-C irradiation is effective in suppressing postharvest diseases of fruits and vegetables and its disease suppressive ability is attributed to induction of resistance in treated fruits (Chap. 3).

Microbial plant pathogens can spread from weeds and wild plants that are infected during the absence of crop plants. The importance of weed control and elimination of diseased plants to reduce the inoculum available for infection of crop plants has been well realized. The role of vectors of plant viral, fungal and bacterial pathogens in introducing the plant diseases into a field and subsequent spread of diseases has been revealed by several investigations. In the case of viruses that are able to multiply in the insect vectors, transovarial transmission to the next generation through eggs was demonstrated as in the case of *Rice dwarf virus*. In such cases, the insect vectors themselves form important sources of virus inoculum. Furthermore, the vectors of the viruses can acquire and transmit the viruses from plants at far way locations. Biological management of these additional sources of inoculum is also an important approach for containing the crop diseases. Some fungal biocontrol agents have been formulated and commercially available for large scale application (Chap. 4).

1.4 Integration of Biological Disease Management Strategies

Several biocontrol agents – biotic and abiotic agents – have been selected based on their efficiency, compatibility with crop production practices and chemicals already in use for disease control. The biocontrol agents by themselves, may significantly reduce disease incidence and severity. But when they are integrated with other disease management tactics, the effectiveness may be enhanced.

1.4.1 Application of Formulated Products of Biological Control Agents

Antagonistic microorganisms, exhibiting high efficiency in the assessments under various greenhouse and field conditions, are selected for formulation and commercialization. Microorganisms selected for their biocontrol activity against the target pathogen(s) have to be mass multiplied in appropriate media, preserved, stored and formulated for commercialization. Two types of formulations are made from microbial antagonists: liquid and dry formulations. Different delivery systems suitable for the crop-pathogen combinations are applied for placing the biocontrol agents at required sites either prior to or after infection by the microbial pathogens. The process of formulation and registration and hurdles in registration of bioproducts are discussed and the fungal and bacterial biocontrol agents available commercially are listed for reference in Chap. 5.

1.4.2 Integration of Biological Control Strategies with Crop Disease Management Systems

Integration of various strategies that are compatible with each other and act on the pathogens synergistically through different mechanisms is essential to develop effective biological disease management systems for agricultural crops. Crop sanitation and proper disposal of infected plants and debris and management of weed populations may result in reduction in pathogen inoculum available for infection of crop plants. In addition, various cultural practices have significant impact on disease incidence, as the pathogen propagules are reduced in number or weakened. Efforts to develop integrated systems of management of disease of agricultural crops have been scarce. The available information on the integration of biological control strategies with other disease management tactics for the diseases of cereals, cotton, pulses, and oilseed crops and the achievement of higher level of disease control through such integration are presented in Chap. 6. Fruit and vegetable crops are high value crops and greater efforts have been taken to contain the diseases caused by microbial pathogens under field conditions, as well as during storage. Prevention of wounds to the plants and vegetables and fruits at pre- and post-harvest stages will have marked effect on the disease incidence and severity. Biological disease management systems have to be developed by integrating several strategies taking into consideration the factors such as pathogen biology, cultivar resistance, and epidemiology. In general, the number of chemicals and frequency of their application are generally more than the optimum and hence, restriction of chemical use is essential to avoid development of resistance in pathogens. Both biotic and abiotic biological control agents have been combined to attain higher level of disease control. Information on compatibility of BCAs with chemicals has helped to reduce the quantity of chemicals applied. Attempts to develop integrated diseases management systems for various vegetable and fruit crops are highlighted in Chap. 7.

The information presented in this book in an easily understandable style, has been distilled from an extensive literature search to include latest findings and concepts on biological management of crop diseases. Various aspects of biotic and abiotic biological control agents and the crop production strategies that have significant impact on the success of crop disease control are discussed, describing some case studies in detail. The information presented in this book is expected to provide a comprehensive knowledge to the upper level undergraduate and graduate students. Researchers and teachers in the Departments of Plant Pathology, Plant Protection, Microbiology, Molecular Biology, Botany, Ecology, Agriculture and Horticulture will find the information presented in the book to be useful for furthering their research investigations. Incorporation of several protocols appended at the end of relevant chapters to assist the researchers in planning their experiments is a unique feature of this book.

References

Austen RA (1657) A treatise of fruit trees. Henry Hall, Oxford

Beirner BP (1967) Biological control and its potential. World Rev Pest Control 6:7–20

Bell CR, Dickie GA, Harvey WLG, Chan JWYF (1995) Endophytic bacteria in grapevine. Can J Microbiol 41:46–53

Chu PWG, Waterhouse PM, Martin RR, Gerlach WL (1989) New approaches to the detection of microbial plant pathogens. Biotechnol Genet Eng Rev 7:45–111

Cook RJ (1987) Research briefing panel on biological control in managed ecosystems. Committee of Science, Engineering and Public Policy, National Academy of Sciences, National Academy of Engineering, Institute of Medicine, and National Academies Press, Washington, DC, 12 pp

Cook RJ, Rovira AD (1976) The role of bacteria in the biological control of *Gaeumannomyces graminis* by suppressive soils. Soil Biol Biochem 8:269–273

Costa AS, Miller GW (1980) Tristeza control by cross-protection: a US-Brazil cooperative success. Plant Dis 64:538–541

Fleming A (1929) On the antibacterial action of cultures of *Penicillium* with special reference to their use in the isolation of *B. influenzae*. Ref Infect Dis 2:129–139

Grigg DB (1974) Agricultural systems of the world: an evolutionary approach. Cambridge University Press, London, 358 pp

Hallman JA, Quadt-Hallman A, Mahafee WF, Kloepper JW (1997) Bacterial endophytes in agricultural crops. Can J Microbiol 43:895–914

Hartley C (1921) Damping-off in forest nurseries. US Dept Agric Bull 934:1–99

Henry AW (1931) The natural microflora of the soil in relation to the foot rot problem of wheat. Can J Res C 4:69–77

Katan J (2010) Cultural approaches for disease management: present status and future prospects. J Plant Pathol 92(Suppl):S4.7–S4.9

Klausner A (1987) Immunoassays flourish new markets. Biotechnology 5:551–556

Marx DH (1969) The influence of ectotrophic mycorrhizal fungi on the resistance of pine roots to pathogenic infections I. Antagonism of mycorrhizal fungi to root pathogenic fungi and soil bacteria. Phytopathology 59:153–163

Millard WA, Taylor CB (1927) Antagonism of microorganisms as the controlling factor in the inhibition of scab by green manuring. Ann Appl Biol 14:202–216

Shipton PJ, Cook RJ, Sitton JW (1973) Occurrence and transfer of a biological factor in soil that suppresses take-all of wheat in eastern Washington. Phytopathology 63:511–517

Weindling R, Fawcett HS (1936) Experiments in the control of the rhizoctonia damping-off of citrus seedlings. Hilgardia 10:1–16

Chapter 2
Cultural Practices Influencing Biological Management of Crop Diseases

Domestication of certain plant species that were once growing in an environment without human interference, to satisfy the need for food, feed, fiber and timber, led to weakening of their ability to overcome the onslaught of adverse environmental conditions that include biotic and abiotic stresses. An agroecosystem contains numerous species of plants, animals, insects and microorganisms. All living organisms detrimental to successful crop production were termed as biotic stresses which include microbial plant pathogens. Crop production systems aim at reducing the adverse effects of microbial plant pathogens by applying both short- and long-term strategies. During the last few decades, crop production is facing several long-term crises such as energy shortage, global warming and environmental pollution. As environmental and ecological issues continue to impact on agriculture, all technologies developed for enhancing crop production should be economically feasible, ecologically sound, environmentally safe and socially acceptable. The philosophy of plant protection has shifted the importance of chemical pesticides to methods that are environment friendly. With the growing awareness for the protection of environment, all practical solutions to crop disease problems must be based on the safety of environment, conservation of natural resources and maintenance of genetic biodiversity of all life forms. Crop production through sustainable agriculture has the goal of providing materials for the existence of humans and animals with least negative effects on the environment. Cultural practices influence the incidence and spread of crop diseases to varying degrees. The effects of cultural practices may vary widely from region to region and it may be difficult to assess them, since beneficial effects become recognizable only over a period of several years or seasons. Cultural practices may be grouped into three types based on the primary aim of their application: (i) practices usually applied for improvement of plant growth and ultimately higher yield and not related to crop protection; (ii) practices used mainly for reduction in disease incidence and/or spread and (iii) practices applied for achieving higher yield level and also providing protection to crops (Katan 2010). The influence of major cultural practices on the pathogen inoculum and consequently disease incidence and subsequent spread within the crop and other fields or locations are discussed in this chapter.

P. Narayanasamy, *Biological Management of Diseases of Crops*,
Progress in Biological Control 16, DOI 10.1007/978-94-007-6377-7_2,

2.1 Crop Sanitation

Crop sanitation aims to prevent the introduction of inoculum into the field/geographical location where the pathogen concerned has not been established and to reduce the amount of initial inoculums, if the pathogen is already present.

2.1.1 Use of Disease-Free Seeds and Propagative Plant Materials

Using disease-free seeds and propagative plant materials is an essential basic disease management strategy recommended for cultivation of all crops especially for organically grown crops. Numerous microbial pathogens are transmitted by infected seeds and asexually propagated planting materials such as tubers, bulbs, corms or setts. Decline in crop yield increases progressively through several generations, if no attention is paid for selecting disease-free seeds or planting materials. Certified seeds are made available by testing the mother plants or crops at different stages employing various techniques for detecting the presence of pathogens in seeds and propagative materials (Narayanasamy 2011). Several abiotic agents like heat and ultraviolet light have been employed to reduce/eliminate the pathogen populations. Quarantines and certification agencies are involved in the process of preventing the introduction of pathogens into the country/area and producing disease-free seeds and propagative materials respectively. Both these agencies adopt seed health standards and tolerance limits for different pathogens as suggested by the International Seed Testing Association (ISTA) through Working Sheets for each pathogen/disease.

2.1.1.1 Production of Propagative Plant Materials Free of Fungal and Bacterial Pathogens

The seeds of different crop plants carry a large number of fungal and bacterial pathogens which can reduce seed germination and adversely affect the seedling vigor. The seed lots are indexed by applying isolation-based methods, using specific or selective medium for isolating the target pathogen or methods based on immunological and/or nucleic acid characteristics of the target pathogens. The infected seed lots with infection levels above the prescribed limits are rejected. Seed lots satisfying the standards are certified and supplied to the growers. Use of seeds free of pathogens which are transmitted primarily through seeds such as sorghum grain smut (*Sphacelotheca sorghi*), rice bakanae (*Fusarium moniliforme*), cabbage blackleg (*Phoma lingam*), cabbage black rot (*Xanthomonas campestris* pv. *campestris*) and common bean bacterial wilt (*Curtobacterium flaccumfaciens* pv. *flaccumfaciens*) has been shown to be a sound strategy to minimize the spread of these diseases in the field or to other countries.

Table 2.1 Biological management of seedborne diseases

Crop/pathogen	Biotic agents	References
Cauliflower/*Pythium aphanidermatum*	*Trichoderma harzianum*	Mukherjee et al. (1989)
Chickpea/*Fusarium oxysporum* and *Rhizoctonia solani*	*Gliocladium virens*	Mukhopadhyay and Mukherjee (1991)
Cotton/*Xanthomonas axonopodis* pv. *malvacearum*	*Streptomyces atroolivaceus*	El-Raheem and Shanshoury (1994)
Cucumber/*Pythium ultimum*	*Pseudomonas fluorescens*	Rip and Seuk (1995)
Mungbean/*Macrophomina* spp.	*Trichoderma viride*	Raguchander et al. (1997)
Tomato/*Fusarium oxysporum* f.sp. *lycopersici*	*Tirchoderma harzianum*	Sivan et al. (1987)
Wheat/*Gaeumannomyces graminis* var. *tritici*	*Pseudomonas fluorescens*	Pierson and Weller (1994)
Ustilago segetum var. *tritici*	*G. virens; P. fluorescens Trichoderma* spp.	Dharam singh and Maheshwari (2001)
Tilletia caries	*Muscodor albus*	Goates and Mercier (2011)
Fusarium culmorum	*Clonostachys rosea*	Jensen et al. (2000)
Carrot/*Alternaria dauci* and *A. radicina*	*C. rosea*	Jensen et al. (2004)
Bean/*Curtobacterium flaccumfaciens* pv. *flaccumfaciens*	*Rhizobium leguminosarum*	Huang et al. (2007)

Several biological control agents (BCAs) such as *Chaetomium* spp., *Gliocladium* spp., *Penicillium* spp., *Pseudomonas* spp. and *Trichoderma* spp. have been employed for treating the seeds for suppressing the development of fungal and bacterial pathogens present in the seeds (Table 2.1). Treatment of cotton seeds with *Gliocladium* (=*Trichoderma*) *virens* or *Bacillus subtilis* reduced the colonization of roots by *Fusarium oxysporum* f.sp. *vasinfectum* and suppressed Fusarium wilt disease incidence and severity also (Zhang et al. 1996). Corn damping-off disease caused by *Pythium ultimum*, *P. arrhenomanes* and *Fusarium graminearum* was more effectively controlled by coating the seeds with *Gliocladium virens* isolate Gl-3, resulting in greater seedling stand, plant height and fresh weight and reduction in root rot severity compared to the fungicide captan (Mao et al. 1997). The potential of *Bacillus* sp. strain L324-92, for control of Pythium root rot caused by *P. irregulare* and *P. ultimum* and Rhizoctonia root rot caused by *Rhizoctonia solani*, in addition to wheat take-all disease caused by *Gaeumannomyces graminis* var. *tritici*, was demonstrated by Kim et al. (1997). Treatment with the bacterial BCA, not only reduced incidence of fungal diseases, but also enhanced grain yields under field conditions in which one or more of the root diseases of wheat were primarily responsible for reduction in yield levels (Kim et al. 1997). In six field experiments, treatment of wheat seeds with *Clonostachys rosea* (IK726) significantly reduced disease caused by *Fusarium culmorum*. The fungal BCA was active against the pathogen at average soil temperatures at sowing ranging from 6.2 to 12.0 °C (Jensen et al. 2000). Treatment with carrot seeds with *Pseudomonas* sp. strain MF416 and *Clonostachys rosea* strain IK726 provided best protection against the fungal pathogens *Alternaria dauci* and

A. radicina. In addition, the effect of hot water treatment was significantly improved, when combined with either of the bacterial or fungal biocontrol agents (Koch et al. 2010). Application of the biofumigant fungus *Muscodor albus* treatment of wheat seeds effectively protected the plants against wheat common bunt disease caused by *Tilletia caries* (Goates and Mercier 2011).

2.1.1.2 Production of Seeds and Propagative Plant Materials Free of Viral Pathogens

Plant viruses are transmitted through true seeds of several crop plants. Development of virus-free seed stocks is a major management strategy for diseases such as lettuce mosaic, barley stripe mosaic, peanut mottle and soybean mosaic diseases. In the case of lettuce mosaic disease, lettuce seedlings are raised in insect-proof glasshouses; infected seedlings are thoroughly rogued out and healthy seedlings are then planted in mosaic-free areas under constant supervision by technical personnel to eliminate plants developing symptoms of infection by *Lettuce mosaic virus* (LMV), as and when they are seen (Grogan et al. 1952). In Salinas Valley, California, USA, following introduction of the program for production of virus-free lettuce seeds, yield increased by 33.5 % consequent to prevention of losses due to LMV (Kimble et al. 1975). Elimination of *Barley stripe mosaic virus* (BSMV) from barley seeds could be achieved by a procedure involving a combination of assaying seedlings for the presence of BSMV by serological technique and roguing infected plants from certified seed plots. Following enforcement of zero tolerance for seedborne BSMV on all certified seeds, loss due to BSMV declined dramatically (Lister et al. 1981; Carroll 1983).

The efficacy of biopriming seed treatment for suppression of root rot pathogens infecting faba bean was assessed. The antagonistic fungi *Trichoderma viride* and *T. harzianum* and the bacteria *Bacillus subtilis* and *Pseudomonas fluorescens* were applied as either growth culture discs or bioprimed faba bean seeds. Under greenhouse conditions, all the tested fresh and stored seeds (for 2 months), bioprimed faba bean seeds protected the plants entirely at both pre- and post-emergence stages, compared with untreated control treatment. Seeds stored for 4 and 6 months showed less protective effect against the root rot disease. After 3 months of storage, incidence of both pre- and post-emergence damping-off disease was effectively controlled under field conditions by biopriming the faba bean seeds. The efficiency of protection declined progressively with increase in storage period. With suitable application technique, the safe and inexpensive seed biopriming technique may be considered for large scale use (El-Mougy and Abdel-Khader 2008). The effectiveness of biopriming soybean seeds for protecting the seedlings against damping-off disease caused by *Colletotrichum truncatum* was investigated, using *Trichoderma harzianum*, *T. virens* and *Pseudomonas aeruginosa*. Biopriming with *P. aeruginosa* was the most effective in protecting the seedlings against pre- and post-emergence damping-off, reduction in disease incidence being 48.6–51.9 % and 65.0–97.2 % respectively. The effectiveness of *P. aeruginosa* treatment was generally comparable to *T. harzianum* and the fungicide Benlate®. The results indicated that biopriming with *P. aeruginosa* or *T. harzianum* had the potential for application as an alternative to fungicide treatment (Begum et al. 2010). Experiments under greenhouse and field conditions were performed to evaluate the effectiveness of seed dressing and soil application of formulations containing *Trichoderma viride*, *T. virens* and *T. harzanum* against wet root rot disease of mungbean caused by *Rhizoctonia solani*. Effects on plant growth and yield were also assessed. Seed dressing formulation Pusa 5SD and soil application formulations Pusa Biogranule 6, Pusa Biopellet 16 G containing *T. virens* was found to be more effective than the other formulations in reducing disease incidence and increasing seed germination and improving plant growth parameters. Seed treatment was more effective than soil application in reducing disease incidence and enhancement of plant growth and yield of mungbean. The results clearly revealed the effectiveness of the treatment with biocontrol agents (Dubey et al. 2011).

Indexing and certification are essential for asexually propagated crops. It may be possible to select propagative materials by avoiding infected mother plants to obtain healthy planting materials in crops like potato, sweet potato, cassava and banana. However, mother plants which are infected late in the season and with latent infection cannot be recognized by visual observations. Hence, the health status of planting materials such as tubers, bulbs, corms and setts has to be verified by infectivity tests using appropriate indicator plants or molecular diagnostic techniques. Serological tests and nucleic acid-based assays are increasingly employed to select virus-free healthy planting materials. The importance of planting certified propagative materials has been emphasized for crops grown in the glass-houses. Chrysanthemum production was seriously hampered by several viral, viroid, fungal and bacterial diseases. The serious losses caused by these diseases could be avoided by planting certified disease-free propagating materials. This approach has played a crucial role in the successful commercial production of chrysanthemum (Narayanasamy 2002, 2011).

2.1.2 Proper Disposal of Infected Plant Debris, Plants and Plant Parts

Proper disposal of infected plant materials is essential for preventing the buildup of pathogen inoculum capable of infecting the standing crops and/or subsequent crops. Infected plant debris should be removed and properly disposed to reduce the conidia and vegetative hyphae which form potential sources of infection for *Botrytis cinerea* on geranium and other greenhouse crops. *B. cinerea* is known to sporulate readily on wounded and senescent plant tissues, providing inoculum for further infection and spread (Hausbeck and Pennypacker 1991). Infected crop plants left in the fields, volunteer (self-sown) plants growing from infected seeds and additional (collateral) hosts of the pathogens form important sources of infection. Removal of all such infected plants and infected plant parts reduces the amount of inoculum of the pathogens, resulting in considerable reduction in incidence of diseases in several crops, especially in perennial crops such as banana, cocoa and sugarcane (Kommedahl and Todd 1991). Removal of infected stumps and windrowing was found to be effective in reducing the root disease due to *Armillaria* spp. in *Pinus radiata* from 22 to 5 % in a period of 5 years from planting (Self and MacKenzie 1995). When the disease spread is slow as in cocoa swollen shoot disease, roguing and replanting with virus-free plants resulted in relatively greater productive stands (Matthews 1991). Regular roguing of infected plants was found to be effective in restricting the spread of *African cassava mosaic virus* in tropical African countries (Robertson 1981) and *Banana bunchy top virus* in eastern Australia (Dale 1987).

Dormant pruning is a conventionally followed practice in apples. If summer pruning is adopted, there is consistent reduction in the incidence of flyspeck diseases of fruits caused by *Schizothyrium pruni* by 50 %, even when no fungicide was applied. It was suggested that summer pruning might lead to a small change in the apple canopy microclimate involving a decrease in the hours of relative humidity (<95 %) in the canopy by 63 % and increase in the evaporative potential and improvement in spray deposition, if fungicide is to be applied (Cooley et al. 1997). Hand removal of all basal growth of hops up to 4 ft from the soil line was found to be highly effective in reducing the incidence of downy mildew disease caused by *Pseudoperonospora humuli* in the United States (Romanko 1965). The spread of maize sheath blight caused by *Thanatephorus sasakii*, was reduced by removing infected leaves as effectively as the chemical application (Jinggangmycin) (Qing et al. 1994). It is important to bear in mind that roguing of infected plants or plant parts should be carried out as soon as the symptoms are visible and before the pathogens produce resting spores/asexual spores. Timing of roguing is very important to realize the maximum benefit. The effectiveness of eliminating alternate host plants and alternative plants that are required for their survival in the absence of the primary crop plant species, has been demonstrated in many pathosystems (Narayanasamy 2002). Plant sanitation involving removal of dead and infected leaves and/or fruits has been shown to have significant effect on the incidence and severity of diseases in susceptible crop cultivars. The effects of leaf and fruit sanitation on the severity of strawberry leaf spot caused by *Mycosphaerella fragariae* and gray mold caused by *Botrytis cinerea* were assessed. Leaf spot severity was significantly reduced by about 90 % due to leaf sanitation and by 50 % due to planting strawberry in one-row system instead of two-row system. Neither leaf sanitation nor fruit sanitation had any beneficial effect on the incidence/severity of gray mold disease. However, the combination of one-row system, leaf- and fruit sanitation reduced the gray mold disease incidence by 50 % in the first crop year, compared to two-row system without leaf- and fruit sanitation. Severity of gray mold disease correlated significantly and positively with plant biomass. Based on the results, it was suggested that for humid central European conditions, adoption of a one-row system combined with leaf sanitation in early spring and fruit sanitation during harvest, if fruit density is high, might be effective in reducing the risk of damages caused by leaf spot and gray mold diseases in strawberry (Schmid et al. 2005).

2.2 Crop Nutrition

Crops to be cultivated in a particular location are selected based on several factors such as soil productivity, water availability, weather, cost/benefit ratio and marketability of the produce harvested. Among the cultural practices, crop nutrition has a marked influence not only on the growth and yield of crop plants, but also on the development of diseases, when other favorable conditions exist. During the second half of the twentieth century, rapid increase in human population enhanced the pressure on the production of food and other essential requirements for comfortable living, resulting in the adoption of intensive agriculture. As a result, high yield could be achieved through heavy application of chemical fertilizers, synthetic pesticides and monoculture of new cultivars with or without crop rotation. The success of control of plant diseases by synthetic chemicals created a wrong impression that chemical control could provide permanent solution to the disease problems in modern agriculture. During this period, agricultural economy was highly dependent on the development and application of chemical fertilizers and fungicides. Development of resistance in plant pathogens to chemicals and growing awareness for protection of environment, drastically altered the approaches for solving plant disease problems from chemical methods to other alternative methods aimed to restrict or replace the use of synthetic chemicals. Disease symptoms often reflect the altered nutritional status of the plant. Many cultural disease control tactics such as crop sequence, organic amendments, soil pH adjustment, tillage and irrigation management influence disease incidence and severity and function through altered biological activity. Response to a particular nutrient may be different, when its level moves from deficiency to sufficiency than from sufficiency to excess. Since each nutrient functions as part of a delicately balanced interdependent system influenced by plant genetic constitution and the environment, it is essential to maintain a nutritional balance for optimum crop response. Through an understanding of the disease interactions with each specific nutrient, the effects on the plant pathogen and environment may be effectively modified to improve the effectiveness of disease management, enhance production efficiency and increase in crop quality. The effects of many nutrients on disease development have been examined incidentally as a consequence of fertilizing to optimize plant growth or yield (Huber and Haneklaus 2007).

2.2.1 *Application of Organic Matter*

Addition of organic matter favorably improves crop yield mainly by enhancing soil fertility rather than through provision of nutrients to the plants. In addition, many microorganisms antagonistic to plant pathogens are activated following addition of organic matter to the soil, leading to suppression of disease development. Organic matter incorporated into the soil is known as amendment, whereas organic matter placed or spread on the soil surface is termed as mulch. Use of organic amendments has been shown to promote biological destruction of pathogen inoculum through germination-lysis mechanism and the intensification of microbiological activity, resulting in the decay of pathogen propagules.

2.2.1.1 Organic Amendments

Organic amendments may exert stimulatory or inhibitory effects on the microbial plant pathogen populations and disease development. They may either prevent infection by activating the soil microflora potentially competitive with or antagonistic to plant pathogens present in the soil or control plant pathogens by producing toxic compounds in the soil, when they decompose in the soil. It is necessary to compost the organic materials properly before incorporation into the soil, since some pathogens may survive in the organic amendments and are likely to be introduced in the field. The need for proper composting may be understood by the investigation on *Phytophthora ramorum*. Three main processes involved in pathogen suppression, while composting plant materials, have been identified: (i) high temperatures (40–70 °C) during thermophilic composting phase; (ii) colonization of compost by a variety of different microorganisms that either antagonize or outcompete the target microorganism(s) and (iii) production of antibiotic compounds by organisms involved in the composting process. *P. ramorum* has a wide host range and it can infect different plant parts of numerous plant species. It is possible that *P. ramorum* may be carried into composting facilities via infected green wastes. In commercial turned windrow composting facilities minimum of 55 °C for 1 h had to be maintained for eliminating *P. ramorum*. Under field conditions, it would be difficult to avoid temperature fluctuations. Hence, it would be preferable to have higher temperature for short duration. After composting, the absence of *P. ramorum* was verified by employing different methods of detection such as baiting, direct plating and PCR-based technique. Pathogen variability in compost was assayed by direct tests capable of differentiating between the pathogen suppression and pathogen elimination. The data suggested that *P. ramorum* was capable of surviving in finished compost, if introduced. Hence, it should not be construed that all compost originating from infected green waste might be safe by virtue of the fact that the USA EPA guidelines were applied for the production of composts (Swain et al. 2006).

Compost-amended substrates offer the potential for the management of diseases caused by soilborne pathogens as well as pathogens infecting aerial plant parts. The efficacy of composted pine bark mix fortified with *Trichoderma hamatum* 382 (FCPB) against bacterial spot disease of radish, lettuce and tomato was assessed. Plants grown in the FCPB mix and inoculated with bacterial leaf spot pathogens were less severely infected than plants grown in commercial peat mix or vermiculite. Disease suppression capacity of FCPB mix was lost by autoclaving and restored by reinoculating the autoclaved FCPB, indicating that the disease suppressive ability of the FCPB was associated with the biological component of the amendment (Aldahmani et al. 2005). The efficacy of wheat bran, saw dust, coffee grounds, chicken manure or mixture of different compost amendments with or without crab shell powder (5 % w/w) in suppressing the development of *Fusarium oxysporum* f.sp. *spinaciae* was assessed. The soil amended with composts became suppressive to disease development on the second and third cropping. The coffee compost lowered soil pH, but became suppressive to the disease after modifying the soil pH. In the field trial using the mixture of the different composts containing 5 % crab shell powder, a combination of 5 % before the first cropping and 2.5 % every second cropping gave stable disease control and promoted plant growth also (Escuadra and Amemiya 2008). The effectiveness of neem cake amendment in sandy loam artificially infested with *Rhizoctonia solani* and a muck soil naturally infested with damping-off pathogens was investigated. Neem cake had no immediate effect on damping-off of cucumber seedlings. However, incubation of amended soil for 7 days before planting radish or cucumber reduced damping-off severity. The results suggested that neem cake was not directly toxic to the damping-off pathogens, but during incubation, neem cake might have created a biological climate that was suppressive to the disease (Abbasi et al. 2005).

The potential of condensed distiller's soluble (CDS), a co-product of ethanol production from corn, rich in organic matter and high in carbon to nitrogen ratio was assessed for its efficacy as a preplant amendment against Verticillium wilt of eggplant and potato scab in potato soils from commercial fields and against damping-off disease of radish and cucumber seedlings in a peat-based mix and muck soil. Eggplants grown in a potato soil amended with CDS showed less Verticillium wilt and increased fresh and dry plant biomass compared with control plants. In muck soil from a commercial field naturally infested with *Pythium* spp., application of CDS provided protection to cucumber seedlings from immediately after incorporation, but maximum protection was observed after 1 week with different rates (0.25, 0.5 and 1 %) of CDS applied. The number of total bacteria was enhanced in the CDS-amended muck soil. In the micro-plots, CDS (0.5 and 1 %) applied as an amendment to muck soil 2 weeks before planting improved the percentage of healthy cucumber seedlings and fresh plant weight compared to the control. The CDS was not toxic to the pathogens and disease suppression was believed to be due to biological activity stimulated by CDS in the substrate (Abbasi et al. 2007). Application of composted cattle manure at the rate equivalent to the 100 % recommended inorganic fertilizer dose, significantly reduced the intensity to diseases caused by *Pseudomonas syringae* pv. *lachrymans* and *Erwinia tracheiphila* in cucumber and by *Colletotrichum piperatum* and *Erwinia carotovora* in pepper (Huelsman and Edwards 1998). The effect of incorporation of soil with lopsided oat or woolly pod vetch as green manure was assessed for the suppression of potato common scab disease caused by *Streptomyces turgidiscabies*. Significantly fewer diseased tubers were harvested from soil incorporated with lopsided oat or wooly pod vetch, compared with those from oat and continuous potato cultivation. Treatment with lopsided oat followed by lopsided oat or wooly pod vetch was significantly more effective ($P<0.001$) in suppressing the disease severity than oat and continuous potato cultivation. An increase in the marketable tuber ratio was also greater in this treatment than for oat and continuous potato cultivation (Sakuma et al. 2011).

Various types of amendments have been incorporated into the soil to assess their effect on the pathogen population and suppression of diseases. Incorporation of barley straw in the top soil significantly reduced populations of *Verticillium albo-atrum*, causing wilt diseases of potato, cotton, tomato, strawberry, watermelon and other cruciferous crops. Reduction in disease incidence was attributed to the direct quantitative reduction in pathogen population, since there was no change in host resistance or fungal pathogenicity (Harrison 1976). Verticillium wilt disease of potato could be effectively suppressed by application of Sudan grass or corn as green manures. Positive correlation was observed between wilt incidence and colonization of apical stem by *V. dahliae* (Davis et al. 1976, 1996). Maximum reduction in the severity of Rhizoctonia stem canker of potato was observed following application of farm yard manure (FYM) and white mustard grown as green manure (Scholte and Lootsma 1998). Development of avocado root rot caused by *Phytophthora cinnamomi* could be effectively checked by continuously planting cover crops and applying abundant poultry manure to maintain high organic matter content and soil pH at neutrality. This approach is now widely practiced in eastern Australia (Utkhade 1992).

Soil amendments of crucifer crop residues such as cauliflower, broccoli or cabbage have been demonstrated to suppress the development of soilborne pathogens, in part due to release of chemical breakdown products of glucosinolates (GLS) (Subbarao et al. 1996). The effect of fresh cauliflower residue amendment alone and with a low dose of metam sodium (MS) combined with soil solarization was assessed for the control of Verticillium wilt disease of artichoke in two commercial fields under artichoke-cauliflower rotation. Inoculum densities (ID) of *Verticillium dahliae* were measured before and after soil treatments as well as determination of disease incidence, symptom severity and yield. Soil solarization with transparent 50 μm thick PE plastic sheets reduced the inoculum of *V. dahliae* and the disease incidence in artichoke. No added benefit was evident, when solarization was combined with cauliflower residue amendment. In addition to toxic volatile compounds, other mechanisms could be involved in disease suppression, because the effects of incorporation of cauliflower residue were not enhanced by solarization. The effect of cauliflower residues on *V. dahliae* population (microsclerotia) in soil was inconsistent, possibly due to varying pretreatment of inoculum levels. Combination of cauliflower residues and low doses of MS reduced inoculum densities significantly in the fields throughout the growing season and significantly reduced the percentage of plants infected by *V. dahliae*. Rotation of artichoke with cauliflower for crop residue incorporation with or without low doses of MS could be an effective disease management strategy for Verticillium wilt in fields with low preplanting inoculum density (Berbegal et al. 2008).

Allelopathy is a direct or indirect biochemical inhibition of one plant or microorganisms on another through production of toxic compounds or allelochemicals released into the environment. Plant residues applied as organic amendments have been reported to contain a variety of compounds which can reduce the severity of several soilborne diseases through their effect on microbial antagonisms, antibiosis and other dynamic mechanisms operating in the soil. Many allelopathic agents are produced during the decomposition of crop residues (Patrick 1986). Residues of cruciferous plants have been shown to have disease suppressive (allelopathic) compounds effective against several pathogens (Gamliel and Stapleton 1993; Keinath 1996). The interactive effects of incorporation of broccoli residue and temperature were investigated in the cauliflower-*Verticillium dahliae* pathosystem. Both fresh and dry broccoli almost entirely eliminated the microsclerotia of *V. dahliae* at 35 °C. The optimum temperature for the broccoli-mediated elimination of *V. dahliae* microsclerotia were between 25 and 30 °C. The results indicated that to obtain maximum reduction of microsclerotia in the soil leading to reduction in wilt disease incidence in cauliflower, broccoli residue should be incorporated at temperatures 20 °C and above. Enhanced growth of cauliflower in broccoli-amended soil was also observed (Subbarao and Hubbard 1996).

2.2.1.2 Organic Mulches

Organic mulches consisting of stubbles and other plant parts are applied on the soil surfaces primarily for the conservation of soil moisture and organic matter and for reducing soil erosion. Infection of apple roots by *Sclerotium rolfsii* in Israel could be markedly reduced by application of mulches which could reduce the soil temperature, though the pathogen was present on the surface of apple roots. The low soil temperature in mulched soil prevented the penetration of roots of apples by the pathogen (Lavee as cited by Avizohar-Hershenzon and Palti 1962). The effectiveness of management system for black sport disease of citrus caused by *Guignardia bidwelli* could be improved by integrating mulching with *Panicum maximum* and fungicidal sprays. There was an increase of 13–16 % in the amount of exportable disease-free fruits, because of mulching. The amount of inoculum appeared to be reduced as reflected by the percentage of rejected fruits (Schutte and Kotze 1997). The effect of application of inert materials on the incidence of soilborne diseases was investigated in sugar beet and cabbage nurseries. Sand and saw dust covering reduced damping-off disease caused by *Pythium debaryanum*, *Rhizoctonia solani*, *Phytophthora* spp. and *Fusarium* spp., infecting seedlings in the nurseries. However, emergence of lettuce seedlings in the nurseries was adversely affected by mulching with these inert materials (Balardin et al. 1994). Organic mulches may introduce both beneficial and harmful microbes into the soil. Most of the isolates of actinomycetes isolated from organic mulch used in avocado plantations in western Australia were inhibitory to *Phytophthora cinnamomi*. But some isolates were harmful to plant growth also (You et al. 1996). Wood-based mulches (cellulose-rich) have been used in avocado production and these mulches were evaluated for their efficacy in reducing Phytophthora root rot disease caused by *P. cinnamomi*. A role for microbial cellulose enzymes in pathogen suppression through effects on the cellulose cell walls of *P. cinnamomi* was examined. A standard curve was developed to correlate cellulose activity in mulches with concentrations of a cellulose product. Sustained exposure of *P. cinnamomi* to cellulose at 10–50 U/ml significantly reduced sporangial production, but the biomass was only reduced at concentrations of cellulose over

100 U/ml. The results indicated that cellulose activity in mulch was sufficient to impair sporangial production of *P. cinnamomi*, but not always sufficient to reduce the vegetative biomass of the pathogen (Richter et al. 2011).

Straw mulches have been reported to be good alternatives to expensive inorganic mulches in a few pathosystems. In the case of *Cucumber vein yellowing virus* infecting cucumber, the vector whiteflies were attracted by the yellow color and killed by the high temperatures, when they alighted on straw mulch (Nitzany et al. 1964). Wheat straw mulch was found to be effective in reducing the incidence of aphid-borne viruses and whitefly populations in cucurbits. In zucchini squash grown over straw mulch, yields were as high and the incidence of the aphid-borne virus diseases was no greater than in plants grown over reflective plastic mulch. Straw mulch also deterred colonization by whitefly and reduced the incidence of squash silver leaf (Summers et al. 2005). However, the possibility of using stubble and debris mulches, leading to higher level of incidence of fungal, bacterial and viral diseases has also been indicated by many investigations. Diseases such as wheat blotch (*Septoria nodorum* and *S. tritici*), barley leaf blotch (*Rhynchosporium secalis*), maize stalk rot (*Diplodia zeae* and *Gibberella zeae*), bacterial wilt (*Clavibacter michiganensis* subsp. *nebraskensis*) and sunflower leaf spot (*Alternaria helianthii*). Many viral and spiroplasmal diseases are spread by vectors that may acquire them from infected stubbles left in the field, due to adoption of no-tillage practice where stubbles of previous crop are left undisturbed. *Maize dwarf mosaic virus*, *Wheat streak mosaic virus*, *Rice tungro-associated viruses* and Rice yellow dwarf phytoplasma may be transmitted from regenerated growth from stubbles left in the field (Narayanasamy 2002).

Polyethylene mulch alters the environmental conditions in the soil and also alters the environmental conditions above the mulch. The actual effects and their magnitude depend on the characteristics of the polyethylene sheet used and how the covering of the soil is carried out (attached to the soil or not). The effectiveness of different kinds of polyethylene mulches was assessed for suppressing or preventing the development of potato tuber blight disease caused by *Phytophthora infestans* under field conditions. A water-permeable agricultural textile treated with copper hydroxide was also tested for its efficacy. Two barriers that covered the potato hills, black polyethylene film and copper hydroxide-treated agricultural textile reduced the incidence of tuber blight relative to appropriate controls. Other barriers (mulches) had little effect on tuber blight incidence. The effectiveness of black polyethylene film might be attributed to the blocking of water resulting in prevention of infiltration of inoculum through potato hill and alteration of the soil environment within the hill, thus markedly affecting inoculum movement and infection capacity. In the case of textile mulch, copper hydroxide coating (6 g/m^2) applied to textile was considered to be responsible for the disease suppression, since copper hydroxide could kill both sporangia and zoospores of *P. infestans* in the soil. The mulch treatments provided only partial protection to potato tubers and hence, this approach is unlikely to be adopted for commercial potato production system (Glass et al. 2001).

The efficiency of covering the soil with polyethylene before planting with or without application of fungicides commonly used by organic growers for suppressing the development of tomato late blight disease caused by *Phytophthora infestans* was assessed. Polyethylene mulch provided consistent, effective and highly significant

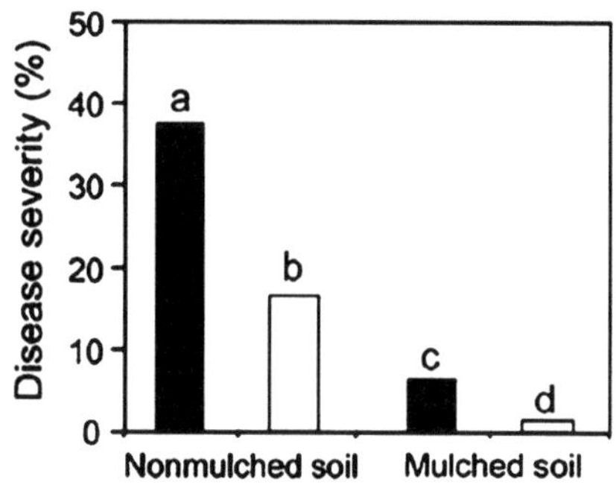

Fig. 2.1 Effect of mulching and chemical control on the severity of tomato late blight disease in walk-in tunnels. *Open bars*: treated with a mixture of Kocide 2000 (0.5 %) and Neemguard (2 %) at weekly intervals; *filled bars*: untreated controls. *Bars accompanied by different letters* are significantly different at P=0.05 (Courtesy of Shtienberg et al. 2010 and with kind permission of The American Phytopathological Society, MN, USA)

suppression of the late blight disease with a control efficacy of 83.6±5.5 %. On the other hand, the effectiveness of fungicides was inconsistent and insufficient to suppress the disease to the desirable level with a control efficacy of 34.5±14.3 %. The combination of mulch and fungicides showed an additive effect, reducing the disease severity by 96 % (Fig. 2.1). There was no significant difference in the effectiveness of different types of polyethylene mulches like biocolor aluminized, clear or black in the suppression of tomato late blight disease development. The polyethylene mulch effectively suppressed the cucumber downy mildew disease caused by *Pseudoperonospora cubensis*, but the suppressive effect was less than that was obtained for tomato late blight disease. The disease suppressive effect of the polyethylene mulch appeared to be due to the reduction in both the leaf wetness and duration, since mulching resulted in decrease in both frequency of nights when dew formed and the number of dew hours per night, when it formed. Mulching also reduced the relative humidity (RH) in the canopy which might reduce sporulation of both fungal pathogens. Other benefits of using plastic mulches are reduction in weed infestation and saving up to 30 % of irrigation water (Shtienberg et al. 2010).

2.2.2 *Use of Inorganic Fertilizers*

As adequate organic matter is not available to meet the requirements of crop plants, inorganic fertilizers have to be applied to increase the growth of yield potential of different crops. The fertilizers may act indirectly through their effects on host plant resistance resulting in increase or reduction in disease incidence and/or severity. They may also have some effect on the environment around the plant and antagonists on or near the plant contributing to the suppression of disease development. Nitrogen favors rapid vegetative growth and excessive supply of nitrogen is known to increase the susceptibility of crop plants to several diseases. Excessive application of nitrogen leads to formation of loose tissues with large thin-walled cells, larger intercellular spaces and stomata remaining wide open. Such conditions facilitate rapid development of pathogens especially bacterial pathogens (Grossmann 1970). The incidence and intensity of wheat powdery mildew caused by *Erysiphe graminis* increased proportionately with increase in levels of nitrogen applied (Heitefuss 1989). Delaying or split application of nitrogen when weather conditions are favorable for the development of rice blast disease caused by *Pyricularia grisea*, may be an effective strategy to reduce the disease incidence. Reduction of nitrogen supply for wheat crops in dry land areas of the Pacific Northwest delayed the onset of the plant water stress that was required for development of Fusarium foot rot disease and consequently the incidence of the disease was reduced (Cook 1980). Development of soilborne diseases is markedly influenced by nitrogenous fertilizers which alter the soil pH considerably. Cotton wilt disease (*Fusarium oxysporum* f.sp. *vasinfectum*) incidence increased with excessive application of nitrogen (Hillocks 1997). Potato tuber rot due to infection by *Phytophthora infestans* was significantly increased at higher levels of nitrogen (Heitefuss 1989).

The nutritional status of the plant species has a marked impact on its susceptibility/resistance to disease(s). Application of optimum amount and proper form of nitrogen may be a potential strategy for disease control. Ammonium nitrogen was found to be suppressive to wheat take-all disease caused by *Gaeumannomyces graminis* var. *tritici* (Colbach et al. 1997). Ammonium fertilization was reported to encourage pseudomonas antagonistic to the pathogen (Smiley 1978a, b). Disease suppression by nitrate fertilizers may be achieved in the case of Fusarium wilt diseases affecting bean, chrysanthemum, cotton, tomato and wheat as well as diseases caused by *Sclerotium rolfsii*. Suppression of cabbage club root disease caused by *Plasmodiophora brassicae* obtained by application of lime could be further enhanced by nitrate fertilizer. The effect of ammonium nitrogen in *Verticillium dahliae* infecting tomato was suppressed by ammonium nitrogen (Smiley 1975), whereas it did not affect the percentage of foliage affected or root and stem colonization by *V. dahliae* infecting eggplant (Elmer and Ferrandino 1994). The effects of many major and minor nutrients on wilt diseases have been investigated (Jones et al. 1989). Tomato seedlings were grown in soilless rockwool system infested with *Fusarium oxysporum* f.sp. *radicis-lycopersici* and they are fertilized with a nutrient solution amended with macro- and micro-elements at different rates. Disease severity was significantly increased by ammonium nitrogen, while disease severity was reduced by nitrate nitrogen. A differential effect of nitrogen form on pH was considered as a major mechanism of action for suppression of soilborne pathogens (Duffy and Défago 1999).

Phosphate fertilization favors plant growth by promoting rapid root development. Enhancement of levels of resistance to diseases in crops receiving adequate phosphotic fertilizers has also been observed. Wheat take-all disease caused by *Gaeumannomyces graminis* var. *tritici* was found to be favored by phosphorus deficiency. The onset and severity of the disease was appreciably reduced by repeated application of phosphates in the long-term field experiments at Rothamsted, England (Mattingly and Slope 1977). Reduction in take-all disease in wheat infected by vesicular-arbuscular mycorrhizal (VAM) fungus *Glomus mosseae* in phosphorus-deficient soil was possibly due to enhanced absorption of phosphorus by wheat roots facilitated by *G. mosseae* (Graham and Menge 1982). Higher levels of phosphorus reduced the incidence and lesion morphology of potato scab (*Streptomyces scabies*) (Davis et al. 1976), potato late blight (*Phytophthora infestans*) (Heitefuss 1989), peanut (groundnut) rust (*Puccinia arachidis*) (Mayee et al. 1983) and cowpea anthracnose (*Colletotrichum lindemuthianum*) (Adebitan 1996) diseases. Application of superphosphate increased the growth of forest trees and reduced root rot infection by *Phytophthora cinnamomi.* This effect was presumed to be due to removal of excessive soil moisture (Nework 1970). Since phosphates are applied to correct the deficiency of this nutrient in the soil, determination of their effect on disease development was not done as the primary aim of the investigation. Hence, the available information is scanty and inadequate to draw any reliable conclusion for the management of plant diseases.

Potassium has a key role in osmoregulation in the cell and maintenance of electrical balance across the plasma membrane. Deficiency of potassium, phosphorus, magnesium and minor nutrients leads to impairment of processes involved in the maintenance of cell structure and function. This situation increases susceptibility to pathogens such as *Gaeumannomyces gramins* var. *tritici* (Reis et al. 1982). Abundant supply of potassium, in addition to optimum levels of other nutrients, has been shown to increase the levels of resistance of plants to several diseases. Resistance to cereal rusts and powdery mildew was increased by higher doses of potassium due to strengthening of the epidermal cell wall. Formation of cellulose and sclerenchymatous structures is accelerated by adequate supply of potassium, resulting in enhancement of resistance to cereal foot rots and stem rots and counteracting effects of excess nitrogen which results in lodging of plants (Heitefuss 1989). Reduction in incidence or severity of diseases following application of abundant potassium has been reported, indicating the possibility of using potassium fertilizers as a management strategy for some crop diseases. Incidence of soybean root rot and mosaic diseases could be reduced by applying complete fertilizer containing N, P, and K in required proportion. The yield of the soybean crop was significantly increased to the maximum extent (Pacumbaba et al. 1997). Resistance of wheat and flax to rust diseases and maize to Stewart's wilt disease may be lost under conditions of potassium deficiency (Huber and Arny 1985), indicating that maintenance of potassium supply to adequate level is necessary to avoid predisposition of plants to diseases caused by microbial pathogens.

Nutrient management is important for the activation of passive and active mechanisms of disease control operating in plants. Mineral nutrients are components of plants and regulate metabolic activity associated with resistance of a plant species and virulence of a pathogen capable of infecting that plant species. Adequate nutrition is generally required to maintain a high level of resistance. Nutrients such as calcium that suppress disease caused by *Erwinia* spp. (bacterial soft rots), *Sclerotium rolfsii*, *Pythium myriotylum*, *Rhizoctonia solani*, *Fusarium solani* and *Sclerotinia minor*, increase the structural integrity and resistance of the middle lamella, cell wall components and cell membranes to the extracellular macerating enzymes produced by these pathogens (Bateman and Basham 1976; Kelman et al. 1989). Incidence of *Phytophthora cinnamomi* causing root rot disease of *Quercus ilax* was at low level in soils with medium to high calcium (Ca^{2+}) content. The possibility of inducing soil suppressiveness by applying calcium fertilizers was examined. The in vitro assays showed that Ca^{2+} fertilizers inhibited significantly production of sporangia, zoospores and chlamydospores at mM concentrations, while mycelial growth was not affected by them. In artificially infested soils, some Ca^{2+} induced significant decrease in viability of chlamydospores. In the greenhouse experiments using soils infested with *P. cinnamomi*, significant reduction in severity of foliar and root symptoms was observed in Holm oak seedlings growing in soils amended with Ca^{2+} fertilizers. Based on the results obtained in this investigation, limestone amendments in oak rangelands were suggested for enhancing suppressiveness of soils to *P. cinnamomi* (Serrano et al. 2012).

Immobilization of manganese (Mn) at infection sites by *Gaeumannomyces graminis* var. *tritici* or *Magnaporthe grisea* predisposed wheat tissues to take-all disease or to rice to blast disease respectively. The ability to oxidize Mn from the reduced, plant available form, to the oxidized, non-available form was identified as a virulence mechanism of the fungal pathogens. The isolates of *G. graminis* and *M. grisea* that could not oxidize Mn were also found to be avirulent. Many cultural disease control practices may function through their effect on mineral availability. Transformation from insoluble Mn^{+3} or Mn^{+4} oxides to plant-available soluble Mn^{+2} is highly dependent on environmental factors so that many factors predisposing plants to disease may act through their effect on Mn availability. A firm seed bed recommended for reducing wheat take-all disease increased Mn uptake by wheat. Plant growth-promoting rhizobacteria (PGPRs) applied to seed or seedlings reduced diseases by modifying the microbial environment and increasing the availability of specific micronutrients such as iron (Fe) and Mn (Huber and Haneklaus 2007).

Silicon added as a supplement to nutrient solutions for hydroponically grown cucumber was effective against the powdery mildew disease caused by *Sphaerotheca fuliginea*. Silicon application (100 mg/l) increased the latent period, reduced the number of colonies per leaf, decreased colony area per leaf and germination rates of conidia of *S. fuliginea* on cucumber plants grown in greenhouses (Bélanger et al. 1995). Susceptible cucumber cultivars irrigated with potassium silicate at 100, 150 or 200 mg/l exhibited a slight but statistically significant reduction in powdery mildew over the course of a 49-day trial. Temperature in combination with silicon was found to alter powdery mildew suppression on cucumber leaves. The greatest effect of temperature on powdery mildew suppression was observed at 20 °C at which cucumber plants irrigated with silicon at 100 mg/l exhibited significant reduction in the number of powdery mildew colonies per leaf. The differences in the number of powdery mildew colonies per leaf between the Si^- and Si^+ treatments decreased at 25 °C or 30 °C, although the number of powdery mildew colonies on Si^+ treated plants remained significantly lower than that on Si^- treated plants (control). The results revealed that temperature could alter the effectiveness of silicon amendments for powdery mildew suppression of cucumber in hydroponic nutrient solution. The effect of temperature on powdery mildew suppression by silicon may explain the differences in the results obtained in Canada and Florida, USA on the influence of silicon (Schuerger and Hammer 2003).

2.2.3 *Use of Biofertilizers*

Different bacterial species that can fix atmospheric nitrogen in the root nodules or in the soil have been produced commercially and applied as biofertilizers. The potential of these bacteria in suppressing the development of crop diseases has been investigated in some pathosystems. Biofertilizers Bio1 and Bio2 were generated from the anaerobic and aerobic digestion of cattle manure respectively. The effect of application of Bio1 and Bio2 was assessed in vitro and in vivo for suppressing the mycelial growth of *Phyllosticta citricarpa* (*Guignardia citricarpa*), causing citrus black spot (CBS) disease. Bio2 produced by aerobic digestion of cattle manure did not have any effect on the mycelial growth of *P. citricarpa*. In contrast, Bio1 produced by anaerobic digestion inhibited the mycelial growth in proportion to its concentration. Under field conditions, the effect of Bio1 in controlling CBS was directly proportional to its concentration. The disease index (DI) values were 0.25 and 0.23 for 10 and 20 % concentration of Bio1, while the untreated control had a value of 0.33, indicating the potential of the biofertilzer in controlling the disease (Kupper et al. 2006). The efficacy of the bio-organic fertilizer was assessed for the control of watermelon Fusarium wilt disease. The bio-organic fertilizer formulation containing organic fertilizer *Paenibacillus polymyxa* and *Trichoderma harziaum* was applied at 0.5 %. The disease incidence at 27 and 63 days following treatment was reduced by 84.9 and 75 % respectively compared with control under growth chamber and greenhouse conditions (Wu et al. 2009).

The potential of *Rhizobium* strains and biofertilizers was evaluated for suppressing the development of bush bean and root rot complex caused by *Fusarium oxysporum*, *F. solani*, *Sclerotium rolfsii*, *Sclerotinia sclerotiorum* and *Rhizoctonia solani*. The seeds were treated with the bioinoculants and sown in the field. The biofertilizers and *Rhizobium* strains improved the seed germination significantly, compared with untreated control. Foot and root rot disease was reduced up to 59.2 % by different strains and biofertilizers. Yield of green pods was also significantly increased by the *Rhizobium* strains and biofertilizer treatments (Khalequzzaman and Hossain 2008). The efficacy of Effective Microorganism (EM), a microbial inoculant comprising of actinomycetes, photosynthetic bacteria, lactic bacteria and yeasts and Vaira, a biofertilizer derived from anaerobic fermentation of cow manure was assessed for suppressing the development of *Xanthomonas campestris* pv. *vesicatoria* (*Xcv*), infecting sweet pepper (*Capsicum annuum* cv. Margareth). Sweet pepper seedlings were transplanted in field plots where the soil had been treated with EM, the biofertilizer or left untreated to serve as control. A suspension of *Xcv* at 10^8 CFU/ml was sprayed on the plants at 60 days after planting. The number of bacterial pathogen (colonies) population isolated from leaves of plants in different treatments and disease incidence were determined. The pathogen populations in diseased leaf in

EM and biofertilizer treatments were 0.5×10^8 CFU and 0.4×10^8 CFU respectively as against 3.5×10^8 CFU in the untreated control. The disease incidence was also significantly reduced by both EM and biofertilizer treatments (Castro et al. 1997).

2.3 Tillage

Tillage is an important operation in the crop management system, carried out for the preservation of soil status to provide optimum conditions for plant growth and to maintain long-term productivity of the soils. As soil is turned, mixed, loosened or compacted using different implements, alteration in water retention, aeration and soil temperatures occurs to varying degrees. These changes have a significant influence on the survival and dissemination of pathogens present in the soil or residues of crops grown earlier. Most of the soilborne facultative fungal pathogens, plant pathogenic bacteria and microorganisms antagonistic to these pathogens complete a part or entire life cycle in the soil or on the plant residues. Obviously the soil conditions altered by tillage may be expected to affect both pathogens and antagonists (Cook and Baker 1983). Tillage reduces disease risk mainly by reducing pathogen inoculum by acceleration of decomposition of infested residues, physically separating the pathogen from host plants by burying the pathogen deep in the soil or by changing soil conditions such as, increasing temperature and reducing moisture) (Yang 2002). Minimum tillage and conservation tillage practices are advocated in the recent years, because of the reduction in costs, improvement of soil structure and water infiltration ensuring reduction of soil erosion. However, these tillage practices result in accumulation of postharvest plant residues in the top layer. The infected residues may be a potential infection source for a range of soilborne diseases (Jirak-Peterson and Esker 2011; Váňová et al. 2011).

Tillage as a crop management practice has a general beneficial effect in destroying the weeds and volunteer plants that may serve as sources of inoculum for pathogens. In addition, it may reduce the amount of pathogen inoculum. Deep ploughing may bury the pathogen for inactivation by soil microflora or the pathogen present in the soil, after ploughing, may be brought to soil surface and inactivated by heat (solar radiation) and drought. Inoculum of *Rhizoctonia solani* infecting many crops and *Verticillium dahliae* infecting cotton may be reduced by deep ploughing (Papavizas and Lewis 1979; Butterfield in Fry 1982). Deep ploughing for the control of bean white mold disease caused by *Sclerotinia sclerotiorum* is recommended, because the sclerotia buried deep in the soil do not germinate (Abawi and Grogan 1979). Deep ploughing may have different effects on the current and successive crops. Lettuce drop disease caused by *Sclerotinia minor* was reduced in the crop planted immediately after ploughing, whereas the disease incidence greatly increased in the second crop. This was attributed to the significant reduction in the number of sclerotia and the altered distribution of sclerotia following ploughing, providing greater chances of infection in the succeeding crop (Subbarao et al. 1996). In the presence of the stubbles from a fallow-sown, no-till wheat cover crop, the dispersal of soil inoculum of *Phytophthora capsici* causing Phytophthora blight of bell pepper was suppressed, resulting in considerable reduction in disease incidence. The final disease incidence ranged from 2.5 to 43.0 % in no-till plots as against 71 to 72 % incidence in pepper planted in a bare field, where all dispersal mechanisms operated without hindrance (Ristaino et al. 1997). On the other hand, the detection frequency of *Phytophthora sojae* infecting soybean near the soil surface was two to three times greater than that of conventional till fields at this depth resulting in higher incidence of damping-off disease (Workneh et al. 1998). Variations in the results reported may possibly be due to the difficulty in precisely assessing the pathogen propagules present in the soils. Hence, no attempt was made to relate the disease incidence with the amount of viable pathogen population capable of inducing disease. Bocus and Shroyer (1998) suggested that the reduced or no-till practice may be coupled with proper crop sequence to achieve the twin objectives of retaining the crop residues and satisfactory level of disease suppression.

The effects of different tillage practices such as reduced tillage, conservation tillage and no-till on crop production differ widely and conflicting results have been reported by agronomists. No-till or nontillage practice was adopted in some annual field crops and perennial crops. Direct drilling of barley was found to reduce incidence of take-all disease in Scotland (Lockhart et al. 1975). In contrast, the incidence and severity of take-all disease of wheat were greater in no-till plots than in tilled plots in Washington, USA (Moore 1978). The tillage practices employing moldboard plough, chisel plough and no-till were evaluated for their efficacy in reducing Fusarium head blight (FHB) disease infecting wheat. Incidence and severity of FHB were lower in moldboard ploughed plots than in either chisel ploughed or no-till plots, although differences among chisel plough and no-till treatments were apparent. Yields of wheat were 10 % greater in moldboard ploughed plots than in either chisel ploughed or no-till treatments. The results suggested that changes in the regional tillage practices, principally the move towards conservation tillage and reduced till systems might have contributed to the FHB epidemics in the Upper Midwest in the United States (Dill-Macky and Jones 2000). The effects of different long-term conventional and conservation tillage treatments on the incidence and diversity of *Fusarium* spp. in soil were investigated. The isolation frequency of *Fusarium* spp. and the total number of CFUs were affected by the sampling years (2000 and 2001) and the cultivated crop (winter wheat, mustard or maize) and showed significant differences between tillage treatments (moldboard plough, chisel plough or rotary tiller). Moldboard plough treatments showed a lower diversity of *Fusarium* species than in chisel plough and rotary tiller treatments. The number of total CFUs per g of soil and *Fusarium* spp. was higher in 2000, when wheat was cultivated (Fig. 2.2) than in 2001 when maize was grown (Fig. 2.3) and showed differences between treatments. Moldboard-based plots at 24 and 17 cm showed higher populations in 2001 than in 2000. Based on the analysis of variance the isolation frequency of total CFUs and *Fusarium* spp. was significantly influenced by tillage treatments. Besides the tillage systems, the tillage depth also appeared to affect the *Fusarium* populations. The deeper the tillage, the lower was the number of *Fusarium* spp. isolated. In conservation tillage, a higher diversity of *Fusarium* species was found than in the moldboard plough-based tillage plots.

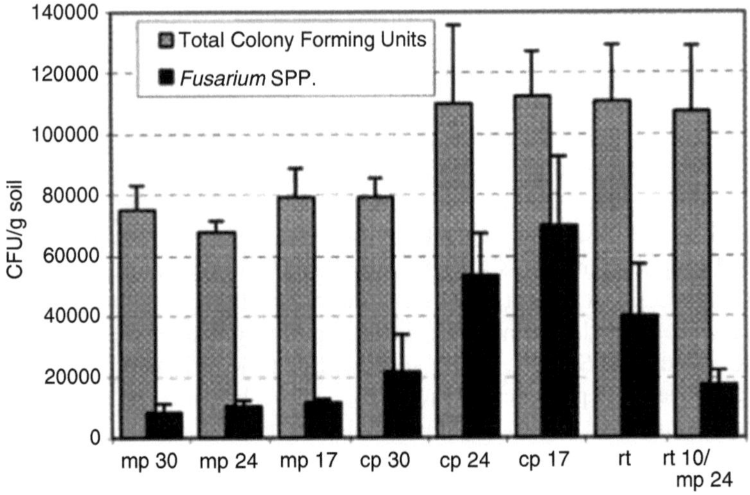

Fig. 2.2 Influence of tillage treatments on isolation frequency of total colony forming units (CFUs) and *Fusarium* spp. in wheat. *mp* moldboard plough, *cp* chisel plough, 30, 24 and 17 depth of treatment in cm (Courtesy of Steinkellner and Langer 2004 and with kind permission of Springer Science + Business B.V., Heidelberg, Germany)

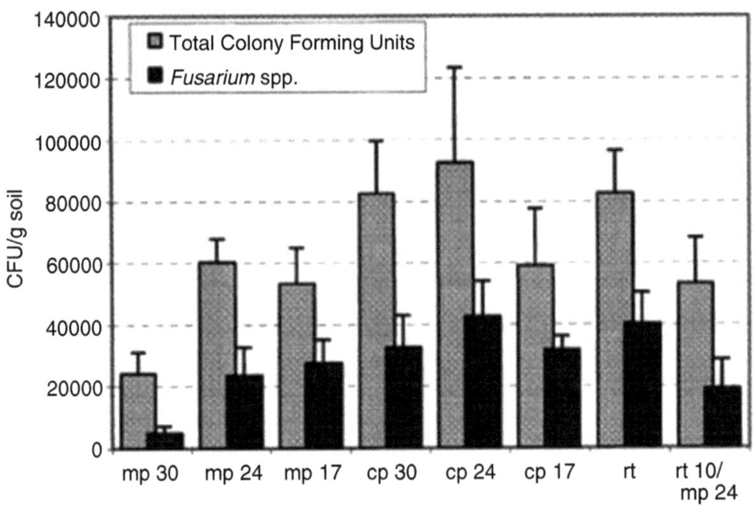

Fig. 2.3 Influence of tillage treatments on isolation frequency of total CFUs and *Fusarium* spp. in maize. *mp* moldboard plough, *cp* chisel plough, *rt* rotary tiller, 30, 24 and 17 depth of treatment in cm (Courtesy of Steinkellner and Langer 2004 and with kind permission of Springer Science + Business B.V., Heidelberg, Germany)

Moldboard ploughing resulted in a lower occurrence of pathogenic *Fusarium* spp. such as *F. culmorum* and *F. graminearum*. The results indicated that conservation soil tillage led to conditions which increased the incidence of *Fusarium* spp. in soil (Steinkellner and Langer 2004).

The influence of conventional and minimum tillage treatments in combination with 2- and 3-year crop rotations on the development of soil suppressiveness to potato canker and black scurf disease caused by *Rhizoctonia solani*, dry rot caused by *Fusarium* spp. and silver scurf disease induced by *Helminthosporium solani* was investigated. Analysis of root zone bacteria recovered from the rhizosphere (exoroot) and potato root tissue (endoroot) showed that antibiosis activity inhibiting the growth of the soilborne pathogens was at the maximum level in bacterial isolates recovered from the endoroot tissues of 3-year rotation crop (barley, red clover and potato) under minimum tillage management. The results suggested that the soil agroecosystems could be modified through rotation and conservation tillage practices to improve disease suppression by enhancing the antibiosis abilities of endophytic and root zone bacteria (Peters et al. 2003). Peanut (groundnut) was raised under four different tillage systems viz., disc harrow, chisel plough, no-till and paratill subsoiler and the effect of the tillage systems on the incidence of brown root rot disease caused *by Fusarium solani* was assessed. Use of paratill subsoiler before sowing peanut in the no-till system generally reduced the subsurface layer of compacted soil which provided a favorable environment for root growth and consequently reduced disease incidence. Reduction of root rot disease incidence was greater in a 2-year rotation including corn-soybean or soybean-corn prior to peanut. Root rot incidence in different tillage systems at 2 weeks before harvest, showed a strong negative correlation with pod yield. Use of paratill subsoiler before sowing peanut in a no-till system had the potential for application as an alternative strategy for the control of root rot disease of peanut in the southern Argentina, since this system could improve peanut root growth and reduce the disease as well (Oddino et al. 2008). Tillage practices namely conventional tillage, no tillage (NT) and minimum tillage (MT) were evaluated for their efficacy in reducing the occurrence of the mycotoxin deoxynivalenol (DON, produced by *Fusarium* spp.), stem base diseases and take-all disease in winter wheat. Tillage practices had statistically significant effect on the values of DON content in the grain. The least amount of DON was present in the grains of plants grown in conventional tillage treatment. DON contents were higher in the treatments using conservation tillage. Highest DON content was recorded in the grains of plants raised in no-tillage treatment. The eyespot disease incidence was higher in plots with conventional tillage treatment. Higher level of risk of getting enhanced level of incidence of stem base diseases, root diseases and Fusarium head blight disease appeared to be associated with conservation tillage practice (Váňová et al. 2011).

2.4 Soil Flooding

Flooding may be feasible as a cultural practice for disease control only in areas where large resources of water are available (Palti 1981). Soil flooding may be considered as a soil disinfestation treatment performed prior to sowing or planting. Flooding may become effective by restricting oxygen availability, increasing CO_2 or other microbial interactions such as production of substance toxic to the soilborne pathogens upon anaerobic processes (Bruehl 1987). A well known example of flooding being effective is that of Panama wilt of banana caused by *Fusarium oxysporum* f.sp. *cubense*. The soil was flooded for 3–4 months or more with a minimum of 30 cm of water. This practice was not effective in soils with very high pathogen populations or under conditions that might favor rapid development of the pathogen (Stover 1962). Flooding was successfully applied as a cultural practice for the control of Verticillium wilt disease infecting cotton. *V. dahliae* was effectively suppressed by long-term summer soil flooding with or without growing rice crops. The control of *V. dahliae* was directly associated with control of Verticillium wilt in subsequent cotton crops and with increased yields. But soil flooding done during winter months and irrigated rice without flooding were not effective. Anaerobic conditions and low oxidation-reduction potentials in floods might be responsible for the pathogen suppression (Pullman and De Vay 1981). The duration of soil saturation was found to be an important factor in both greenhouse and field testing where population densities declined after 6–8 weeks of flooding. Suppression of *V. dahliae* development might be due to anaerobic conditions and low oxidation-reduction potentials in flooded soils (Katan 2000). The effect of soil flooding was investigated also on other soilborne fungal pathogens. The sclerotia of *Sclerotinia sclerotiorum* were entirely decayed after 24–45 days in three types of flooded soils at temperatures of 22–26 °C (Moore 1949). Likewise, sclerotia of *S. minor* and *S. sclerotiorum* were no longer viable after 3 weeks in flooded soils at mean temperatures ranging from 30 to 33 °C (Matheron and Porchas 2005). Summer flooding treatment for 60 days of soil infested with *F. oxysporum* f.sp. *vasinfectum* significantly reduced the occurrence of Fusarium wilt on cotton crops grown subsequently (Allen and Lonergan 2000). The effectiveness of flooding, for suppressing the development of *F. oxysporum* f.sp. *lactucae* infecting lettuce, was assessed. Flooding treatment was applied by completely saturating the soil with water and maintaining a 2-cm layer of water on the soil surface by adding water daily to replace that was lost due to evaporation in experiments conducted in microplots. Growth of lettuce in flooded soil containing the pathogen occasionally was significantly higher than in non-flooded soil. However, the effect on plant growth was not as consistent as that recorded for soil solarization treatment with which flooding treatment was compared (Matheron and Porchas 2010).

2.5 Planting Date and Density

Date of sowing/planting of a crop depends on several factors such as crop duration, availability of irrigation facility/rainfall and other requirements for good growth of plants to realize the maximum yield potential of cultivar selected. From the disease management point of view, date of sowing/planting may be determined with the primary objective of reducing the period for which the crop may be exposed to the virulent pathogen or propagules available in the environment/substrate. In the case of viruses or other vector-borne pathogens, the date of sowing has to be so adjusted that the crop is not exposed to the period, when the vector population is high with active movements. The pattern of incidence of the disease(s) on a crop in a given location has to be studied over several seasons or years to plan a realistic proposal for avoiding the disease by suitable manipulation of date of sowing/planting. Although this approach may appear to be simple, some problems imposed by crop requirements and seasons have to be overcome.

By selecting appropriate dates of sowing, some diseases could be significantly controlled as in the case of wheat foot rot disease caused by *Pseudocercosporella herpotrichoides*. When wheat was sown in late September or early October, the incidence of foot rot disease was markedly reduced in Washington and other adjacent states of USA. The take-all disease of wheat caused by *Gaeumannomyces graminis* var. *tritici* could be reduced by early sowing of winter wheat (late October to early November) in irrigated areas (Cook and Baker 1983). Likewise, Sclerotinia crown and stem rot disease of alfalfa caused by *Sclerotinia trifoliarum* could be appreciably reduced by combining early sowing and no-till practice (Sulc and Rhodes 1997). Some of the viral diseases can be managed by properly adjusting sowing dates. Early sowing of sugar beet permits its germination and growth, when temperatures are too low for infection of roots by the fungal vector *Polymyxa betae*. On the other hand, late sowing leads to the additional advantage of avoiding the rhizomania disease caused by *Beet necrotic yellow vein virus* for which *P. betae* is the vector (Blunt et al. 1992; Rush and Heidel 1995). The effect of planting dates on the incidence of rice tungro disease (RTD) in the Cauvery delta of Tamil Nadu State, India was studied in detail. Rice crops planted before the first fortnight of August generally showed minimal infection by RTD, while the disease incidence increased markedly in rice crops planted in October and thereafter. The higher disease incidence had a significant positive relationship with the buildup of population of green leafhoppers (Narayanasamy 2002).

Planting density, as determined by seed rate and spacing adopted, has a distinct effect on the amount of foliage canopy and consequently on the microclimate around the crop. It is well known that high rate of seeds sown in nurseries of tobacco, tomato and other vegetable crops along with high soil moisture favor the incidence of damping-off disease due to *Pythium* spp. and *Rhizoctonia* spp., resulting in the death of large number of seedlings in the nurseries. High planting density favors the development of diseases also in the main field after transplanting as in the case of downy mildew of tobacco, sugar beet and pepper (chilli) (Palti 1981). Coffee rust (*Hemileia vastatrix*) and tea blister blight (*Exobasidium vexans*) disseminated by wind may spread rapidly in densely planted crops. Severity of potato silver scurf (*Helminthosporium solani*) and black scurf (*Rhizoctonia solani*) disease increased with increase in plant density (Firman and Allen 1995). Canola seeding rates, likely to affect the microclimate beneath the canopy, were evaluated for their impact on the incidence of stem rot disease caused by *Sclerotinia sclerotiorum*. A significant relationship between Sclerotinia stem rot disease incidence (DI) and seed rate was observed. With an increase in seeding rate, the DI was significantly increased. Lodging of plants of all four cultivars increased, when seeding rates exceeded the recommended seeding rate (6.5 kg/ha). Both plant density and lodging resistance varied in having larger influence on DI depending on the year (2001–2003) and cultivar analyzed. The results indicated that higher seeding rate modified the micro-environment and might increase the potential for lodging, resulting in plant-to-plant spread of the disease (Jurke and Fernando 2006). The effects of plant density on the susceptibility of medium and medium-late maturity maize hybrids to ear rot and to mycotoxin contamination in natural infection conditions were determined. The plant density significantly influenced the percentage of kernels infected by *Fusarium* infection (+24 %) and higher fungal ear rot severity (+43 %) than plots with lower plant densities. However, the plant density did not influence the type of mycotoxin found in the kernel which depended mainly on the climatic conditions during the season and their influence on the infection and the development of different fungal genera and species associated with ear rot disease. The natural occurrence of the mycotoxins found each year (2001–2004) was always significantly higher in crops with higher plant density. The use of a medium-late maturity hybrid instead of a medium maturity hybrid could lead to higher contamination of mycotoxins. Maize crops cultivated in temperate areas at high plant populations might exhibit a significant increase in mycotoxin contamination risk in the kernels (Blandino et al. 2008). Sowing time, plant density and nitrogen fertilization were evaluated for their effect on Fusarium ear rot incidence and severity, kernel infection by *F. verticillioides* and *F. graminearum* and contamination of kernel with fumonisin and deoxynivalenol (mycotoxins). The fumonisin levels were less affected by nitrogen fertilization, plant density and hybrid maturity. The presence of deoxynivalenol was related to late sowing, high plant density and N fertilization (Blandino et al. 2009).

The influence of plant density on the incidence of virus diseases has been investigated. Reduction of incidence of *Groundnut rosette virus* in groundnut (peanut) by adopting closer spacing was suggested as a practical measure of rosette disease management. Closer spacing (2.5 × 90 cm) reduced the rosette disease significantly compared to that in widely spaced (30 × 90 cm) plots (Farrell 1976). The incidence of bud necrosis virus in groundnut could be reduced by adopting closer spacing (15 × 15 cm) followed by elimination of infected plants up to 6 weeks after sowing (Narayanasamy et al. 1976). Similar effect of closer spacing on the incidence of bud necrosis disease was observed also by Mali (1992). Adoption of narrow row spacing (17.5 cm) resulted in reduction of *Bean yellow mosaic virus* incidence in narrow-leaved lupine (*Lupinus angustifolius*) (Jones 1994). The effect of varying plant density and crop variety on the incidence of virus diseases of cowpea (*Vigna unguiculata*) was investigated with the aim of identifying tolerant cowpea cultivars and determining optimum planting density that could lower viral disease incidence and increase cowpea yield. Planting density and variety had significant effect on the incidence of *Cowpea aphid-borne mosaic virus*, *Cowpea mild mottle virus*, *Cowpea mosaic virus*, *Cowpea chlorotic mottle virus* and *Cowpea golden mosaic virus*. The lowest mean disease incidence was recorded in plots with planting density of 25 × 75 cm, compared with planting density of 50 × 75 cm and 75 × 75 cm. However, planting at a mid-level density 50 × 75 cm with variety IT89KD-288 showed lowest disease incidence and outperformed the other combinations for yield. Hence, this combination was recommended for wider application (Aliyu and Balogun 2011).

2.6 Irrigation Practices

Irrigation, probably, has greater influence on the development of both host plant and microbial plant pathogens than other agricultural practices. Irrigation is essential to maintain required soil moisture levels for the absorption of plant nutrients from the soil and for maintaining turgidity of plan cells/tissues. Both high moisture and water stress may predispose plant to infection by different kinds of pathogens. Four kinds of irrigation techniques are used for various crops depending on the quantum of water available/required. Flooding is done, when the quantity of water is abundantly available, but this system is likely to spread the soilborne inoculum throughout the field. Furrow irrigation saves some water, but spreads the pathogen along the furrows. The sprinkling system restricts the amount of water used further, but causes enormous splash dissemination of the pathogen inoculum, while surface dissemination is limited. The trickling (drip) system conserves water requirement and aids pathogen dissemination to some extent. A combination of subsurface drip irrigation and minimum tillage practices more effectively reduced the incidence of lettuce drop disease caused by *Sclerotinia minor* and the severity of corky root disease caused by *Rhizomonas suberifaciens*, compared with furrow irrigation and conventional tillage. Differential moisture and temperature effects were considered to be responsible for disease suppression (Bell et al. 1998). The effect of three irrigation methods, daily drip, 3-day drip and alternate row furrow irrigation on the incidence of Phytophthora root rot disease caused by *Phytophthora capsici* in pepper (*Capsicum annuum*) was compared. Root rot incidence in the infested plots was higher under alternate furrow irrigation than in other two systems of irrigation. Drip irrigation increased the yield of marketable green fruits by providing either favorable soil moisture conditions or unfavorable condition for the pathogen development (Xie et al. 1999). *Phytophthora capsici* produces zoospores that initiate infection of roots primarily. Conditions of wet and dry cycles in soil are required for completion of different stages in its life cycle. Rainfall and periodic furrow irrigation provide the wet-dry cycle in the soil, favoring sporangia production during dry period and zoospore release during wet period. The role of irrigation methods (drip, basin and furrow) in the development of pepper crown rot disease caused by *P. capsici* was studied. Disease percentage was higher (78.39 %) in drip irrigation system than in basin (56.27 %) and furrow irrigation (55.92 %) methods (Sağir et al. 2005).

The type of irrigation can be very important in terms of disease risk. Sprinkler irrigation results in wetting of the entire crop canopy and soil surface, compared to furrow or flood irrigation and as a consequence may increase the risk of foliar diseases. Frequent irrigation provides a favorable microclimate for downy mildews which are almost absent in areas with low rainfall and relative humidity. Favorable conditions for the incidence of potato late blight disease caused by *Phytophthora infestans* in northwestern United States and Israel were created by sprinkler irrigation system (Rotem and Palti 1969). In dry lands of Washington, USA, use of overhead sprinkler irrigation upset the biological balance existing before the introduction of this practice. In these fields, increase in the incidence of diseases due to *Verticillium dahliae*, *Gaeumannomyces graminis* var. *tritici* and *Fusarium* spp. was observed over the years (Baker and Cook 1974). Peanut (groundnut) late leaf spot, rust and *Tomato spotted wilt virus* (TSWV) diseases affected peanut crops more severely in plots irrigated by sprinkler system than in plots receiving surface irrigation. Splash dissemination of spores and disturbance to thrips (vector of TSWV) by sprinkling water may account for the increased disease incidence/spread in plots receiving water through sprinkler irrigation system (Ganapathy 1985). Supplemental irrigation under drought stress conditions may improve the growth and yield of potatoes. But it can influence the incidence of tuber diseases. Surface sprinkler irrigation was applied based on tensiometer or moisture block readings deployed in field plots. Black scurf (*Rhizoctonia solani*), black dot (*Colletotrichum coccodes*), silver scurf (*Helmithosporium solani*) and common scab (*Streptomyces scabei*) were quantified on potato tubers randomly sampled at harvest and kept at storage temperature of 7.2 °C, before visual assessment. The mean incidence of black scurf, silver scurf and black dot ranged from 3 to 18 %, 2 to 33 % and 4 to 7 % respectively in best, unirrigated and reduced irrigation respectively. The results suggested that irrigation might lead to increase in disease incidence. However, crop rotation and application of amendment may reduce incidence of potato tuber diseases selectively to some extent (Olanya et al. 2010).

2.7 Effects of Other Crops

Many other crops may be grown in rotation with the primary crop in different seasons or as intercrop along with the primary crop in the same season. Incidence of diseases may be influenced by other crops to varying degrees with either positive or negative effects.

2.7.1 Crop Rotation

Crop rotation or sequence refers to a system of inclusion of appropriate crops in a sequence for a definite period of time (usually 2 or 3 years in most cases). The success of crop rotation as a biological method of disease control depends on several factors including time, the environment, and the nature of the pathogen(s) to be managed and characteristics of the host crop. Rotation to a non-host(s) for the pathogen concerned for a sufficient period allows enough time for the decomposition of infected crop residues and/or a reduction in the viability of pathogen survival structures and the ability of the pathogen to produce inoculum, thus eliminating a potential source of disease. Crop rotation is likely to be an effective disease management strategy under certain conditions only. The source of the inoculum must be from the field concerned or the potential movement of the pathogen from other adjacent fields is limited. Crop rotation may not be effective for seed-borne pathogens such as cereal smut pathogens. The host range of the pathogen has to be narrow and crop rotation may not be effective for diseases like Sclerotinia stem rot of canola, because the pathogen has an extensive host range which includes several crops and weed species. Another requirement for the effectiveness of crop rotation is the inability of the pathogen to survive for long time in the absence of the crop plant species. If the pathogen has the ability to produce resting spores such as sclerotia or chlamydospores which can remain viable for several years, crop rotation may not be a suitable option for the management of such pathogens. Crop rotation may be the most effective for the diseases caused by pathogens that are soilborne or plant residue-borne.

Inclusion of legumes in crop rotations provides double benefits because of their ability to fix atmospheric nitrogen that can be used by the succeeding crops and to reduce the incidence of the target disease. Lupin included in the rotation either in the preceding year or even 2 years previously, significantly reduced the incidence of wheat take-all disease caused by *Gaeumannomyces graminis* var. *tritici* and increased the yield compared with monoculture of wheat (Reeves et al. 1984). For the management of wheat take-all disease and winter wheat Cephalosporium stripe disease caused by *Cephalosporium graminearum*, a 2-year or 3-year rotation was found to be effective. The sequence of wheat-potato, wheat-corn, wheat-beans or wheat-sugar beet for wheat take-all disease and wheat followed by spring wheat, spring barley, peas or lentils for winter wheat stripe disease was recommended (Bruehl 1968; Cook and Baker 1983). The inclusion of a legume cover crop was found to be useful for improving soil fertility as well as disease suppression. Hairy vetch (*Vicia villosa*), when incorporated into the soil, exerted differential influence on the pathogens causing cotton disease complex. The incidence of black root rot disease caused by *Thielaviopsis basicola*, soil populations of *T. basicola* and the frequency of isolation of the pathogen from diseased cotton seedling were significantly reduced, following cultivation of hairy vetch as a cover crop in winter compared to fallowing the field. On the other hand, soil populations of *Rhizoctonia solani* and *Pythium* spp. forming the components of the disease complex increased after hairy vetch as a cover crop (Rothrock et al. 1995). This differential effect was attributed to the increase in soil atmospheric ammonia levels in hairy vetch amended soil. *T. basicola* was more sensitive to the ammonia than *R. solani* and *Pythium ultimum* (Candole and Rothrock 1997).

Incidence and severity of common root rot caused by *Cochliobolus sativus* was reduced in wheat following lupine, compared to wheat following wheat. Significant increase in yield of wheat following lupine was also obtained and this enhanced yield was considered to be due to suppression of disease and higher nitrogen status of the soil (Doyle et al. 1988). In another investigation, crown root rot of wheat caused by *Fusarium graminearum* was reduced and wheat yield increased by a chickpea-wheat versus wheat-wheat rotation. The mean incidence of crown rot was 12 % for chickpea-wheat rotation, compared with 30 % infection in wheat-wheat system (Felton et al. 1998). The effect of crop rotation including canola, lupine, medic, medic-clover mixture on the incidence of wheat crown root rot disease was investigated. The incidence of *F. graminearum*-induced disease was significantly higher on wheat for the rotation treatment that included wheat and canola. Highest disease incidence and severity were recorded for this treatment. Medic-clover mixture as the rotation component reduced the incidence of the crown root rot to the maximum extent and enhanced the yield of wheat grains (Lamprecht et al. 2006). Rhizoctonia bare patch disease caused by *Rhizoctonia solani* AG-8 increased in no-till cereals and the disease was considered as a major limiting factor for adoption of no-till technology. In an 8-year experiment, crop rotation had no effect on the bare patch incidence during the first 5 years and from years six to eight, both soft white and hard white classes of spring wheat (*Triticum aestivum* L.) grown in a 2-year rotation with spring barley had an average of only 7 % of total land area with bare patches, compared with 15 % in continuous annual soft white wheat or hard white wheat (monoculture). Both classes of wheat had less bare patch area and produced higher grain yield, when grown in rotation with barley. But monoculture of hard white was more severely affected by *Rhizoctonia solani* than soft white wheat. The results indicated that Rhizoctonia bare patch disease could be suppressed in no-till system by adopting rotation of cereal crops (Schillinger and Paulitz 2006).

The effect of mixed cropping on disease suppressiveness of soils was tested for two cropping systems, Brussels sprouts-barley and triticale-white clover. No significant effects of mixed cropping on general disease suppressiveness of three different soils to three different soilborne pathogens, *Rhizoctonia solani*, *Fusarium oxysporum* f.sp. *lini* and *Gaeumannomyces graminis* var. *tritici* could be observed. However, for Fusarium wilt disease, soils cropped with barley had the highest level of disease suppressiveness, compared with soils cropped with Brussels sprouts and the mixed crop. The results of the bioassay GGT1 for *G. graminis* var. *tritici* indicated a weak, but similar trend for suppressiveness of soils for the cropping systems and soilborne pathogens tested in this investigation (Hiddink et al. 2005). Take-all decline is considered a consequence of the development of antagonistic microbial community in the soil. The population of take-all pathogen *G. graminis* may be much reduced by a break crop which is other than a cereal that is not susceptible to the target pathogen. The introduction of barley into a sequence of wheat crops may result in less take-all than would be expected in another crop of wheat, as barley is more resistant to the disease. Effects of different cultivated or weed grasses as pure stands or in combination with wheat on take-all and its suppression in subsequent wheat crops were evaluated. Annual brome grasses maintained take-all inoculum in the soil as efficiently as wheat (grown as a continuous sequence) and much better than the cultivated species with a perennial habit. Take-all developed more in wheat grown after *Anisantha sterilis* (barren brome) or *Bromus secalinus* (rye brome) with or without wheat, than in continuous grass-free wheat in the same year, whereas take-all decline was apparently occurring. While annual grass species were found to be more effective hosts of the take-all pathogen, some of the perennial grasses were more effective obstacle to the development of suppression (Gutteridge et al. 2006).

Corn anthracnose disease caused by *Colletotrichum graminicola* occurs in two phases as leaf blight and stalk rot. The effect of soybean-corn crop rotation on the incidence and severity of leaf blight and stalk rot diseases was assessed. Continuous corn cropping increased the leaf blight incidence and severity of 91 % and 24–78 % respectively over that of soybean-corn rotation. The corn residue was found to be an important source of primary inoculum and the buildup of inoculum was accelerated by continuous corn cropping. Managing residue levels through crop rotation could be expected to reduce anthracnose leaf blight (Jirak-Peterson and Esker 2011). Rotation of pepper (*Capsicum annuum*) with peanut or sesame resulted in reduction of the incidence of blight caused by *Phytophthora capsici* by 39 and 11 % respectively (Kim 1989). Nonhosts include in the rotation and/or fallowing reduced the selection pressure for a given soilborne pathogen and 'starve' it out, preventing the buildup of large populations. The frequency and composition of crops in a potato rotation influence the incidence and development of soilborne diseases affecting potato crops. *Phytophthora erythroseptica* causing pink rot disease of potato can survive for many years in soil as oospores and propagules of the pathogen which are endemic in most cultivated soils where potatoes are grown repeatedly (Lambert and Salas 2001). Crop rotation with at least one alternate crop was found to reduce the pink rot disease incidence significantly (Lambert et al. 2001). In a later investigation, the effect of 2- and 3-year crop rotations and conservation tillage practices on the severity of pink rot disease was assessed. Barley was included in 2-year rotation, while red clover was also included additionally in the 3-year rotation. The development of pink rot disease due to *P. erythroseptica* was significantly less pronounced in potatoes from 3-year rotational soils than from 2-year rotational soils. Potato plants grown in greenhouse using soils from field plots, had less disease in the 3-year rotation than in 2-year rotation ($P=0.05$). The results suggested that potatoes grown in soils managed under a 3-year rotation appeared to be intrinsically more resistant to infection by *P. erythroseptica* than those managed under a 2-year rotation (Peters et al. 2005).

A system of cultivar rotation was suggested for the management of Verticillium wilt of cotton caused by *Verticillium dahliae*. The cotton cultivar Acala SJ-4 was relatively tolerant to Verticillium wilt and could yield better than susceptible cv. SJ-2 in fields with high inoculum densities (9–13 microsclerotia/g of soil). The cv. SJ-2 had higher yield potential in soils with lower inoculum densities. It was recommended that cv. SJ-4 could be grown in fields until the pathogen population was reduced below the threshold level and then cv. SJ-2 might be grown to reduce the damage to this cultivar (Ashworth and Huisman 1980). The efficiency of crop rotation in eliminating *Verticillium dahliae* causing cauliflower wilt disease was assessed. Incorporation of broccoli residue resulted in significant reduction in number of propagules of *V. dahliae* and the decline continued throughout the cauliflower season. Rotation of broccoli with cauliflower followed by incorporation of broccoli residue was suggested as a practical measure of control for this disease (Xiao et al. 1998). Adoption of a rotation system of 2 years of cotton and 1 year of sorghum reduced the density of microsclerotia of *V. dahliae* (2/cm^3 of soil), compared with continuous cotton (23/cm^3 of soil). Crop rotation avoiding excessive irrigation and using a partially resistant cultivar reduced incidence of Verticillium wilt and improved net returns (Wheeler et al. 2012).

As crop rotation with broccoli has been shown to be effective in reducing Verticillium wilt disease infecting cauliflower, similar approach was adopted for the management of Verticillium wilt disease in strawberry. The effects of broccoli and lettuce rotations on population densities of *Verticillium dahliae* and *Pythium* spp. in soil and on strawberry (*Fragaria* × *ananassa*) growth and yield were determined in conventional and organic production systems. Strawberry was planted after two successive crops of broccoli or lettuce. In the control treatment in the conventional field, the soil was fumigated with methyl bromide + chloropicrin prior to strawberry planting. No difference in the inoculum densities of *V. dahliae* was seen in organic and conventional plots following lettuce rotations. But in plots following broccoli rotations, decrease in the inoculum densities of *V. dahliae* in organic plots was greater than in conventional plots. Crop rotation treatments did not have any consistent effect on the inoculum densities of *Pythium* spp. Verticillium wilt incidence in strawberry was lower by 12–24 % in fields rotated with broccoli, compared with fields rotated with lettuce. Likewise, wilt severity was also reduced to a greater extent in fields rotated with broccoli, compared with the fields rotated with lettuce. The usefulness of employing broccoli rotation coupled with postharvest incorporation of broccoli residue was indicated by this investigation (Njoroge et al. 2009).

The effect of rotation with broccoli or Brussels sprouts on the inoculum densities of *Verticillium dhaliae*, infecting strawberry was assessed. Inoculum densities of *V. dahliae* were reduced to a greater extent, when broccoli or Brussels sprouts were grown as rotation crop, compared with lettuce. The yield increase was proportional to the reduction in inoculum densities of the pathogen which had direct effect on disease incidence (Fig. 2.4a, b). The mechanism of broccoli-mediated disease control was suggested to be biofumigation, due to degradation of glucosinolates in the crop residue (Martin and Bull 2002). The effectiveness of rotation of strawberry with broccoli, Brussels sprouts and lettuce in reducing soilborne inoculum of *Verticillium dahliae* and *Pythium* spp. and severity of wilt disease was assessed. Rotation did not alter the total population levels of *Pythium* spp. at two sites tested. On the other hand, *V. dahliae* microsclerotia were significantly reduced with broccoli and Brussels sprouts rotations, compared with lettuce rotations at *V. dahliae*-infected site. Reduction in propagules was reflected by lower Verticillium wilt severity on strawberry plants in the broccoli and Brussels sprouts rotations, compared with lettuce rotations. Further, strawberry vigor and fruit yield were significantly lower in lettuce-rotated plots. None of the rotation treatments was better than the standard chemical treatment with methyl bromide + chloropicrin for all the variables quantified. Rotation with broccoli and Brussels sprouts could be an effective cultural practice for managing strawberry wilt disease, in the absence of chemical fumigation (Subbarao et al. 2007). Adoption of a rotation system of Chinese leek-banana reduced the Fusarium wilt of banana caused *by F. oxysporum* f.sp. *cubense* (*Foc*) incidence and severity index by 88–97 % and 91–96 %

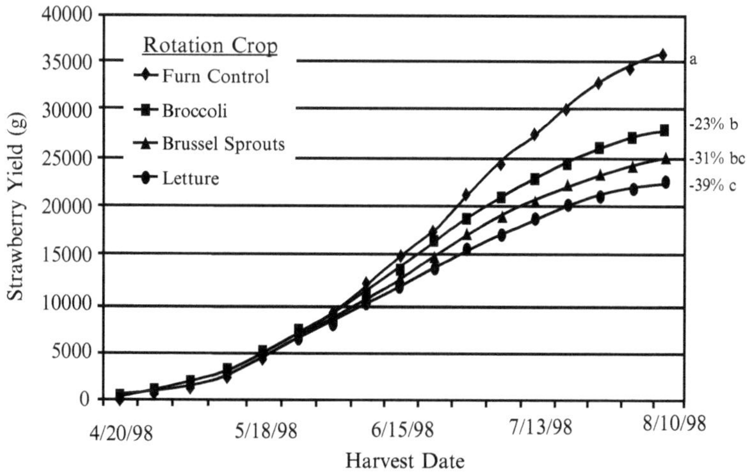

Fig. 2.4 Effect of crop rotation on yield of fruits of strawberry from fields infested by *Verticillium dahliae*. *Numbers followed by same letters* are not significantly different (P = 0.05) (Courtesy of Martin and Bull 2002 and with kind permission of The American Phytopathological Society, MN, USA)

respectively and improved the crop value by 36–86 % in an area heavily infested by *Foc* during 2007–2009 (Huang et al. 2012).

Sclerotinia minor, causative agent of lettuce drop disease, has a wide host range. Use of crop rotation as a disease management strategy for reducing the incidence in lettuce was examined. The efficiency of broccoli as a rotation crop was assessed for reducing the number and viability of sclerotia of *S. minor* and the incidence of lettuce drop in the field. Continuous lettuce did not increase the lettuce drop disease incidence for at least 2 years, although an increase in soilborne sclerotia was observed annually. Rotation with broccoli resulted in small reduction in disease incidence only in the first year. The density of sclerotia was the lowest in plots with lettuce, followed by a fallow period (LFL) and highest in plots with continuous lettuce (LLL). In another location, rotation with broccoli significantly reduced both sclerotia and lettuce drop disease incidence. The number of broccoli crops rather than the sequence of lettuce rotations with broccoli was critical for reducing the numbers of sclerotia of the pathogen in the soil. Fallowing after a lettuce crop caused marginal reduction in sclerotial number and disease incidence. Incorporation of broccoli residue reduced the numbers of sclerotia, possibly because of volatiles formed during the breakdown of glucosinolates present in broccoli. The results indicated that broccoli as a nonhost of *S. minor* and the glucosinolates present in its tissues could be desirable candidate(s) for use as a component of crop rotation for the control of lettuce drop disease (Hao et al. 2003). The effect of growing *Brassica* crops as green manure or cover crops on the incidence of potato stem rot disease caused by *Sclerotinia sclerotiorum* was assessed. *Brassica napus*, *B. juncea* and *B. campestris* were raised in three naturally infected potato fields during three cropping seasons of 2008–2010. *Brassica* crops significantly reduced the disease incidence and mean percentage of dead plants. The extent of disease reduction varied in different field sites. However, growing *B. juncea* was the most effective in reducing the disease incidence by 55.6 % (average over all years and sites), whereas average of disease reduction percentage for *B. napus* and *B. campestris* were 31.6 and 45.8 % respectively. The results showed that the cultivation of *Brassica* crops could be considered as a component of potato production management system (Ojaghian et al. 2012).

The effects of a cropping system designed specifically for suppression of soilborne diseases of potato and also for their effect on soil microbial community characteristics were assessed. The efficacy of crop rotation, as a disease management strategy for the control of bacterial diseases has been evaluated in a few pathosystems. The effect of previous crop in the incidence of tomato bacterial wilt disease caused by *Burkholderia* (*Ralstonia*) *solanacerum* was assessed. The population of *B. solanacearum* declined after cowpea and rice, but not after eggplant (brinjal) and a similar effect was also observed on disease incidence, when these crops were raised prior to tomato. The yield of tomato was adversely affected by cultivation of eggplant prior to tomato (Michel et al. 1996). The disease suppressive (DS) system included diverse crops with known disease suppressive capability. The DS system consisting of *Brassica* and Sudan grass as green manure crops, fall cover crops and high crop diversity resulted in greatest reductions in stem and stolon canker and black scurf (*Rhizoctonia solani*) and common scab (*Streptomyces scabies*) under both irrigated and non-irrigated conditions, compared with other cropping systems designed for soil conservation (SC), soil improvement (SI) and a status quo standard rotation (SQ) as control. All cropping systems reduced Rhizoctonia stem and stolon canker of potato caused by *R. solani*, relative to the continuous potato (PP) with individual yearly reductions of 10–50 % and overall average reductions of 20–30 % (all years considered together). However, no significant differences among the cropping systems in any year or average over all 3 years could be seen (Fig. 2.5). DS also produced significant shifts in soil microbial community characteristics different from all other rotations. Biofumigation was considered to be the mechanism of actions for these crops. However, additional mechanisms including specific changes in soil microbial communities not related to levels of glucosinolates or other toxic metabolites could be important in the reduction of soilborne diseases by *Brassica* crops. Soil microbial population and microbial diversity may be increased due to interactions between the components of cropping system. Fatty acid methylester (FAME) profiles revealed distinct differences in soil microbial community characteristics associated with DS system. Significant differences in the bacterial and fungal communities were indicated by FAME analysis. A cropping system incorporating disease suppressive rotations, cover crops and crop diversity might provide opportunities for enhanced sustainable production and disease management (Larkin et al. 2011).

Aspergillus flavus causes the aflaroot disease in peanut and the kernels contaminated with aflatoxins produced by *A. flavus* pose potential danger to human and animal health. Investigations on crop sequences (rotation) that could reduce *A. flavus* populations and identification of microbial groups that could reduce the population

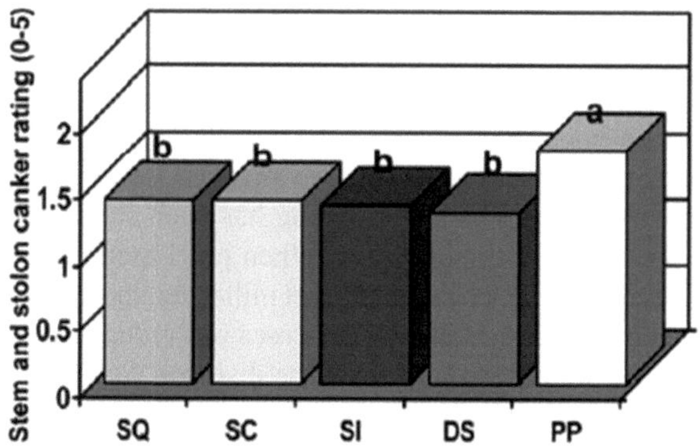

Fig. 2.5 Effect of cropping system on average severity of stem and stolon canker disease of potato under field conditions during 2006–2008. *SQ* status quo system, standard 2-year rotation (barley/clover-potato), *SC* soil-conservation system, 3-year rotation, limited tillage (barley/timothy-timothy-potato), *SI* soil-improving system, 3-year rotation, same as SC, but with yearly compost amendments, *DS* disease suppressive system, 3-year rotation (mustard blend GM/rapeseed cover crop-sudan grass GM/rye cover crop-potato), *PP* continuous potato, non-rotational control. *Bars with same letter* are not significantly different based on analysis of variance and Fisher's protected least significant difference test (P=0.05) (Courtesy of Larkin et al. 2011 and with kind permission of The American Phytopathological Society, MN, USA)

levels of the aflatoxin-producing fungi in peanut soils were carried out. Peanut cropping sequences, continuous peanut, 4 years of continuous bahia grass followed by peanut, peanut-corn-cotton and peanut-cotton were evaluated. Ribosomal intergenic spacer analysis (RISA) provided greater solution and analysis of soil microbial diversity than other nucleic acid-based techniques. Soil enumeration indicated that using bahia grass in rotation with peanut resulted in lower *A. flavus* population levels, compared with other cropping sequences. Fungal RISA profiles showed similarities among communities for replicate plots of the same crop rotation. The presence of common bands in all cropping sequences might be attributed to the presence of common saprophytes in these soils. The minimum population density of 2.6×10^6 CFU/g soil was necessary for detection of *A. flavus* in the soil. The molecular technique like RISA to determine microbial community structure with emphasis on aflatoxin-producing fungi may provide a better understanding and consequent development of management approaches for the aflatoxin problem (Sudini et al. 2011).

2.7.2 Multiple Cropping

The system of growing two or more crops as intercrop, mixed crop or barrier crops has been a traditional practice followed in several countries. Two or more crops may be sown simultaneously or in a staggered manner so that the growth periods of the crops may overlap. Generally, the crops that are not infected by the same pathogen(s) are selected. Crops which are planted as intercrop or mixed crop may (i) modify the microclimate of another crop; (ii) intercept or filter out the spores of the pathogens; (iii) restrict the movement of the vectors of viruses or other pathogens and (iv) reduce the infectivity of stylet-borne viruses which are lost, when the vectors feed on nonhosts. Intercropping or mixed cropping has been shown to be effective in reducing the incidence of several diseases. When pearl millet was grown as intercrop with peanut (groundnut) (one row of pearl millet and three rows of peanut), the severity of early and late leaf spot and rust diseases was reduced in six peanut varieties with varying degrees of resistance to these diseases. Sorghum as an intercrop with peanut reduced the incidence of rust disease (ICRISAT 1981). Rust disease was less severe on bean, when it was intercropped with corn (Moreno 1979). In Kenya, bean cultivated along with corn as intercrop had much less incidence of several diseases such as anthracnose, common blight, halo blight, powdery mildew and scab (Van Rheenan et al. 1981).

Wheat (*Triticum aestivum*) and field beans (*Vicia faba*) were grown in an organic system as sole crops (SC) and additive intercrops (IC). All of the intercrops accumulated more nitrogen than the sole-cropped wheat, but did not exceed that accumulated by sole-cropped beans. Both powdery mildew (*Erysiphe graminis*) and brown rust (*Puccinia recondita*) were at very low levels. The plant density had no significant effect on either powdery mildew or brown rust. In contrast, increasing the density of beans increased the incidence of mildew on the lowest leaf of wheat. The presence of beans at 25 % (recommended density, RD) also significantly increased ($P<0.05$) the incidences of brown rust on flag leaf, though bean densities had no significant effect. The increase in powdery mildew with increasing bean density probably related to the increase in the nitrogen content (N %) of wheat. Increase in bean density significantly increased the incidence of chocolate spot disease caused by *Botrytis cinerea* on bean. Wheat density did not have any significant effect on the incidence of chocolate spot disease (Bulson et al. 1997). Planting shade trees to shield cocoa plants against infection by *Verticillium dahliae*, reduced the incidence of disease (Trocmé 1972). Intercropping of cotton with sorghum or *Phaseolus aconitifolius* reduced root rot caused by *Rhizoctonia solani* and *R. bataticola* by reducing soil temperature and maintaining high humidity. The Phytophthora blight of red pepper (*Capsicum annuum*) caused by *Phytophthora capsici* was effectively suppressed by growing sesame (*Sesamum indicum*) or peanut (*Arachis hypogaea*) as intercrops in Korea (Kim 1989). Extracts of soils on which peanut or sesame was cultivated and the root exudates of these intercrops inhibited mycelial growth, sporangium formation and release of zoospores by the pathogen. In soils amended with these extracts, the incidence of Phytophthora blight was markedly reduced (Lee et al. 1990, 1991). The effect of intercropping in three consecutive cropping seasons on the incidence of barley diseases was assessed by including dual grain legume (pea, faba bean and lupine)-barley intercropping (IC), as compared to the respective sole crops (SC). The stability was not greater than that of grain legume SC. The ICs were effective at suppressing weeds, capturing a greater share of available resources than SC. Reduction in disease incidence was observed in all IC systems, compared to the corresponding SC with a general disease reduction in the range of 20–40 %. Barley net blotch disease caused by *Pyrenophora teres* was the most serious during the 3 years of experimentation, in addition to the infection by brown rust caused by *Puccinia hordei*. Incidence of Ascochyta blight (*Mycosphaerella pinoides*), chocolate spot (*Botrytis cinerea*) and brown spot (*Peronospora*) on grain legumes was also reduced in intercrops. In the case of brown spot, disease incidence was drastically reduced by almost 80 % by intercropping (Hauggaard-Nielsen et al. 2008).

Plasmodiophora brassicae causing club root disease of crucifers produces resting spores that can persist in the soil for more than 15 years (Wallenhammer 1996). Germination of the resting spores is presumed to be stimulated in the presence of some nonhost plant species. The effect of four nonhost plants on *P. brassicae* was investigated in a 3-year field experiment and a 14-month glasshouse experiment. None of the plant species, leek (*Allium porrum*), winter rye (*Secale cereale*) and perennial rye grass (*Lolium pernne*) and clover (*Trifolium pratense*) reduced the concentration of *P. brassicae* in soils, when tested with bioassay host Chinese cabbage. In the glasshouse experiment, there was a reduction in disease incidence in all plant treatments, compared with plant-free controls, following incorporation and decomposition of plant roots. The results indicated that use of nonhost plant species for reducing club root disease incidence would not be an effective approach (Friberg et al. 2006). On the other hand, intercropping field vegetables with other crops has been shown to have potential to be used as alternative to the application of chemicals for reducing the disease incidence. Further, intercropping with commonly used vegetable crops fits into environmentally acceptable and sustainable crop production practices adopted by various farms. Vegetable intercrops, carrot, spider plant (*Cleome gynandra*) and French bean (*Phaseolus vulgaris*) were evaluated for their efficacy in reducing the incidence of downy mildew (*Peronospora destructor*) and purple blotch (*Alternaria porri*) diseases on onion. Each vegetable was intercropped with three onion cultivars Bombay Red, Red Creole and Orient (F1) and development of downy mildew and purple blotch diseases was determined until physiological maturity. The vegetable intercrops reduced severity and incidence of both diseases on all three onion varieties. However, only spider plant showed consistent and significant ($P \geq 0.05$) reduction of disease incidence, severity and the area under disease progress curve (AUDPC) up to 24.5 % on all onion varieties. Although intercropping onion with vegetables reduced onion bulb yield, it improved the gross return per unit area. The results showed that intercropping bulb onion with vegetables could be beneficial in reducing foliar diseases and improving gross return per unit area (Narla et al. 2011).

Intercropping practice has been reported to be effective in reducing the incidence and/or severity of bacterial diseases of field crops. The effect of intercropping cowpea (*Vigna unguiculata*) with cassava or maize on bacterial blight disease of cowpea caused by *Xanthomonas axonopodis* pv. *vignicola* was determined in five cropping systems viz., cowpea monoculture at high density and at low density, cowpea-maize ‘within row’ and cowpea-maize ‘in alternate rows’. Disease incidence was higher in cowpea monoculture at high density than in cowpea-cassava ‘in alternate rows’ in

1996 and in cowpea-maize 'in alternate rows' in1997. The disease severity index was reduced by 50 % in cowpea-cassava 'in alternate rows' pattern in 1997, but not in 1996. Highest percentages of leaves with spots or blight occurred only in the cowpea monoculture in high density. Comparing the land efficiency, use of cowpea monoculture at high density to intercropping, the land equivalent ratio (LER) was similar in the intercropping systems with 'alternate row' patterns and the yield loss of cowpea in intercropping was compensated by the additional yield of intercrops (Sikirou and Wydra 2008). The effectiveness of intercropping tobacco with garlic in reducing tobacco bacterial wilt disease caused by *Ralstonia solanacearum* (*Rs*) was assessed in 2008 and 2009. Tobacco bacterial wilt disease development was significantly inhibited by intercropping with garlic in both years. The appearance of the disease was delayed for about 15 days in intercropped fields. The population of *Rs* in the rhizosphere was significantly lower in intercropped fields (138×10^4–161×10^4 CFU/g dry soil) than in monocultured fields. The results indicated the potential of intercropping of tobacco with garlic for reducing the bacterial wilt disease and obtaining higher monetary return for the grower (Lai et al. 2011).

Any form of plant diversification like mixed cropping, cover crops, border plants, intercrops and trap crops, used to protect a primary crop from insect-transmitted viral diseases may be called as barrier cropping. The influence of intercrops and barrier crops on the incidence of *Tomato yellow leafcurl virus* (TYLCV) was reduced, when tomato was interplanted with cucumber (Al Musa 1982) or with pigeonpea (Yassin and Abu Salih 1976). Incidence of peanut bud necrosis disease caused by *Tomato spotted wilt virus* (TSWV) was significantly reduced by growing sorghum or pearl millet as intercrop or barrier crop (Ganapathy and Narayanasamy 1991). A wide barrier of oats reduced the incidence of *Bean yellow mosaic virus* and *Pea mosaic virus* from 99 to 13 %. Oats as a barrier crop significantly reduced the incidence *of Watermelon mosaic virus* (WMV) also (Chalfont et al. 1977). Cultivation of other crops as inter-, mixed- or barrier crops has been shown to reduce the incidence of several viral diseases. Infection of capsicum (pepper) by *Potato virus Y* (PVY) was reduced by >50 % by growing sunflower as border crop to protect pepper (Simmons 1957) and in potato with borders of sorghum, soybean or wheat (DiFonzo et al. 1996). Likewise, the incidence of *Broadbean yellow mosaic virus* was reduced by raising barley as border strips (Jayasena and Randles 1985). Incidence of PVY and *Cucumber mosaic virus* (CMV) in pepper was significantly reduced by growing maize or sunflower as barrier crops. Reduction in disease incidence was due to decrease in infectivity of vectors that landed on maize or sunflower before reaching capsicum crop (Avilla et al. 1996). In a later study, the effectiveness of growing sorghum, corn or vetch as border crop for protecting peppers against PVY and CMV was assessed. Sorghum and vetch borders were found to be effective in only 2 of 4 years tested and the level of reduction in disease was less than 15 % (Fereres 2000). The influence of intercropping with grain sorghum, peanut, soybean or corn on the incidence of *Water melon mosaic virus* (WMV) and *Papaya ringspot virus* type-W (PRSV-W) in pumpkin was assessed in five field trials over 3 years. Reduction in disease incidence achieved by intercropping with grain sorghum ranged from 43 to 96 % ($P\leq0.05$). Surrounding pumpkin plots with borders of peanut, soybean or corn was not effective. Borders of grain sorghum were effective, but disease control was generally less than for the intercrop treatment. Intercropping soybean and peanut with pumpkin reduced disease incidence to a lesser extent than intercropping with grain sorghum (Damicone et al. 2007).

The effectiveness of intercropping as a management strategy for virus diseases and its relationship with vector, compatibility of intercrops with the main crop and the types of cultural practices followed commonly was determined. Three intercropping models viz., cassava-pepper, plantain-pepper and maize-pepper were evaluated for their efficacy in reducing the incidence of *Pepper vein mottle virus* during 2004–2006. Based on the averages of disease incidence for 3 years, the viral disease incidence and severity were significantly reduced by all three intercrops raised with pepper. Viral disease incidence was relatively higher in sole crop of pepper with disease incidence of 42 % and severity of 23.3 %. Plantain-pepper combination recorded the maximum disease incidence of 5–8.33 % and severity of 4–7 %. Maize-pepper fruit combination gave the highest yield (5.98 t/ha), while the sole crop yield was far low (1.54 t/ha). The effectiveness of the intercrops was determined based on the extent of reduction in number of diseased plants, enhanced plant vigor and yield and quality of pepper fruits (Fajinmi and Fajinmi 2010).

The exact mechanisms by which the barrier crops reduce the incidence of virus diseases transmitted by aphids in a nonpersistent manner are not clearly known. The aphids cannot distinguish hosts from nonhosts until they make brief, shallow exploratory probes with their mouthparts. During this process, the virus particles present on their stylets/mouth parts may be released into the protector/barrier plant. Thus the barrier plants may function as virus-sink for nonpersistent viruses. Alternatively, the barrier plants may act as physical barriers and reduce the total number of aphids entering the primary crop resulting in reduction of disease incidence. The barrier plants may protect the primary crop by camouflaging them from vector insects. Another possibility suggested is that the aphids may be attracted by the secondary plants used as trap crops. The protector plants may act as a "decoy" by attracting aphids away from the primary crop plants. The competition between the primary crop and barrier crop for nutrients should not become a limiting factor for its use to protect the primary crop against virus diseases (Hooks et al. 2007).

2.8 Effects of Cultural Practices on Soil Microbial Communities

The effects of crop management practices on microbial populations in the soil have been investigated. Management practices are primarily adopted for the improvement of plant growth and yield potential. Tillage and crop rotations have exerted indirect beneficial effects by reducing pathogen populations and consequently resulting in reduction in disease incidence. Soil disturbance by tillage is a major factor affecting microbial communities, leading to reduction in soil microflora due to desiccation, mechanical destruction, soil compaction, reduced pore volume and disruption of access to food resources. However, soil dwelling organisms are able to improve crop production by releasing available forms of nutrients from soil organic and inorganic sources, fixing nitrogen within plant roots and increasing phosphorus uptake (Carpenter-Boggs et al. 2003). Crop rotation and residue management may regulate soil microbial biomass which mediates residue decomposition, nutrient cycling and organic matter turnover. Microbial communities associated with the rhizosphere may vary with different plant species and component of crop rotation. Cover crops may affect the soil microbial communities and may suppress the development of plant pathogens (Mazzola 1999; Smalla et al. 2001). The effects of 2- and 3-year crop rotation with conventional and minimum tillage treatments on the severity of soilborne diseases of potato canker and black scurf (*Rhizoctonia solani*), dry rot (*Fusarium* spp.) and silver scurf (*Helminthosporium solani*) and the root zone bacteria were investigated. The 2-year rotation included spring barley, while the 3-year rotation had barley and red clover in addition to potato. The disease incidence/severity was reduced to greater extent in 3-year rotational soils compared with 2-year rotational soils. Analysis of root zone bacteria recovered from the rhizosphere (exoroot) and potato root tissues (endoroot) showed that the greatest antibiosis activity inhibiting the growth of soilborne pathogens in vitro occurred in bacterial isolates recovered from the endoroot tissues of 3-year rotation crops under minimum tillage management. The soil agroecosystems can be modified through rotation and conservation tillage practices to improve disease suppression by enhancing antagonistic activities of endophytic and root zone bacteria (Peters et al. 2003, 2005).

The influence of mixed cropping on disease suppressiveness of soils using two cropping systems, Brussels sprouts-barley and triticale-white clover was assessed for the soilborne pathogens *Rhizoctonia solani*, *Fusarium oxysporum* f.sp. *lini* and *Gaeumannomyces graminis* var. *tritici*. Diversity of the indigenous bacterial and fungal microflora was determined with denaturing gradient gel electrophoresis (DGEE) of amplified 16S- and 18S rDNA fragments respectively and related to disease severity determined in different bioassays applied for the three different pathogens. For both cropping systems, mixed cropping did not enhance disease suppressiveness of the soil, except in the case of Fusarium wilt disease. Soils cropped with barley had the highest level of disease suppressiveness, compared with soils cropped with Brussels sprouts and the mixed crop. In some cases, soil cropped to barley alone was significantly more suppressive to *F. oxysporum* f.sp. *lini* than soils cropped to Brussels sprouts or the mixture of Brussels sprouts and barley. Analyses of the diversity of the indigenous bacterial and fungal microflora by DGEE of amplified 16S- and18S- rDNA fragments respectively revealed in most cases, no significant difference between mixed and mono-cropped soils (Hiddink et al. 2005).

Tillage and crop rotation practices have been shown to influence the population densities of soilborne pathogens and also beneficial microflora, especially actinomycetes, *Trichoderma* spp. and *Gliocladium* spp. (Rojo et al. 2007). A long-term field study was taken up to analyze the effect of tillage and crop rotation on the

Table 2.2 Effect of tillage systems and preceding crops on the potential biocontrol agents (PABs) (Gil et al. 2008)

		log 10 CFU/g soil of PABs		
Treatments		Actinomycetes[a]	*Trichoderma* spp.[b]	*Gliocladiumy* spp.[b]
Tillage systems	No tillage	7.00a	4.20a	2.90a
	Disc harrow	5.90b	3.70b	1.90b
	Mouldboard plough	4.80c	3.10c	1.50c
Previous crop	Maize	7.00a	5.40a	4.00a
	Soybean	6.00b	2.80b	1.40b

The values are average of samples taken after peanut sowing and before harvest
For each biocontrol agent, the numbers followed by the same letter are not significantly different according to LSD test at $P<0.05$
[a]Actinomycetes are expressed as $\times 10^4$
[b]Fungi as $\times 10^2$

abundance of actinomycetes, *Trichoderma* spp. and *Gliocladium* spp. and the incidence of peanut (groundnut) root rot caused by *Fusarium solani.* Isolation on specific media and quantification of microbial population were followed. The tillage intensity affected organic matter which in turn was negatively correlated with the abundance of potential biocontrol agents. Higher populations of potential biocontrol agents (PBAs) were associated with reduction of peanut root rot disease incidence (Table 2.2). The biotic soil quality factors might be inversely related to tillage intensity and no-till systems appeared to increase microbial activity. Maize and soybean residues had a strong effect on populations of actinomycetes, *Trichoderma* spp. and *Gliocladium* spp. which were higher, when maize was the previous crop and lower, when it was soybean. The microbial populations could be an index of soil conditions and crop health and their assessment might be useful for the development of sustainable cropping systems (Gil et al. 2008).

The influence of intercropping of lentil with different crops on the incidence/severity of damping-off and root rot diseases caused by *Rhizoctonia solani* and *Fusarium solani* was investigated under greenhouse and field conditions. Intercropping cumin, anise, onion and garlic significantly reduced damping-off and root rot disease and also increased the yield. Highest positive influence was exerted by anise, while onion as intercrop had the least effect. The effect of preceding crop on the disease severity was also assessed. Severity of damping-off and root rot diseases was reduced to the maximum extent, when cowpea and gaur were the preceding crops. Lentil crop following cowpea produced the highest seed yield followed by gaur or millet as preceding crops. In addition, cultivation of cowpea prior to lentil resulted in the lowest population of *R. solani*, whereas sorghum as the preceding crop, most effectively reduced the populations of *F. solani.* The root exudates of crops effective in reducing pathogen populations, reduced mycelial dry weight of the fungal pathogens in in vitro assays (Abdel-Monaim and Abo-Elyousr 2012).

References

Abawi GS, Grogan RG (1979) Epidemiology of diseases caused by *Sclerotinia* species. Phytopathology 69:899–904

Abbasi PA, Riga E, Conn KL, Lazoravits G (2005) Effect of neem cake soil amendment on reduction of damping-off severity and population densities of plant-parasitic nematodes and soilborne plant pathogens. Can J Plant Pathol 27:38–45

Abbasi PA, Conn KL, Lazarovits G (2007) Managing soilborne diseases of vegetable crops with a pre-plant soil or substrate amendment of corn distillation product. Biocontrol Sci Technol 17:331–334

Abdel-Monaim MP, Abo-Elyousr KAM (2012) Effect of preceding and intercropping crops on suppression of lentil damping-off and root rot disease in New Valley-Egypt. Crop Prot 32:41–46

Adebitan SA (1996) Effects of phosphorus and weed interference on anthracnose of cowpea of Nigeria. Fitopatol Brasil 21:173–179

Al Musa A (1982) Incidence, economic importance and control of tomato yellow leafcurl in Jordan. Plant Dis 66:561–563

Aldahmani JH, Abbasi PA, Sahin F, Hoitink HAJ, Miller SA (2005) Reduction of bacterial leaf spot severity on radish, lettuce and tomato plants grown in compost-amended potting mixes. Can J Plant Pathol 27:186–193

Aliyu TH, Balogun OS (2011) Effects of variety and planting density on the incidence of common viral diseases of cowpea (*Vigna unguiculata*) in a Southern Guinea Savannah agro-ecology. Asian J Plant Pathol 5:126–133

Allen SJ, Lonergan PA (2000) Control strategies for Fusarium wilt of cotton in Australia. In: Proceedings of beltwide cotton conferences. National Cotton Council America, Memphis, pp 136–138

Ashworth LJ Jr, Huisman OC (1980) New hope for Verticillium control in cotton. Calif Agric 34:19–20

Avilla C, Collar JL, Duque M, Hernaiz P, Martin B, Fereres A (1996) Barrier crops as a method of control of nonpersistent viruses in pepper. Bol San Veg Plagas 22:301–307

Avizohar-Hershenzon Z, Palti J (1962) *Sclerotium rolfsii* in Israel. Plant Dis Rep 46:418–423

Baker KF, Cook RJ (1974) Biological control of plant pathogens. WH Freeman, San Francisco

Balardin RS, Verona LA, Rebelo JA (1994) Damping-off control in vegetable nurseries with inert materials. Ciência Rural 24:230–241

Bateman DF, Basham HG (1976) Degradation of plant cell walls and membranes by microbial enzymes. In: Heitefuss R, Williams PH (eds) Encyclopedia of plant physiology: physiological plant pathology, vol 4. Springer, New York, pp 124–136

Begum MM, Sariah M, Putch AB, Zainal Abidin MA, Rahman MA, Siddiqui Y (2010) Field performance of bio-primed seeds to suppress *Colletotrichum truncatum* causing damping-off and seedling stand of soybean. Biol Control 53:18–23

Bélanger RR, Bowen PA, Ehret DL, Menzies JG (1995) Soluble silicon: its role in crop and disease management of greenhouse crops. Plant Dis 79:329–336

Bell AA, Liu L, Reidy B, Davis RM, Subbarao K (1998) Mechanisms of subsurface drip irrigation-mediated suppression of lettuce drop caused by *Sclerotinia minor*. Phytopathology 88:252–259

Berbegal M, García-Jiménez J, Armengol J (2008) Effect of cauliflower residue amendments and soil solarization on Verticillium wilt control in artichoke. Plant Dis 92:595–600

Blandino M, Reyneri A, Vanara F (2008) Effect of plant density on toxigenic fungal infection and mycotoxin contamination of maize kernels. Field Crop Res 106:234–241

Blandino M, Reyneri A, Vanara F, Tamietti G, Pietri A (2009) Influence of agricultural practices on *Fusarium* infection, fumonisin and deoxynivalenol contamination of maize kernels. World Mycotoxin J 2:409–418

Blunt SJ, Asher MJC, Gilligan CA (1992) The effect of sowing date on infection of sugar beet by *Polymyxa betae*. Plant Pathol 41:148–153

Bockus WW, Shroyer JP (1998) Impact of reduced tillage on soilborne plant pathogens. Annu Rev Phytopathol 36:485–500

Bruehl GW (1968) Ecology of Cephalosporium stripe disease of winter wheat in Washington. Plant Dis Rep 52:590–594

Bruehl GW (1987) Soil-borne plant pathogens. Macmillan, New York

Bulson HAJ, Snaydon RW, Stopes CE (1997) Effects of plant density on inter-cropped wheat and field beans in an organic farming system. J Agric Sci Camb 128:59–71

Candole BL, Rothrock CS (1997) Characterization of the suppressiveness of hairy vetch-amended soils to *Thielaviopsis basicola*. Phytopathology 87:197–202

Carpenter-Boggs L, Stahl PD, Lindstrom MJ, Schumacher TE (2003) Soil microbial properties under permanent grass, conventional tillage and no-till management in South Dakota. Soil Tillage Res 71:15–23

Carroll TW (1983) Certification scheme against barley stripe mosaic. Seed Technol 11:1033–1042

Castro CM, Motta SD, Pereira DS, Akiba F, Ribeiro RLD (1997) Effective microorganisms for control of *Xanthomonas campestris* pv. *vesicatoria* in sweet pepper. www.infrc.or.jp/english/KNF_Data_Base

Chalfont RB, Jaorskic A, Johnson AN, Sumber DR (1977) Reflective film mulches, millet barriers and pesticides: effects on watermelon mosaic virus, insects, nematodes, soilborne fungi and yield of yellow summer squash. J Am Soc Hortic Sci 102:11–15

Colbach N, Lucas P, Maynard JM (1997) Influence of crop management on take-all development and diseases cycles on winter wheat. Phytopathology 87:26–32

Cook RJ (1980) Fusarium foot rot of wheat and its control in the Pacific Northwest. Plant Dis 64:1061–1066

Cook RJ, Baker KF (1983) The nature and practice of biological control of plant pathogens. The American Phytopathological Society, St. Paul

Cooley DR, Gamble JW, Antonio WR (1997) Summer pruning as a method for reducing flyspeck disease on apple fruit. Plant Dis 81:1123–1126

Dale JL (1987) Banana bunchy top: an economically important tropical plant virus disease. Adv Virus Res 33:301–325

Damicone JP, Edelson JV, Sherwood JL, Myers LD, Motes JE (2007) Effects of border crops and intercrops on control of cucurbit virus diseases. Plant Dis 91:509–516

Davis JR, Dole RE, Callihan RH (1976) Fertilizer effects on common scab of potato and the relation of calcium and phosphate-phosphorus. Phytopathology 66:1236–1241

Davis JR, Huisman OC, Westerman DT, Hafez SL, Everson DO, Sorensen LH, Schneider AT (1996) Effects of green manures on Verticillium wilt of potato. Phytopathology 86:444–453

Dharam Singh, Maheshwari VK (2001) Biological seed treatment for the control of loose smut of wheat. Indian Phytopathol 54:457–460

DiFonzo CD, Ragsdale DW, Radcliffe EB, Gudmestad NC, Secor GA (1996) Crop borders reduce *Potato virus Y* incidence in seed potato. Ann Appl Biol 129:289–302

Dill-Macky R, Jones RK (2000) The effect of previous crop residues and tillage on Fusarium head blight of wheat. Plant Dis 84:71–76

Doyle AD, Moore KJ, Herridge (1988) The narrow-leaved lupin (*Lupinus angustifolius* L.) as a nitrogen-fixing rotation crop for cereal production. III. Residual effects of lupins on subsequent cereal crops. Aust J Agric Res 39:1029–1037

Dubey SC, Bhavani R, Singh B (2011) Integration of soil application and seed treatment formulation of *Trichoderma* species for management of wet root rot of mungbean caused by *Rhizoctonia solani*. Pest Manag Sci 67:1163–1168

Duffy BK, Défago G (1999) Macro- and microelement fertilizers influence the severity of Fusarium crown and root rot of tomato in a soilless production system. Hortscience 34:287–291

Elmer WH, Ferrandino FJ (1994) Comparison of ammonium sulfate and calcium nitrate fertilization effects on Verticillium wilt of eggplant. Plant Dis 78:811–816

El-Mougy NS, Abdel-Khader MM (2008) Long-term activity of bio-priming seed treatment for biological control of fababean root rot pathogens. Australas Plant Path 37:464–471

El-Raheem REL, Shanshoury A (1994) *Azotobacter chroococcum* and *Streptomyces atroolivaceus* as biocontrol agents of *Xanthomonas campestris* pv. *malvacearum*. Acta Microbiol Pol 16:635–642
Escuadra GME, Amemiya Y (2008) Suppression of Fusarium wilt of spinach with compost amendments. J Gen Plant Pathol 74:267–274
Fajinmi AA, Fajinmi OB (2010) Evaluation of pepper intercropped with tall companion crop in the management of pepper veinal mottle virus potyvirus disease and its vectors on cultivated pepper (*Capsicum annuum* L.) in Nigeria. Agric J 5:205–210
Farrell JAK (1976) Effects of groundnut sowing date and plant spacing on rosette virus disease in Malawi. Bull Entomol Res 66:159–171
Felton WL, Marcellos H, Alston C, Martin RJ, Backhouse D, Burgess LB, Herridge DF (1998) Chickpea in wheat-based cropping systems of northern New Southwales. II. Influence on biomass, grain yield and crown rot in the following wheat crop. Aust J Agric Res 49:401–407
Fereres A (2000) Barrier crops as a cultural control measure of non-persistently transmitted aphid-borne viruses. Virus Res 71:221–231
Firman DM, Allen EJ (1995) Effects of seed size, planting density and planting pattern on the severity of silver scurf (*Helminthosporium solani*) and black scurf (*Rhizoctonia solani*) diseases of potatoes. Ann Appl Biol 127:73–85
Friberg H, Lagerlöf J, Rämert B (2006) Usefulness of nonhost plants in managing *Plasmodiophora brassicae*. Plant Pathol 55:690–695
Fry WF (1982) Principles of plant disease management. Academic, New York
Gamliel A, Stapleton JJ (1993) Characterization of antifungal volatile compounds evolved from solarized soil amended with cabbage residues. Phytopathology 83:899–905
Ganapathy T (1985) Studies on the management of major diseases of groundnut (*Arachis hypogaea* L.). Master's thesis, Tamil Nadu Agricultural University, Coimbatore, India
Ganapathy T, Narayanasamy P (1991) Effect of barrier- and intercropping on the incidence of *Tomato spotted wilt virus* on groundnut. Madras Agric J 78:22–24
Gil SV, Pedelini R, Oddino C, Zuza M, Marinelli A, March GJ (2008) The role of potential biocontrol agents in the management of peanut root rot in Argentina. J Plant Pathol 90:35–41
Glass JR, Johnson KB, Powelson ML (2001) Assessment of barriers to prevent the development of potato tuber blight caused by *Phytophthora infestans*. Plant Dis 85:521–528
Goates BJ, Mercier J (2011) Control of common bunt of wheat underfield conditions with the biofumigant fungus *Muscodor albus*. Eur J Plant Pathol 131:403–407
Graham JH, Menge JA (1982) Influence of vesicular-arbuscular mycorrhizae and soil-phosphorus on take-all disease of wheat. Phytopathology 72:95–98
Grogan RG, Welch JE, Bardin R (1952) Common lettuce mosaic and its control by the use of disease-free seed. Phytopathology 42:573–578
Grossmann F (1970) Einflub der ernährung der pflanze auf den befall durch krankheitserreger and schädlinge. Landwirtsch Forsch 23:79–91
Gutteridge RJ, Jenkyn JF, Bateman GL (2006) Effects of different cultivated or weed grasses, grown as pure stands or in combination with wheat, on take-all and its suppression in subsequent wheat crops. Plant Pathol 55:696–704
Hao J, Subbarao KV, Koike ST (2003) Effects of broccoli rotation on lettuce drop caused by *Sclerotinia minor* and on the population density of sclerotia in soil. Plant Dis 87:159–166
Harrison MD (1976) The effect of barley straw on the survival of *Verticillium albo-atrum* in naturally infested field soil. Am J Potato Res 53:385–394
Hauggaard-Nielsen H, Jørnsgaard B, Kinane J, Jensen ES (2008) Grain legume-cereal intercropping: the practical application of diversity, competition and facilitation in arable and organic cropping systems. Renew Agric Food Syst 23:3–12
Hausbeck MK, Pennypacker SP (1991) Influence of grower activity and disease incidence on concentration of airborne conidia of *Botrytis cinerea* among geranium stock plants. Plant Dis 75:798–803
Heitefuss R (1989) Crop and plant protection: the practical foundations. Ellis Horwood Ltd., Chichester

Hiddink GA, Termorshuizen AJ, Raaijmakers JM, van Bruggen AHC (2005) Effect of mixed and single crops on disease suppressiveness of soils. Phytopathology 95:1325–1332

Hillocks RJ (1997) Cotton and tropical fibres. In: Hillocks RJ, Waller JM (eds) Soil-borne diseases of tropical crops. CAB International, Wallingford

Hooks CR, Fereres A, Wang K-H (2007) Using protector plants to guard crops from aphid-borne nonpersistent viruses. Soil Crop Manage 18:1–7

Huang HC, Erickson RS, Hsieh TF (2007) Control of bacterial wilt of bean (*Curtobacterium flaccumfaciens*) by seed treatment with *Rhizobium leguminosarum*. Crop Prot 26:1055–1061

Huang YH, Wang RC, Li CH, Zuo CW, Wei YR, Zhang L, Yi GJ (2012) Control of Fusarium wilt in banana with Chinese leek. Eur J Plant Pathol 134:87–95

Huber DM, Arny DC (1985) Interactions of potassium with plant disease. In: Muson RD (ed) Potassium in agriculture. American Society of Agronomy, Madison, pp 467–488

Huber DM, Haneklaus S (2007) Managing nutrition to control diseases. Landbauforsch Volk 57:313–322

Huelsman MF, Edwards CA (1998) Management of diseases in cucumbers (*Cucumis sativus*) and peppers (*Capsicum annuum*) by using compost as fertility inputs. Brighton Crop Protect Conf 3:881–886

ICRISAT (1981) Annual report of the international Crops Research Institute for semi-arid tropics, Patancheru, India. www.icrisat.org/SoilSolarization.pdf

Jayasena KW, Randles JW (1985) The effects of insecticides and a plant barrier row on aphid populations and spread of bean yellow mosaic potyvirus and subterranean clover red leaf luteovirus in *Vicia faba* in South Australia. Ann Appl Biol 107:355–364

Jensen B, Knudsen IMB, Jensen DF (2000) Biological seed treatment of cereal with fresh and long-term stored formulations of *Clonostachys rosea*: biocontrol efficacy against *Fusarium culmorum*. Eur J Plant Pathol 106:233–242

Jensen B, Knudsen IMB, Madsen M, Jensen DF (2004) Biopriming of infected carrot seed with an antagonist *Clonostachys rosea* selected for control of seedborne *Alternaria* spp. Phytopathology 94:551–560

Jirak-Peterson JC, Esker PD (2011) Tillage, crop rotation and hybrid effects on residue and corn anthracnose occurrence in Wisconsin. Plant Dis 95:601–610

Jones RAC (1994) Effect of mulching with cereal straw and row spacing on spread of bean yellow mosaic potyvirus into narrow-leaved lupins (*Lupinus angustifolius*). Ann Appl Biol 124:45–58

Jones JP, Engelhard AW, Woltz SS (1989) Management of Fusarium wilt of vegetables and ornamentals by macro- and micronutrients. In: Englehard AW (ed) Soilborne plant pathogens: management of diseases with macro- and microelements. APS Press, St. Paul, pp 18–32

Jurke CJ, Fernando WGD (2006) Effects of seeding rate and plant density on Sclerotinia stem rot incidence in canola. Nature 444:323–329. doi:10.1080/03235400600679743

Katan J (2000) Physical and cultural methods for the management of soilborne pathogens. Crop Prot 19:725–731

Katan J (2010) Cultural approaches for disease management: present status and future prospects. J Plant Pathol 92:S4.7–S4.9

Keinath AP (1996) Soil amendment with cabbage residue and crop rotation to reduce gummy stem blight and increase growth and yield of watermelon. Plant Dis 80:564–570

Kelman A, McGuire RG, Tzeng KC (1989) Reducing severity of bacterial soft rot by increasing the concentration of calcium in potato tubers. In: Engelhard AW (ed) Soilborne plant pathogens: management of diseases with macro- and microelements. APS Press, St. Paul, pp 102–123

Khalequzzaman KM, Hossain I (2008) Efficacy of *Rhizobium* strains and biofertilizers for controlling foot and root rot and increasing green pod yield of bush bean. Bangladesh J Agric Res 33:617–622

Kim CH (1989) Phytophthora blight and other diseases of red pepper in Korea: disease and pest problems from continuous cropping. II. Soil-borne diseases. FFTC Ext Bull 302:10–17

Kim DS, Cook RJ, Weller DM (1997) *Bacillus* sp. L324-92 for biological control of three root diseases of wheat grown with reduced tillage. Phytopathology 87:551–558

Kimble KA, Grogan RG, Greathead AS, Paulus AO, House HK (1975) Development of application and comparison of methods for indexing lettuce seeds for mosaic in California. Plant Dis Rep 59:461–464

Koch E, Schmitt A, Stephan D, Kromphardt C, Jahn M, Karuthausen H-J, Forsberg G, Werner S, Amein T, Wright SAI et al (2010) Evaluation of nonchemical seed treatments for the control of *Alternaria dauci* and *A. radicina* on carrot seeds. Eur J Plant Pathol 127:99–112

Kommedahl T, Todd LA (1991) The environmental control of plant pathogens using eradication. In: Pimentel D (ed) Handbook of pest management in agriculture. CRC Press, Boca Raton

Kupper KC, Bettiol W, de Goes A, de Souza PS, Bellotte JAM (2006) Biofertilizer for control of *Guignardia citricarpa*, the causal agent of citrus black spot. Crop Prot 25:569–573

Lai R, You M, Jiang L, Lai B, Chen S, Zeng W, Jiang D (2011) Evaluation of garlic intercropping for enhancing the biological control of *Ralstonia solanacearum* in flue-cured tobacco fields. Biocontrol Sci Technol 21:755–764

Lambert DH, Salas B (2001) Pink rot. In: Stevenson WR, Loria R, Franc GD, Weingartner DP (eds) Compendium of potato diseases. APS Press, St. Paul, pp 33–34

Lambert DH, Giggie E, Currier AI, Mahoney K (2001) Dynamics of mefenoxam resistance in pink rot populations. In: Proceedings of the Northeast Potato Technology Forum, McCain Foods Ltd, New Brunswick, p 70

Lamprecht SC, Marasas WFO, Hardy MB, Calitz FJ (2006) Effect of crop rotation on crown rot and the incidence of *Fusarium graminearum* in wheat in the Cape, South Africa. Australas Plant Path 35:419–426

Larkin RP, Honeycutt CW, Griffin TS, Olanya OM, Halloran JM, He Z (2011) Effects of different potato cropping system approaches and water management on soilborne diseases and soil microbial communities. Phytopathology 101:58–67

Lee HU, Kim CH, Lee FJ (1990) Effect of pre- and mixed cropping with nonhost plants on incidence of Phytophthora blight of red pepper. Korean J Plant Pathol 6:440–446

Lee HU, Kim CH, Nam KW (1991) Suppression of Phytophthora blight incidence of red pepper by cropping system. Korean J Plant Pathol 7:140–146

Lister RM, Carroll TW, Zaske SK (1981) Sensitive serologic detection of *Barley stripe mosaic virus* in barley seeds. Plant Dis 65:809–814

Lockhart DAS, Heppel VAF, Holmes JC (1975) Take-all [*Gaeumannomyces graminis* (Sacc.) Arx and Oliver] incidence in continuous barley by growing and effect of tillage method. EPPO Bull 5:375–383

Mali VR (1992) Integrated virus disease management. In: Verma JP, Verma A (eds) Farming systems and integrated pest management. Malhotra Publishing House, New Delhi, pp 197–233

Mao W, Lewis JA, Hebbar PK, Lamsden RD (1997) Seed treatment with a fungal or bacterial antagonist for reducing corn damping-off caused by species of *Pythium* and *Fusarium*. Plant Dis 81:450–454

Martin FN, Bull CT (2002) Biological approaches for control of root pathogens of strawberry. Phytopathology 92:1356–1362

Matheron ME, Porchas M (2005) Influence of soil temperature and moisture on eruptive germination and viability of sclerotia of *Sclerotinia minor* and *S. sclerotiorum*. Plant Dis 89:50–54

Matheron ME, Porchas M (2010) Evaluation of soil solarization and flooding as management tools for Fusarium wilt of lettuce. Plant Dis 94:1323–1328

Matthews REF (1991) Plant virology, 3rd edn. Academic, San Diego

Mattingly GEG, Slope DB (1977) Phosphate fertilizer and "take-all" disease of wheat and barley. J Sci Food Agric 28:658–659

Mayee CD, Gangwal KU, Ekbote AU (1983) Contributions to the groundnut rust epidemiology from Marathwada Region of Maharashtra. III. Cultural practice and disease progress. In: Narayanasamy P (ed) Management of diseases of oilseed crops. Tamil Nadu Agricultural University, Coimbatore, p 106

Mazzola M (1999) Transformation of soil microbial community and structure and Rhizoctonia-suppressive potential in response to apple roots. Phytopathology 89:920–927

Michel VV, Hartman GL, Midmore DJ (1996) Effect of previous crop in soil populations of *Burkholderia solanacearum*, bacterial wilt and yield of tomatoes in Taiwan. Plant Dis 80:1367–1372

Moore WD (1949) Flooding as a means of destroying the sclerotia of *Sclerotinia sclerotiorum*. Phytopathology 39:920–927

Moore KJ (1978) The influence of no-tillage on take-all of wheat. Doctoral thesis, Washington State University, Pullman

Moreno RA (1979) Crop protection implication of cassava intercropping-intercropping with cassava. In: Proceedings of the International Workshop, Trivandrum, India

Mukherjee PK, Upadhyay JP, Mukhopadhyay AN (1989) Biological control of Pythium damping-off of cauliflower by *Trichoderma harzianum*. J Biol Control 3:119–124

Mukhopadhyay AN, Mukherjee PK (1991) Innovative approaches in biological control of soilborne diseases in chickpea. Petria 1:146

Narayanasamy P (2002) Microbial plant pathogens and crop disease management. Science Publishers, Enfield

Narayanasamy P (2011) Microbial plant pathogens – detection and disease diagnosis, vol 1–3. Springer Science + Business B.V., Heidelberg

Narayanasamy P, Ramiah M, Kandaswamy TK (1976) Influence of elimination of virus sources and insecticides on the incidence of groundnut ring mosaic disease. In: Proceedings of the symposium on modern concepts of plant protection, University of Rajasthan, Udaipur, India, pp 94–95

Narla RD, Muthomi JW, Gachu SM, Nderitu JH, Olubayo FM (2011) Effect of intercropping bulb onion and vegetables on purple blotch and downy mildew. J Biol Sci 11:52–57

Nework FJ (1970) *Phytophthora cinnamomi* in New Zealand. In: Toussoun TA, Bega RV, Nelson PE (eds) Root diseases and soil-borne pathogens. University of California Press, Berkeley, pp 173–176

Nitzany FE, Giesenberg H, Koch B (1964) Tests for the protection of cucumbers from whitefly-borne virus. Phytopathology 54:1059–1061

Njoroge SMC, Kabir Z, Martin FN, Koike ST, Subbarao KV (2009) Comparison of crop rotation for Verticillium wilt management and effect on *Pythium* species in conventional and organic strawberry production. Plant Dis 93:519–527

Oddino CM, Marinelli AD, Zuza M, March GJ (2008) Influence of crop rotation and tillage on incidence of brown root rot caused by *Fusarium solani* in Argentina. Can J Plant Pathol 30:575–580

Ojaghian MR, Cui Z, Xie G-L, Li B, Zhang J (2012) Brassica green manure rotation crops reduce potato stem rot caused by *Sclerotinia sclerotiorum*. Australas Plant Pathol 41(4):347–349, Online First™

Olanya OM, Porter GA, Lambert DH, Lakin RP, Starr GC (2010) The effects of supplemental irrigation and soil management on potato tuber diseases. J Plant Pathol 9:65–72

Pacumbaba RP, Brown GF, Pacumbaba RO Jr (1997) Effects of fertilizers and rates of application on incidence of soybean diseases. Plant Dis 81:1459–1460

Palti J (1981) Cultural practices and infectious crop diseases. Springer, Berlin

Papavizas GC, Lewis JA (1979) Integrated control of *Rhizoctonia solani*. In: Schippers B, Grams W (eds) Soil-borne plant pathogens. Academic, New York, pp 415–424

Patrick ZA (1986) Allelopathic mechanisms and their exploitation for biological control. Can J Plant Pathol 8:225–228

Peters RD, Sturz AV, Carter MR, Sanderson JB (2003) Developing disease-suppressive soils through crop rotation and tillage management practices. Soil Tillage Res 72:181–192

Peters RD, Sturz AV, Carter MR, Sanderson JB (2005) Crop rotation can confer resistance to potatoes from *Phytophthora erythroseptica* attack. Can J Plant Sci 85:523–528

Pierson EA, Weller DM (1994) Use of mixtures of fluorescent pseudomonads to suppress take-all and improve the growth of wheat. Phytopathology 84:940–947

Pullman GS, De Vay JE (1981) Effect of soil flooding and paddy rice culture on the survival of *Verticillium dahliae* and incidence of Verticillium wilt in cotton. Phytopathology 71:1285–1289

Qing JL, Huang CW, Huang RJ, Li QC (1994) Study on the control of maize sheath blight disease by removing infected leaves. Plant Protect 20:4–6

Raguchander T, Rajappan K, Samiyappan R (1997) Evaluating methods of application of biocontrol agent in the control of mungbean root rot. Indian Phytopathol 50:229–234
Reeves TG, Ellington A, Brooke HD (1984) Effects of lupin-wheat rotation on soil fertility by crop disease and crop yields. Aust J Exp Agric Anim Husb 24:595–600
Reis EM, Cook RJ, McNeal BL (1982) Effect of mineral nutrition on take-all of wheat. Phytopathology 72:224–229
Richter BS, Ivors K, Shi W, Benson DM (2011) Cellulase activity as a mechanism for suppression of Phytophthora root rot in mulches. Phytopathology 101:223–230
Rip YJ, Seuk PC (1995) Enhancement of plant growth and suppression of damping-off of cucumber by low temperature growing *Pseudomonas fluorescens*. Korean J Plant Pathol 11:252–257
Ristaino JB, Parra G, Campbell CL (1997) Suppression of Phytophthora blight in bell pepper by a no-till wheat cover crop. Phytopathology 87:242–249
Robertson AD (1981) The whitefly *Bemisia tabaci* (Gennadium) as a vector of *African cassava mosaic virus* at Kenya coast and ways in which the yield losses caused in cassava, *Manihot esculenta* Crantz. can be reduced. Appl Insect Sci 8:797–801
Rojo FG, Reynoso MM, Ferez M, Chulze SN, Torres AM (2007) Biological control by *Trichoderma* species of *Fusarium solani* causing peanut brown root rot under field conditions. Crop Prot 26:549–555
Romanko RR (1965) Factors affecting the spread of downy mildew in arid climate. Plant Dis Rep 49:247
Rotem J, Palti J (1969) Irrigation and plant disease. Annu Rev Phytopathol 7:267–288
Rothrock CS, Kirkpatrick TL, Frans RE, Scott HD (1995) The influence of winter legume cover crops on soil-borne pathogens and cotton seedling diseases. Plant Dis 79:161–171
Rush CM, Heidel GB (1995) Furovirus diseases of sugar beets in the United States. Plant Dis 79:868–875
Sağir A, Eylen M, Pirinc V (2005) Effect of irrigation methods on pepper crown rot disease caused by *Phytophthora capsici*. Int J Agric Biol 7:804–806
Sakuma F, Maeda M, Takahashi M, Kondo N (2011) Suppression of common scab of potato caused by *Streptomyces turgidiscabies* using lopsided oat green manure. Plant Dis 95:1124–1130
Schillinger WF, Paulitz TC (2006) Reduction of Rhizoctonia bare patch in wheat with barley rotations. Plant Dis 90:302–306
Schmid A, Daniel C, Weibel F (2005) Effect of cultural methods on leaf spot (*Mycosphaerella fragariae*) and gray mold (*Botrytis cinerea*) damage in strawberries. Biocontrol 50:179–194
Scholte K, Lootsma M (1998) Effect of farm yard manure and green manure crops on populations of mycophagous soil fauna and Rhizoctonia stem canker of potato. Pedobiologia 42:223–231
Schuerger AC, Hammer W (2003) Suppression of powdery mildew on greenhouse-grown cucumber by addition of silicon to hydroponic nutrient solution is inhibited at high temperature. Plant Dis 87:177–185
Schutte G, Kotze JM (1997) Grass mulching as part of an integral control programme for the control of citrus black spot. Citrus J 7:18–20
Self M, MacKenzie M (1995) Intensive site preparation to control Armillaria root disease in second rotation *Pinus radiata*. N Z J For Sci 25:111–116
Serrano MS, De Vita P, Fernandez-Rebollo P, Sanchez-Hernandez ME (2012) Calcium fertilizers induce soil suppressiveness to *Phytophthora cinnamomi* root rot of *Quercus ilex*. Eur J Plant Pathol 132:271–279
Shtienberg D, Elad Y, Bornstein M, Ziv G, Grava A, Cohen S (2010) Polyethylene mulch modifies greenhouse microclimate and reduces infection of *Phytophthora infestans* in tomato and *Pseudoperonospora cubensis* in cucumber. Phytopathology 100:97–104
Sikirou R, Wydra K (2008) Effect of intercropping cowpea with maize or cassava on cowpea bacterial blight and yield. J Plant Dis Protect 115:145–151
Simmons JN (1957) Effects of insecticides and physical barriers on field spread of *Pepper vein banding mosaic virus*. Phytopathology 47:139–145

Sivan A, Ucko O, Chet I (1987) Biological control of Fusarium crown rot of tomato by *Trichoderma harzianum* under field conditions. Plant Dis 71:587–592
Smalla K, Wieland G, Buchner A, Parzy J, Kaiser S, Roskot N, Heuer H, Berg G (2001) Bulk and rhizosphere soil bacterial communities studied by denaturing gradient gel electrophoresis plant-dependent enrichment and seasonal shifts revealed. Appl Environ Microbiol 67:4742–4751
Smiley RW (1975) Forms of nitrogen and the pH in the root zone and their importance to root infections. In: Brueh GW (ed) Biology and control of soilborne pathogens. APS, St. Paul, pp 52–62
Smiley RW (1978a) Antagonists of *Gaeumannomyces graminis* from the rhizoplane of wheat in soils fertilized with ammonium- or nitrate nitrogen. Soil Biol Biochem 10:169–174
Smiley RW (1978b) Colonization of wheat roots by *Gaeumannomyces graminis* inhibited by specific soils, microorganism and ammonium nitrogen. Soil Biol Biochem 10:175–179
Steinkellner S, Langer I (2004) Impact of tillage on the incidence of *Fusarium* spp. in soil. Plant Soil 267:13–22
Stover RH (1962) Fusarium wilt (Panama disease of bananas) and other *Musa* species. Phytopathol paper no. 4, CMI, Surrey
Subbarao KV, Hubbard JC (1996) Interactive effects of broccoli residue and temperature on *Verticillium dahliae* microsclerotia in soil and on wilt in cauliflower. Phytopathology 86:1303–1310
Subbarao KV, Koike ST, Hubbard JC (1996) Effects of deep plowing on the distribution and density of *Sclerotinia minor* sclerotia and lettuce drop incidence. Plant Dis 80:28–33
Subbarao KV, Kabir Z, Martin FN, Koike ST (2007) Management of soilborne diseases in strawberry using vegetable rotations. Plant Dis 91:964–972
Sudini H, Arias CR, Liles MR, Bowen KL, Huettel RN (2011) Comparison of soil fungal community structure in different peanut sequences using ribosomal intergeneric spacer analysis in relation to aflatoxin-producing fungi. Phytopathology 101:52–57
Sulc RM, Rhodes LM (1997) Planting date, fungicide and cultivar effects on Sclerotinia crown and stem rot severity in alfalfa. Plant Dis 81:13–17
Summers CG, Mitchell JP, Stapleton JJ (2005) Mulches reduce aphid-borne viruses and whiteflies in cantaloupe. Calif Agric 52:90–94
Swain S, Harnik T, Mejia-Chang M, Hayden K, Bakx W, Creque J, Garbelotto M (2006) Composting is an effective treatment option for sanitization of *Phytophthora ramorum*-infected plant material. J Appl Microbiol 101:815–827
Trocmé O (1972) Contribution á l'étude d'une maladie du cacoyer en Ouganda: le déssechement écofongique des branches. Cafe Cacao The 16:219–225
Utkhade RS (1992) Biological control of *Phytophthora* on fruit trees. In: Mukherji KG, Tewari JP, Arora DK, Saxena G (eds) Recent developments in biocontrol of plant diseases. Aditya Books Pvt Ltd, New Delhi, pp 1–16
Van Rheenan HA, Hasslebach OE, Muigai SGS (1981) The effect of growing beans together with maize on the incidence of bean diseases and pests. Neth J Plant Pathol 87:193–199
Váňová M, Matušinsky P, Javůrek M, Vach M (2011) Effect of soil tillage practices on severity of selected diseases in winter wheat. Plant Soil Environ 57:245–250
Wallenhammer AC (1996) Prevalence of *Plasmodiophora brassicae* in a spring oilseed rape growing area in central Sweden and factors influencing soil infestation level. Plant Pathol 45:710–719
Wheeler TA, Bordovsky JP, Keeling JW, Mullinix BG Jr (2012) Effects of crop rotation, cultivar and irrigation and nitrogen rate on Verticillium wilt in cotton. Plant Dis 96:985–989
Workneh F, Yang XB, Tylka GL (1998) Effect of tillage on vertical distribution of *Phytophthora sojae*. Plant Dis 82:1258–1263
Wu HS, Yang XN, Fan J-Q, Miao W-G, Ling N, Xu Y-C, Huang Q-W, Shen Q (2009) Suppression of Fusarium wilt of watermelon by a bio-organic fertilizer containing combinations of antagonistic microorganisms. Biocontrol 54:287–300
Xiao CL, Subbarao KV, Schulbach KF, Koike ST (1998) Effect of crop rotation and irrigation on *Verticillium dahliae* microsclerotia in soil and wilt in cauliflower. Phytopathology 88:1046–1055

Xie J, Cardenas ES, Sammis TW, Wall MM, Lindsey DL, Murray LW (1999) Effects of irrigation method on Chile pepper yield and Phytophthora root rot incidence. Agric Water Manag 42:127–142

Yang XB (2002) Timing tillage for disease control. www.G:Biocontrol-culturalPractices/Yang-BC-CultPract(F).html

Yassin AM, Abu Salih HS (1976) Annual report 1975/76. Gezira Res Station Agric Res Corp, Jordan

You MP, Sivasithamparam K, Kurtboke DI (1996) Actinomycetes in organic mulch used in avocado plantations and their ability to suppress *Phytophthora cinnamomi*. Biol Fertil Soils 22:237–242

Zhang JX, Howell CR, Starr JL (1996) Suppression of Fusarium colonization of cotton roots and Fusarium wilt by seed treatment with *Gliocladium virens* and *Bacillus subtilis*. Biocontrol Sci Technol 6:175–187

Chapter 3
Physical Techniques for Biological Crop Disease Management

3.1 Heat Treatments

The sensitivity to high temperature of microbial pathogens varies considerably, because of their ability to adapt themselves to the unfavorable environments.

3.1.1 Fungal Pathogens

3.1.1.1 Field Crop Diseases

Heat may be applied as hot air or hot water to treat the seeds or asexually propagative planting materials such as tubers or setts. Hot water was used for the control of smuts of barley and oats (Jensen 1888). Application of heat for the elimination of *Ustilago tritici* and *U. nuda* causing loose smut of wheat and barley respectively was found to be effective. The seeds were immersed in water at 20–30 °C to stimulate the dormant mycelium of the pathogen present in them and the pathogen could be killed, when the seeds were immersed in hot water at 50–52 °C (Appel and Riehm 1911). A useful modification of this principle that can be adopted in tropical countries was developed by Luthra (1953) in Punjab, India. The seeds are soaked in cold water for about 4 h and then they are spread as a thin layer under the sunlight (at temperatures around 50 °C) for 4 h. They are then stored under dry conditions. Maize seeds were first pretreated in water at 18–22 °C for 4 h and then they were treated with hot water at 60 °C for 5 min to eliminate *Fusarium moniliforme* (Daniels 1983). Soaking onion seeds in hot water at 50 °C for 20 min was found to be more effective than fungicidal application in reducing the seedborne infection by *Alternaria porri* and *Stemphylium vesicarium* though seed germination and seedling emergence were reduced (Aveling et al. 1993). Hot water treatment at 60 °C for 10 min was the most effective in suppressing the colonization of rice seeds by *Drechslera oryzae* (Krishnamurthy et al. 2001).

P. Narayanasamy, *Biological Management of Diseases of Crops*,
Progress in Biological Control 16, DOI 10.1007/978-94-007-6377-7_3,

Hot air treatment was applied for the control of anthracnose pathogen *Colletotrichum capsici* infecting pepper (chilli) seeds. Exposure of pepper seeds to hot air at 85 °C for 5 days reduced seedborne infection with some reduction in seed germination (Kobabyashi 1990). Likewise, *Phytophthora infestans* causative agent of late blight disease, carried in discolored seeds from infected tomato fruits was eliminated by drying seeds in an oven at 29.5–37.5 °C for 6 h (Vartanian and Endo 1985). A high precision method, involving the use of warm air which is in moisture equilibrium with the grains to be treated was employed for the control of fungal pathogens in wheat seeds. Sanitation of wheat grains using warm air resulted in reducing seedborne infection by *Tilletia caries* and *Microdochium nivale* to a level comparable with chemical seed treatment. No adverse effect on plant development following warm air treatment was observed. The warm air treatment is suitable for treatment of seeds with thick layers at short durations and has the potential for large scale application at a low cost (Forsberg 2001). Heat treatment of wheat and barley seeds with high level of infection by *Fusarium graminearum* (84 and 23 % respectively) was evaluated. The pathogen was eliminated from Canada Western red spring wheat seeds treated at 60 °C for15 days or at 70 °C for 5 days or at 80 °C for 2 days. Similarly, the barley seeds were freed of the pathogen after treatment at 60 °C for 21 days, or at 70 °C for 9 days or at 80 °C for 5 days. Rates of germination in most samples were not reduced due to heat treatments (Clear et al. 2003). Aerated steam treatment of carrot seeds was more effective than hot water or electron treatment in eliminating *Alternaria dauci* and *A. radicina* from carrot seeds. Hot water combined with *Pseudomonas* sp. MF416 or *Clonostachys rosea* IK726 treatment showed enhanced performance against the seedborne pathogens of carrot (Koch et al. 2010).

The microclimate of the crop plants grown in the glasshouses may be manipulated to some extent, so that temperature and humidity become unfavorable to the development of pathogens without causing any obvious ill effects on the host plants. By applying forced heated air continuously from beneath an open bottom bench, the incidence of geranium stem blight disease caused by *Botrytis cinerea* could be significantly reduced. The conidial concentrations were lower in the treated area and the temperature increase was not more than 1.5 °C that of the control area (Hausbeck et al. 1996a). A similar approach of using hot water was shown to be effective for the control of pineapple heart rot disease caused by *Phytophthora cinnamomi*. Pineapple crown buds, lateral bud and suckering buds and rhizomes were exposed to sunlight for 4–8 h at a temperature of 33 °C to eliminate the pathogen present in these tissues (Yang and Chang 1998). Heat has been applied as steam for soil disinfestations. It was introduced in Europe mainly in greenhouses by steaming the soil. The temperature should reach a level that is lethal for most heat tolerant soilborne pathogens existing in the soil. The temperatures above 70 °C or approaching 100 °C become detrimental to soil biota including fungi and bacteria antagonistic to plant pathogens. Drastic reduction in soil microbial activity may lead to rapid reinfestation of the heated soil by a contaminating inoculum resulting in higher level of disease incidence/severity than that in the non-treated soil due to 'biological vacuum' (Baker 1962). Aerated steam at 60 °C for 30 min applied to glasshouse

Table 3.1 Application of different forms of heat for the control of postharvest diseases of fruits and vegetables (Adapted from Narayanasamy 2006)

Fruit/vegetable	Fungal pathogen	Form of heat	Temperature (°C)/ treatment time (min)
Apple	*Moilinia fructicola*	HW	40/5–10
	Gloeosporium sp.	HW	45/10
	Penicillium expansum	HA	45/15
Banana	*Colletotrichum musae*	HW	45/30
Mango	*Colletotrichum gloeosporioides*	HW	51/15
Orange	*Phomopsis citri*	HW	53/5
	Penicillium digitatum		
Papaya	*Colletotrichum* sp.	HW	43/20
Pear	*Botryosphaeria*	HA	30/35
	Berengeriana	HW	50/5–7
Strawberry	*Botrytis cinerea*	HA	44/40
Bean	*Pythium butler*	HW	52/0.5
	Sclerotinia sclerotiorumi		
Pepper (chilli)	*Alternaria alternata*	HW	50/2
		HA	38/48 h
Lemon	*Penicillium digitatum*	HW	52/5
Melon	*Fusarium* spp.	HW	57/0.5

HA, hot air; HW, hot water

soils was found to be thermally selective by eliminating plant pathogens, while leaving a part of the established nonpathogenic microflora to multiply and compete with pathogenic organism(s) that may be introduced (Baker and Cook 1974).

3.1.1.2 Postharvest Diseases

Heat treatments in different forms have been demonstrated to be effective in providing disease-free environment for fresh fruits and vegetables against postharvest diseases. Postharvest decay control through the use of heat differs essentially from other uses of heat on agricultural commodities such as curing to promote wound healing or heating to eliminate viruses, insects or nematodes. Postharvest heat treatment for disease management is restricted to short periods, whereas longer periods of heat treatments are required for other purposes. Various factors are to be considered, while heat treatments are opted for different fruits and pathogen combinations for successful suppression of disease(s) during storage. Many pathogens are present on the surface or within a few coat cell layers of the infected produce and many of them have thermal sensitivity much lower than the fruits and vegetables. The location of the target pathogen in or on the fruit and the thermal sensitivity of the target organism and the fruit are the major factors determining the effectiveness of heat treatment (Narayanasamy 2006). Some of postharvest diseases of fruits that could be effectively controlled are presented in Table 3.1.

Application of heated water has been shown to be effective against certain fungal pathogens causing postharvest diseases. The mango stem-end rot disease is caused by *Dothiorella dominicana, Lasiodiplodia theobramae* and *Phomopsis mangiferae.* Infection by *D. dominicana* could be effectively reduced by vapor heat treatment of mango fruits cv. Kensington Pride (fruit seed surface temperature of 46.5 °C) for 10 min (Coates et al. 2003). The efficacy of vapor heat treatment was assessed with a view to replacing the use of SO_2 for the control of gray mold pathogen *Botrytis cinerea* infecting table grapes. Vapor heat treatment at 52.5 °C for 21 or 24 min and at 55 °C for 18 or 21 min provided most effective protection against the gray mold disease (Lydakis and Aked 2003). Hot water dip (HWD) treatment was effective against anthracnose diseases caused by *Colletotrichum musae* and *C. gloeosporioides* infecting banana and mango (Loepez-Escudero and Blanco-Lopez 2001). By wrapping fruits in plastic bags prior to HWD, the temperature requirement could be reduced to achieve the level of disease control expected (Acedo et al. 2001). Hot water brushing (HWB) was developed to control green mold pathogen *Penicillium digitatum* infecting citrus fruits. The method involves rinsing hot water and brushing of fruits. HWB treatment at 56, 59 and 62 °C for 20 s reduced decay development in the infected wounds to 20, 5 and less than 1 % respectively of that in untreated control fruits (Porat et al. 2000). *Alternaria alternata* infection of mangoes was effectively reduced by HWB treatment. The level of disease reduction was equal to that in fungicide treatment (Prusky et al. 2003).

The effect of hot water (HW), *Bacillus subtilis* species complex CPA-8 and sodium bicarbonate (SBC) treatments applied separately or in combination for the control of *Monilinia laxa* infecting peaches was assessed. The treatments with HW at 60 °C for 40 s, SBC at 2 % for 40 s and CPA-8 at 10^7 CFU/ml did not affect fruit quality. The combinations of these treatments were evaluated in three cultivars of peaches and nectarines artificially inoculated with *M. laxa*. When the fruits were incubated for 5 days at 20 °C, a significant additional effect to control of *M. laxa* was observed with the combination of HW followed by the antagonist CPA-8. Only 8 % of the fruits were infected in the combined treatment, compared to 84, 52 or 24 % of the fruits among control, CPA-8 and HW treatments respectively. When fruits were incubated for 21 days at 0 °C plus 5 days at 20 °C, the significant differences between separate or combined treatments were reduced and generally the incidence of brown rot was higher than when the fruits were incubated for 5 days at 20 °C. The effects of the treatments were similar, when naturally infected fruits were tested (Casals et al. 2010a, b).

Pineapple suffers markedly due to infection by *Chalara paradoxa*, causing black rot disease. Hot water dip treatment at 54 °C for 3 min suppressed the development of *C. paradoxa* in fruits inoculated with 10^4 spores/ml, when the fruits were stored at 10 °C for 21 days followed by 48 h at an ambient temperature (28 ± 2 °C). Inoculated dip-treated fruits held at 28 ± 2 °C for 6 days also remained healthy. No observable symptoms were produced on treated fruit, while characteristic disease symptoms developed in untreated fruit under similar storage conditions. No significant difference was recorded between hot water-dipped fruits in respect of flesh and shell color of the fruits, ascorbic acid contents and titrable acidity and other quality parameters (Wijeratnam et al. 2005). Hot air treatment was applied as a tool to induce resistance in citrus. Oranges were inoculated with *Penicillium digitatum* and 1 day after inoculation, they were exposed to hot air treatment at 37 °C for 3 days. Activities of phenylalanine ammonia lyase (PAL), peroxidase, ß-1,3-glucanase and chitinase increased in treated fruit in parallel with increased resistance, especially in the albedo. Generally highest activities were recorded in the flavedo. Expression of the gene encoding PAL and that of the genes coding for the basic, rather than for the acidic isoforms of the PR proteins was also induced in both tissues, but most dramatically in the albedo (Ballester et al. 2010). Treatment of *Monilinia fructicola*-infected apples with hot water at 40 °C for 5–10 min was preferred for effective suppression of the development of the brown rot pathogen (Barkai-Golan and Phillips 1991). Direct antimicrobial effects on pathogen propagules as well as induction of host defense responses have been considered to have a role in restricting the decay development. The effect of heat treatment (HT, hot water treatment at) 40 °C for 5 or 10 min on *M. fructicola* causing brown rot was assessed. Hot water treatment inhibited spore germination and germ tube elongation of *M. fructicola* in in vitro assays. Further, hot water treatment triggered the accumulation of reactive oxygen species (ROS), collapse of mitochondrial membrane potential and a decrease in intracellular ATP in *M. fructicola*. In treated peach, the expression of defense-related genes including chitinase, ß-1,3-glucanase and phenylalanine ammonia lyase was induced. Enhanced activity of these enzymes could be detected in treated peach fruit. The results indicated that hot water treatment exerted both direct inhibitory effects of enhancing the resistance in the fruits against infection by *M. fructicola* (Liu et al. 2012).

The effects of heat, in combination with other alternative methods of fruit decay control were investigated using 'Montenegrina' tangerines. Different treatment combinations using hot water (60 °C), brushing and immersion in chloride dioxide, imazalil (fungicide), sodium bicarbonate and hand-applied wax were used. The tangerines were then placed in cold storage for 20 days at 5 °C and retrieved to ambient conditions for seven more days. Heat treatments significantly reduced the number of tangerines with decay symptoms and enhanced the efficacy of the tested products. Application of carnauba wax significantly increased the number of rotten fruits, exerting a protective effect on the fungi present on the fruit surface by covering their structures. Sodium bicarbonate in combination with heat treatments effectively reduced the incidence of fruit decay. Heat treatments partially removed the hyphae and spores of fungal pathogens on the fruit surface and also melted the cuticular waxes that covered stomata and cracks, reducing possible entry points for the pathogens (Montero et al. 2010). The efficacy of hot air treatment at 38 °C for 36 h or the yeast *Pichia guillermondii* applied alone or in combination for suppressing the development of anthracnose rot caused by *Colletotrichum acutatum* in loquat fruit was assessed. Combined treatment significantly reduced natural decay and disease incidence and lesion diameter in artificially inoculated fruit. Low activities of catalase (CAT) during the early storage period were recorded, but CAT and superoxide dismutase (SOD) activities both increased later in storage. This condition corresponded with a high level of H_2O_2 in the early storage period and lower level of H_2O_2 in later stages in storage. At the same time, induced activity of PAL and ß-1,3-glucanase stimulated synthesis of lignin, thus eliciting disease resistance. Furthermore, the combined treatment significantly inhibited the mycelial growth and spore germination of *C. acutatum*, reducing anthracnose rot of loquat fruit (Liu et al. 2010).

3.1.2 Bacterial Diseases

Thermotherapy involving the use of heat for eliminating the bacterial and phytoplasmal pathogens from propagative materials has been successfully applied in a few cases. The bacterial pathogen *Clavibacter xyli* subsp. *xyli*, causative agent of sugarcane ratoon stunting disease was efficiently eliminated by treating the setts in hot water at 50 °C for 2–3 h (Steindl 1961). Treatment with hot air (Steindl 1961) and moist hot air treatment (Shukla et al. 1974) was also found to be effective in eliminating the bacterial pathogen from infected setts. The adverse effect on germination of setts was found to be a limiting factor. However, large scale use of heat therapy as a disease control strategy is followed in several sugarcane growing countries. For the control of grassy shoot and white leaf diseases caused by phytoplasmas, treatment of setts with hot water, hot air, moist-hot air and aerated steam have been reported to be successful in eliminating the pathogens to varying degrees (Edison and Ramakrishnan 1972; Srivastava et al. 1977). Hot water treatment of dormant cuttings of grapevine was applied as a means of freeing the cuttings from the Pierce's disease caused by *Xylella fastidiosa* (Goheen et al. 1973). Treatment of cuttings at 50 °C for 20 min was effectively applied on imported cuttings in Australia (Barlass et al. 1987). Hot air treatment was reported to be effective in reducing the cabbage black rot pathogen infection in seeds (Shiomi 1992) and bacterial wilt pathogen infection of ginger rhizome (Tsang and Shintaku 1998). *Agrobacterium tumefaciens* biovar (AT3) is a systemic pathogen and spreads through propagative materials to different locations or generations. The effect of heat therapy on the elimination of the pathogen AT3 from grape cuttings was assessed, since excluding the pathogen by shoot-tip culture was found to be time-consuming. AT3 was eradicated or reduced to below the level of detection in dormant cuttings of grape rootstocks K5140 (*Vitis schampini* × *V. riparia*) and Ramsey (*V. champini*) and in artificially infested Cayuga (interspecific hybrid) by a hot water treatment at 50 °C for 30 min. Strains of AT3 exhibited different levels of sensitivity to heat. But they were generally more sensitive than strains of biovar1 or biovar2. Populations of about 10^3 CFU/ml of AT3 in broth were killed by a 30-min treatment. The treated cuttings were assayed by taking the vascular extracts and plating them on artificial medium. As few as 50 CFU of AT3 in the triturated callus tissues were detected (Table 3.2). The hot water treatment was found to be a simple, effective, economical and environmentally safe method of eradicating crown gall pathogen from dormant grape cuttings (Burr et al. 1989). Immersion of bud sticks of sweet cherry in hot water at 52 °C, followed by immediate cooling by immersion in water at 21 °C eliminated

Table 3.2 Effect of hot water treatment at 50 °C on the persistence of *Agrobacterium tumefaciens* biovar 3 (AT3) in dormant grape cuttings[a] (Burr et al. 1989)

	No. of cuttings with AT3/no. of cuttings treated			
	Time (min)			
Rootstock or cutivar	0	10	20	30
Ramsey[b]	6/10	–	–	0/10
K5140	8/12	–	0/10	0/10
K5140	6/10	–	0/10	0/10
Ramsey	7/10	–	0/10	0/10
K5140[c]	8/10	6/10	0/10	0/10
Ramsey	7/10	0/10	0/10	0/10
K5140	5/10	1/10	0/10	0/10
K5140	4/10	0/10	0/10	0/10
K5140[c]	8/20	–	0/50	0/50
Cayuga White[b]	20/20	–	–	0/20

[a]All cuttings after hot water treatment were indexed by plating vascular extract on RS medium

[b]Ramsey and K5140 with natural infection; Cayuga White artificially inoculated with AT3

[c]Calluses generated from K5140 were indexed for the presence of AT3

Pseudomonas syringae pv. *syringae* (*Pss*), causative agent of bacterial canker disease. The treated bud sticks were grafted onto established Mazzard rootstocks. Bud blast and canker development were monitored relative to the untreated bud sticks. The heat treatment eliminated the pathogen with minimal damage to the bud sticks (Hall et al. 2002).

The synergistic effect of relatively less efficient disease control strategies when combined together, was investigated for the control of tomato bacterial leaf speck disease caused by *Pseudomonas syringae* pv. *tomato* (*Pst*). Treatment of tomato seeds infected by *Pst*, with a combination of mild heat (42–45 °C in a circulating water bath for 2 h), inoculation with *Azospirillum brasilense* (a biocontrol agent) and later a single application of commercial bactericide (a mixture of 0.5 % copper hydroxide and 0.3 % copper oxychloride), almost eliminated bacterial leaf speck pathogen. The treated plant grew normally as the untreated control plants and remained without disease symptoms in a mist chamber for 6 weeks (Bashan 2000). Angular leaf spot (ALS) disease seriously affects strawberry seedlings in the nurseries and the infected seedlings are the principal mode of spread of the disease to the main field. Disease control strategies are directed to obtain disease-free strawberry seedlings that are to be exported from the USA to European countries. Populations of the pathogen *Xanthomonas fragariae (Xf)* exposed to 56 and 52 °C were killed entirely after 15 and 60 min of exposure respectively, but both treatments killed the plants also. Strawberry seedlings of six cultivars were treated with hot water at 44 or 48 °C for 0, 60, 120, 180 and 240 min. Bacterial populations exposed to 44 °C for 4 h or 48 °C for 2 h were reduced by 10^5 or 10^6 CFU/ml. These treatments minimally

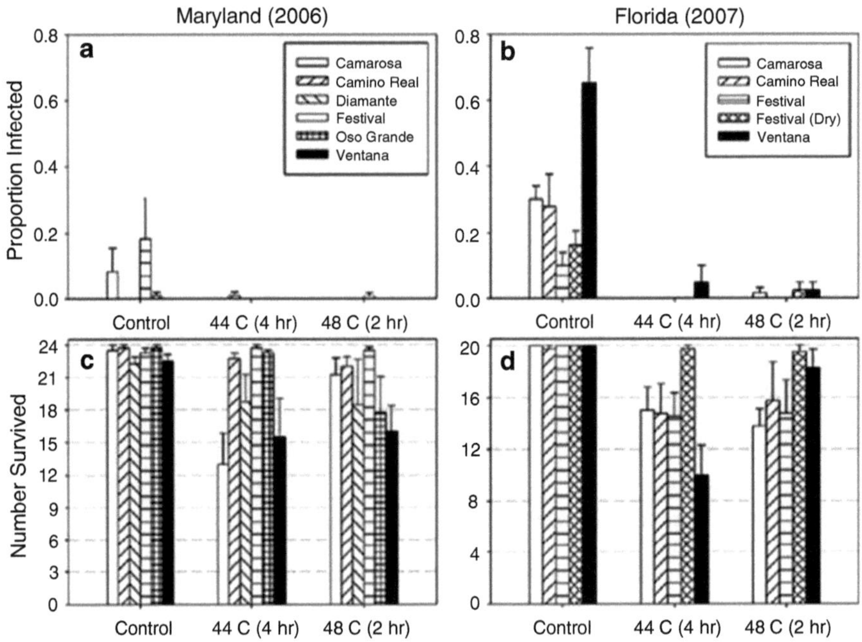

Fig. 3.1 Assessment of thermosensitivity and survival of five cultivars of strawberry for selection of angular leaf spot disease-free strawberry seedlings under field conditions at two locations in 2 years (2006 and 2007) (Courtesy of Turechek and Peres 2009 and with kind permission of The American Phytopathological Society, MN, USA)

affected the vegetative growth of plants bagged dry or wet, but flowering was adversely affected. The strawberry cultivars varied in their sensitivity to heat. By selecting heat tolerant strawberry cultivar, heat treatment for nursery stock might be feasible and it could be applied to supplement standard production practices for producing pathogen-free nursery stock (Fig. 3.1) (Turechek and Peres 2009).

Thermosensitivity of phytoplasmas (earlier known as mycoplasmas) was reported as early as 1936 by Kunkel who demonstrated that peach yellows-infected plants could be cured by heat or hot water treatment. Attempts to eliminate the phytoplasma from Cabot high bush blueberry by exposing the whole plant to high temperature (38 °C) in growth chamber failed, when the plants were grown at ambient CO_2 level. But at an enhanced level of CO_2 (1,200 ppm), Cabot plants produced new shoots after 6 weeks. Soft wood cuttings (10–20 mm long) were taken from new shoots and propagated. These cuttings were free from the phytoplasma, as revealed by the test using diaminophenylindole (DAPI) fluorescent stain. The number of DAPI-positive plants grown from heat-treated cuttings decreased with increase in the duration of treatment. At 5 weeks after treatment, all cuttings were free from infection by the phytoplasma. Three clones of blueberry free of the phytoplasma were established by this high carbon-dioxide-thermotherapy technique (Converse 1987). Sweet potato plants could be freed from the little leaf phytoplasma (*Candiatus*

Phytoplasma aurantifolia) by applying cryotherapy treatment. Shoot tips with 3–4 leaf primordia were excised from in vitro grown infected plants. All plants regenerated from cryo-treated shoot tips were free of phytoplasma (100 %), whereas shoot tip culture or dehydration of shoot tips without subsequent cryotherapy resulted in plants free of phytoplasma at a frequency of only 7–10 %. Light and transmission electron microscopy (TEM) indicated that cryotherapy was lethal to all cells, except those in the apical dome of the meristem and the two youngest leaf primordia. No phytoplasma bodies could be seen in the surviving parts of the shoot tip. In contrast, abundant phytoplasmal bodies were present in the sieve elements located in the lower parts of the untreated shoot tip. Cryotherapy holds promise for long-term storage of plant germplasm and also for production of pathogen-free plants effectively from infected plants (Wang and Valkonen 2008).

3.1.3 Virus Diseases

Higher temperatures have been demonstrated to be useful to create variability in the virulence of viruses, resulting in the selection of mild strains that could be employed to cross-protect the susceptible plants against the severe strains of the same virus or related viruses or strains. This aspect of biological control of virus diseases through cross-protection has been discussed earlier. Heat treatments (thermotherapy) has been applied to effectively eliminate the pathogens especially viruses from asexually propagated planting materials using different forms of heat based on their thermosensitivity. Heat treatments have also been used as the first step to reduce the virus concentration or to eliminate the virus present in the apical meristem to enhance the percentages of meristematic tissue free of virus. Consequently, the plants generated from the tissue culture may be free from viruses that cannot be eliminated by other methods.

3.1.3.1 Elimination of Plant Viruses by Thermotherapy

The possibility of eliminating over 100 plant viruses from infected plants was indicated by Nyland and Goheen (1969). Heat may be applied as hot air, hot water or aerated steam. Growing plants at temperatures of 35–40 °C for several days or weeks was effective in eliminating the viruses from the host plant species. The plants growing from cuttings from *Cassava mosaic virus*-infected plants grown at 35 °C for 30 days, failed to develop any visible symptoms of the disease (Kartha and Gamborg 1975). Grapevine plants infected by *Grapevine fan leaf virus* were entirely freed of the virus, by growing them at 38 °C for 163 days (Rives 1970). Heat treatments have been applied as a practical means for the control of sugarcane diseases transmitted through setts. *Sugarcane mosaic virus* (SMV) and its strains A, B, D, H and I were eliminated from sugarcane setts by serial hot water treatment on four successive days at 54.8, 57.3, 57.3 and 57.3 °C for 7 min on each day (Benda 1972;

Benda and Ricaud 1978). Physiological changes in the virus-infected plants following heat treatments were studied. An isolate of *Tomato aspermy virus* (TAV) was inactivated both in vitro and in vivo at 36 °C. Virus inactivation was not entirely due to loss of infectivity following heat treatment. The concentration of polyphenoloxidase (PPO) increased greatly in tobacco plants grown at 36 °C. The results indicated that PPOs either directly or indirectly enhanced the inactivation of TAV during heat treatment. In addition, the concentration of ribonucleases also increased in heat-treated tissues and these might supplement the inactivation of the virus by heat treatment. Reduction in the pH and ionic strength of the sap in heat-treated plants might affect the optimal requirement of pH and molarity for TAV infectivity. The alterations in cellular metabolism during heat-induced stress may have a role in the viral inactivation (Johnstone and Wade 1974).

Stable viruses like *Tobacco mosaic virus* (TMV) can remain viable on the seed surface and the externally seedborne virus can be inactivated by various mechanical, chemical and physical treatments (Hall et al. 2002). *Cucumber green mottle mosaic virus* (CGMMV), a member of the genus *Tobamovirus* is transmitted through cucumber seeds. Virus-contaminated seeds were dried at 35 and 50 °C for 24 h each and then incubated at 72, 75, 79 or 82 °C for 72 h. Purified CGMMV was also subjected to these temperature treatments. The mechanism of virus inactivation by high temperatures was studied. Electron microscopy and reverse transcription (RT)-polymerase chain reaction (PCR) assay were employed to determine the changes caused by heat treatments. Heat treatment at 75 °C for 72 h completely inactivated CGMMV on contaminated seeds as shown by the bioassay of heat-treated seeds and the treatment at 85 °C for 24 h also inactivated the virus. Observations under the electron microscope revealed that virus particles purified from heat-treated seeds were broken down into a variety of fragments as the temperature of treatment increased. Virus purified from untreated control seeds infected *Nicotiana benthamiana*, while the preparation from any of the heat-treated seeds did not infect the assay host plants. Analysis of viral RNA revealed that there were heat-sensitive and heat-resistant regions of the virions of CGMMV. The heat-sensitive regions were located in the RNA-dependent RNA polymerase (RdRp) gene. The results indicated that RT-PCR assay of CGMMV-specific primers corresponding to the 2–2.5 kb and/or 4–4.8 kb regions could be employed for rapid detection of virus inactivation by heat treatment. The technique developed in this investigation would allow detection of virus inactivation on heat-treated seeds in large number of seed samples facilitating development of virus-free seed stocks (Kim et al. 2003).

Carnation plants are seriously damaged by *Carnation latent virus* (CLV). Heat therapy approach was evaluated for its effectiveness in producing virus-free plants for further propagation. Carnation plants infected by CLV were placed in heat therapy chamber and expose to two heat regimes of 35 ± 2 °C or 40 ± 2 °C for 1, 2, 3 weeks and more than 3 weeks. The surviving plants were indexed for the presence of CLV by applying enzyme-linked immunosorbent assay (ELISA) test. Plants exposed to either 35 ± 2 °C for 4 weeks or 40 ± 2 °C for 3 weeks were found to be free of the virus, while other temperature treatments were ineffective in eliminating the virus from carnation plants. Furthermore, all plants regenerated from the axillary buds which were heat-treated at 35 ± 2 °C for 4 weeks or at 40 ± 2 °C for 3 weeks were free of CLV as revealed by ELISA technique (Mangal et al. 2004). In order to obtain virus-free lily plants, different techniques including thermotherapy were employed in combination. *Lily symptomless virus* (LSV) adversely affected plant growth and quality of cut flowers. Stock bulbs of pollenless Asiatic hybrid lily (*L.* x *elegans*) lines (409 and 509) were used as explants. Shoot meristems were excised and micropropagated. Thermotherapy treatment (35 °C for 42 days) was applied to in vitro growing bulblets and a second meristem cut was made from heat-treated material. ELISA or RT-PCR techniques were employed to check the leaf tissues from bulblets formed before or post-heat treatments for the presence of LSV. The line 409 produced LSV-free plants without heat treatments, but line 599 produced LSV-free plants only after heat treatment. The growth of virus-free lily bulblets was at a faster rate and acclimatized promptly. The results showed that thermotherapy was effective in virus elimination and also promoted plant growth resulting in vigorous, virus-free lily plants (Nesi et al. 2009).

Propagative materials from three clones of grapevine cv. Chardonnay infected by multiple virus combinations of *Grapevine leafroll-associated virus* (GLRaV) and *Grapevine rupestris stem pitting virus* (GRSPaV) and virus-like diseases (vein mosaic and vein necrosis) were propagated. Selected progenies were obtained after virus elimination by heat therapy and in vitro apex grafting. The effect of heat therapy was assessed by graft indexing and double antibody sandwich (DAS)-ELISA format. The effect of selective virus elimination on vigor, yield and fruit quality was assessed. Elimination of GLRaV-2 provided the highest beneficial impact with a marked increase in cumulative weight growth (21 %), fresh fruit yield (22 %) and sugar concentration of fruit juice (9 %). Selective elimination of GLRaV-1 and GLRaV-3 also showed beneficial effect on fruit quality. The results indicated that heat therapy has the potential for wide application in view of the enhancement of desirable yield and quality parameters (Komar et al. 2007). The effectiveness of in vitro thermotherapy at 35–37 °C for 20 days in obtaining propagative materials of apricot (*Prunus armeniaca*) cv. Bebecou free from *Plum pox virus* (PPV) was compared to the conventional heat treatment of potted plants at 30–35 °C for 8 weeks. The in vitro thermotherapy was combined with shoot tip culture. The combination resulted in the enhancement of effectiveness of virus elimination by more than five times, while requiring only half the time needed for conventional thermotherapy (Koubouris et al. 2007). In another investigation, thermotherapy and chemotherapy treatments were combined for the elimination of *Apple chlorotic leaf spot virus* (ACLSV) and *Prunus necrotic ringspot virus* (PNRSV) from myrobalan (*Prunus cerasifera* var. *diversicata*), PNRSV from 'Empress' plum (*Prunus domesticata*) and Prune dwarf virus from 'Early Rivers' sweet cherry (*Cerasus avium*) plants. The cultured shoots were subjected to heat therapy in a growth chamber wherein the temperature was gradually increased from 28 to 36 °C within a week and maintained at 36 °C for the following 4 weeks. The treated and untreated (maintained at 22 °C) shoots were assayed for the presence of the viruses by DAS-ELISA tests for 1 year after heat therapy treatment. PNRSV and ACLSV were effectively eliminated by heat treatment, while it was not efficient in eliminating PDV from sweet cherry.

Table 3.3 Efficacy of thermotherapy and chemotherapy in eliminating viruses infecting *Prunus* spp. (Cieślińska 2007)

	No. of ELISA negative plants/total no. of plants tested			
	Myrobalan		Plum	Sweet cherry
Treatment	PNRSV	ACLSV	PNRSV	PDV
Control	0/30	0/30	0/30	0/25
Thermotherapy	14/24	16/24	15/20	0/25
Chemotherapy-Virazole				
10 mg/l	0/27	10/27	10/26	0/25
25 mg/l	0/26	16/26	12/24	0/25
50 mg/l	0/26	23/26	7/10	0/22
100 mg/l	0/21	18/21	8/8	0/14
Thermotherapy + Chemotherapy Virazole				
10 mg/l	16/25	14/25	22/22	0/25
25 mg/l	17/25	20/22	20/20	3/25
50 mg/l	19/19	19/19	10/10	8/21
100 mg/l	18/18	15/18	2/2	14/15

Combination of thermotherapy and chemotherapy with Virazole® (a synthetic broad spectrum antiviral nucleoside, 1-ß-D-ribofuranesyl-1,2,4-triazole-3-carboxamide) at a concentration of 50–100 mg/l effectively eliminated PDV from sweet cherry shoots. However, phytotoxic symptoms were observed in plum shoots (Table 3.3) (Cieślińska 2007).

Indian citrus ringspot virus (ICRSV) disease is a serious problem affecting citrus production in India. In order to obtain virus-free plants, the effect of hot water and moist hot air treatments on the elimination of ICRSV was assessed. Nodal segments from mother plants were treated at different temperature (40–50 °C for 30 min). The percentage of virus elimination achieved varied from 40.74 to 60.0 % and 48.15–73.33 % as assessed by RT-PCR and ELISA tests respectively. The results indicated that ELISA test was less efficient in deleting low virus concentrations, whereas RT-PCR technique could detect the virus in such ELISA-negative plant materials, resulting in lower percentages of virus elimination by the treatments (Sharma et al. 2008). It was difficult to eliminate *Raspberry bushy dwarf virus* (RBDV) from raspberry plants using conventional meristem-tip culture. Subjecting plants to thermotherapy prior to meristem-tip culture, reduced drastically the concentrations of viral RNA2 and RNA3 and the coat protein in the shoot tips. But the plants were not virus-free. Hence, a novel method involving thermotherapy followed by cryotherapy was developed for efficient virus elimination. Heat treatment induced subcellular alterations such as enlargement of vacuoles in the more developed virus-infected cells which were largely eliminated after application of cryotherapy. Using this procedure, 20–36 % of treated shoot tips survived; 30–40 % regenerated and up to 35 % of regenerated plants were virus-free as revealed by ELISA and reverse transcription loop-mediated isothermal amplification techniques. This study enabled to study the factors influencing virus elimination, including invasion of shoot tips and meristematic tissues by RBDV, enhanced viral RNA degradation and increased sensitivity to freezing caused by thermotherapy and subcellular changes and subsequent death of cells caused by cryotherapy (Wang et al. 2008). The feasibility of applying thermotherapy as a virus disease management strategy was assessed under field condition. Trees of two cultivars in each of plum, apricot and peach were artificially inoculated with two strains of *Plum pox virus* (PPV-D, PPV-M). Two cycles of thermotherapy in vivo were adopted. During the first cycle, 16 trees of individual cultivars of plum, apricot and peach were treated for 15 days at 37 °C. In the second cycle 10 trees of individual cultivars of plum, apricot and peach were treated for 22 days at 37 °C. DAS-ELISA and RT-PCR assays were employed to check the trees for freedom from virus. Out of a total of 26 trees treated, four trees (two plum and two apricot) recovered from virus infection. Thermotherapy was not successful in the case of peach trees. The results indicated that though the chemotherapy efficiency in vivo was only 15 %, it would be possible to enhance the efficiency by improving treatment application (Polák and Hauptmanova 2009).

3.1.3.2 Electrotherapy for Elimination of Plant Viruses

Application of electric current for producing virus-free plants has been demonstrated as a new approach for virus disease management. *Cactanucia* tree stakes showing mosaic symptoms were treated with electric current at a voltage of 500 V for 5–10 min. Up to 90 % of the treated plants were found to be virus-free (Quacquerrelli et al. 1980). Later, the electrotherapy was comprehensively employed for the elimination *Potato virus X* (PVX) from potato plants. The infected plants were exposed to an electric current of 15 mA for 5 min, leading to 60–100 % elimination of PVX in different potato cultivars (Lozoya-Saldaña et al. 1996). In addition to elimination of PVX, electrotherapy was successfully used for the elimination of *Potato virus Y* (PVY), *Potato virus A* (PVA), *Potato virus S* (PVS) and *Potato leafroll virus* (PLRV) (Pazhoushande and Mozafari 2001). Furthermore, electrotherapy has been reported to be effective for the elimination of *Onion yellow dwarf virus* (OYDV) and *Leek yellow stripe virus* from garlic (Hernández et al. 1997), *Banana streak virus* (BDV) from banana (Hernández et al. 2002), *Dasheen mosaic virus* (DaMV) from cocoyam (Igarza-Castro et al. 2001) and *Tomato yellow leafcurl virus* (TYLCV) (a DNA virus) from tomato plants (Fallah et al. 2007). The effect of electrotherapy involving the use of continuous electric field for elimination of *Grapevine leafroll-associated virus* (GLRaV) was assessed in comparison to chemotherapy. Infected grapevine plants raised from one bud woody cuttings were subjected to continuous electric field and in vitro culture. The efficacy of virus elimination of GLRaV 1+3 from grapevine cv. Ranâi Magaraci was achieved by electrotherapy to an extent of 66.66 %. No clear correlations could be established between explants positions and virus-free plant regeneration percent. When Oseltamivir was applied at 120 µmol/l for 30 days, the virus elimination rate obtained was 71.42 %. However, when the exposure time was doubled, the virus

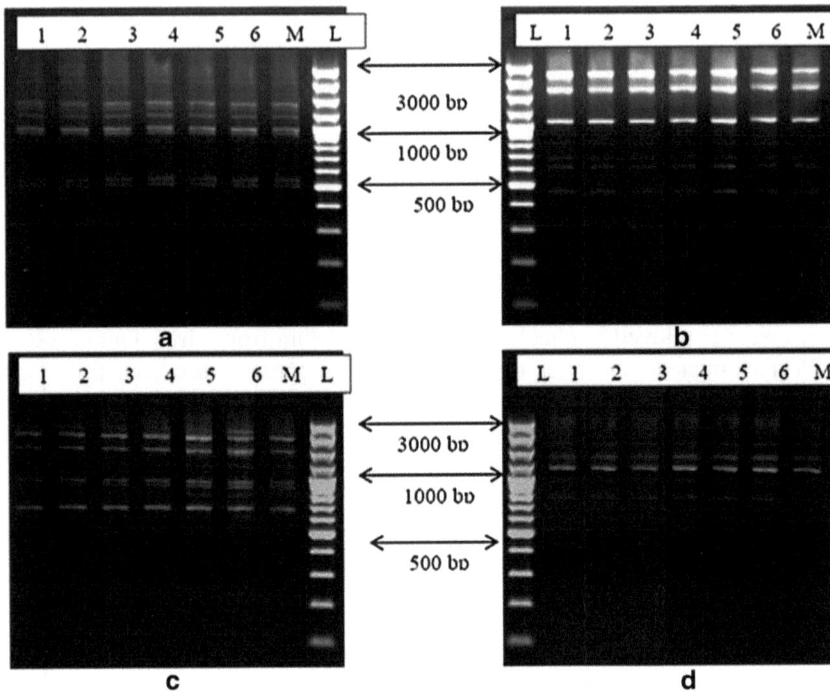

Fig. 3.2 RAPD patterns amplified by primers OPC05 (**a**), OPC06 (**b**), OPC01 (**c**) and OPA02 (**d**) for determining the genetic stability of virus-free plants generated after electrotherapy treatment Generants did not show any deviation from the mother plant (Courtesy of Gută et al. 2010 and with kind permission of Romanian Biotechnological Letters, Romania)

elimination was reduced to 66.66 %. Another chemotherapeutant ribavirin was ineffective in eliminating the virus. For assessing the genetic fidelity of virus-free plants, random amplified polymorphic DNA (RAPD) techniques was efficient in generating reproducible results. No variation was observed between mother plant and randomly tested regenerants indicating the maintenance of genetic uniformity among regenerants (Fig. 3.2) (Gută et al. 2010).

Use of virus-free germplasm is a prerequisite for production of certified seeds as in the case of common bean (*Phaseolus vulgaris*) in which *Bean common mosaic virus* (BCMV) is transmitted through seeds. The effectiveness of electrotherapy in eliminating BCMV and producing virus-free plants in two common bean cultivars, was evaluated. Nodal cuttings of infected bean plants were exposed to electric currents of 5, 10, and 15 mA for 10 min and planted in vitro. There was no significant difference in the response of the bean cultivars to electrotherapy. One month old regenerated plantlets were tested by double antibody sandwich (DAS)-ELISA format for detecting BCMV infection. The three intensities of electric current resulted in mean virus elimination percentages of 37.2, 45.5 and 71.9 % and mean plant regeneration of 79.5, 68.6 and 58.4 % respectively (Table 3.4). The therapy efficiency

Table 3.4 Effects of electrotherapy treatments on the regeneration of in vitro plantlets and elimination rate of BCMV from two bean cultivars (Hormozi-Nejad et al. 2010)

Cultivar/treatment (mA/min)	Regeneration rate		Elimination rate	
	Regenerated[a]	Percentage treated[b]	BCMV-free[c]	Percentage regenerated[c]
Khomein				
0/0	17/20	85.0	–	–
5/10	38/47	80.8	11/38	40.0
10/10	27/41	65.8	12/27	44.4
15/10	26/49	53.6	19/26	73.0
Capsouli				
0/0	18/20	90.0	–	–
5/10	29/37	78.3	10/29	34.4
10/10	30/42	71.4	14/30	46.6
15/10	30/49	63.2	22/30	70.9

[a]Number of regenerated plants
[b]Number of explants treated
[c]Number of virus-free plants

index (TEI) for the three electrotherapy treatments was estimated as 28.9, 29.2 and 39.1 in cv. Khomein and 26.9, 33.3 and 44.8 in cv. Capsouli respectively. Based on the TEI, an electric current of 15 mA for 10 min was found to be the most effective treatment for eliminating BCMV from both cultivars. The regeneration rate of virus-free plants obtained after electrotherapy was higher than that of plants exposed to more conventional virus elimination techniques including meristem culture, thermotherapy and chemotherapy. Regenerated plantlets after exposure to electrotherapy were morphologically identical to those generated from non-treated control explants. The results suggested that electrotherapy might denature viral nucleoproteins, by increasing the temperature inside the tissues. There is no need for any chamber as required for thermotherapy treatments (Hormozi-Nejad et al. 2010).

3.1.3.3 Tissue Culture Technique for Virus Elimination

Tissue culture technique has been demonstrated to be useful for the preservation of germplasm of several crops. Another application of tissue culture is rescuing valuable plant varieties from virus diseases, when all plants are systemically infected by the viruses. Propagation of virus-free plants from meristem cultures has been successful, if the viruses spread relatively slowly to the newly formed tissues. Virus-free stocks of seeds and asexually propagated plant materials have been established in many countries by ensuring freedom from virus infection by employing immunoassays and/or nucleic acid-based detection techniques (Narayanasamy 2011). The meristem tip culture is an aseptic culture of apical meristem dome plus the first pair of leaf primordia. The length of meristem pieces may be about 0.5–1.0 mm and this length may vary with plant species. The chances of getting virus-free plants are more with shorter pieces of tissue, but chances of tissue survival are less than that of longer pieces. By meristem tip culture, chlorotic virus infecting two jute species *Corchorus olitorius* and *C. capsularis* could be eliminated using 0.2–0.8 mm meristem dome. The smaller the dome size, greater was the chance of getting virus-free plant. The percentages of obtaining virus-free plants through direct plant regeneration and through callus to plant formation methods were 40–50 % and 50–60 % respectively (Das et al. 1983). In order to enhance the chances of obtaining virus-free plants, two kinds of pre-treatments to source plants have been applied. By exposing the infected plants to high temperatures, it may be possible to get virus-free plants even with longer pieces of meristematic tissue. Cassava plants infected by *Cassava mosaic virus* were exposed to 35 °C for 30 days and meristems up to 8 mm in length were cultured. Mosaic-free plants could be regenerated from such longer pieces of meristem (Kartha and Gamborg 1975). Similar procedure was adopted to eliminate *Tomato apsermy virus* from chrysanthemum and a wide range of other viruses from host plant species (Garrett et al. 1985). Another approach for pre-treating the infected plants for virus elimination, is to accelerate the elongation of internodes of plants by applying auxins. Using different combination of kinetin and indole acetic acid (IAA) for treating virus-infected sweet potato plants, it was possible to get sweet potato plants free of leaf spotting, internal cork and feathery mottle virus complex by culturing meristematic tips (0.4–0.8 mm long). The auxin-treated plants elongated at a rate faster than the viruses could invade the newly developed tissues, thereby increasing the possibility of having higher percentages of plants free of viruses infecting sweet potato (Alconero et al. 1975).

The potential of tissue culture method for the elimination of viruses infecting lily plants was assessed. The bulbils from stem, bulblets from scale, basal pieces of leaves and basal pieces of petals from plants infected by *Lily symptomless virus* (LSV) were used as explants and they were placed on Murashige and Skoog (MS) medium. For Asiatic hybrids, methods using adventitiously formed meristems on bulbscale explants resulted in 50–70 % LSV-free plants of *Lolium longifolium* cultivars. Detection of LSV in the regenerated plants by electron microscope was performed to confirm the freedom of regenerated plants from infection by LSV. Plantlets from basal parts of the petals and of basal pieces of leaves, used as explants, were free from virus to the maximum extent (Dapkūniene et al. 2004). Microshoot-tip culture technique has shown to be an effective method of eliminating viruses infecting grapes. In microshoot-tip culture method, a growing tip (less than 0.5 mm) is excised from shoot tip. It includes the meristematic dome and two to four leaf primordia. Rapidly growing shoots in the spring and early summer provide the best tissue for excision. Terminal buds are preferred, since they are larger, easier to excise and more vigorous than axillary buds. Normally 10–30 % of the meristem pieces survive and become rooted plants. Of these surviving plants usually 70–100 % were found to be virus-free, depending on the virus and grapevine cultivar. Microshoot-tip culture eliminated *Grapevine leafroll-associated virus, Grapevine fan leaf virus, Grapevine virus A, B and D* (associated with rugose wood disease) and *Rupestris stem pitting-associated virus* and *Tomato ringspot virus*. In addition, *Agrobacterium vitis* causing crown

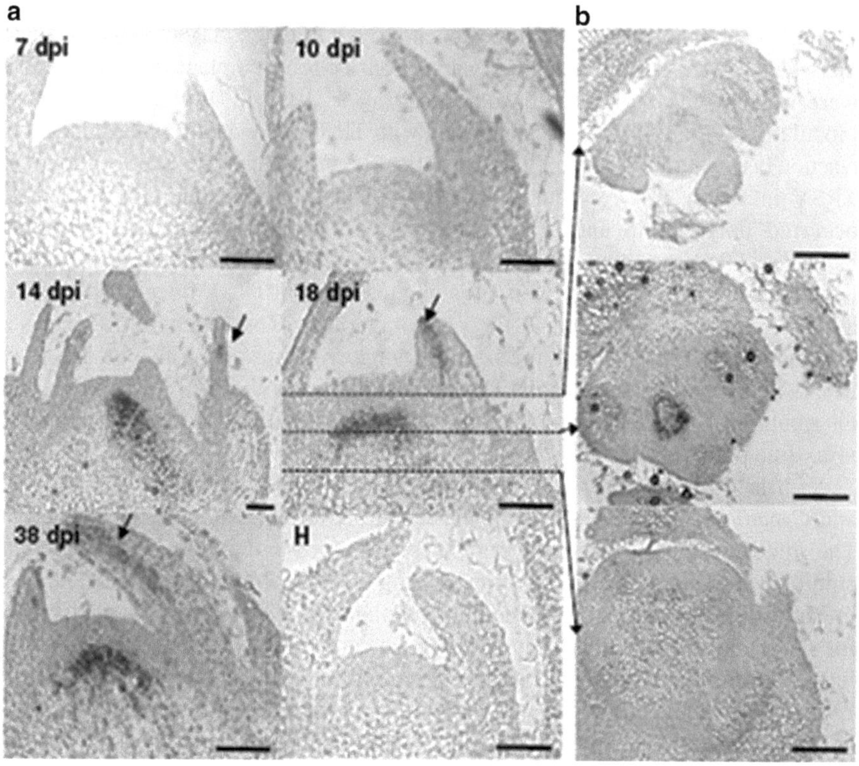

Fig. 3.3 Distribution of *Tobacco ringspot virus* (TRSV) in shoot meristem of infected tobacco plants (**a**) longitudinal sections of tobacco plant meristem tissues at 7, 10, 14, 18 and 38 days post-inoculation (dpi), treated with an anti-TRSV serum. *Dotted lines* in the 18 dpi image indicate the location of the transverse sections in (**b**). H: healthy, negative control; note absence of TRSV signal (Courtesy of Dong et al. 2010 and with kind permission of Springer Science + Business B. V., Heidelberg, Germany)

gall disease was also eliminated from infected grapevine plants. The microshoot-tip technique has several advantages over other tissue culture techniques. It avoids the production of plants from callus, resulting in regeneration of an off-type plant. Many pathogens including the bacterial pathogen can be eliminated by applying microshoot-tip culture technique (Sim 2006; Sim and Golino 2010) (Appendix A.3).

The absence of viruses in the shoot apical meristems (SAM) tissues is of practical importance, because virus-free plant clones can be generated by culturing meristem tips. *Cucumber mosaic virus* (CMV) and *Tobacco rattle virus* (TRV) were able to invade the shoot meristem. But the SAM tissues recovered from virus infection later. These shoots became virus-free in due course (Mochizuki and Ohki 2000; Martin-Hernández and Baulcombe 2008). The kinetics of distribution of *Tobacco ringspot virus* (TRSV) in infected tobacco plants were analyzed, since asymptomatic leaves were formed above the symptomatic leaves that exhibited ring spots. Immunohistochemical microscopy was employed to detect TRSV in shoot apical meristem (SAM) tissues. Longitudinal sections of tobacco plant meristem tissues were treated with an anti-TRSV antiserum at 7, 10, 14, 18 and 38 days post-inoculation with TRSV. The cells infected with TRSV were darkly stained due to the reaction between the virus and the virus-specific antibodies (Fig. 3.3). In contrast, TRSV infection was transient in the root apical meristem that TRSV infection occurred only within and around the tissues exhibiting ring spot symptoms. However, TRSV was not or hardly detected in the asymptomatic leaves of inoculated tobacco plants. The recovery of young leaves from TRSV infection becoming aymptomatic was attributed to the phenomenon of RNA silencing. Decreased distribution of TRSV in root meristem of tobacco plants could also be due to an RNA silencing-like mechanism. The results suggested the need for studying the distribution of viruses in meristematic tissues before utilizing them for obtaining virus-free plants (Dong et al. 2010).

Vanilla (*Vanila planifolia*) is infected *by Cucumber mosaic virus* (CMV) and *Cymbidium mosaic virus* (CyMV) accounting for marked losses in production. The infected plants were freed from these two viruses through apical meristem culture technique. Apical meristem measuring 0.1–0.25 mm in length were cultured in Murashige and Skoog (MS) medium supplemented with 0.45 μM thiadiazuron for 40–45 days to initiate growth. Following enlargement of meristem, it was transferred to MS medium supplemented with 4.43 μM 6-benzyl aminopurine (BAP) and 2.68 μM α-naphthalene acetic acid (NAA) for regeneration. The regenerated plantlets were hardened in insect-proof glasshouse. The plants were indexed for virus infection using RT-PCR assay using virus-specific primers. The frequency of CMV elimination was 79.4 %, whereas CyMV was eliminated from 82.4 % of regenerated plants, when tested individually. Both viruses were eliminated by meristem culture procedure in 75 % of the regenerated vanilla plants (Retheesh and Bhat 2010). Purple passion fruit plants are infected by *Passion fruit woodiness virus* resulting in marked reduction in yield. Apical meristem culture technique was adopted for production of virus-free plants. Shoot tips (2 mm) were excised from infected plants and shoots were regenerated by culturing on MS medium supplemented with different concentrations of benzyladenine (BA) and NAA. Root formation was promoted by indole-3-butyric acid. The regenerated shoots were assayed for the presence of woodiness virus using ELISA or a test strip kit. The virus-free shoots were used for further propagation leading to production of virus-free plants (Prammanee et al. 2011).

3.1.3.4 Combination of Thermotherapy and Tissue Culture for Virus Elimination

Tissue culture of apical meristem of plants infected by viruses has been shown to be effective for eliminating some viruses from infected plants and to develop virus-free stocks. But several viruses are able to invade the meristematic tissues and it is difficult to eliminate such viruses only by tissue culture methods. Heat therapy either prior to or after excision of meristems has been useful to reduce the virus titer (concentration) appreciably. Somatic embryogenesis technique is based on the production of somatic plantlets via callus tissue from anthers and/or ovaries. Elimination of grapevine viruses using in vitro somatic embryogenesis was successfully achieved, but this procedure was not sufficient to eliminate *Grapevine fan leaf virus* (GFLV) (Goussard et al. 1991). Somatic embryos were successfully regenerated from callus tissue of anthers and ovaries excised from inflorescences of grapevines infected with GFLV. Production of pro-embryogenic masses (PEMs) was controlled by specific growth regulators and culture conditions. The presence of GFLV particles in plantlets regenerated by somatic embryogenesis without heat incubation indicated that the virus was translocated from infected via proliferating callus to embryoids. The cultures were incubated in darkness and constantly agitated on an orbital shaker at 35 °C for 60 days. Somatic embryos (containing roots and cotyledons) and plantlets were analyzed by immunosorbent electron microscopy (ISEM) and ELISA test. The observations showed that GFLV was effectively eliminated by the combination of embryogenesis and heat therapy (Goussard and Wiid 1992).

In grapevines, infection by individual viruses causes losses to different degrees and combined infections become more destructive. Production of virus-free plants is considered to be desirable in the case of asexually propagated crops. Virus elimination could be achieved through thermotherapy and meristem- or shoot tip culture, but the combination of both methods was found to be more effective (Grammatikakk et al. 2005). In a later investigation, meristem or shoot-tip cultures coupled with thermotherapy were used to successfully eliminate *Grapevine leafroll-associated virus-1* (GLRaV-1) and *Grapevine rupestris stem-pitting-associated virus-1* (GRSPaV-1), resulting in a relatively high percentage of virus-free plantlets. GRSPaV-1 was found to be earlier recalcitrant to elimination, whereas GLRaV-1 was more easily eliminated. The effectiveness of two different virus elimination techniques including meristem tip culture was evaluated. Meristem tip culture combined with thermotherapy [26 °C (day) and 23 °C (night) gradually raised by 3 °C/week to reach the final temperature of 40 °C (day) and 37 °C (night) over a period of 6 weeks] was the most effective in eliminating both viruses. The freedom from viruses was checked by nested RT-PCR assays. Success rate for GLRaV-1 (91.2 %) was greater than for GRSPaV-1 (67.6 %). The ratio of virus elimination to survival was greater for meristem tip culture than for shoot tip culture. The technique developed in this investigation allowed to have bigger explants such as shoot tips and the survival rate was significantly greater than in meristem tip culture. The plants were virus-free, making the adoption of this technique feasible for wider application (Skiada et al. 2009).

Thermotherapy was applied for recovering healthy citrus plants from trees infected by *Citrus tristeza virus* (CTV) and *Citrus psorosis virus* (CPsV). It was found to be effective for the elimination of most graft-transmissible viruses, but not viroids and spiroplasmas (Grant 1976; Calavan et al. 1972). Shoot tips (0.14–0.18 mm long) were aseptically isolated from diseased citrus plants and they were grafted (inverted T graft) onto young etiolated seedlings grown in vitro. One month after grafting, plants were transferred to in vitro conditions. Bigger shoot tips were collected from thermo-treated buds at 32–35 °C. This combination reduced virus replication and increased meristematic cell activity and the shoot tips were free of *Citrus psorosis virus* which was found to be difficult to be eliminated by shoot-tip grafting alone (Navarrot et al. 1975; Carvalho et al. 2002). Citrus somatic embryogenesis from stigma and style in vitro culture is the efficient sanitation technique for total recovery of the most infected *Citrus* spp. Styles and stigmas are dissected from freshly collected closed flowers under laminar flow. Explants are vertically placed on the medium in petridishes. Green somatic embryos developed in 2–7 months and embryos developed into plantlets which are in vivo transferred. The embryos formed are totally free from CTV. Somatic embryogenesis is user-friendly and can be applied for the sanitation of *Citrus* spp. (Meziane et al. 2009).

Potato virus Y (PVY) is an important virus infecting potato crops and it is considered to be primarily responsible for high yield losses and degeneration of seed tubers. The effect of tissue culture technique, thermo- and chemo- and electro-therapies on elimination of PVY was assessed. Direct antigen coating (DAC)-ELISA format was employed to detect the presence of PVY in the treated materials. Plants developed from tubers stored for 3 months in the cold room were subjected to two cycles of thermotherapy at approximately 37 °C during 40 and 30 days resulted in elimination of virus in 33.3 and 14.2 % of plantlets respectively, indicating the moderate efficiency of virus elimination. Application of an antiviral compound, ribavirin (RBV) at 20 mg/l was more effective, providing elimination efficiency of 30–43 %, when the medium was amended with ribavirin. Stems with axillary buds were excised from potato plants; grown for 45 days and treated with electric shock. Stems were directly connected to the electrodes of power supply or they were immersed in electrified water in a wide range of time and power. Axillary buds excised from electric-shocked stems or tissues immersed in electrified water, were transferred into the medium supplemented with or without ribavirin. Electro-therapy treatments at 5, 10, and 15 mA electric current, eliminated PVY in 66.7, 90 and 100 % of the plantlets respectively. Among the therapeutic treatments, electrotherapy was the most efficient in elimination of PVY from treated plantlets (Mahmoud et al. 2009). Sweet potato, a vegetatively propagated crop, is infected by several viruses which contribute to the decline in yield and quality of tubers, because of virus accumulation following many cycles of vegetative propagation. Stem cuttings of sweet potato cv. Abees were grown under controlled greenhouse conditions and indexed for the presence of *Sweet potato feathery mottle virus* (SPFMV), using dot-ELISA procedure. Infected plants were exposed to 42 °C/day and 30 °C/night for 3 weeks. After heat treatment, meristems were excised, cultured and plants were regenerated. Plantlets derived from tissue culture were analyzed by RT-PCR using specifically designed primers based on the conserved regions present in all strains of SPFMV. All plantlets were found to be SPFMV-free. This RT-PCR procedure was efficient in detecting the specific band (~300 nucleotides) used as the molecular marker for SPFMV. Tissue-culture derived plants were routinely tested for 2 years subsequently by dot-ELISA. None of the plants showed SPFMV infection in in vitro propagation (El Far and Ashoub 2009).

3.1.3.5 Combination of Tissue Culture and Ultraviolet Irradiation for Virus Elimination

Ultraviolet (UV)-irradiation has been employed to induce mutations in plant viruses, resulting in the selection of mild strains that could be used to cross-protect plants against infection by severe strains of the parent virus/strain. Ultraviolet-B (UV-B) irradiation was evaluated for its efficiency in protecting plants regenerated from calli which were exposed to UV-B irradiation. The calli produced from tomato cv. PKM1 were treated with UV-B irradiation. Plantlets regenerated from treated calli were challenge-inoculated with *Cucumber mosaic virus* (CMV). All treated plants had remarkably high levels of soluble phenolics, salicylic acid (SA) and peroxidase compared with untreated controls. The challenge-inoculated plants showed disease severity ranging from 32 to 78 %, while the control plants exhibited disease severity of 92 %. Accumulation of CMV in young uninoculated leaves was significantly reduced in treated plants. The results suggested that exposure of calli to UV-B might activate defense responses in plantlets resulting in enhancement of resistance to CMV (Sudhakar and Murugesan 2011).

3.2 Solar Energy Treatments

Solar energy derived from sunlight provides heat that can be harvested to enhance the temperature, resulting in improvement of sanitation of the soil and plant materials. Ultraviolet light, a component of the sunlight has also been shown to have germicidal property. Soil solarization is a method of heating the soil by covering it with transparent polyethylene sheeting during summer season to control soilborne diseases affecting several crops. This technique has been commercially exploited for growing high-value crops in environments with hot summer, when air temperatures exceed 35 °C. This clear polyethylene sheets are used to cover the soil, after irrigation and solar energy is trapped by plastic sheets which prevent the loss of heat by evaporation and convection currents. Solarization results in a layer of pasteurized or disinfected soil, but not sterilized soil. The effect of solarization is similar to that obtained by application of aerated steam. It is important that high soil moisture levels are maintained during solarization. The important benefit of soil solarization appears to be its long-term benefits, as it is a physical treatment which has neither phytotoxic effect nor leaves any harmful residues as the chemical treatments. Solarization selectively eliminates pathogens, while high biological activity is maintained during treatment. Another, probably, most important advantage of employing solarization for elimination of pathogens, is that this approach is simple, economical and safe with none of the hazards associated with chemical treatments. Though there are several advantages, use of solarization as a disease management strategy, is limited to areas where hot conditions exist. In addition, the high cost of plastic sheets may prohibit its application for many economically important crops (Narayanasamy 2002).

3.2.1 Fungal Diseases

Elimination of propagule-like microsclerotia and sclerotia of soilborne pathogens such as, *Verticillium dahliae* (Pullman et al. 1981), *Sclerotium rolfsii* (Katan et al. 1976), *Pythium* spp., *Thielaviopsis basicola* and *Rhizoctonia solani* (Pullman et al. 1981) has been achieved to a satisfactory level by covering the soil with polyethylene sheets. The inoculum level of *V. dahliae* was reduced to trace amounts even at a soil depth of 60 cm in the unshaded areas in a 4-year old pistachio orchard in Southern California, by covering the soil with plastic sheets for a period of 6 weeks (Ashworth et al. 1982). In a greenhouse experiment with tomato, *V. dahliae* could not be isolated from soil solarized using transparent polyethylene sheets for 10 weeks, whereas high levels of pathogen populations were present in nonsolarized soil. The yield increase due to solarization was 112.4 % over control treatment. No infected plant was present in the treated area and the percentage of infected roots was as low as 0.3–0.4 %, as against 60 % in nonsolarized soil, indicating clearly the beneficial effect of soil solarization (Bourboos and Skoudridakis 1996). *Verticillium dahliae* causes serious damages to olive trees in Spain. The effectiveness of single (1 year) or double (two consecutive years) soil solarization treatments was assessed for the control of Verticillium wilt in young olive orchards. Plantations had different initial inoculum densities of *V. dahliae*. Soil solarization treatments were applied to lines of trees for either one or two consecutive years. Solarization significantly reduced the pathogen population in the top 20 cm of soil for at least 3 years in relation to untreated control plots. Pathogen reduction after the single solarization obscured effects of the second solarization treatment. Decrease of inoculum density in soil by solarization did not correspond to a similar reduction in disease severity. Disease severity was reduced only in orchards with medium or high inoculum densities. A second soil solarization treatment did not improve the effect of single solarization on Verticillium wilt control. In orchards with low inoculum densities, solarization did not result in significant difference in disease incidence and severity. However, recovery of the trees from disease was improved. Solarized plots remained free of weeds and no growth improvement due to solarization was evident (Loepez-Escudero and Blanco-Lopez 2001). Effects of soil solarization for the suppression of Verticillium wilt disease affecting artichoke (*Cynara cardunculus* var. *scolymus*) were investigated. Soil solarization reduced the inoculum of *V. dahliae* and also the incidence of Verticillium wilt disease in artichoke. However, no added benefit could be realized, when solarization was combined with addition of cauliflower residue amendments. Solarization had a significant effect on yield, producing an increase in yield by 16–40 % relative to the uncovered plots. Soil solarization would be compatible with normal agronomic practices adopted for artichoke production (Berbegal et al. 2008).

Long-term benefit offered by soil solarization was indicated by the investigation on cotton wilt disease caused by *Fusarium oxysporum* f.sp. *vasinfectum*. Effective control of the cotton wilt disease could be achieved by adopting a combination of solarization and proper crop sequence (rotation). Drastic reduction in pathogen population was obtained with soil solarization and following appropriate crop sequence, the availability of susceptible cotton plants could be avoided, resulting in the significant reduction in disease incidence in cotton, when it was cropped again (Katan et al. 1983). The effectiveness of soil solarization using transparent polyethylene sheets was assessed for the control of watermelon wilt disease caused by *F. oxysporum* f. sp. *raphani* and China aster wilt disease caused by *F. oxysporum* f.sp. *chrysanthemi*. Solarization for 3–4 weeks prior to planting reduced the wilt disease incidence by 39–74 %. The maximum temperatures were 50.5 and 44 °C at soil depths of 0–15 cm and 15–30 cm respectively. Plastic sheet mulching (covering the soil surface) controlled the weeds and improved the crop growth also (Huang 1993). Soil solarization using photoselective low density polyethylene film for a period of 32–49 days prior to planting tomatoes significantly reduced the density of *Fusarium oxysporum* f.sp. *lycopersici* and *F. oxysporum* f.sp. *radicis-lycopersici* in the upper 5 cm of soil, whereas the treatment significantly decreased the density of *Phytophthora nicotianae (= P. parasitica*) and *Burkholderia solanacearum (= Pseudomonas solanacearum*) up to depths of 25 and 15 cm respectively. The increase in temperature in solarized soils was 5.7, 7.1 and 5.0 °C over non-solarized soils at depths of 5, 15 and 25 cm respectively. The sensitivity of the pathogens to higher temperature may be an important factor determining the effectiveness of soil solarization in eliminating the pathogens present at different soil depths (Chellemi et al. 1994). Incidence of Fusarium wilt of tomato following solarization for 40–55 days was reduced to a level comparable to fumigation with a mixture of methyl bromide + chloropicrin (Chellemi et al. 1997). Soil solarization for 15 days resulted in significant reduction in the incidence of Fusarium wilt disease in cumin (Lodha 1995). Soil solarization and semi-solarization were evaluated alone or in combination with calcium cyanamide as soil amendment or a nonpathogenic suppressive strain of *Fusarium oxysporum* to develop alternative and ecologically compatible methods to control *F. oxysporum* f.sp. *melonis (Fom)*. The mean soil temperatures at 25 cm depth were increased by 8.6–12.6 °C and 12.6–16.3 °C respectively by semi-solarization and full solarization treatments. Solarization reduced the population of *Fom* grown in sterile soil samples from 3.8–60 × 10^5 CFU/g to 0–60 × 10^5 CFU/g soil in those placed at 15 and 25 cm depth. Solarization reduced Fusarium wilt disease incidence by 82–90 % in three out of five experiments. Calcium cyanide did not improve the effectiveness of soil solarization (Tamietti and Valentino 2006).

The efficacy of soil solarization in suppressing the development of *Rhizoctonia solani* AG-1 causing lettuce bottom rot disease and *Sclerotinia minor* causing lettuce drop disease was assessed. Soil solarization was applied for 60 days during the summer in lettuce fields. Incidence of both diseases was very low in the first year after solarization and crop cycle was shortened by 10 days as compared to the non-solarized plots. In non-solarized plots a loss of 25–40 % of lettuce heads was recorded. In addition to reduction in disease incidence due to soil solarization, additional advantages such as preservation of fluorescent pseudomonads and increased nutrient availability were also realized (Patricio et al. 2006). Biodegradable plastics and plastic films employed for soil solarization were evaluated for their comparative

Fig. 3.4 Solarization using clear polyethylene sheeting (100 μm thick) under field conditions (With kind permission of the International Crops Research Institute for Semi-Arid Tropics (ICRISAT), Patancheru, India)

efficiency in suppressing the development of tomato wilt disease pathogen *Fusarium oxyporum* f. sp. *lycopersici* and *Sclerotinia minor* causing lettuce drop disease. Plastic cover was more effective and consistent in increasing soil temperature compared to biodegradable plastics. Both plastic and biodegradable solarizing materials were effective in reducing the lettuce drop disease. Solarization with plastic films significantly suppressed weed development, while biodegradable plastic offered limited weed control. The results indicated that biodegradable solarizing materials may have to be considered for control of soilborne diseases, in view of the difficulties in the disposal of used plastic materials (Bonanomi et al. 2008).

The efficacy of soil solarization in reducing the incidence of wilt diseases was assessed in Fusarium sick plots of chickpea and pigeonpea during 3 years (1984–1985, 1985–1986 and 1986–1987). In these fields all plants of the susceptible cultivars were entirely (100 %) infected. Clear polyethylene sheeting (400 gauge) was laid in plots (Fig. 3.4). Soil was placed on the edges of each polyethylene sheet to secure it and sheets were removed after 6 weeks, followed by sowing of seeds of pigeonpea and chickpea. At 5 cm depth, soil solarization increased the temperature by about 10 °C. Heavy reduction in the population of both pathogens and wilt disease incidence in pigeonpea, due to solarization in both dry and irrigated conditions was observed (Fig. 3.5). Wilt susceptible cultivars did not show mortality in solarized plots, in spite of the presence of pathogen propagules. In contrast, all plants of wilt susceptible cultivars of chickpea and pigeonpea were killed before maturity in non-solarized plots. In addition to protection against wilt diseases, stimulatory effects of solarization on plant growth were also observed (ICRISAT 1980). The effects of soil solarization on the incidence of Fusarium wilt disease caused by *Fusarium oxysporum* f.sp. *lactucae* (FOL) and the growth of lettuce plants were assessed. In four trials conducted in the fields infested with FOL, the incidence of Fusarium

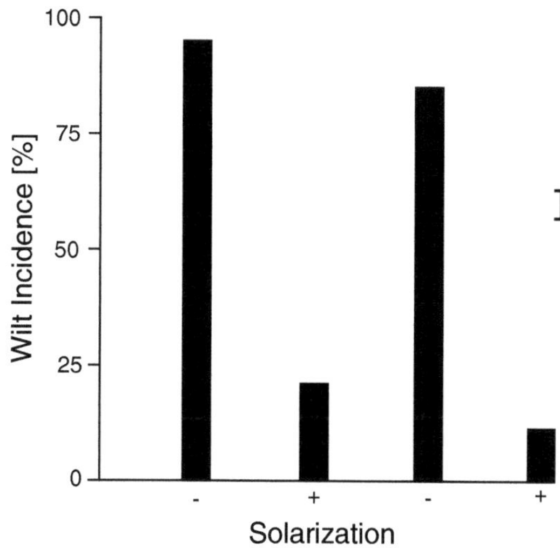

Fig. 3.5 Effect of solarization on the incidence of Fusarium wilt disease in chickpea genotype ICCC4 (With kind permission of the International Crops Research Institute for Semi-Arid Tropics (ICRSAT), Patancheru, India)

wilt on lettuce sown in soil after solarization was reduced from 91 to 42 %, compared with non-solarized plots. No significant benefit was observed due to a 2-over a 1-month solarization period under natural field conditions, where the mean soil temperature at a depth of 5 cm during 1-month solarization period was 47 and 49 °C in 2005 and 2006 respectively. The growth of lettuce plants in solarized plots was consistently greater than the plants in the non-solarized plots. This growth response might be due to more than the direct effect of solarization on the reduction of pathogen populations, since solarization was able to induce plant growth response in soil free of known soilborne pathogens. Retention of soil moisture by the plastic film during solarization process was considered as an important factor contributing to the reduction in pathogen population, since microorganisms are much more resistant to heat under dry conditions. Soil solarization significantly reduced the number of plants affected by Fusarium wilt disease in a subsequent planting of susceptible lettuce cultivar, indicating the long-term benefit of solarization (Matheron and Porchas 2010).

Phytophthora fragariae var. *fragariae* and *P. fragariae* var. *rubi* are the pathogens primarily involved in root diseases infecting respectively strawberry and raspberry crops. The effect of soil solarization for 2 months on the incidence and severity of root rot diseases was assessed. Increase in temperatures at 10 cm depth of the soil was between 7 and 17 °C over that prevailing in non-solarized soils. Soil collected after solarization was assayed for growing bait plants cv. Totem strawberry or cv. Qualicum raspberry respectively to detect the presence of the fungal pathogens at 15 °C for 6 weeks in saturated soil to promote infections. Solarization significantly reduced root necrosis ($P>0.05$) and increased root weight of bait plants compared to plants grown in soil from non-solarized plots. Infection of strawberry roots by *P. fragariae*, *Pythium*, *Rhizoctonia* and *Cylindrocarpon* spp. was reduced significantly ($P<0.05$) by solarization in sampled soil. Disease incidence was reduced in solarized field plots also. Red raspberry plants yielded more, when planted in solarized than in non-solarized soil. The beneficial effect of solarization in suppressing root rot was observed in strawberry and raspberry plantings for two or more years after solarization (Pinkerton et al. 2002). Soil solarization using transparent polyethylene sheets to cover tobacco seedbeds for 6 weeks effectively reduced the incidence of damping-off disease caused by *Pythium aphanidermatum* and black shank disease caused by *Phytophthora parasitica* var. *nicotianae*, due to enhancement of soil temperature (Wajid et al. 1995). Solarization with clear low-density polyethylene mulch decreased the final incidence and progress of Rhizoctonia crown rot and blight infection by *Pythium* spp. in roots and root discoloration in impatiens. In addition, shoot biomass was also increased by solarization (McGovern et al. 2002). The efficacy of soil solarization and biofumigation with cabbage residue to reduce the damping-off disease of cucumber crops grown in greenhouse was assessed. Both solarization (for 21 days) and biofumigation with organic amendment (5 kg/m^2) reduced the damping-off disease significantly, relative to the control treatment. Disease incidence was reduced more effectively by biofumigation compared with soil solarization, at the final count of disease incidence at 35 days after planting. Solarization and biofumigation enhanced plant growth as revealed by stem height and stem diameter. Reduction in pathogen population levels and disease incidence and enhancement of plant growth were greater during summer season than in winter season (Deadman et al. 2006). *Phytophthora cinnamomi* causes the major root rot disease of avocado worldwide. The effectiveness of soil solarization for the control of the Phytophthora root rot disease was evaluated under field conditions. Solarization was applied in a naturally infested plot before planting avocado (*Persea americana*) and viñatigo (*P. indica*) seedlings and also in established avocado trees for four consecutive summers. Soil solarization provided long-term positive effects, when applied as a preplanting treatment. Five years after solarization, disease severity in avocado and viñtago planted in solarized soil was 2.03 and 0.71 respectively, compared with 4.05 and 4.84 (in 0–5 disease rating scale) in the controls. Eleven years after solarization, the percentage of dead plants in solarized soils was 73 % for avocado and 43 % for viñtago, while all trees (100 %) were killed in control plot. In contrast, the level of control achieved in established orchards was insufficient, probably due to lower temperature reached during solarization under the shade of tree canopy (Gallo et al. 2007).

Phytophthora cactorum, a causative agent of leathery rot disease of strawberry fruit and *Rhizoctonia* spp. are soilborne fungal pathogens. The effectiveness of raised bed-solarization (RBS) alone or in combination with chicken manure (CM) amendment, methyl bromide (MB), TeloDrip (1,3-dichloropropene + chloropicrin) was assessed for suppressing the soilborne diseases of strawberry caused by these pathogens. Raised bed solarization for 7 weeks or combined with CM amendment (10 t/ha), metham sodium (50 ml/m^2) after 2-week RBS and MB (50 g/m^2) significantly reduced the soilborne diseases. Treatment with TeloDrip was less effective, compared with other treatments in the field experiments conducted during two cropping seasons of 2002–2004. All treatments effectively controlled four weed species commonly found in strawberry fields. Total marketable yields from RBS with or without CM and 2-week solarization plus MS were enhanced to a level equal to that of methyl bromide treatment in the first year. On the other hand, only raised bed solarization and CM amendment resulted in increase in yield in equivalence to MB treatment in the second year. The results indicated that raised bed solarization had the potential as an alternative to MB which has to be phased out in due course (Benlioğlu et al. 2005). The efficacy of soil solarization and *Trichoderma* spp. in suppressing the soil population of *Phytophthora cactorum*, when applied alone or in combination was determined. Clear low density (50 μm) polyethylene sheets were used to cover the plots, while *Trichoderma* spp. (10^8 conidia/ml) was applied via dip and drip, adding to the soil at 7 days before planting. Solarization reduced the population of *P. cactorum* by 100 % in the first year, by 47 % in the second year and by 55 % in the third year, relative to the untreated control. Combined application of solarization and *Trichoderma* spp. reduced the population of the pathogen which remained at low levels in a sustained manner for all the 3 years. In addition, the combined application increased strawberry yield appreciably (Porras et al. 2007). Effect of soil solarization for the control of raspberry root rot caused by *Phytophthora rubi* was investigated under field conditions using raised bed plots, flat bed plots or hilled bed plots. Soil solarization could be an effective component of integrated management of Phytophthora root rot disease of raspberry, especially, when combined with raised beds and gypsum amendments (Pinkerton et al. 2009).

Macrophomina phaseolina causing charcoal rot disease infects several crop plants. The effectiveness of soil solarization alone or combined with millet residue or paunch contents as amendments was assessed for the control of disease affecting cowpea (*Vigna unguiculata*) in naturally infested soil. Solarization increased the soil temperature to 50 °C for at least 4 h/day during June, resulting in a significant reduction of soil inoculum of *M. phaseolina* by 44 %. Paunch contents and millet residues reduced the initial inoculum density by 46 and 66 % respectively. Combination of soil solarization with millet residues and paunch contents reduced the disease severity as expressed by the area under disease progress curve (AUDPC) by 78 and 96 % respectively. The results suggested that in the Sahelian region, combination of soil solarization with organic amendment could be effective as an alternative to fungicides for the management of charcoal rot disease caused by *M. phaseolina* (Ndiaye et al. 2007). The efficiency of soil solarization for the control of *Macrophomina phaseolina* infecting strawberry was assessed using polyethylene films (30 μm thick) containing different additives ultraviolet (UV), UV + infrared (UV + IR), UV + antifog (AF) + IR + antidust (UV + IR + AF + AD) and also applying used polyethylene film (260 μm thick). The highest soil temperatures reached at the depth of 10 cm under polyethylene sheet containing UV + IR + AF + AD were 54 °C in 2007 and 50.7 °C in 2008. *M. phaseolina* sclerotia survived more than 18 days at 45 °C. There was a sharp decline in *M. phaseolina* at 50 °C, where it survived for 19 h, but was completely killed at 20 h. In the field, solarization did not reduce the viability of *M. phaseolina* at a soil depth of 10 or 20 cm. However, a significant reduction (66 %) in survival of *M. phaseolina* occurred at a soil depth of 5 cm was recorded, indicating the loss of effectiveness of soil solarization with increase in soil depth. Marketable fruit yield was increased to a maximum extent in the plots treated with UV + IR. All treatments effectively eliminated all weeds except *Cyperus rotundus* by 100 % as an additional advantage of contributing to the increases in yield of marketable fruits (Yildiz et al. 2010).

Soil solarization alone or residue incorporation, summer irrigation and biocontrol agent in combination was evaluated for the ability to adversely affect the survival of *Macrophomina phaseolina* and *Fusarium oxysporum* f.sp.*cumini (Foc)*, causative agents of root rot and wilt disease of cluster beans and cumin respectively. Combining amendments and soil solarization elevated the soil temperatures by 0.5–5.0 °C and 2.5–13.0 °C, compared to non-amended solarized and non-solarized plots respectively. The treatment combinations significantly reduced *M. phaseolina* and *Foc* propagules compared to control. The effect of surviving propagules of *M. phseolina* and *Foc* on the incidence was assessed in the subsequent rainy and winter seasons. Significant reduction in the incidence of both diseases was observed with or without soil solarization, compared to control (Israel et al. 2005). Combination of soil solarization with biocontrol agents *Bacilus subtilis* or *Trichoderma harzianum* reduced infection of table beet by *Rhizoctonia solani*, increased the plant stand by 25–95 % and fresh weight of beet by 32–41 %, when compared to individual treatment (Gasoni et al. 2008).

Pyrenochaeta lycopersici causes tomato corky root disease which increases in severity annually, reaching the maximum after 7 years under protected monoculture in greenhouses. The effectiveness of soil solarization for controlling tomato corky root disease in closed greenhouses was demonstrated by Garibaldi and Tamietti (1983). In a later investigation, the performance of plastic films used for soil mulching in greenhouses in solarization treatments for suppression of corky root disease was assessed, in addition to comparison of soil solarization with chemical fumigation for their effectiveness. Tests were carried out with clear, traditional and innovative plastic films and fumigant application. Soil solarization was effective in controlling corky root disease, relative to the untreated control. Use of different greenhouse covering and mulching films for solarization was found to be effective in reducing corky root disease severity, compared to the control. Solarization reduced disease incidence to a level comparable with application of methyl bromide and more effective than metham sodium and metham potassium. Among the films tested, green coextruded film could be preferred, because it is possible to use it as mulch after solarization (Vitale et al. 2011).

Polyethylene mulch has been demonstrated to offer beneficial influence for the control of fungal pathogens infecting aerial plant organs. Black polyethylene mulch reduced cucumber downy mildew (*Pseudoperonospora cubensis*) disease index and also increased the yield (Milvoj and Osvald 1994). Colonization of cucumber leaves by *P. cubensis* and sporangial production were significantly inhibited using polyethylene film with added blue pigment for solarization of soil. Use of plastic foil mulch for soil solarization was found to be effective for the control of gray mold disease of strawberry caused by *Botrytis cinerea.* The plastic foil prevented loss of radiation heat during the night, resulting in reduction of dew formation and also effectively prevents the fruits coming into contact with soilborne inoculum (Palti 1981). Green-pigmented polyethylene sheet screened visible light mainly from 560 to 800 nm and

UV radiation up to 380 nm. Use of green-pigmented sheet in the greenhouse reduced the conidial load of *B. cinerea* and gray mold by 35–75 %. Filtration of solar light by green-pigmented polyethylene sheet was also effective against disease caused by *Sclerotinia sclerotiorum* on cucumber *Fulvia fulva* on tomato and powdery mildew on cucumber (Elad 1997). The effect of plastic mulch and forced heated air on *B. cinerea* infecting geranium plants was investigated. The extent of sporulation of *B. cinerea* on necrotic lower leaves of mature geranium plants was reduced significantly. The combined effect of both plastic mulch and forced heated air was still greater. However, the stem blight phase of the disease was not reduced by combined treatments (Hausbeck et al. 1996b). Reflecting film mulching combined with rain-shelter treatment appreciably reduced the incidence of *Monilinia fructicola* and *Gleosporium laeticolor (= Glomerella cingulata)* on peach. Fruits from plants subjected to combined treatment matured early and were of better quality compared to fruits from untreated control trees (Hyun et al. 1995).

3.2.2 Bacterial Diseases

The potential of soil solarization for the suppression of bacterial plant diseases has been examined. The effectiveness of soil solarization using plastic films was assessed for the control of crown gall bacteria *Agrobacterium tumefaciens*. Solarization significantly reduced the populations of *Agrobacterium* spp. populations in the soil. The reduction in the pathogen population was positively correlated with the average temperature during the trial. The maximum population reduction obtained was 99 %, when the temperature ranging from 39 to 51 °C prevailed. However, *A. tumefaciens* population was still present after solarization. The treatment resulted in 89 and 94 % reduction in the incidence of crown gall in the rootstocks GF677 and bitter almond, compared to seedlings planted in non-solarized plots (Khlaif 2003). Transparent polyethylene sheet were used to cover the soil in commercial tomato production areas for the control of bacterial canker disease caused by *Clavibacter michiganensis* subsp. *michiganensis (Cmm)*. Soil mulching with the plastic film was applied for 6 weeks in plastic houses. Symptoms (disease intensity) induced by *Cmm* were reduced drastically throughout the growing season. The bacterial species of *Pseudomonas, Bacillus* and *Streptomyces* present in the tomato rhizosphere were able to survive the enhanced temperatures following solarization. The fluorescent pseudomonads isolated from the rhizosphere soil of tomato raised in solarized soils, induced resistance to the bacterial canker disease, when applied as tomato seed treatment. Tomato growers in Greece appeared to be convinced about the beneficial effect of soil solarization using polyethylene sheets (Antoniou et al. 1995). Soil solarization using transparent plastic sheet as mulch during summer months was found to effectively reduce the bacterial wilt disease of tomato caused by *Ralstonia solanacearum*. The bacterial biocontrol agent *Pseudomonas fluorescens* was incorporated uniformly in soils infested by *R. solanacearum* followed by solarization for 8 and 10 weeks at two locations in Himachal

Pradesh, India. The soil temperature was increased in solarized soils by 8.9 and 10.0 °C in two locations where the experiments were conducted. The bacterial antagonist proliferated in the solarized soils in the first year and the population declined subsequently. The incidence of bacterial wilt disease in the solarized soil was reduced by 43–63 %. There appeared to be no correlation between disease suppression and duration of soil solarization. Possibly, the population of *Pseudomonas* spp. present after soil solarization might be an important factor to be reckoned with (Ambadar and Sood 2010).

3.2.3 Virus Diseases

Reflective plastic mulches have been used to prevent/reduce the incidence of some of the virus diseases affecting crop plants (Loebenstein and Raccah 1980). These mulches reflect short-wave length light which confuses the incoming alate aphids, resulting in reduction of number of aphids alighting on plants. This in turn reduces the incidence of virus diseases transmitted by them. Plastic foils of different colors have been tested for their effect on the vectors of virus diseases (attraction/repellance) and consequently on the incidence of virus diseases. Significant reduction in the incidence of *Cucumber mosaic virus* (CMV) and *Potato virus Y* (PVY) was achieved by placing yellow polyethylene sheets spread with sticky substances for trapping aphids which were attracted by yellow color. Yellow polyethylene sheets delayed infection of tomatoes by *Tomato yellow leafcurl virus* (TYLCV), because of the stronger attraction of whitefly vectors to them over aluminum or blue-colored polyethylene sheets and death of whiteflies due to heat (Cohen 1981; Keren et al. 1991).

The repellant action of reflective surfaces has been considered as the principal reason for their use for the control of nonpersistent viruses transmitted by aphid vectors (Loebenstein and Raccah 1980). Infection of bell pepper by *Cucumber mosaic virus* (CMV) and *Potato virus Y* (PVY) was significantly reduced by using gray plastic mulch and the winged aphids trapped in the mulched plots were only 6–12 % of that in unmulched plots (Loebenstein et al. 1975). The efficacy of silver reflective plastic mulches in reducing aphid population and mosaic virus diseases infecting yellow crookneck summer squash (*Cucurbita pepo* var. *melopepo*) was assessed. Silver reflective mulch alone and silver reflective mulch with insecticide were superior to other colors of plastic mulches (white, yellow and black with yellow edges) in reducing aphid populations. Silver reflective plastic mulch with or without insecticide, resulted in a delay of 10–13 days in the onset of mosaic diseases caused by CMV, *Watermelon mosaic virus* I and II, *Zucchini yellow mosaic virus* and *Squash mosaic virus*. In addition, marketable yields were higher in plants grown on silver plastic mulch compared to controls. The silver mulch was also found to be cost-effective, due to efficient virus disease control (Brown et al. 1993).

Silver gray mulch and silver plastic mulch reduced the incidence of *Watermelon mosaic virus* and *Cucumber mosaic virus* on zucchini melon. In the crop raised in fall, 100 % of plants grown in unmulched plots, with and without insecticide application, were infected by viruses at first harvest, whereas only 10 % of plants grown over silver mulch exhibited virus infection. Silver-pigmented mulches were more effective in repelling aphids and delaying disease incidence than white-pigmented mulches. Yields of marketable fruits increased by 75 and 80 % due to silver polyester film and silver spray mulches respectively (Summers et al. 1995). The symptom development in squash following infection by *Zucchini yellow mosaic virus* (ZYMV) and CMV was delayed with application of aluminum foil mulch. The number of marketable fruits increased significantly due to mulching. Infection by these viruses on Courgette was reduced by reflective mulches (Ansanelli et al. 1997). Use of white and silvery mulches was found to be effective in reducing the incidence of viral diseases transmitted by other kinds of insect vectors with persistent relationship with respective viruses. White and silvery plastic mulches reduced thrip injury and delayed virus epidemics on hot pepper (*Capsicum* spp.) in West Java, Indonesia. The plastic mulch also had a favorable effect on plant growth (Vos et al. 1995). Whitefly-borne *Tomato yellow leafcurl virus* (TYLCV) incidence was reduced by reflective aluminum colored plastic mulches, resulting in increase in tomato fruit yields (Mauromicale et al. 1996). However, the efficiency of aluminum-colored plastic mulch was reported to be inferior to yellow polyethylene sheets (Cohen 1981).

Cantaloupes are seriously infected by aphid-transmitted viruses viz., *Cucumber mosaic virus* (CMV), *Watermelon mosaic virus* (WMV) and *Zucchini yellow mosaic virus* (ZYMV). Reflective, metalized plastic mulch, formed by adhering a thin coat of aluminum ions has been demonstrated to reduce the incidence of aphid-borne viruses. The number of winged aphids per leaf was significantly higher on plants grown over bare soil (control) than those grown over plastic mulch. The lower aphid counts on plants grown over plastic mulch resulted in lower virus disease incidence. The reflective plastic mulch maintained virus incidence below 10 % through mid-September 2002. The incidence of virus diseases closely followed that of winged aphids per leaf with the highest percentage of infected plants grown over bare soil (Summers et al. 2005). The effect of integration of UV-reflective plastic mulch and commercially available resistance-inducing compound BioYield™ on the incidence of *Watermelon mosaic virus* (WMV) infecting summer squash (*Cucurbita pepo* var. *melopepo*) was assessed. Incidence of WMV was significantly reduced in plants grown on silver-on-black (UV-reflective) mulch, compared with plants grown on black mulch (control). In the fall trial, highly UV-reflective silver mulch was also included for testing. WMV incidence and ELISA values were significantly lower for squash plants in the silver mulch treatment, compared with silver-on-black and black mulch treatments. However, yields of squash plants were significantly greater for plants in the silver-on-black mulch treatment than for those in the silver or black mulch treatments. The treatments with BioYield™ neither reduced WMV incidence nor increased the squash fruit yields in both spring and fall trials (Murphy et al. 2008).

Whiteflies form another major group of insects involved in the transmission of viruses infecting several economically important crops. The sweet potato whitefly *Bemisia tabaci* B biotype (also known as *B. argentifolii*) is an important pest of zucchini squash (*Cucurbita pepo*) and also functions as the vector of *Cucurbit leafcrumple virus* (CuLCrV). The efficacy of UV-reflective mulch and the living mulch, buckwheat (*Fagopyrum esculentum*) was assessed for the control of the whitefly and the cucurbit leafcrumple disease, including white mulch as the control treatment. The reflective mulch and the living mulch were more effective than control treatment in reducing the whitefly densities and incidence of CuLCrV on zucchini plants. The white mulch (control) treatment had the highest virus disease incidence and number of whiteflies. Reflective mulch alone gave protection to zucchini plants as the insecticide imidacloprid, addition of which did not provide any additional advantage. Hence, the reflective mulch alone was considered to be satisfactory. The effectiveness of UV-reflective mulch in reducing the incidence of CuLCrV was attributed to its ability to repel whiteflies, preventing them from alighting on zucchini plants (Nyoike et al. 2008). The effect of root zone temperature (RZT) as influenced by plastic film mulch on symptom expression in tomato inoculated with *Tomato spotted wilt virus* (TSWV) or in naturally infected plants was assessed. The root zone temperatures attained were 27.5, 27.0, 25.8 and 24.8 °C in plants grown respectively on black, gray, silver and white mulches. Plastic mulches that created high RZT stress resulted in reduced plant growth and yield and predisposed plants to rapid expression of symptoms of TSWV under artificial and natural infections (Diaz-Perez et al. 2007).

3.3 Radiation Treatments

3.3.1 Ultraviolet Irradiation

Radiation treatments under suitable conditions, may reduce populations of, or eliminate microbial pathogens and retard physiological processes such as ripening or senescence and sprouting. The radiation systems such as gamma rays (Cobalt-60 or Caesium-137), fast electrons (linear accelerators) and ultraviolet (UV) rays have been employed as radiation sources. UV light has been used to treat several fruits and vegetables as a postharvest disease management strategy. Three ranges of wave lengths of UV light, present in sunlight, have been designated UV-A (310–390 nm), UV-B (280–310 nm) and UV-C (190–280 nm). All ranges of UV light can damage the plant DNA and alter physiological processes depending on the duration of exposure (Luckey 1980; Stapleton 1992). A wide array of fruits and vegetables has been reported to respond positively to treatment with UV light and the treated commodities became resistant to the respective pathogens (Narayanasamy 2006).

3.3.1.1 Postharvest Diseases

Treatment of grapefruits marsh Seedless and Star Ruby and orange cultivars Washington Navel, Biondo Commune, Tarocco and Valencia Late with UV-C light at 0.5 kJ/m^2 reduced significantly the decay due to *Penicillium digitatum*, causing green mold disease. UV-C irradiation markedly reduced the growth of *P. digitatum* at the fruit wound sites and enhanced the resistance of treated fruits (Porat et al. 1999). The cultivars of orange responded to UV-C light differently based on treatment dose and harvest date, indicating their differential response to UV-C treatment. Higher dose of UV-C at 3.0 kJ/m^2 reduced the decay intensity on Washington Navel to a greater extent, compared with lower dose (0.5 kJ/m^2) (D'hallewin et al. 1999). However, in Star Ruby grapefruit, UV-C irradiation at 0.5 kJ/m^2 was effective in reducing the decay and increase in the dose above this level did not improve the decay control any further (D'hallewin et al. 2000). The efficacy of UV-type C (254 nm) irradiation, two yeast species *Candida saitoana* and *C. oleophila* (antagonists), chitosan and harpin (resistance inducer) was assessed for protecting apple cv. Red Delicious fruit against the blue mold pathogen *Penicillium expansum*. Treated fruits were inoculated with *P. expansum* at 24, 48 or 96 h following treatment and stored at 24 °C in the dark. All treatments were effective in reducing the area under disease progress curve (AUDPC). Treatment with UV-C was the most effective among the treatments used, followed by harpin, chitosan and the antagonistic yeast. There was a clear time-dependent response of the fruit to the treatments, in which treatments applied at 96 h before challenge inoculation with the pathogen provided best protection to apples. UV-C induced resistance within the first 24 h following treatment. However, greater control was achieved, when fruits were treated at 48 or 96 h before inoculation. UV-C irradiation besides being germicidal and causing delayed ripening of fruit, elicited biochemical responses in host tissue (de Capdeville et al. 2002).

The efficacy of treatment of table grapes with chitosan alone or in combination with UV-C radiation in suppressing the development of gray mold disease caused by *Botrytis cinerea* was assessed. The influence of these treatments on catechin and resveratrol contents and chitinase activity in grape berry skins was also investigated. Clusters of cvs. Thompson Seedless, Autumn Black and Emperor were sprayed in the vineyard with 1 % chitosan and at 5 days after harvest, they were inoculated with *Botrytis cinerea*. Decay incidence and disease severity were significantly reduced by chitosan treatment which was the most effective on berries harvested 1 or 2 days after treatment. In another set of bunches, grape berries were sprayed with chitosan, harvested 2 days later, irradiated for 5 min with UV-C (0.36 J/cm^2) and inoculated with the pathogen 2 days later. Combined application of chitosan and UV-C to cv. Autumn Black or Selection B36-55 was synergistic in reducing the gray mold disease incidence and severity, compared with individual treatment. Berries treated with chitosan and exposed to UV-C had higher catechin and trans-resveratrol induction, compared with the same treatments applied separately. The integrated strategy of combining chitosan and UV-C treatments exploiting the synergistic effects of the components could be an effective alternative for postharvest disease control. Further, induction of trans-resveratrol by the treatments may be beneficial, because of its anticarcinogenic and antioxidant properties. UV-C irradiation was shown to be a rapid, simple and economical postharvest treatment with potential for large scale application (Romanazzi et al. 2006).

Treatment of *Monilinia fructigena*, infecting strawberry with UV-C at 0.50 J/cm^2 inhibited the pathogen growth significantly. When UV-C exposure was combined with heat treatment (45 °C for 15 min), the intensity of disease symptoms was reduced, indicating the synergistic action of both these treatments (Marquenie et al. 2002a). In the case of *Botrytis cinerea*, exposing strawberry fruits to UV-C light at > 0.01 J/cm^2 resulted in significant reduction in pathogen growth. Heat treatment had no appreciable effect on the development of *B. cinerea*, but the highest temperature (48 °C) tested, damaged the berry surface and reduced fruit firmness (Marquenie et al. 2002b). Papaya anthracnose disease caused by *Colletotrichum gloeosporioides* is an important postharvest disease. The effects of UV-C and gamma irradiation on pathogen development and disease progress were investigated. Gamma irradiation at 0.75 and 1.0 kGy doses inhibited conidial germination and mycelial growth in vitro, but all doses increased fungal sporulation. Anthracnose disease incidence and severity were reduced by 0.75 and 1.0 kGy doses applied post-inoculation. But there was no effect of gamma irradiation, if applied prior to pathogen inoculation. In the case of UV-C, all doses inhibited conidial germination, while doses greater than 0.84 kJ/m^2 inhibited the mycelial growth and sporulation also in vitro. There was no effect of UV-C doses (0.2–2.4 kJ/m^2) and time intervals between treatment and inoculation on anthracnose disease control and fungal sporulation on fruit lesions. The results indicated that gamma irradiation could be an alternative for reducing the use of fungicides (Cia et al. 2007).

Postharvest decay of strawberries is caused by *Botrytis cinerea* which is widely distributed and is capable of infecting several fruit and vegetable crops. UV-C treatment at 0.5 and 1.0 kJ/m^2 followed by storage of strawberry fruit at 20 ± 1 °C or 3 °C significantly reduced the development of the gray mold pathogen. The results indicated that reduction in disease incidence might be due to a direct germicidal effect of UV-C light by disinfecting the external contamination, in addition to induction of resistance by activating host defense systems (Nigro et al. 2000). Strawberry fruit exposed to a hermetic dose of UV-C prior to storage showed less disease incidence and severity. Hence, the gene expression and enzymatic activity of a set of strawberry genes that are related to plant defense systems against *B. cinerea*, were studied. The expression and activity of phenylalanine ammonia lyase (PAL) increased over the level found in control after 4 and 24 h of storage. The activity of ß-1,3-glucanase and peroxidase showed a biphasic pattern. Higher polyphenoloxidase activity was detected in treated fruit compared with control from 10 to 48 h post-irradiation. The chitinase genes *FaChi2-2* and *FaChi3* were induced immediately after exposure to UV-C and the activity of the corresponding chitinase was increased at 10 h post-treatment. Gene expression profile of ß-1,3-glucanases was different from those of chitinases. The *FaPR1* gene expression was also stimulated by UV-C treatment at 4 and 24 h of storage. The results indicated that the reduction in strawberry fruit decay by UV-C treatment at harvest could be related to the enhancement of transcription and activity of a set of enzymes and proteins involved in the defense against microbial pathogens causing postharvest disease in strawberry (Pombo et al. 2011).

Treatment of carrots with UV-C radiation inhibited the development of *Botrytis cinerea* causing gray mold diseases in several vegetables. UV irradiation does not appear to have any systemic effect, as disease resistance could be induced only in tissues directly exposed to the irradiation. UV radiation induced local accumulation of 6-methoxymellein (6-MM) which reached a concentration near or higher than the ED_{50} (50 % effective dose) for inhibiting *Botrytis cinerea*. It is possible that 6-MM could be involved in inducing resistance. Freshly harvested carrots had a number of constitutive chitinases and ß-1,3-glucanases which were not affected by UV irradiation. When challenged with *B. cinerea*, induction of a 24-kDa chitinase was enhanced in UV-treated carrots. UV radiation had only a local effect in priming the chitinase response. Treatment of vegetables with UV-C irradiation for the control of postharvest diseases has been performed successfully (Mercier et al. 2000). Application of UV-C (254 nm) hormesis on tomato fruits stimulated beneficial responses, leading to reduction in disease development initiated by *Rhizopus stolonifer*, incitant of soft rot disease. The delay in ripening of tomatoes treated with UV-C was considered to be partly due to the high level of putrecine and spermine polyamines as well as accumulation of tomatine (Stevens et al. 1998). Further, mature green tomato fruits irradiated with UV-C light (24–36 kJ/m^2) exhibited delayed ethylene production and reduced respiration rate. Infection of *Alternaria alternata* was reduced, in addition to retarded ripening and color development in tomato fruits (Rong and Feng 2001).

Physiological, biochemical and structural changes induced by UV-C (3.7 kJ/m^2) in tomato fruits resulting in reduction in gray mold rot caused by *Botrytis cinerea* were investigated. Treated fruits were more susceptible to disease immediately after treatment, but they became gradually resistant to infection by *B. cinerea*. The resistance was maintained until the end of storage duration of 35 days. Synthesis and accumulation of the phytoalexin rishitin were induced by prestorage treatment with UV-C radiation. Rishitin concentration reached the peak at 15 days (46.23 mg/kg) after treatment and declined at the end of storage (3.5 mg/kg). A significant correlation was observed between rishitin accumulation in UV-treated fruit both before and after inoculation and disease resistance. The level of rishitin present at the time of inoculation with *B. cinerea* appeared to be the primary, but not the only factor in the expression of resistance, while its accumulation after inoculation might have a reinforcing role in resistance to gray mold pathogen (Charles et al. 2008d). Changes in the topography and fine structure of tomato fruit pretreated with UV-C light were monitored using scanning electron microscope (SEM). The formation of an operculum over broken trichomes was a common feature of ripened control fruit, while this structure was incompletely formed in UV-C-treated fruits. Colonization of the fruit surface was sparse on treated fruit. UV-C treatment resulted in alteration in the amount of epicuticular wax and its ultrastructural arrangement. In addition, the UV-C-induced physical and chemical modifications of tomato fruit surface could be an improved capacity of the tissues to resist infection by *B. cinerea* (Charles et al. 2008c).

UV-C induced changes in tomato fruit treated prior to storage were investigated to reveal the ultrastructural modifications of the pericarp leading to development of resistance to gray mold pathogen *Botrytis cinerea*. UV-induced plasmolysis of the epicarp cells as well as some cell layers of the mesocarp. Collapse of these cells, resembling HR-like cell death, led to the formation of the cell wall stacking zone (CWSZ). In the UV-treated fruit, pathogen development was primarily restricted to the outer most part of the fruit and further invasion of the deeper tissues seemed to be blocked by the CWSZ. This zone, through physical obstruction to pathogen spread, seemed to be an important factor in restricting the development of gray mold pathogen on tomato fruit (Charles et al. 2008a). Stimulation of biosynthesis of phenolic compounds in the epicarp and mesocarp cells of tomato exposed to UV-C prior to storage was detected. Biochemical reinforcement of cell wall through lignifications and suberization was induced. These responses were primarily localized in the CWSZ induced by UV-C treatment. These responses were further intensified in treated fruit following challenge inoculation with *B. cinerea* (Charles et al. 2008b). The role of constitutive defense enzymes and inducible pathogenesis-related (PR)-proteins was studied in the development of resistance induced by UV-C treatment prior to storage. Polyacrylamide gel elelctrophoresis (PAGE) analysis revealed (i) enhanced repression of expression of some proteins, (ii) enhanced expression of constitutive proteins like ß-1,3-glucanase, chitinases (acidic and basic) and (iii) synthesis of new proteins. The results suggested that PR-proteins with glucanohydrolase activities induced by UV-C might be an integral part of the long-term resistance exhibited by UV-C treated tomato fruit (Charles et al. 2008e).

Botrytis cinerea causing postharvest petal speckling is one of the major factors adversely affecting marketability of cut flowers of *Freesia hybrida*. The germicidal and inducible effects of UV-C irradiation on cut flowers of *F. hybrida* were assessed. UV-C irradiation of freesia inflorescences after artificial inoculation with *B. cinerea* was more effective in reducing petal speckling, compared with UV-C treatment before inoculation with the pathogen, indicating its germicidal effect. Cut freesia inflorescences exposed to 1.0 kJ/m^2 UV-C after initial inoculation with conidial suspension (10^4/ml) displayed reduction in disease severity scores, lesion numbers and lesion diameter by 74, 68 and 14 % respectively, compared to untreated controls. In contrast, UV-C irradiation with 1.0 kJ/m^2 before artificial inoculation reduced lesion numbers and lesion diameter by 13 and 24 % respectively, compared to the non-irradiated controls. Higher UV-C doses of 2.5 or 5.0 kJ/m^2 reduced disease severity scores, lesion number and lesion diameter, when applied after inoculation, but increased disease severity, when applied before inoculation with the pathogen. Irradiation with UV-C at 0.5–2.5 kJ/m^2 did not adversely affect the vase life of freesia flowers. But at a higher dose of 5.0 kJ/m^2, irradiation caused phytotoxicity, as inferred from petal discoloration and reduced vase life compared to non-irradiated inflorescences (Darras et al. 2010).

3.3.1.2 Field Crop Diseases

The beneficial effects of UV light in reducing the diseases affecting certain field crops have also been reported. Exposure of cabbage seeds to UV-C light at 3.6 kJ/m^2 was effective in reducing black rot disease caused by the bacterial pathogens *Xanthomonas campestris* pv. *campestris (Xcc)* and the density of bacterial pathogen population. In addition, the cabbage plants growing from treated seeds had the most desirable color, highest weight, largest head diameter and delayed maturity (Brown et al. 2001). Treatment of mungbean (*Vigna radiata*) and groundnut (peanut) seeds with UV-C radiation for 5, 10, 20, 30 and 60 min was carried out to assess the effects on plant growth and the development of fungal pathogens such as *Macrophomina phaseolina, Rhizoctonia solani* and *Fusarium* spp. causing root diseases. Infection of roots by the fungal pathogens was significantly reduced in mungbean plants growing from seeds treated with UV-C light for different periods (5–20 min). Likewise, treatment of groundnut seeds for 20, 30 and 60 min resulted in maximum reduction in infection or roots by *R. solani* and *M. phaseolina.* In addition to disease control, UV-C treatment of seeds improved the growth parameters such as shoot weight, shoot length, root weight and number of root nodules formed. The results indicated that treatment with UV-C light exerted beneficial effects on plant growth as well as suppression of pathogen development (Siddiqui et al. 2011).

The effects of UV-B radiation on the development of fungal diseases infecting plants have been assessed in certain cases. The defense signaling pathways share components with those involved in plant responses to UV radiation which can induce expression of plant genes required for resistance to infection by microbial pathogens. These intriguing links suggest that UV treatment might activate resistance to pathogens in normally susceptible plant species/cultivars. Pre-inoculative UV (254 nm) radiation of *Arabidopsis thaliana* (model plant) susceptible to infection by the obligate pathogen *Hyaloperonospora prasitica* causing downy mildew disease induced dose- and time-dependent resistance to the pathogen detectable up to 7 days after exposure to UV. The UV irradiation may initiate the development of resistance to *H. parasitica* in plants normally susceptible to the pathogen and UV-induced DNA damage also may have a role in the process of resistance development (Kunz et al. 2008). The efficacy of UV-B radiation as an alternative strategy of management of strawberry powdery mildew disease caused by *Spherotheca aphanis* var. *aphanis* was assessed. The suppressive effect of UV-B radiation against powdery mildew pathogen was evaluated in a vinyl house using three strawberry cultivars. All cultivars were effectively protected by UV-B radiation. RT-PCR analysis revealed that UV-B radiation induced the expression of a disease-resistance gene (encoding PR-protein). The results indicated that the mechanism of suppression of powdery mildew development was through induction of disease resistance in strawberry plants as one of its possible mode of action (Kanto et al. 2008).

Ultraviolet radiation in the environment has increased due to depletion of ozone. Investigations on plant responses to UV were taken, since solar UV wave lengths can reduce plant genome stability, growth and productivity. These detrimental effects may result from damage to cell components, including nucleic acids, proteins and membrane lipids. Plants are obligate phototrophs and plants have to counter the onslaught of cellular damage due to prolonged exposure to sunlight. The plants overcome the obstacles by attenuating the UV dose received through accumulation of UV-absorbing secondary metabolites, neutralizing UV-induced pyrimidine dimers by photoreactivation, extracting UV-photoproducts from DNA via nucleotide excision repair and perhaps transiently tolerating the presence of DNA lesions via replicative bypass of the damage. The signaling mechanisms determining these responses suggested that exposure to UV may also be beneficial to plants by increasing cellular immunity to pathogens. Thus, resistance to microbial plant pathogens may be enhanced by UV treatment (Kunz et al. 2006).

3.3.2 Gamma Radiation

The effect of gamma radiation on the development of fungal flora in dry onion bulbs has been assessed. The average storage moisture maintained at room temperature and at 4 °C were 75 and 85 % respectively. A positive correlation between the reduction in initial fungal flora and dose of irradiation was observed. The combination of chilling and ionization was the most effective in reducing fungal flora at a dose of 0.31 kGy (Benkeblia and Selselet-Attou 1997). Bunches of grapes cv. Italia were irradiated at different doses of gamma rays and stored at 0 °C and 95 % RH for 30 days and then transferred to 25 °C and 85 % RH to evaluate the effect of treatment on the development of gray mold disease caused by *Botrytis cinerea.* Rate and severity of infection decreased at a dose of 1.5 or 2.0 kGy on inoculated fruit. The spread of the disease to adjacent and distant fruits also was significantly reduced. Irradiation treatment did not adversely affect the color and composition of fruits. Favorable effects following irradiation were at low rates of berry splitting and natural shattering, but there were greater losses of fruit weight and firmness (Cia et al. 2000).

3.3.3 Short-Wave Infrared Radiation

Short wave infrared radiation (IR) has been demonstrated to be a highly efficient form of heating. The efficacy of IR was compared with the commercially used hot water treatment for the management of postharvest diseases of mango. Exposure of mango fruits to IR for 3 min was found to be as effective as the hot water dip at 50 °C for 5 min followed by a 20 s ambient temperature and prochloraz dip in controlling anthracnose disease caused by *Collletotrichum gloeosporioides* and soft brown rot caused by *Nattrassia mangiferae* on seven mango cultivars. The fruit quality was similar to those treated with hot water. The rapidity and lower treatment cost are additional benefits of using IR treatment for the control of postharvest diseases of mangoes (Saaiman 2003). The efficacy of red light irradiation in suppressing the development of Corynespora leaf spot disease of cucumber caused by *Corynespora cassiicola* was assessed under greenhouse conditions. In the greenhouse with red light (+Red) lesion appearance was delayed relative to that under –Red and its development was also significantly suppressed. Disease suppression under red light was observed in the glasshouse grown pathogen inoculated cucumbers. Red light did not show any direct effect on infection behavior of the pathogen. The results indicated that the delay and suppression of Corynespora leaf spot of cucumber under +Red were due to induction of resistance in cucumber and not due to differences in the environmental conditions or fungal population between the +Red and –Red greenhouses. Use of red light has the potential for adoption as an alternative strategy for the management of Corynespora leaf spot disease of cucumber (Rahman et al. 2010).

3.3.4 Microwave Treatment

The possibility of employing microwaves for suppressing the development of postharvest diseases of peach was investigated. Peach fruits inoculated with *Botrytis cinerea* were treated with microwave for 2 min. Percentages of infection and lesion diameter were significantly reduced ($P \leq 0.05$), compared with control fruits. The storage experiments demonstrated the effectiveness of use of microwave power as a prestorage treatment in controlling natural infection by fungal pathogens. Furthermore, the fact that the microwave treatment neither caused surface damage to fruits nor impaired fruit quality, is an additional advantage for the wider application of this approach for the management of postharvest diseases of various fruits (Karabulut and Baykal 2002). The effectiveness of radiofrequency (RF) heating in controlling brown rot disease caused by *Monilinia fructicola* in peaches and nectarines was assessed. RF treatment was applied at 27.12 MHz with 17 nm distance between fruit and upper electrode and 18 min exposure time. The fruits were inoculated with the pathogen at 0, 24, and 48 h before RF treatment using inoculum concentration of 10^3, 10^4, and 10^5 conidia/ml. The average brown rot incidence ranged from 44–82 % to 63–100 % in Summer Rich and Placido peaches respectively. RF treatment reduced the infection in Summer Rich more effectively than in Placido peach. On the other hand, RF treatment completely inhibited the development of brown rot in naturally infected peaches of both cultivars. The treatment with RF was not effective in reducing brown rot disease in nectarines infected by natural or artificial inoculum (Casals et al. 2010a, b).

3.4 Ozone Treatment

Ozone, the triatomic form of oxygen (O_3) has been recognized as being generally safe (GRAS) for food contact applications in the USA (Graham et al. 1997). Storage of lemons and oranges under 1.0 ppm ozone was reported to effectively prevent sporulation of *Penicillium digitatum*, causing green mold disease on infected fruit and moderately reduce the incidence of the disease (Harding 1968). Valencia oranges were continuously exposed to 0.3 ± 0.05 ppm (v/v) ozone at 50 °C for 4 weeks, whereas Eureka lemons were exposed to an intermittent day/night ozone cycle (0.3 ± 0.01 ppm ozone only at night) in a commercial cold storage room at 4.5 °C for 9 weeks. Exposure to ozone did not reduce final incidence of green mold (*Penicillium digitatum*) and blue mold (*P. italicum*), although incidence of both diseases was delayed about 1 week and infection developed more slowly under ozone (Palou et al. 2001). In a later investigation, ozone gas penetration through packaging materials and its effectiveness in reducing sporulation of *P. digitatum* and *P. italicum* was evaluated using artificially inoculated and commercially packed Lanelate oranges stored at 12.8 °C and exposed to an average ozone concentration of 0.72 ppm (v/v) for 14 days. Ozone penetration was strongly dependent on the vented area of each type of package. Sporulation inhibition of both pathogens was clearly related to ozone penetration and it was satisfactory only on oranges packed naked in returnable plastic containers (RPC). Since ozone was not able to penetrate through fiberboard cartons or plastic bags which are used commonly for packaging, the practical use of ozone gas exposure for treatment of fresh produce during storage is limited (Palou et al. 2003). The effectiveness of gaseous ozone exposure on the development of *Botrytis cinerea* causing stem-end rot of kiwifruit (*Actinidia delicosa*) cv. Hayward was assessed. Kiwifruits were inoculated with *B. cinerea* and placed in conventional storage conditions for 4 months (0 °C, RH 95 %) where catalytic oxidation of ethylene was applied (control) or exposed to continuous supply of ozone (0.3 μl/l). Ozone treatment delayed and simultaneously decreased disease incidence by 56 %, while disease severity on infected fruit did not show any change. Sclerotia were produced and there was no sporulation of the pathogen on the treated fruits. Pre-inoculation exposure of fruit to ozone, at increasing exposure time intervals led to significant suppression of disease incidence, while post-inoculation exposure did not have any effect on disease development. The results suggest that ozone treatment might induce resistance in kiwifruit against *B. cinerea*, as strong negative correlation between disease incidence or severity and phenol content in treated fruits (Minas et al. 2010).

Appendix Elimination of Viruses from Infected Plants Using Microshoot Tip Culture Technique (Sim 2006; Sim and Golino 2010)

1. Harvest shoot tips about 2 cm long from greenhouse- or field-grown plants; rinse under running tap water for 1 h with addition of a drop of detergent at 20 min interval and surface sterilize the tissues by submersing in10 % commercial bleach plus one drop of detergent for 10 min.
2. Excise the microshoot tips aseptically in a transferhood under 10–50× magnification with the aid of a zoom binocular dissecting microscope; remove

leaf scales individually to expose the shoot tip using a sterile forceps and scalpel; after each cut, sterilize in flame and cool and cut at the base of the last of several leaf primordia and place the microshoot tips (about 0.4–0.5 mm) gently on the surface of the initiation medium and include 1–3 pairs of leaf primordia.

3. Initial and maintenance medium, Murashige and Skoog (MS) salts and vitamins with 1.0 mg/l of cytokinin growth hormone 6-benzylaminopurine (BA), 3 % sucrose and 6.0 g/l gum agar adjusted to pH 5.8 (MSB); the rooting medium has half-strength MS salts and vitamins with 1.0 mg/l of the auxin growth hormone indole acetic acid (IAA), 1.5 % sucrose and 6.0 g/l gum agar adjusted to pH 5.8 (RM).
4. Incubate the explants at 2.5 °C in a growth chamber at 70 % RH for 16 days under cool white fluorescent and incandescent light; transfer the explants to fresh medium every 3 weeks and transfer them to rooting medium every 3 weeks and transfer them to rooting medium, when shoots 2 cm long with 4 or 5 well developed leaves are formed.
5. After rinsing of the roots to remove the medium, plant them in pots containing sterilized potting mix; place the pots in clear plastic box with lid-on for 2 weeks for gradual acclimatization and transplant them in bigger pots and transfer them to the greenhouse.

References

Acedo AL Jr, Benetisz MM, Milan LA (2001) Heat treatment effects on ripening and disease infection of Philippines banana and mangoes. Acta Hortic 553:417–419

Alconero R, Santiago AG, Morales F, Rodriguez F (1975) Meristem tip culture and virus indexing of sweet potatoes. Phytopathology 65:769–773

Ambadar VK, Sood AK (2010) Effect of solarization on tomato wilt incidence and dynamics of *Pseudomonas fluorescens* and *Ralstonia solanacearum*. J Mycol Plant Pathol 40:120–123

Ansanelli C, di Giovannantonio C, Tomassoli L (1997) Tolerant varieties and reflective soil covering to control virus of courgette. Inform Agrario 53:43–45

Antoniou PP, Tjmos EC, Andreou MT, Panago-Poulos CG (1995) Effectiveness, modes of action and commercial application of soil solarization for control of *Clavibacter michiganensis* subsp. *michiganensis* of tomato. Acta Hortic 382:119–128

Appel O, Riehm E (1911) See Martin H and Woodcock D (1983) The scientific principles of protection, 7th edn. Edward Arnold, London

Ashworth LJ Jr, Morgan DP, Gaona SA, McCain AH (1982) Polythene tarping for control of Verticillium wilt in pistachios. Calif Agric 36(5–6):17–18

Aveling TAS, Snyman HG, Maude SP (1993) Evaluation of seed treatments for reducing *Alternaria porri* and *Stemphylium vesicarium* on onion seed. Plant Dis 77:1009

Baker KF (1962) Principles of heat treatment of soil and planting material. J Aust Inst Agric Sci 28:118–126

Baker KF, Cook RJ (1974) Biological Control of Plant Pathogens. W.H. Freeman, San Francisco

Ballester AR, Izquierdo A, Lafuente MT, González-Candelas L (2010) Biochemical and molecular characterization of induced resistance against *Penicillium digitatum* in citrus fruit. Postharvest Biol Technol 56:31–38

Barkai-Golan R, Phillips DJ (1991) Postharvest heat treatment of fruits and vegetables for decay control. Plant Dis 75:1085–1089

Barlass M, Newman HP, Possingham JV (1987) Grapevine quarantine at Merbein, Victoria. Australas Grapevine Winemaker 286:15–17

Bashan Y (2000) Reduction of bacterial leaf speck disease (*Pseudomonas syringae* pv. *tomao*) in tomatoes treated with a combination of *Azospirillum barasilense*, bactericides and mild heat. In: Proceedings of the 10th International Conference, Charlottetown, pp 296–300

Benda GTA (1972) Control of sugarcane mosaic by serial heat treatment. Proc Int Soc Sugar Technol 14:955–960

Benda GTA, Ricaud C (1978) The use of heat treatment for sugarcane disease control. Proc Int Soc Sugar Technol 16:483–486

Benkeblia N, Selselet-Attou G (1997) Evaluation of fungal flora of irradiated dry onions (*Allium cepa* L.) during storage. Microbiol Aliments Nutri 15:71–77

Benlioğlu S, Boz Ö, Yildiz A, Kaşkavalci G, Benlioğlu K (2005) Alternative soil solarization treatments for the control of soilborne diseases and weeds of strawberry in the Western Anatolia. J Phytopathol 153:423–430

Berbegal M, García-Jiménez J, Armengol J (2008) Effect of cauliflower residue amendments and soil solarization on Verticillium wilt control in artichoke. Plant Dis 92:595–600

Bonanomi G, Chiurazzi M, Caporaso S, Del Sorbo G, Moschetti G (2008) Soil solarization with biodegradable materials and its impact on soil microbial communities. Soil Biol Biochem 40:1989–1998

Bourboos VA, Skoudridakis MT (1996) Soil solarization for the control of Verticillium wilt of greenhouse tomato. Phytoparasitica 24:277–280

Brown JE, Dangleer JM, Woods FM, Titt KM (1993) Delay in mosaic virus onset and aphid vector reduction in summer squash grown on reflective mulches. Hort Sci 28:895–896

Brown JE, Ty L, Stevens C, Khan VA, Lu JY, Wilson CL, Collins DJ, Wilson MA, Igwebe ECK, Chalutz E, Droby S (2001) The effect of low dose ultraviolet-C seed treatment on induced resistance in cabbage to black rot (*Xanthomonas campestris* pv. *campestris*). Crop Protect 20:873–883

Burr TJ, Ophel K, Katz BH, Kerr A (1989) Effect of hot water treatment on systemic *Agrobacterium tumefaciens* biovar 3 in dormant grape cuttings. Plant Dis 73:242–245

Calavan EC, Roistacher CN, Nauer EM (1972) Thermotherapy of citrus for inactivation of certain viruses. Plant Dis Report 56:976–980

Carvalho A, Santos FA, Machado MA (2002) Psorosis virus complex elimination from citrus by shoot-tip grafting associated to thermotherapy. Fitopat Brasil 27:306–308

Casals C, Teixidó N, Viñas I, Slivera E, Lamarca N, Usall J (2010a) Combination of hot water, *Bacillus subtilis* CPA-8 and sodium bicarbonate treatments to control postharvest brown rot on peaches and nectarines. Eur J Plant Pathol 128:51–63

Casals C, Viñas I, Landl A, Picouet P, Torres R, Usall J (2010b) Application of radio frequency heating to control brown rot on peaches and nectarines. Postharvest Biol Technol 58:218–224

Charles MT, Benhamou N, Arul J (2008a) Physiological basis of UV-C-induced resistance to *Botrytis cinerea* in tomato fruit: III. Ultrastructural modifications and their impact on fungal colonization. Postharvest Biol Technol 47:27–40

Charles MT, Goulet A, Arul J (2008b) Physiological basis of UV-C-induced resistance to *Botrytis cinerea* in tomato fruit: IV. Biochemical modification of structural basis. Postharvest Biol Technol 47:41–53

Charles MT, Makhloul J, Arul J (2008c) Physiological basis of UV-C-induced resistance to *Botrytis cinerea* in tomato fruit: III. Modification of fruit surface and changes in fungal colonization. Postharvest Biol Technol 47:21–26

Charles MT, Mercier J, Makhloul J, Arul J (2008d) Physiological basis of UV-C-induced resistance to *Botrytis cinerea* in tomato fruit: I. Role of pre- and post-challenge accumulation of the phytoalexins-rishitin. Postharvest Biol Technol 47:10–20

Charles MT, Tano K, Asselin A, Arul J (2008e) Physiological basis of UV-C-induced resistance to *Botrytis cinerea* in tomato fruit: V. Constitutive defence enzymes and inducible pathogenesis-related proteins. Postharvest Biol Technol 51:414–424

Chellemi DO, Olson SM, Mitchell DJ (1994) Effects of soil solarization and fumigation on survival of soilborne pathogens of tomato in northern Florida. Plant Dis 78:1167–1172

Chellemi DO, Olson SM, Mitchell DJ, Secker I, McSorley R (1997) Adaptation of soil solarization to the integrated management of soilborne pests of tomato under humid conditions. Phytopathology 87:250–258
Cia P, Beanto FA, Anjos VD, Vietes RL (2000) Effect of irradiation on storage of grapes cv. Italia. Revista Brasil Fruitic 22:62–67
Cia P, Pacholati SF, Benato EA, Camili EC, Santos CA (2007) Effects of gamma and UV-C irradiation on the postharvest control of papaya anthracnose. Postharvest Biol Technol 43:366–373
Cieślińska M (2007) Application of thermo- and chemotherapy in vitro for eliminating some viruses infecting *Prunus* sp. fruit trees. J Fruit Ornam Plant Res 15:117–124
Clear RM, Patrick SK, Turkington TK, Wallis R (2003) Effect of heat treatment on seedborne *Fusarium graminearum* and other cereal pathogens. Can J Plant Pathol 25:489–498
Coates LM, Cooke AW, Dean JR (2003) The response of mango stem-end rot pathogens to heat. In: Vth International Mango Symposium. Acta Horticulture, vol 455
Cohen S (1981) Control of whitefly vectors of viruses by non-conventional means. In: Proceedings of the International Workshop, Pathogens Transmitted by Whiteflies. Association of Applied Biologists, Wellesbourne, p 51
Converse RH (1987) Elimination of mycoplasma-like organisms in Cabot highbush blueberry with high carbondioxide thermotherapy. Plant Dis 71:36–38
D'hallewin G, Schirra M, Manueddu E, Piga A, Ben-Yehoshua S (1999) Scoparone and scopoletin accumulation and ultraviolet-C induced resistance to postharvest decay in oranges as influenced by harvest date. J Am Soc Hortic Sci 124:702–707
D'hallewin G, Schirra M, Pala M, Ben-Yehoshua S (2000) Ultraviolet-C irradiation at 0.5 kJm^{-2} reduces decay without causing damage or affecting postharvest quality of Star Ruby grapefruit (*Citrus paradisi* Macf.). J Agric Food Chem 48:4571–4575
Daniels BA (1983) Elimination of *Fusarium moniliforme* from corn seeds. Plant Dis 67:609
Dapkūniene S, Indrišinūnaite G, Juodkaite R, Navalinskiene M, Samuitiene M (2004) Tissue culture for elimination of lily viruses depending on explant type. Act Univ Latvi Biol 676:163–166
Darras AI, Joyce DC, Terry LA (2010) Postharvest UV-C irradiation on cut *Freesia hybrida* L. inflorescences suppress petal specking caused by *Botrytis cinerea*. Postharvest Biol Technol 55:186–188
Das K, Sinha RR, Sen SK (1983) Recovery of virus-free plantlets of cultivated jute species. In: Sen SK, Giles KC (eds) Plant cell culture in crop improvement. Plenum Press, New York, pp 459–464
de Capdeville G, Wilson CL, Beer SV, Aist JR (2002) Alternative disease control agents induce resistance to blue mold in harvested 'Red Delicious' apple fruit. Phytopathology 92:900–908
Deadman M, Al Hasani H, Al Sa'di A (2006) Solarization and biofumigation reduce *Pythium aphanidermatum* induced damping-off and enhance vegetative growth of greenhouse cucumber in Omani. J Plant Pathol 88:335–337
Diaz-Perez J, Gitaitis R, Mandal B (2007) Effects of plastic mulches on root zone temperature and on the manifestation of tomato spotted wilt symptoms and yield of tomato. Sci Hortic 114:90–95
Dong F, Mochizuki T, Ohki ST (2010) Tobacco ringspot virus persists in the shoot apical meristem but not in the root apical meristem of infected tobacco. Eur J Plant Pathol 126:117–122
Edison S, Ramakrishnan K (1972) Aerated steam therapy for the control of grassy shoot disease (GSD) of sugarcane. Mysore J Agric Sci 6:492–494
El Far MMM, Ashoub A (2009) Utility of thermotherapy and meristem tip for freeing sweet potato from viral infection. Aust J Basic Appl Sci 3:153–159
Elad Y (1997) Effect of filtration of solar light on the production of conidia by field isolates of *Botrytis cinerea* and on several disease of greenhouse-grown vegetables. Crop Protect 16:635–642
Fallah M, Mozafari M, Sokhandan-Bashir N, Hashemi M (2007) Elimination of a DNA virus associated with yellow leaf curl disease in tomato using an electrotherapy technique. In: Second International Symposium on Tomato Diseases, Kusadasi, 8–12, October 2007
Forsberg G (2001) Heat sanitation of cereal seeds with a new efficient, cheap and environmentally friendly method. In: Biddle AJ (ed) Seed treatment: challenges and opportunities. British Crop Protection Council, Farnham, pp 69–72

Gallo L, Siverio F, Rodriguez-Perez A-M (2007) Thermal sensitivity of *Phytophthora cinnamomi* and long-term effectiveness of soil solarization to control avocado root rot. Ann Appl Biol 150:65–73

Garibaldi A, Tamietti G (1983) Attempts to use soil solarization in closed glasshouses in Northern Italy for controlling corky root of tomato. Acta Hortic 152:237–243

Garrett RG, Cooper JA, Smith PR (1985) Virus epidemiology and control. In: Francki RIB (ed) The plant viruses, vol 1. Academic Press, New York, pp 269–297

Gasoni L, Kahn N, Yossen V, Cozzi J, Kobayashi K, Babbitt S, Barrera V, Zumelzu G (2008) Effect of soil solarization and biocontrol agents on plant stand and yield on table beet in Córdoba (Argentina). Crop Protect 27:337–342

Goheen AC, Nyland G, Lowe SK (1973) Association of rickettsia-like organism with Pierce's disease of grapevine and alfalfa dwarf and heat therapy of the disease in grapevines. Phytopathology 63:341–345

Goussard PG, Wiid J (1992) The elimination of *Fan leaf virus* from grapevines using in vitro somtic embryogenesis combined with heat therapy. South Afr J Enol Vitic 13:81–83

Goussard PG, Wiid J, Kasdorf GGF (1991) The effectiveness of in vitro somatic embryogenesis in eliminating *Fan leaf virus* and leafroll-associated viruses from grapevines. South Afr J Enol Vitic 12:77–81

Graham DM, Pariza M, Glaze WH, Newell GW, Erdman JW, Borzelleca JF (1997) Use of ozone for food processing. Food Technol 51:72–76

Grammatikakk G, Argirakis E, Avgelis A (2005) Virus elimination in infected clones of three grapevine wine cultivars (Roditis, Xinomavro and Savatiano) using in vitro thermotherapy and tip meristem culture. Geotech Sci Sub 4:25–30

Grant T (1976) Identification of heat treatments on tristeza and psorosis viruses in citrus. Plant Dis Reporter 41:232–234

Gută IC, Buciumeanu E-C, Gheorghe RN, Teodorescu A (2010) Solutions to eliminate *Grapevine leafroll-associated virus* serotype 1+3 from *Vitis vinifera* cv. Ranâi Magaraci. Rom Biotechnol Lett 15:72–78

Hall TW, Heidenreich MC, Cicciarelli R, Andersen RL, Turechek WW (2002) Eradication of *Pseudomonas syringae* pv. *syringae* in sweet cherry bud sticks. Phytopathology 92:S33

Harding PR Jr (1968) Effect of ozone on Penicillium mold decay and sporulation. Plant Dis 52:245–247

Hausbeck MK, Pennypacker SP, Stevenson RE (1996a) The use of forced heated air to manage Botrytis stem blight of geranium stock plants in a commercial greenhouse. Plant Dis 80:940–943

Hausbeck MK, Pennypacker SP, Stevenson RE (1996b) The effect of plastic mulch and forced heated air on *Botrytis cinerea* on geranium stock plants in a research greenhouse. Plant Dis 80:170–173

Hernández PR, Fontanella J, Noa JC, Pichardo T, Igarza Y, Cardenas H, Manso R (1997) Electrotherapy: a novel method for eliminating viruses from garlic. Hortic Argent 16:68–71

Hernández PR, Bertrand PH, Lepoiver P, González JE, Rojas X, Pairol A, González Y, González GY, Cortés C (2002) Diagnostico y saneamiento de *Banana streak virus* (BSV) en *Musa* spp. Centro Agric 2:42–47

Hormozi-Nejad MH, Mozafari J, Rakhshandehroo F (2010) Elimination of *Bean common mosaic virus* using an electrotherapy technique. J Plant Dis Protect 117:201–205

Huang SH (1993) Application of solarization for controlling soilborne fungal disease. Spl Publ Taichung Dis Agric Improv Sta 32:247–256

Hyun CD, Soo KJ, Tak YJ, Yong CS, Sull CB (1995) Effect of rain shelter and reflecting film mulching on fruit quality and disease infection on peach. RDA J Agric Sci Hortic 37:456–460

ICRISAT (1980) International Crops Research Institute for Semi-Arid Tropics. www.icrisat.org/what-we-do/learning opportunities/lsu-pdfs/Soil Solarization

Igarza-Castro J, Hernandez PR, Castellanos BC (2001) Electrotherapy as an alternative for elimination of the *Dasheen mosaic virus* in Tania. Span Manej Integr de plagas 60:57–60

Israel S, Mawar R, Lodha S (2005) Soil solarization, amendments and biocontrol agents for the control of *Macrophomina phaseolina* and *Fusarium oxysporum* f.sp. *cumini* in arid soils. Ann Appl Biol 146:481–491

Jensen JL (1888) The propagation and prevention of smut in oats and barley. J R Soc Engl 24:397–415

Johnstone GR, Wade GC (1974) Therapy of virus-infected plants by heat treatment I. Some properties of *Tomato aspermy virus* and its inactivation at 36 °C. Aust J Bot 22:437–450

Kanto T, Matsuura K, Yamada M, Usami T, Ameniya Y (2008) UV-B radiation for control of strawberry powdery mildew. Acta Hortic 842, VI Internatl Strawberry Symp 2008

Karabulut OA, Baykal N (2002) Evaluation of the use of microwave power for the control of post-harvest diseases of peaches. Postharvest Biol Technol 26:237–240

Kartha KK, Gamborg OL (1975) Elimination of cassava mosaic disease by meristem culture. Phytopathology 65:826–828

Katan J, Greenberger A, Alon H, Grinstein A (1976) Solar heating by polyethylene mulching for control of diseases caused by soilborne pathogens. Phytopathology 66:683–688

Katan J, Fishler G, Grinstein A (1983) Short- and long-term effects of soil solarization and crop sequence on Fusarium wilt and yield of cotton in Israel. Phytopathology 73:1215

Keren J, Decco Z, Cohen S, Bar-Yossef R (1991) Reduction of damage from yellow top virus and tobacco whitefly in tomatoes by yellow plastic mulch. Hassadeh 71:864–868

Khlaif H (2003) Effect of soil solarization on total *Agrobacterium* spp. population, inoculated *Agrobacterium tumefaciens* and on the development of crown gall. J Plant Pathol 85:117–122

Kim S-M, Nam S-H, Lee J-M, Yim K-O, Kim K-H (2003) Destruction of *Cucumber green mottle mosaic virus* by heat treatment and rapid detection of virus inactivation by RT-PCR. Mol Cells 16:338–342

Kobabyashi T (1990) Impact of seedborne pathogen on quarantine in Japan. Seed Sci Technol 18:427

Koch E, Schmitt A, Stephan D, Kromphardt C, Jahn M, Krauthausen H-F, Forsberg G, Werner S, Amein T, Wright SAI et al (2010) Evaluation of non-chemical seed treatment methods for the control of *Alternaria dauci* and *A. radicina* on carrot seeds. Eur J Plant Pathol 127:99–112

Komar V, Vigne E, Demangeat G, Fuchs M (2007) Beneficial effect of selective virus elimination on the performance of *Vitis vinifera* cv. Chardonnay. Am J Enol Vitic 58:202–210

Koubouris GC, Maliogka VI, Efthimiou K, Katis NI, Vasilakakis MD (2007) Elimination of *Plum pox virus* through in vitro thermotherapy and shoot tip culture compared to conventional heat treatment in apricot cultivar Bebecou. J Gen Plant Pathol 73:370–373

Krishnamurthy CD, Lokesh S, Shetty HS (2001) Occurrence, transmission and remedial aspects of *Dreschlera oryzae* in paddy (*Oryza sativa* L.). Seed Res 29:63–70

Kunz BA, Cahill DM, Mohr PG, Osmond MJ, Vonarx EJ (2006) Plant responses to UV radiation and links to pathogen resistance. Int Rev Cytol 255:1–40

Kunz BA, Dando PK, Grice DM, Mohr PG, Schenk PM, Cahill DM (2008) UV-induced DNA damage promotes resistance to the biotrophic pathogen *Hyaloperonospora parasitica* in Arabidopsis. Plant Physiol 148:1021–1031

Liu F, Tu K, Shao X, Zhao Y, Tu S, Su J, Hou Y, Zou X (2010) Effect of hot air treatment in combination with *Pichia guillermondii* on postharvest anthracnose rot of loquat fruit. Postharvest Biol Technol 58:65–71

Liu J, Sui Y, Wisniewski M, Droby S, Norelli J, Herskovitz V (2012) Effect of heat treatment on inhibition of *Monilinia fructicola* and induction of disease resistance in peach fruit. Postharvest Biol Technol 65:61–68

Lodha S (1995) Soil solarization, summer irrigation and amendments for the control of *Fusarium oxysporum* f.sp. *cumini* and *Macrophomina phaseolina* in arid soils. Crop Protect 14:215–219

Loebenstein G, Raccah B (1980) Control of nonpersistently transmitted aphid-borne viruses with aluminum foil or plastic mulch. Phytoparasitica 8:221–235

Loebenstein G, Alper M, Levy S, Palevitch D, Menagemi E (1975) Protecting peppers from aphid-borne viruses with aluminum foil or plastic mulch. Phytoparasitica 3:43–53

Loepez-Escudero FJ, Blanco-Lopez MA (2001) Effect of a single or double soil solarization to control Verticillium wilt in established orchards in Spain. Plant Dis 85:489–496
Lozoya-Saldaña H, Abello FJ, García G (1996) Electrotherapy and shoot tip culture eliminate *Potato virus X* in potatoes. Am Potato J 73:149–154
Luckey TD (1980) Hormesis with ionizing radiation. CRC Press, Boca Raton
Luthra JC (1953) Solar energy treatment of wheat loose smut *Ustilago tritici* (Pers.) Rostr. Indian Phytopathol 6:49–56
Lydakis D, Aked J (2003) Vapor heat treatment of Sultania table grapes. I. Control of *Botrytis cinerea*. Postharvest Biol Technol 27:109–116
Mahmoud SYM, Hosseny MH, Abdel-Ghaffar MH (2009) Evaluation of some therapies to eliminated potato Y potyvirus from potato plants. Int J Virol 5:64–76
Mangal M, Bharadwaj SV, Handa A (2004) Production of virus-free carnation plants through heat therapy. Defence Sci J 54:53–56
Marquenie D, Lammertyn I, Geeraerd AH, Soontjens C, van Impe JF, Nicolai BM, Michiels CW (2002a) Inactivation of conidia of *Botrytis cinerea* and *Monilinia fructigena* using UV-C and heat treatments. Int J Food Microbiol 74:27–35
Marquenie D, Schenk A, Nicolai B, Michiels CW, Soontjens C, van Impe JF (2002b) Use of UV-C and heat treatment to reduce storage rot of strawberry. Acta Hortic 567:779–782
Martin-Hernández AM, Baulcombe DC (2008) *Tobacco rattle virus* 16-kilodalton protein encodes a suppressor of RNA silencing that allows transient viral entry in meristems. J Virol 82:4064–4071
Matheron ME, Porchas M (2010) Evaluation of soil solarization and flooding as management tools for Fusarium wilt of lettuce. Plant Dis 94:1323–1328
Mauromicale G, Davino M, Nucifora S, Sortino O, Raccuia S, D'Urso F (1996) Effects of mulching with polyethylene sheets of different characteristics on infestation by *Bemisia tabaci* TYLCV infection and yield of tomato grown in greenhouses. Colture Protette 25:73–79
McGovern RJ, McSorley R, Bell ML (2002) Reduction of landscape pathogens in Florida by soil solarization. Plant Dis 86:1388–1395
Mercier J, Roussel D, Charles MT, Arul T (2000) Systemic and local responses associated with UV-C and pathogen-induced resistance to *Botrytis cinerea* in stored carrot. Phytopathology 90:986–986
Meziane M, Frasheri D, Carra A, Djelouah K, Carimi F, D'Onighia AM (2009) Citrus sanitation methods for the elimination of *Citrus tristeza virus* (CTV). Options Mediterran 65:177–180
Milvoj L, Osvald J (1994) The effects of drip irrigation on the occurrence of downy mildew of cucumber [*Pseudoperonospora cubensis* (Berk.et Court) Rostow]. Meded Facult Landbou Taege Biol Wetensch Univ Gent 59:855–858
Minas IS, Karaoglanidis GS, Manganaris GA, Vasilakakis M (2010) Effect of ozone application during cold storage of kiwifruit on the development of stem-end rot caused by *Botrytis cinerea*. Postharvest Biol Technol 58:203–210
Mochizuki T, Ohki ST (2000) Shoot meristem tissue of tobacco inoculated with *Cucumber mosaic virus* is infected with the virus and subsequently recovers from infection by RNA silencing. J Gen Plant Pathol 70:363–366
Montero CRS, Antes RB, Schwarz LL, dos Santos LC, dos Santos RP, Bender RJ (2010) Complementary physical and chemical treatments as an alternative to fungicide use to control postharvest decay incidence and fruit quality of Montenegrina tangerines. Crop Protect 29:1076–1083
Murphy JF, Eubanks MD, Masiri J (2008) Reflective plastic mulch but not a resistance-inducing treatment reduced *Watermelon mosaic virus* incidence and yield losses in squash. Int J Veg Sci 15:3–12
Narayanasamy P (2002) Microbial plant pathogens and crop disease management. Science Publishers, Enfield
Narayanasamy P (2006) Postharvest pathogens and disease management. Wiley, Hoboken
Narayanasamy P (2011) Microbial plant pathogen detection and disease diagnosis, vol 3. Springer Science+Business Media B.V., Heidelberg

Navarrot L, Roistacher CN, Murashige T (1975) Improvement of shoot-tip grafting in vitro virus-free citrus. Am Soc Hortic Sci 100:471–479

Ndiaye M, Termorshuizen AJ, Van Bruggen AHC (2007) Combined effects of solarization and organic amendment on charcoal rot caused by *Macrophomina phaseolina* in the Sahel. Phytoparasitica 35:392–400

Nesi B, Trinchello D, Lazzereschi S, Grassotti A (2009) Production of *Lily symptomless virus*-free plants by shoot meristem tip culture and in vitro thermotherapy. HortScience 44:217–219

Nigro F, Ippolito A, Lattanzio V, Venere DD, Salerno M (2000) Effect of ultraviolet-C light on postharvest decay of strawberry. J Plant Pathol 82:29–37

Nyland G, Goheen AC (1969) Heat therapy of virus diseases of perennial plants. Annu Rev Phytopathol 7:331–354

Nyoike TW, Liburd OE, Webb SE (2008) Suppression of whiteflies *Bemisia tabaci* (Hemiptera: Aleyrodidae) and incidence of *Cucurbit leaf crumple virus*, a whitefly-transmitted virus of zucchini squash new to Florida, with mulches and imidacloprid. Florida Entomol 91:460–465

Palou L, Smilanick JL, Crisosto CH, Mansour M (2001) Effect of gaseous ozone exposure on the development of green and blue molds on cold-stored citrus fruit. Plant Dis 85:632–638

Palou L, Smilanick JL, Crisosto CH, Mansour M, Plaza P (2003) Ozone gas penetration and control of the sporulation of *Penicillium digitatum* and *Penicillium italicum* with commercial packages of oranges during cold storage. Crop Protect 22:1131–1134

Palti J (1981) Cultural practices and infectious crop diseases. Springer, Berlin

Patricio FRA, Sinigaglia C, Barros BC, Freitas SS, Neto JT, Cantarella H, Ghini R (2006) Solarization and fungicides for the control of drop, bottom rot and weeds in lettuce. Crop Protect 25:31–38

Pazhoushande M, Mozafari J (2001) In vitro development of potato virus-free germplasm. Dissertation, Tarbiat Modarres University

Pinkerton JN, Ivors KL, Reeser PW, Bristow PR, Windom GE (2002) The use of soil solarization for the management of soilborne plant pathogens in strawberry and red raspberry production. Plant Dis 86:645–651

Pinkerton JN, Bristow PR, Windom GE, Walters TW (2009) Soil solarization as component of an integrated program for control of raspberry root rot. Plant Dis 93:452–458

Polák J, Hauptmanova A (2009) Preliminary results of in vivo thermotherapy of plum, apricot and peach cultivars artificially infected with PPV-M and PPV-D strains of *Plum pox virus*. Hortic Sci (Prague) 36:92–96

Pombo MA, Rosli HG, Martinez GA, Civello PM (2011) UV-C treatment affects the expression and activity of defense genes in strawberry fruit (*Fragaria* x *ananassa* Duch.). Postharvest Biol Technol 59:94–102

Porat RB, Lers A, Dori S, Cohen L, Weiss B, Ddaus A, Wilson CL, Droby S (1999) Induction of chitinase and ß-1,3-glucanase proteins by UV-irradiation and wounding in grapefruit peel tissue. Phytoparasitica 27:233–238

Porat R, Daus A, Weiss B, Cohen L, Fallik E, Droby S (2000) Reduction of postharvest decay in organic citrus fruit by a short hot water brushing treatment. Postharvest Biol Technol 18:151–157

Porras M, Barrau C, Arroyo FT, Santos B, Blanco C, Romero F (2007) Reduction of *Phytophthora cactorum* in strawberry fields by *Trichoderma* spp. and soil solarization. Plant Dis 91:142–146

Prammanee S, Thumjamras S, Chiemsombat P (2011) Efficient shoot regeneration from direct apical meristem tissue to produce virus-free purple passion fruit plants. Crop Protect 30:1425–1429

Prusky D, Falik E, Kobiler I, Fuchs Y, Zauberman G, Pesis E, Roth I, Weksler A, Akerman M, Ykutiely O, Waisblum A, Keinan A, Ofek G (2003) Hot water brush: a new method for the control of postharvest disease caused by Alternaria rot in mango fruits. In: Vth International Mango Symposium. Acta Horticulture, vol 455

Pullman GS, De Vay JE, Garber RH, Weinhold AR (1981) Soil solarization: Effects on Verticillium wilt of cotton and soilborne populations of *Verticillium dahliae,* Pythium spp. Rhizoctonia solani and Thielaviopsis basicola. Phytopathology 71:952–959

Quacquerrelli A, Gallitelli O, Savinow V, Pizzola PC (1980) The use of electrical current (RACE) for obtaining mosaic free almonds. Acta Phytopathol Sci Hung 15:251–255

Rahman MZ, Khanam H, Veno M, Kihara J, Honda Y, Arase S (2010) Suppression by red light irradiation of Corynespora leaf spot of cucumber caused by *Cercospora cassiicola*. J Phytopathol 158:378–381

Retheesh ST, Bhat AI (2010) Simultaneous eliminaton of *Cucumber mosaic virus* and *Cymbidium mosaic virus* infecting *Vanilla planifolia* through meristem culture. Crop Protect 29:1214–1217

Rives M (1970) Complete cure of a whole plant of *Vitis rupestris* cv. DuLt (St. Georges) of grapevine fan leaf by thermotherapy. Plant Dis Reporter 54:915–916

Romanazzi G, Milkota Gabler F, Smilanick JL (2006) Preharvest chitosan and postharvest UV irradiation treatments suppress gray mold of table grapes. Plant Dis 90:445–450

Rong RF, Feng SQ (2001) Effect of UV-C light irradiation on ripening and disease infection of postharvest tomato. J China Agric Univ 6:68–73

Saaiman WC (2003) Short-wave infrared radiation as an alternative to the hot water bath to control postharvest decay in mangoes. In: Vth International Mango Symposium. Acta Horticulture, vol 455

Sharma S, Singh B, Rani G, Zaidi AA, Hallan VK, Nagpal AK, Virk GS (2008) In vitro production of *Indian citrus ringspot virus* (ICRSV)-free Kinnow plants employing thermotherapy coupled with shoot tip grafting. Plant Cell Tiss Org Cult 92:85–92

Shiomi T (1992) Black rot of cabbage seeds and its disinfection under hot air treatment. Jpn Agric Res Q 26:13–18

Shukla US, Ram RS, Tripathi RC (1974) Effect of moist air treatments on the control of GSD and RSD. Annual Report, Indian Institute of Sugarcane Research, Lucknow, pp 72–74

Siddiqui A, Dawar S, Zaki MJ, Hamid N (2011) Role of ultraviolet (UV-C) radiation in the control of root infecting fungi on groundnut and mungbean. Pak J Bot 43:2221–2224

Sim ST (2006) Virus elimination from grape selections using tissue culture. www.ucanr.org/sites/intvit/files/24507.pdf

Sim ST, Golino D (2010) Micro- vs macro-shoot tip tissue culture therapy for disease elimination in grapevines. www.ucanr.org/sites/natipn/files/63882.pdf

Skiada FG, Grigoriadou K, Maliogka VL, Katis NI, Eleftheriou EP (2009) Elimination of *Grapevine leaf roll-associated virus-1* and *Grapevine rupestris stem pitting-associated virus* from grapevine cv. Agiorgitiko and micropropagation protocol for mass production of virus-free plantlets. J Plant Pathol 91:177–184

Srivastava NSL, Singh K, Sharma MP (1977) The IISR moist hot air treatment plant for seed cane. Sugar News 9:96–102

Stapleton AE (1992) Ultraviolet radiation and plants: Burning questions. Plant Cell 4:1353–1358

Steindl DRL (1961) Ratoon stunting disease. In: Martin JP, Abbott EV, Hughes CG (eds) Sugarcane diseases of the world, vol 1. Elsevier, Amsterdam, pp 433–459

Stevens C, Lui J, Khan VA, Lu JY, Wilson CL, Igwegbe ECK, Kabwe MK, Chalutz E, Droby S (1998) Application of hormetic UV-C for delayed ripening and reduction of Rhizopus soft rot in tomatoes: the effect of tomatine on storage rot development. J Phytopathol 146:211–221

Sudhakar N, Murugesan K (2011) Induction of systemic resistance in *Lycopersicon esculentum* cv. PKM1 (tomato) against *Cucumber mosaic virus* by UV-B irradiation on calli. Arch Phytopathol Plant Protect 44:460–474

Summers CG, Stapleton JJ, Newton AS, Duncan RA, Hart D (1995) Comparison of sprayable and film mulches in delaying the onset of aphid-transmitted virus diseases in zucchini squash. Plant Dis 79:1126–1131

Summers CG, Mitchell JP, Stapleton JJ (2005) Mulches reduce aphid-borne viruses and whiteflies in cantaloupe. Calif Agric 59:90–94

Tamietti G, Valentino D (2006) Soil solarization as an ecological method for the control of Fusarium wilt of melon in Italy. Crop Protect 25:389–397

Tsang MMC, Shintaku M (1998) Hot air treatment for the control of bacterial wilt in ginger root. Appl Eng Agric 14:159–163

Turechek WW, Peres NA (2009) Heat treatment effects on strawberry plant survival and angular leaf spot caused by *Xanthomonas fragariae* in nursery production. Plant Dis 93:299–308

Vartanian VG, Endo RJ (1985) Survival of *Phytophthora infestans* in seeds extracted from tomato fruits. Phytopathology 75:375

Vitale A, Castello I, Cascone G, D'Emilio A, Mazzarella R, Polizzi G (2011) Reduction of corky root infections on greenhouse tomato crops by soil solarization in South Italy. Plant Dis 95:195–201

Vos JGM, Ulhan TS, Sutarya R (1995) Integrated crop management of hot pepper (*Capsicum* spp.) under tropical lowland conditions effect of rice straw and plastic mulches on crop health. Crop Protect 14:445–462

Wajid SA, Shenoi MM, Sreenivas SS (1995) Seed bed soil solarization as a component of integrated disease management in FCU tobacco nurseries of Karnataka. Tobacco Res 21:58–65

Wang QC, Valkonen JPT (2008) Efficient elimination of sweet potato little leaf phytoplasma from sweet potato by cryotherapy of shoot tips. Plant Pathol 57:338–347

Wang QC, Cuellar WJ, Rajamäki M-L, Hirata Y, Valkonen JPT (2008) Combined thermotherapy and cryotherapy for efficient virus eradication: relation of virus distribution, subcellular changes, cell survival and viral RNA degradation in shoot tips. Mol Plant Pathol 9:237–250

Wijeratnam RSW, Hewajulige IGN, Abeyratne N (2005) Postharvest hot water treatment for the control of Thielaviopsis black rot of pineapple. Postharvest Biol Technol 36:323–327

Yang RG, Chang Z (1998) Study on the control of pineapple heart rot disease by sunning the propagative materials. South China Fruits 27:35

Yildiz A, Benlioğlu S, Boz Ö, Benlioğlu K (2010) Use of different plastics for soil solarization in strawberry growth and time-temperature relationships for the control of *Macrophomina phaseolina* and weeds. Phytoparasitica 38:463–473

Additional References for Further Reading

Kunkel LO (1936) Heat treatments for the cure of yellows and other virus diseases of peach. Phytopathology 26:809–830

Lopez-Cabrera JJ, Marrero-Dominguez A (1998) Use of hot water dips to control the incidence of banana crown rot. Acta Hortic 490:563–569

Tamietti G, Valentino D (2001) Soil solarization: a useful tool for control of Verticillium wilt and weeds in eggplant crops under plastic in the Po Vallley. J Plant Pathol 83:173–180

Chapter 4
Biological Control of Microbial Plant Pathogens in Alternative Sources of Infection

4.1 Biological Control of Weeds

4.1.1 Biotic Agents-Based Control Strategies

Biological control of weeds using microbial pathogens may be achieved by applying two distinct strategies. In classical biocontrol systems, pathogens are isolated from regions where both the pathogen and weed species are indigenous and where the pathogen exhibits some degree of control of the weed species concerned. These exotic pathogens are released into the areas where the weed species is to be controlled. These pathogens have to establish and reach a sustainable population level to suppress further spread of weed population. They should be capable of self-dissemination generally through wind. Such biocontrol agents should be environmentally safe and host-specific, since the spread of the biocontrol agent to nontarget plants including crops cannot be limited. In bioherbicide strategy, potential antagonistic microorganisms are multiplied artificially and applied on weeds to be controlled at required concentrations like chemical herbicides. The bioherbicides selected for weed control containing different species/strains of fungi are known as mycoherbicides (Charudattan 1991). A large number of pathogenic microorganisms have been isolated from various weed species. Pathogenicity tests are performed to select effective isolates of microorganisms. The biology of the pathogens has to be studied to determine their optimal requirements for growth, sporulation and production of resistant spore forms. The ability to sporulate in submerged culture is a desirable attribute, since liquid fermentation is more practical and economical than other methods. Solid-state fermentation methods were found to be more difficult and expensive (Boyette et al. 1991). The type of parasitism pursued by the pathogenic fungus affects its ability to serve as a mycoherbicide. Obligate parasites like rusts typically cause less damage on the weed hosts, because of the requirement of live host plants for their survival. On the other hand, facultative parasites or facultative saprophytes can cause serious damages by producing toxic metabolites that kill

P. Narayanasamy, *Biological Management of Diseases of Crops*,
Progress in Biological Control 16, DOI 10.1007/978-94-007-6377-7_4,

plant tissues with differing rapidity and these pathogens can derive their nutrition from the partially killed or dead plant tissues/cells. Further, difficulties in multiplying the obligate parasites in artificial media make them less preferred candidates for biological control of weeds. A pathogen that can become endemic, is capable of being more or less constantly present from year to year in a moderate to severe form may be chosen for further development. The host specificity of mycoherbicides is like a double-edged sword. If the host range of the selected pathogen is narrow, its potential market will be limited by the distribution of the susceptible weed species. But the highly specific mycoherbicides will pose minimal risk to non-target plants including crops. In contrast, the pathogen with wide host range can be effectively used with consequent larger market potential. However, these pathogens may pose substantial risk to other economically important plant species. Hence, the choice of a mycoherbicide has to depend on several factors with suitable follow up measures (Pilgeram and Sands 1999).

4.1.1.1 Isolation and Identification of Biotic Agents

Various isolates of biotic agents obtained from weed plant species are carefully studied to determine their morphological, physiological and genetic characteristics using appropriate protocols. Simultaneously the efficacy of the pathogen species/strains in suppressing the development of target weed species is assessed. Precise identification of the strains/races/biotypes of the pathogens is important, since host-pathogen specificity may be governed by single gene differences or by a small number of genes, especially at sub-specific level (Yandoc et al. 2006). *Valdensinia heterodoxa*, an ascomycete fungus considered as a potential biocontrol agent was examined for genetic diversity. Three geographically separate populations were cultured and single-spore cultures were prepared for DNA isolation. Amplified fragment length polymorphism (AFLP) technique was used to generate individual DNA fingerprints for each of the three *V. heterodoxa* isolates. These isolates exhibited low genetic diversity, with just 30 of 214 loci being polymorphic. There were many shared haplotypes within each population. The results suggested that with low diversity and high population differentiation, the effectiveness of *V. heterodoxa*, as a biocontrol agent might be limited to local populations or to use it in combination with other control methods (Wilkin et al. 2005). The fungal pathogen *Sclerotinia minor* IMI344141 has been developed as a bioherbicide for the control of broad-leaved weeds. A strain specific molecular marker was generated for the detection and monitoring this specific strain. This method was based on polymerase chain reaction (PCR) amplification of two sequence-characterized amplified regions (SCAR) primer pairs for a first round PCR and another two sets of nested primers was employed for a second round PCR, if higher sensitivity was required. The DNA of *S. minor* bioherbicide isolate IMI344141 could be detected in the soil even 2 months after application. This molecular detection protocol provided a mechanism to distinguish this isolate from other related strains and a tool to monitor the presence of the BCA in nature, particularly in soil (Pan et al. 2010).

The study of biology of the fungi with potential for use as bioherbicide is important, as exemplified by the rust pathogens. An understanding of the life cycle of the pathogen is desirable, before its release as a biocontrol agent to assess the risks involved in such a release. The life cycle and infection parameters of the neotropical rust pathogen (*Prospodium tuberculatum*) infecting *Lantana camara* were studied. This rust pathogen was found to be autoecious and microcyclic producing only urediniospores and teliospores on infected *L. camara*. The pathogen survived entirely through several cycles of urediniospores on the target weed species throughout most of its native range. The urediniospores and teliospores could survive for extended periods in liquid nitrogen. Under glasshouse conditions, the urediniospores exhibited wide temperature tolerance (10–25 °C) for infection on *L. camara* with an optimum near 20 °C. The minimum dew period required for plant infection was 9 h with the optimum close to 15 h (Ellison et al. 2006). In order to determine the host range of *P. tuerculatum*, 40 Australian *Lantana camara* forms and 52 closely related non-target plant species were inoculated under glasshouse conditions. The Brazilian rust strain was pathogenic only to two color forms of the flowers viz., pink and pink-edged red. Macro- and microsymptoms were recorded based on 11 categories of symptom types and four susceptibility ratings. The results showed that none of the non-target plant species was susceptible to the rust pathogen, indicating the specificity of host-pathogen interaction (Thomas et al. 2006).

Three rusts *Puccinia nassellae*, *Uromyces pencanus* and *Puccinia graminella* were investigated under field conditions. *Uromyces pencanus* was identified as the most promising candidate. It caused significant damage to Chilean needle grass (*Nassella neesiana*) in the field. An isolate that could infect most population of the weed species in Australia was identified. The uredinospores could be stored for over 12 months. The pathogen was host-specific for use as classical biocontrol agent (Anderson et al. 2010). In a later investigation, *P. nassellae* produced several cycles of urediniospores in a season. The teliospores were dormant and incapable of infecting Chilean needle grass. *U. pencanus* also produced cycles of urediniospores capable of infecting the target weed species, but the teliospores failed to germinate. On the other hand, *P. graminella* produced only aecia and telia. The aeciospores were repetitive and infected the weed plants. These rust species appeared to be heteroecious and their respective alternate hosts are needed for the completion of their life cycles. It is essential to maintain the pathogens in the repetitive phase, so that large amounts of inoculum for infection of Chilean needle grass are available (Anderson et al. 2011). *Puccinia jacea* var. *solstitialis* is an autoecious rust that produces all stages of its life cycle in the same weed host yellow starthistle, an invasive alien weed in California. This pathogen was released throughout California, but long-term establishment rates were generally found to be low, because of its inability to adapt to the climate where the weed was more invasive (Fisher et al. 2011).

Phomopsis amaranthicola was described as a new species capable of infecting weeds belonging to the genus *Amaranthus*. The range of plant species susceptible to *P. amaranthicola* was determined by inoculating 45 accessions of 21 species of *Amaranthus* and 46 species belonging to the genera other than *Amaranthus* which included some crop plants in which fungal BCA might be applied as bioherbicide.

P. amaranthicola was highly pathogenic to several *Amaranthus* spp. But it was unable to infect plant species outside the genus *Amaranthus*. The mortality levels induced by *P. amaranthicola* varied depending on the species of *Amaranthus*. This BCA was found to be confined to *Amaranthus* genus. The genetic level specificity rendered it not only a safe bioherbicide, but also one that could be applied against several weedy *Amaranthus* spp. (Rosskopf et al. 2006). In another study, the ability of *Phomopsis amaranthicola* and *Microsphaeropsis amaranthi* to infect eight species of *Amaranthus* was determined by applying alone or in combination. The fungal pathogens infected most of the *Amaranthus* spp. tested and reduced their growth and survival. However, the responses of the weeds in the greenhouse were less than in field experiments. Treatments with the fungal pathogens caused mortality of *A. albus* and *A. blitoides* to the extent of 80–100 % when the mixture of *P. amaranthicola* and *M. amaranthi* was applied both in the greenhouse and field experiments at 2 weeks after treatment. But in the field trials, the fungal pathogens induced severe disease symptoms with mortality of all weed species ranging from 74 to 100 %. The biomass of treated weeds was also significantly reduced, indicating the effectiveness of treatment with fungal pathogens as mycoherbicides (Ortiz-Ribbing and Williams 2006).

4.1.1.2 Formulations and Delivery Systems

Although numerous fungal pathogens of weeds have been found to significantly reduce the weed population, only a few have been developed and registered as bioherbicides for large scale application. Fresh preparations containing conidial/spore suspensions of the BCA form the simplest formulation. *Phytophthora palmivora* was formulated as a fresh liquid with a 6-week shelf-life and marketed by Abbott Labs., IL, USA for the control of weeds in citrus orchards (Kennedy 1986). In the case of rust pathogens, the urediniospores are applied as inoculum for the control of different weed species like *Lantana camara* (Ellison et al. 2006; Thomas et al. 2006). Conidial suspensions of *Myrothecium verrucaria*, a soil fungus were sprayed on the Old World Climbing Fern *Lygodium microphyllus* which became one of the most invasive and destructive weeds in southern Florida. The fungal BCA retarded the growth of the weed severely (Clarke et al. 2007). Organic (plant) materials infested with the fungal BCA like *Sclerotinia sclerotiorum* were found to be useful in the case of pathogens that did not sporulate readily in liquid cultures. However, large quantity of infested materials (830 kg/ha of infested wheat seeds) would be required to bring about acceptable level of weed control (Brosten and Sands 1986). *Sclerotinia sclerotiorum* was employed for the control of *Ranunculus acris*. Wheat grains were inoculated with the mycelium of the BCA and they were milled and broadcast onto the weed plants. The efficiency of inoculum types-mycelium or ascospores of *S. sclerotiorum* in suppressing the weed development was assessed. The ascospores were found to be more effective, when their concentrations were maintained at 2.5×10^6 spores/ml (Pottinger et al. 2008).

The microsclerotia of the fungus *Colletotrichum truncatum*, an effective mycoherbicide against the weed *Sesbania exalta* were formulated in wheat gluten-kaolin granules called 'Pesta' and applied to the soil. Weed control averaged 84–88 % respectively in plots treated with pre-plant incorporated (PPI) at 'Pesta' rates of 168 or 336 kg/ha over a 2-year period of testing. Postemergence (POE) treatment was less effective. Soybean yields were significantly greater in plots treated with *C. truncatum* 'Pesta' granules (Boyette et al. 2007b). *Myrothecium verucaria* isolate was evaluated for its potential against all serious weed species present in commercial tomato fields in southeastern USA. *M. verrucaria* was found to be highly virulent against all the weeds tested, when applied as conidial sprays formulated in 0.2 % Silwet L-77 surfactant, even in the absence of dew. In the field plots, application of *M. verrucaria* (2×10^7 conidia/ml) in 0.2 % Silwet killed 90–95 % of both purslane species and 85–95 % of both spurge species at 7 days after application. Tomatoes in plots treated with *M. verrucaria* remained healthy and vigorous throughout the growing season. This fungal BCA has the potential for large scale application as an effective bioherbicide for pre-plant weed control in production systems with transplanted tomatoes (Boyette et al. 2007a).

Stagonospora spp. induced severe disease symptoms in *Convolvulus arvensis* (field bind weed) and *C. sepium* at all stages of growth. Formulation of *Stagonospora* spp. in a 10 % rapeseed oil-in-water emulsion significantly improved its biocontrol efficacy and reduced its dependence on extended dew periods. Inoculation of the weed with 10^7 spores/ml in the oil emulsion applied to run-off produced a high level of leaf necrosis even in the absence of 100 % relative humidity (Pfirter and Défago 1998). Solid-state fermentation on couscous (cracked hard wheat) produced up to 4×10^8 spores/g of substrate. These spores were as virulent as those grown on V8 juice agar. After air-drying on kaolin, followed by storage at 3 °C, the spores were viable and infective for 140 days (Pfirter 2000). *S. convolvuli* strain LA39 was applied in maize field heavily infested with *Convolvulus sepium*. Severe infection by the pathogen and heavy defoliation of the weed were observed. The potential of the strain LA39 was also demonstrated in a non-crop situation (Guntli et al. 1999). The soilborne fungal pathogen *Sclerotium rolfsii* isolate SC64 was evaluated for its efficacy in suppressing broad-leaved weeds *Cyperus difformis*, *Lindernia procumbens*, *Rotala rotundifolia*, *Ammannia baccifera* and *Eclipta prostrate* present in rice fields. Pathogen infested solid substrates (composed of mixtures of rice hulls and bran) at 60–140 g/m^2 were applied in two sites in China. Percentage of plant mortality ranged from 50 to 89 % and 30 to 71 % respectively in 2 years of testing. Weed density, prevailing temperature and humidity at the time of pathogen application influenced the effectiveness of the treatment (Tang et al. 2011). Another bioherbicidal fungus *Colletotrichum gloeosporioides* f.sp. *aeschynomene* (*Cga*) was formulated as an invert emulsion (MSG 8.25) with dried formulated spores. This pathogen was highly pathogenic to the leguminous weed *Aeschynomene virginica* (northern jointvetch), but it did not infect another more serious leguminous weed *Sesbania exaltata* (hemp sesbania). A 1:1 (v/v) fungus/invert emulsion mixture caused 100 % infection and mortality of inoculated hemp sesbania seedlings at 21 days after inoculation under greenhouse conditions. In replicated field experiments, the fungal

BCA in invert emulsion controlled 85–90 % in two locations (Stuttgart, AR and Stoneville, MS). The results suggested that the host range of *Cga* could be enlarged by using suitable formulation with a concomitant improvement in the biocontrol potential of the fungal pathogen (Boyette et al. 2011).

The effectiveness of biocontrol of *Amaranthus rudis* (waterhemp) and *Amaranthus* spp. (pigweeds) by the application of *Microsphaeropsis amaranthii* and *Phomopsis amaranthicola* was assessed in irrigated and non-irrigated pumpkin and soybean crops for over 2 years. The bioherbicide was applied with lecithin and vegetable oil at 187 l/ha (in 2008) and 374 l/ha (in 2009). The treatments were spore suspensions of the fungal BCA alone, a mixture of both BCAs and sequential treatments of BCAs with halosulfuron-methyl (Sandea Herbicide) in pumpkin or glyophosphate (Roundup Original Max Herbicide) in soybean. Observations were made to record disease incidence and disease severity, percent weed control and weed biomass reduction. The bioherbicides significantly reduced weed biomass. The results indicated the possibility of tank-mixing *M. amaranthii* with halosulfuron-methyl resulting in enhancement of effectiveness of weed control (Ortiz-Ribbing et al. 2011).

Fungal pathogens are known to produce various kinds of metabolites that are toxic to plant hosts. These toxins may be selective in their action on plants which develop distinctive symptoms similar to those caused by the pathogen itself. Some toxins cause general non-specific symptoms. *Aschochyta caulina* showed herbicidal activity on the weed *Chenopodium album*. This pathogen produced in liquid cultures three main metabolites proposed as possible natural herbicide. Experiments were performed for maximizing toxin production by *A. caulina*, reducing production costs, scaling up a large scale purification procedure and identifying fast and inexpensive chemical methods to quantify toxin yields. *A. caulina* grew well and produced up to 230 mg of toxins/l culture, when grown for 5–10 days in shake cultures, provided the initial inoculum was at least 10^5–10^6 conidia/ml of culture. Addition of yeast extract to the medium could be useful for mass production of toxin by *A. caulina*. The investigation indicated that use of the toxin in place of the fungus producing it, could be a better option for the control of *Chenopodium album* (Vurro et al. 2012).

4.1.1.3 Commercialization of Bioherbicides

Several pathogens have been demonstrated to have potential for effective suppression of weeds infesting agricultural fields where various crops are cultivated. However, very few fungi have been developed as commercial bioherbicides, because of formidable constraints that have to be overcome. Host variability, host range, availability of optimal environmental conditions, mass production of viable, infective and genetically stable pathogen propagules and inconsistency in their performance under varied field environments are the major constraints that have to be addressed. As the bioherbicides contain live propagules, safety to human beings, animals and other non-target plants including crops in which the bioherbicides are to be applied have to be clearly established. The bioherbicides should be fairly fast-acting, provide high levels of control

Table 4.1 Mycoherbicides developed as commercial products (Adapted from Charudattan 1991 and Chutia et al. 2007)

Pathogenic fungus	Product name/company	Country of registration
Phytophthora palmivora	DeVine, Abbott Laboratories	United States
Colletotrichum gloeosporioides f.sp. *aeschynomene*	COLLEGO Univ. Arkansas, USDA UP John Company	United States
C. gloeosporioides f.sp. *malvae*	BioMal™	Canada
C. gloeosporioides f.sp. *cuscutae*	LUBOAII	China
Cylindrocladium leaveae	Stumpout ™	South Africa
Chondrostereum purpureum	Biochon™	Netherlands
	Myco-Tech™	Canada

similar to chemical herbicides and easy to use and offer consistent performance for the acceptance by industries and users. Information on the availability of bioherbicides that have been registered is either scarce or lacking (Table 4.1).

4.1.2 Abiotic Agents-Based Biocontrol Strategies

Soil solarization has been demonstrated to offer several benefits in addition to suppression of soilborne microbial plant pathogens. Soil solarization, an alternative to fumigation with chemicals such as methyl bromide and metham sodium, is a hydrothermal soil-disinfestation process that utilizes clear plastic mulch to trap solar radiation in moist soils. Solarization during hot seasons can enhance soil temperatures to levels that can kill soilborne pathogens and weeds. Thin clear polyethylene sheets are used to cover the soil, after irrigation. By this, the solar energy is trapped by plastic sheets which prevents loss of heat by evaporation and convection currents. Solarization results in a layer of pasteurized or disinfested soil. Solarization is a physical treatment which has neither any phytotoxic effects nor leaves any harmful residue as in chemical treatments. In addition, solarization is a simple and safe method with none of the hazards associated with chemical treatments. Organic growers are at an additional disadvantage, because of the restrictions on pesticides allowable under organic certification programs. The effectiveness of soil solarization may be enhanced by combining it with other treatments such as fertilization and cultivation of or amendment with cruciferous crops. Plants often grow faster and produce higher yields, when grown in solarized soils, as a result of weed and disease suppression, increased solubility of nutrients and changes in the microbial composition of the soil (Stapleton et al. 2000).

The effects of soil solarization on six important weed species were investigated by conducting under laboratory, small plots and on-farm conditions. The seeds of barn-yard grass (*Echinochloa crus-galli*), London rocket (*Sisymbrum irio*), common purslane (*Portulaca oleracea*), black nightshade (*Solanum nigrum*), annual sowthistle (*Sonchus oleraceus*) and tumble pigweed (*Amaranthus albus*), present commonly in parsley and strawberry fields were killed within 20 min at 70 °C and in 3 h at 60 °C. Solarization reduced the weed numbers by 86–94 % and weed biomass by 94–99 %, compared with untreated control plots. Solarization treatments reduced labor time necessary to maintain commercial weed control in the plots by 92–97 % as compared to timed hand-weeding treatment. In terms of parsley foliage yield, solarization treatment provided an economic yield of foliage ranging from 6.7 to >20-fold increase over untreated control. Soil solarization for 4 weeks effectively reduced weed density and increased marketable strawberry fruit yield by 32–42 %. Solarization did not provide satisfactory control of yellow nutsedge (*Cyperus esculentus*) as in the case of application of metham sodium, methyl bromide or chloropicrin. The cost of solarization ($150–$300 per acre) was found to be much cheaper than methyl bromide fumigation. Combination of solarization with metham sodium (where permissible) appeared to be an attractive option for commercial production systems (Stapleton et al. 2005).

Soil solarization has been reported to offer beneficial effect for disease suppression, enhancement of plant growth and weed control as well. Soil solarization using transparent polyethylene sheets reduced the incidence of water melon wilt disease caused by *Fusarium oxysporum* f.sp. *raphani* by 39–74 %. The weeds present in the plots covered with polyethylene sheet were controlled effectively (Huang 1993). Raised bed solarization (RBS) for 7 weeks alone or in combination with chicken manure (CM) amendment effectively reduced the incidence of strawberry diseases caused by *Phytophthora cactorum* and *Rhizoctonia* spp. In addition, four weed species commonly found in strawberry fields were effectively controlled by RBS treatment alone and also in combination with CM amendment. Raised bed solarization increased the fruit yield significantly in the field trials conducted during two cropping seasons of 2002–2004 (Benlioğlu et al. 2005). Soil solarization treatments using polyethylene films (30 μm thick) containing different additives viz., ultraviolet (UV), UV + infrared (IR), UV + IR + antifog (AF) + antidust (AD) were evaluated for their efficacy in suppressing the development of *Macrophomina phaseolina* infecting strawberry and the weeds. The survival of the sclerotia of *M. phaseolina* and weed growth were significantly reduced by solarization. All weeds except *Cyperus rotundus* were effectively eliminated by soil solarization treatments (Yildiz et al. 2010).

4.2 Biological Control of Insect Vectors by Microbial Pathogens

Microbial plant pathogens, especially viruses depend on different kinds of vectors for their spread from infected plants to healthy plants. Some viruses are transmitted by the vectors from one generation to the next generation through eggs of infected females as in the case of *Rice dwarf virus*. Reduction of vector population by different methods may be expected to result in decrease in the incidence of such diseases.

Biological management of insects is considered to be a desirable option for reducing or replacing the chemical application. Plant viruses are transmitted by various insects belonging to different taxonomic groups of which aphids, whiteflies, leafhoppers and thrips are important. Mycopathogens parasitizing these groups of insects that occur as pests of different crops have been identified. Some of the insect species can also be efficient transmitters of plant viruses. Various species of entomopathogenic fungi such as *Lecanicillium* (syn. *Verticillium*) sp., *Beauveria bassiana*, *Metarhizium anisopliae*, *Paecillomyces* sp. and *Nomuraea rileyi* have been evaluated for their potential to be employed for the management of insect pests. The effectiveness of the pathogens infecting insects and prospects of developing them as commercial products for large scale application for managing the insects are discussed in this section.

4.2.1 Aphids

4.2.1.1 Fungal Biocontrol Agents

Among the aphids, *Myzus persicae*, the green peach aphid is considered as the champion of the vectors of plant viruses, since it can transmit over 25 different plant viruses with different efficiencies (Narayanasamy and Duraiswamy 2003). *Aphis gossypii* also transmits many plant viruses infecting different crops. Twelve strains of entomopathogenic fungi (EPF) such as *Lecanicillium lecanii*, *Paecilomyces farinosus*, *Beauveria bassiana*, *Metarhizium anisopliae*, *Cordyceps scarabaeicola* and *Nomuraea rileyi* were screened for their efficacy in reducing the population of two aphid species viz., *Myzus persicae* and *Aphis gossypii* which were commonly observed in vegetable crops grown in greenhouses. The time required to kill 50 % of aphids (lethal time, LT_{50}) was calculated for each strain of the fungal species. The LT_{50} values of *L. lecanii* strain 41,185, 6,543 and 6,541 were lower than those of other fungal strains tested (Table 4.2). Two strains of *L. lecanii* 41,185 and 6,541 caused 100 % mortality of *M. persicae* in 4 days at 25 °C and 75 % RH, while other fungal strains could reach mortality of 72–100 % only at 6–8 days after treatment. The LT_{50} values of the three strains 41,185, 6,541 and 6,543 indicated that they were more effective against *A. gossypii* also, compared with other fungal strains. *L. lecanii* strain 41,185 required only 2 days to reach the mortality level of 100 %. *A. gossypii* was more sensitive to the fungal pathogens than *M. persicae*. The 50 % lethal concentration (LC_{50}) of the conidial suspension of *L. lecanii* 41,185 was 6.55×10^5 conidia/ml. This strain grew well in a wide temperature range (15–30 °C). The conidia of this strain could germinate at a wide range of temperatures (15–30 °C) (Vu et al. 2007).

Efficient application of an entomopathogen as a mycoinsecticide is considered to involve hitting the target insect species directly with a lethal dose of conidia. However, secondary pick-up of pathogen propagule (conidia) from surrounding vegetation may be an important source of inoculum. Increasing aphid mobility

Table 4.2 Efficacy of entomopathogenic fungal strains on two aphid species (Vu et al. 2007)

	LT_{50} values[a] (days for mortality)	
Pathogen strains	*M. persicae*	*A. gossypii*
Lecanicillium lecanii 41,185	1.58	1.40
6,543	2.43	1.50
6,541	2.38	1.49
L. fusisporum 4078	4.24	1.60
Cordyceps scarabaeicola J94	4.78	2.50
Paecilomyces farinosus J301	3.06	2.81
Beavueria bassiana J57	4.75	2.70
P. fumosoroseus J50	4.11	1.97
P. farinosus J302	4.58	5.47
P. farinosus J67	3.75	2.16
Nomuraea rileyi J125	4.80	2.18
Metarhizium anisopliae J88	5.66	1.64

[a]LT50 values of *M. persicae* and *A. gossypii* based on probit analysis program using detached leaves of Chinese cabbage and cucumbers at 25 °C and 75 % RH

with (E)-ß-farnesene (EBF) (aphid alarm pheromone) was shown to improve the pick-up of pesticides and control of the cotton aphid *Aphis gossypii* (Griffiths and Pickett 1987). The methods of increasing acquisition of conidia of *Lecanicillium* (*Verticillium*) *lecanii* by *Myzus persicae* were explored. E-ß-farnesene application significantly increased mortality of *M. persicae* that were exposed for 24 h to discs of green pepper leaf sprayed with conidia of *L. lecanii* and then transferred to fresh untreated leaf discs to allow disease development. The effect of using sub-lethal doses of the chloronicotinyl insecticide imidacloprid on the conidia pick-up by *M. persicae* was assessed. Systemic application of 1 % of the recommended dose of imidacloprid, dramatically increased aphid movement which was quantified by image analysis of videotaped aphid behavior. This resulted in greater mortality from mycosis in the experiments where aphids were exposed to insecticide-treated leaf discs that had been sprayed with conidia of *L. lecanii*. High mortality of the aphids on insecticide treated surface appeared to account for increased aphid mortality. The results provided a clear link between increased movement of aphids and enhanced secondary pick-up of fungal conidia. The combined use of two control agents with different mechanisms of action might reduce the chances of simultaneous development of resistance to both agents (Roditakis et al. 2000). The effectiveness of the endophytic fungus *Chaetomium globosum* in reducing the population of *Myzus persicae* was assessed. *C. globosum* strain YY-11 with antifungal activities was isolated from rape seedlings. The agglutinin (*pta*) gene from *Pinellia ternata* was cloned into the strain YY-11 adopting the procedure of *Agrobacterium tumefaciens*-mediated transformation. The positive transformants, as selected by antibiotic resistance, were evaluated for their efficacy using polymerase chain reaction (PCR) and Western blot assays. The recombinant endophytes colonized different crops. The resistance of rape inoculated with the transformants was enhanced, resulting in the inhibition of growth and reproduction of the aphid *Myzus persicae*. The results indicated the possibility of employing recombinant endophytes expressing *Pinella ternata* agglutinin (PTA) for the control of aphids such as *M. persicae* which is an important pest, as well as an efficient vector of several plant viruses (Qi et al. 2011).

The efficacy of a strain of *Lecanicillium lecanii* in causing mortality of the aphid *Schizaphis graminum* was determined. The fungal pathogen was inoculated onto the fourth instar nymphs and to alate and apterous adult morphs as a ground rice-kernel formulation. *L. lecanii* formulation reduced the survival of the aphids and interacted differently with morphs, the lethal time (LT) values being lower for alate compared to apterous morphs and nymphs. Treatment with *L. lecanii* reduced the fecundity of three treated aphid groups significantly. The pathogen hyphae invaded the hemocoel of a limited number of alate and apterous morphs as revealed by histological observations. Scanning electron microscope (SEM) observations showed that *L. lecanii* adhered to the body surface of both nymphs and adults without any variation due to stage of aphid species. The results indicated the potential of *L. lecanii* for use as a biocontrol agent for the management of *S. graminum* and other aphid species (Ganassi et al. 2010). *Lecanicillium longisporum* formulated and developed as the commercial product Vertalec® was evaluated for its efficacy against the aphid *Aphis gossypii* and the powdery mildew pathogen *Sphaerotheca fuliginea* infecting cucumber under greenhouse conditions with fluctuating temperature and relative humidity (RH) within normal operating ranges. Cucumber plants (5–6 weeks old) were inoculated with cotton aphids or powdery mildew pathogen or both. Vertalec, Vertalec containing irradiation-inactivated blastospores (II Vertalec) or sterilized water was applied to cucumber plants at 1, 4 and 7 days after inoculation. Vertalec treatment provided complete control of aphids at 16 days after inoculation. II Vertalec was ineffective and it was not significantly different from water-treated control. Powdery mildew colonies developing on cucumber leaves and sporulation were significantly reduced by Vertalec. The results showed that the commercial product Vertalec® containing *L. longisporum* had the potential for the effective management of cotton aphids and the powdery mildew disease on cucumber under greenhouse conditions (Kim et al. 2010).

Isolates (23) of *Metarhizium anisopliae* and *M. acridum* were evaluated for their biocontrol potential against *Myzus persicae*. The apterous adults of *M. persicae* were inoculated by spraying conidial suspensions at concentrations of 11.5, 99 and 1,179 conidia/mm^2 in leaf-dish bioassays. Most of the isolates were pathogenic to *M. persicae* at 21 ± 1 °C and 14:10 h light:dark photoperiod, causing corrected mortalities of 10.1–95.3 % at high conidial concentration. Ten isolates caused >50 % mortality and the estimates of their LC_{50} and LT_{50} values were worked out. Four isolates of *M. anisopliae* (ARSEF759, 4,132, 2,080 and 57) had LC_{50} values of 44–80 conidia/mm^2 at 8 days after inoculation and LT_{50} values of 4.9–6.8 days at 100 conidia/mm^2 with 91–98 % of the killed aphids showing mycosis after death. The results indicated that four isolates with high level of virulence and ability to produce high concentrations of conidia could be considered for further development of formulation and commercialization (Shan and Feng 2010). The efficacy of four native isolates of *Metarhizium anisopliae*, LIM1, LIM2, LIM3 and V275 and one native isolate (LIB1) and one exotic isolate of *Beauveria bassiana* (Botaniard) was evaluated for the control of *Myzus persicae*. All isolates tested, were able to cause 100 % mortality of *M. persicae* at 7 days post-inoculation. However, significant differences in their virulence could be observed at 2 days post-inoculation. High mortalities of 80, 81 and 87 % were recorded for native *B. bassiana* LIB1 isolate, Botanigard and *M. anisopliae* V275 respectively. All native strains of *M. anisopliae* and *B. bassiana* (LIB1) induced higher percentage of mycosis (80 %) developed on aphid cadavers at 7 days post-inoculation. A dose-dependent relationship in mortality of *M. persicae* was indicated by LC_{50} and LT_{50} assessments. The estimated LC_{50} for LIM1 isolate was significantly higher than that for LIB1. The results indicated that the isolates of entomopathogenic fungi present in a location have to be tested for their pathogenicity to the target insects along with the commercial products (Ibrahim et al. 2011).

4.2.1.2 Bacterial Biological Control Agents

The plant growth-promoting rhizobacteria (PGPR) are known to be efficient biocontrol agents against fungal, bacterial and viral pathogens infecting a wide range of crops. The effects of the rhizobacterium *Bacillus subtilis* strains BS3A25 on the incidence of *Cucumber mosaic virus* (CMV) on tomato and the vector *Aphis gossypii* were assessed under natural conditions of high levels of vector-virus pressure on tomato plants. Tomato seeds treated with the strain BS3A25 showed enhanced seed germination (99 %), compared with untreated control (78 %). When seed treatment was combined foliar application of the BCA strain, the level of protection of tomato plants against CMV was increased, in addition to improvement in plant growth. Furthermore, the strain BS3A25 had many inherent positive traits such as phosphate solubilization ability, ACC utilization as the sole source of nitrogen and production of significantly greater concentrations of IAA and cytokinin. Treatment with the strain BS3A25 influenced the growth parameters of *A. gossypii* such as development time, time of birth to adult (tD) and preproduction time ((td) which were longer due to bacterial treatment. In addition, the relative growth rate (RGR) and intrinsic rate of natural increase (r_m) were lower following treatment with BS3A25 supernatant compared with the effect of insecticide imidacloprid. In tomato plants treated with BS3A25, higher activities of phenylalanine ammonia lyase (PAL), peroxidase, and polyphenoloxidase and total phenols were recorded, suggesting the possible induction of defense responses. Under field conditions, seed and foliar treatment with BS3A25 effectively reduced the aphid population and CMV incidence compared with imidacloprid treatment. The cost:benefit ratio was the highest in BS3A25 formulation treated plots (1:2.5), followed by the insecticide (1:2.0) and control (1:1.4) treatments. The results showed that the strain BS3A25 could be considered for commercial development (Sudhakar et al. 2011).

4.2.2 Whitleflies

Development of resistance to chemical insecticides in whiteflies has accelerated the search for alternative methods for the management of this group of pests that are also involved in the spread of many economically important virus diseases. Among alternative methods, biological control approach has drawn the attention of a large number of researchers. Natural infections of *Bemisia* spp. by *Paecilomyces*, *Beauveria* and *Lecanilcillium* (=*Verticillium*) have been commonly observed. The ability of fungal pathogens to provide effective control of *Bemisia tabaci* has been investigated under a wide range of conditions. Several in vitro and field experiments have shown that the high ambient humidity conditions required for development of natural epizootics are not necessarily required for fungal infections. Many fungal pathogens find required moisture for spore germination and host penetration within the leaf or insect microclimate boundary layer. This phenomenon was observed for infection of whitefly nymphs by *Paecilomyces fumosoroseus* and *Beauveria bassiana* (Wraight et al. 2000). Temperature has significant effects on the physiology and development of both the insect host and fungal pathogen (host susceptibility and pathogen virulence) and these effects are influenced, in turn, by factors related to the host plants. The potential of fungal pathogens has been assessed using all life stages of whiteflies. Young instars of *B. tabaci* tend to be more susceptible to fungal infections than the later stages (Faria and Wraight 2001). The effects of relative humidity levels on the infectivity of two entomopathogenic fungi *Verticillium lecanii* (Mycotal) and *Beauveria bassiana* (Naturalis-L) on the whitefly *Trialeurodes vaporariorum* were determined by two series of bioassays. In one series, a non-excised bean leaf was sandwiched to create a chamber above the whitefly infested area. In another series, three potted whole bean plants were maintained in an air-tight plastic box. Relative humidity in both series was controlled by injecting a circulating constant flow of humidity-regulated air which had passed over a salt solution used to maintain targeted air humidity levels (from 53 to 98.5 % and 75 to 98.5 % respectively). Microclimatic measurements demonstrated that non-excised leaves significantly increased the air humidity in the air circulating in the immediate environment of the leaf surface as well as that of the whole plant. In the first series of assays under all RH conditions tested (70–98 %) after passage over the leaves, the mycoparasite induced high mortality rates of whiteflies (92–100 %). The fungal pathogenicity occurred early, since at least 70 % of whiteflies died as second instars, before next molt occurred. Fungus-induced mortality appeared to be independent of the humidity level existing around the leaves of bean plants (Vidal et al. 2003).

Fungal infection of whiteflies depends on many biological events which are initiated by adhesion of fungal spores to the insect cuticle, followed by spore germination and hyphal growth. The hyphae exert mechanical pressure and produce cuticle-degrading enzymes such as proteases and chitinases to breach the insect cuticle. *Paecilomyces fumosoroseus* strain 612 was exposed to UV-irradiation to generate mutants. Selection of colonies displaying clearing zones on colloidal chitin agar medium was found to be an efficient qualitative screening method for rapid examination of a large number of colonies after UV mutagenesis. One mutant M84 exhibited a large and stable chitin hydrolysis halo. This stable mutant that over-produced chitinases compared with parent strain was selected. Glucose consumption and biomass production were similar for M84 and the parent strain. Chitinase was inducible by chitin and repressed by glucose in both strains. But when they were grown on minimal medium plus colloidal chitin as sole carbon source, the parent and M84 strains yielded 198 and 690 μmol *N*-acetylglucosamine respectively, indicating enhanced synthesis of a chitinase with higher activity by the mutant M84. Bioassays performed with *Bemisia tabaci* nymphs showed that the strain M84 incited a two-fold mortality of *B. tabaci*, compared with the parent strain. The results suggested that chitinase activity was directly involved in higher virulence of the fungal pathogens of whiteflies (Hernández-Torres et al. 2004). Infestation of tomatoes and spread of whiteflies and thrips that vector *Tomato yellow leafcurl virus* (TYLCV) and *Capsicum chlorosis virus* (CaCV) constitute major risks for tomato production. *Paecilomyces fumosoroseus* was found to be effective against both groups of insects. *P. fumosoroseus* strain FWA3 caused 76.7 % mortality of *Bemisia tabaci*. White rice, brown rice, broken-milled rice, corn and sorghum grains were evaluated for their ability to serve as solid substrate for fungal mass multiplication. White-rice and broken-milled rice were more efficient in enhancing conidial production and the maximum yield of conidia was obtained after an incubation period of 10 days. The infection process of the fungal pathogen was visualized using scanning electron microscope. The conidia germinated by forming a germ tube which gained entry into insect body cavity or haemocoel by penetration through the cuticle and epidermal regions of insect integument or by passing through insect spiracle. The potential of *P. fumosoroseus* for use as a bioinsecticide was assessed by bioassays. Pest-infested tomato plants were sprayed in insect screen cages to determine effective dose rates for spray application of the BCA in greenhouses. The LC_{50} value of the strain against *B. tabaci* was 9.41×10^4 conidia/ml. A simple low-cost method with less expensive and locally available materials was developed for mass multiplication of *P. fumosoroseus* strain FWA3 for the biocontrol of *B. tabaci* occurring in tomato crops (Panyasiri et al. 2007).

Mass multiplication and production of large quantities of biomass of entomopathogenic fungi is a critical step for their successful formulation and further development as commercial products. Solid-state fermentation (SSF) is an effective method of obtaining fungal spores produced by aerial hyphae. Aerial conidia produced by SSF are similar to those produced by naturally on the surface of insect cadavers and are superior to mycelia and blastospores produced under submerged fermentation conditions (Wraight et al. 2001; Roberts and Leger 2004). Synchronous conidia production procedure using a series of solid-state fermentation (SSF) chamber system was developed. The usefulness of the SSF chamber system was verified by employing the method for the conidia production of *Metarhizium anisopliae*, *Paecilomyces fumosoroseus* and *Beauveria bassiana* which have been reported to be effective against several insects. Rice grains were used as a solid substrate and the fungi were grown in a series of chambers at 25 °C. The conidial powder harvested from 5 kg rice cultures varied from 40.0 to 147.1 g and conidial concentrations differed from 1.3×10^{12} to 5.4×10^{12} depending on the pathogen species/strain. The conidial viability of the strains was up to 95 %. The results indicated that the chamber system developed in this investigation had the potential for wider application, after necessary modification to suit the specific requirement of the entomopathogen concerned (Chen et al. 2009). Native and exotic isolates of the entomopathogens *Metarhizium anisopliae* and *Beauveria bassiana* were evaluated for their potential to cause mortality of the first instar larvae of whiteflies. Six isolates LIM1, LIM2, LIM3 and V275 of *M. anisopliae* and LIB1 and Botanigard® of *B. bassiana* showed significant differences in their virulence reflected by the cumulative mortality of whiteflies *Bemisia tabaci* at 7 days post-inoculation. The duration required to cause 50 % mortality also differed for each isolate. The highest mortality of *B. tabaci* was caused by exotic isolate Botanigard® and it required 3.4 days to induce 50 % mortality of *B. tabaci* and 7 days for 96.5 % mortality. In the case of *M. anisopliae*, the native isolate LIB1 alone killed more than 30 % of larvae within 3 days, 50 % within 4.5 days and 88 % mortality in 7 days. Based on the LC_{50} and LT_{50} values, *B. bassiana* appeared to be more efficient than *M. anisopliae* in causing the mortality of whiteflies. Native *B. bassiana* LIB1 isolate, applied in combination with the insecticide diflubenzuron, reduced the population of first instar larvae of whiteflies under greenhouse conditions, indicating an additive effect of the BCA and insecticide (Ibrahim et al. 2011).

Entomopathogenic fungi possess a contact mode of action on target whitefly species. Infectious fungal spores have to reach the target insect or onto substrates in the habitat from which secondary inoculation may be possible through the insect movement or feeding. As the whitefly nymphs are small and do not move about actively, the most effective means of achieving rapid, high rates of infection is by direct inoculation under field conditions. For this, the lower sides of leaves have to be thoroughly covered with sprays of conidial inoculum which is a difficult challenge to be faced effectively. Good coverage is achievable with conventional hydraulic technologies like using drop-tubes with swivel-mounted nozzles. The mycoinsecticides are generally compatible with a wide range of insecticides, resulting in enhancement of their effectiveness against target insect. Commercial products containing the pathogenic fungi have been developed for use in greenhouse crops. *Paecilomyces fumosoroseus* is marketed as PreFeRal (Thermo Triology/Biobest NV) in Belgium, while *Beauveria bassiana* has been commercially developed as Botanigard (Emerald BioAgriculture Corporation, USA), as AgoBiocontrol Beauveria (AgoBiocontrol, Colombia), as Bea-Sin (Agrobiologicos, Mexico) and as Boveril PM (Itaforte BioProducts, Brazil). The efficacy of the entomopathogenic fungi against whiteflies is not generally effective under field conditions, possibly because of harsh climatic conditions in open fields (Faria and Wraight 2001).

4.2.3 Leafhoppers

Leafhoppers constitute an important group of pests on several crops and they are capable of transmitting many plant viruses and phytoplasmas infecting various crops in all countries. Fungal pathogens belonging to the genera *Beauveria*, *Metarhizium* and *Lecanicillium* (= *Verticillium*) have been reported to be pathogenic to leafhoppers and planthoppers (Aguda et al. 1984; Toledo et al. 2007; Jin et al. 2008). The green leafhoppers are efficient transmitters of rice tungro and rice yellow dwarf diseases, while rice grassy stunt disease is transmitted by brown planthoppers. In a few cases, bacterial biological control agents have also been reported to have adverse effects on the development of leafhoppers. The bacterial strains could cause about 90 % mortality of *N. virens*, when they fed on rice leaves treated with the bacterial strains for 7 days (Islam et al. 2010).

Beauveria bassiana and *Metarhizium anisopliae* effectively parasitized the brown planthopper (BPH) *Nilaparvata lugens* obtained from various locations. The local isolates of these pathogens were more virulent than the exotic isolates. The most effective concentration of *B. bassiana* causing high mortality of *N. lugens* was 6×10^{12} spores/ha. *M. anisopliae* and *B. bassiana* significantly reduced the population of BPH at 7 days after spraying. Rice bran medium was found to be the suitable medium for mass multiplication of *B. bassiana*, as it produced greater amounts of dried biomass than other media tested (Van Nghiep et al. 1999). Thirty-five isolates of *Metarhizium anisopliae* and *M. flavoviride* were bioassayed for their potential in causing mortality of *N. lugens*. Third instar-nymphs were inoculated with a concentration of 1,000 conidia/mm^2 of the fungal pathogens at 25 °C with a 14:10 light:dark photoperiod under in vitro conditions. At 9 days after inoculation, mortality of the nymphs due to mycosis ranged from 6.5 to 64.2 %, depending on the virulence of the isolates. Only the two BPH-derived *M. anisopliae* isolates from Philippines (Ma450) and Indonesia (Ma576) could cause mortality of >50 % of the nymphs with a standard suspension of 1×10^8 conidia/ml. These two isolates, however, caused mortality close to 70 % at highest concentration of about 1,000 conidia/mm^2. The two isolates of *M. anisopliae* were superior to *Beauveria bassiana* in causing nymphal mortality. The LC_{50} values were estimated as 731 and 1,124 conidia/mm^2 at 7 days after inoculation respectively for Ma456 and Ma576 (Jin et al. 2008). The bacterial strains could cause about 90 % mortality of *N. virens*, when they fed on rice leaves treated with the bacterial strains for 7 days (Islam et al. 2010).

The efficacy of 17 isolates of *Beauveria bassiana*, *Lecanicillium muscarium*, *Metarhizium anisopliae*, *Isaria farinosa*, and *I. fumosorosea* in causing mortality of three hemipterans, *Peregrinus maidis*, *Delphacodes kuscheli and Dalbulus maidis*, vectors of corn diseases was assessed. Three isolates of *B. bassiana* CEP147, CEP150 and CEP189 were found to be effective against *Peregrinus maidis*. No consistent differences in the susceptibility of males and females of *P. maidis* to fungal pathogens were detectable. There was no correlation between the percentage of conidial germination and percentage of mortality caused by fungal infection of *P. maidis* in any of the treatments. The adults of *P. maidis*, *D. kuscheli* and *D. maidis* were inoculated with the fungal pathogens. The most effective *B. bassiana* isolate CEP147 caused a cumulative mortality of 69.8±6.4 % after 7 days post-inoculation. At 2 weeks after inoculation with the isolate CEP147, the levels of susceptibility of the insect species were determined. Both *D. kuscheli* with cumulative mortality of 73.3±9.0 % and *P. maidis* with a cumulative mortality of 68.6±6.7 % were significantly more susceptible than *D. maidis* with a cumulative mortality of 49.9 ± 9.7 %. The results indicated the potential of *B. bassiana* for the management of leafhoppers as pests and vectors of corn virus diseases (Toledo et al. 2007).

4.2.4 Thrips

The economic importance of thrip species as pests of several crops and vectors of destructive viruses like *Tomato spotted wilt virus* (TSWV) has been well recognized. *Paecilomyces fumosoroseus*, the fungal pathogen of many insects was evaluated for its efficacy in causing mortality of thrips. A strain of *Lecanicillium lecanii* originally isolated from glasshouse whitefly and later commercially produced in the Netherlands was also effective against thrip species *Frankliniella occidentalis* (van der Schaaf et al. 1991). Later, isolates of *L. lecanii*, *Metarhizium anisopliae* and *Beauveria bassiana* were also reported to be pathogenic to *F. occidentalis* (Vestergaard et al. 1995; Sengonca et al. 2006). Isolates of *P. fumosoroseus* were reported to be pathogenic to *F. occidentalis* and an isolate of *P. fumosoroseus* was developed as a commercial product to be used against whiteflies and thrips (Saito and Sugiyama 2005). In a later investigation, *P. fumosoroseus* isolates FWA3 and FWA5 were found to be highly virulent against thrips species *Ceratothripoides claratris* infesting tomato plants. The LC_{50} of *P. fumosoroseus* isolate FWA3 against *C. claratris* and *B. tabaci* were 9.51×10^2 conidia/ml and 9.41×10^4 conidia/ml respectively. The white-rice and broken-milled rice produced the highest concentration of *P. fumosoroseus* conidia after an incubation period of 10 days. The isolate FWA3 with high level of pathogenicity against thrips and whiteflies possessed many attributes required for the candidate for commercialization (Panyasiri et al. 2007).

Paecilomyces lilacinus, a soil inhabiting nematophagous fungus produces chitinases and proteases capable of breaking down egg shell, facilitating penetration into insect body. This mechanism was demonstrated to operate effectively against the western flower thrip *Frankliniella occidentalis* (Fiedler and Sosnowska 2007). *P. lilacinus*, at a concentration of 2.3×10^9 conidia/ml was blended with neem (*Azidirachta indica*) leaf, and seed extract (10 ml/l) and diatomaceous earth formulation (PyriSec) (3 g/l) and the mixture was tested for its effect on *Thrips tabaci*, using cotton leaf disc bioassay technique. The bioassays were performed at 25 ± 1 °C and >70 % relative humidity with 16:8 light:dark photoperiod conditions. *P. lilacinus* was very effective for the control of *T. tabaci*, when applied alone and in combination with neem on cotton. Application of this mixture resulted in significant reduction in thrips population both in in vitro assays and in semi-natural conditions compared with control. The combination of *P. lilacinus* and diatomaceous earth formulation was less effective, compared with fungus + neem extract combination. The results indicated the advantages of combining two naturally derived organic materials for the efficient control of thrips in cotton (Wakil et al. 2012).

In another investigation, five strains of *Beauveria bassiana* were evaluated for their efficacy against *F. occidentalis*. The strain RSB was the most virulent causing 69–96 % mortality at concentrations of $1 \times 10^4 - 1 \times 10^7$ conidia/ml, at 10 days after inoculation of first instar larvae. In greenhouse evaluation, RBS strain applied to broccoli foliage significantly reduced adult and larval populations of *F. occidentalis* (Gao et al. 2012b). The insecticidal property of *Beauveria bassiana* on *Orius sauteri*, an important predator of western flower thrips *Frankliniella occidentalis* was assessed under in vitro conditions. *B. bassiana* strain RSB (Bb-RSB) was not insecticidal against *O. sauteri*, irrespective of the concentration applied to the first instars. The developmental rate of *O. sauteri* was also not affected by direct treatment with Bb-RSB. However, significant differences in the development rates and adult longevity were noted between *O. sauteri* that fed on Bpb-RSB infected *F. occidentalis* cadavers and those fed on untreated thrips. Developmental time (from first instar to adult) increased from 0.3 to 0.7 days for predators-fed thrips treated with low and high concentration of strain Bb-RSB respectively, compared with predators fed on untreated thrips. But these differences were only 3–13 % of mean values for the controls. The results suggested that the observed adverse effects on *O. sauteri* were relatively minor. The usefulness of combination of *B. bassiana* and *O. sauteri* for the control of *F. occid*entalis has to be demonstrated under field conditions (Gao et al. 2012a, b).

References

Aguda RM, Litsinger JA, Roberts DW (1984) Pathogenicity of *Beauveria bassiana* in brown planthopper (BPH), whitebacked planthopper (WBPH) and green leafhopper (GLH). Int Rice Newsl 9(6):20

Anderson FE, Barton J, McLaren DA (2010) Studies to assess the suitability of *Uromyces pencanus* as a biological control agent for *Nassella neesiana* (Poaceae) in Australia and New Zeland. Australas Plant Pathol 39:69–78

Anderson FE, Diaz ML, Barton J, Flemmer AC, Hansen PV, McLaren DA (2011) Exploring the life cycles of three South American rusts that have potential as biocontrol agents of the stipoid grass *Nassella neesiana* in Australasia. Fungal Biol 115:370–380

Benlioğlu S, Boz Ö, Yildiz A, Kaşkavalci G, Benlioğlu K (2005) Alternative soil solarization treatments for the control of soilborne diseases and weeds of strawberry in the western Anatolia of Turkey. J Phytopathol 153:423–430

Boyette CD, Quimby PC Jr, Connick WJ Jr, Daigle DJ, Fulgham FC (1991) Progress in the production, formulation and application of mycoherbicides. In: Te Beest DO (ed) Microbial control of weeds. Chapman & Hall, New York, pp 209–222

Boyette CD, Hoagland RE, Abbas HK (2007a) Evaluation of the bioherbicide *Myrothecium verrucaria* for weed control in tomato (*Lycopersicon esculentum*). Biocontrol Sci Technol 17:171–178

Boyette CD, Jackson MA, Bryson CT, Hoagland RE, Connick WJ, Daigle DJ (2007b) *Sesbania exaltata* biocontrol with *Colletotrichum truncatum* microsclerotia formulated in 'Pesta' granules. BioControl 52:413–426

Boyette CD, Gealy D, Hoagland RE, Vaughn KC, Bowling AJ (2011) Hemp sesbania (*Sesbania exaltata*) control in rice (*Oryza sativa*) with the bioherbicidal fungus *Colletotrichum gloeosporioides* f.sp. *aeschynomene* formulated in an invert emulsion. Biocontrol Sci Technol 21:1399–1407

Brosten BS, Sands DC (1986) Field trials of *Sclerotinia sclerotiorum* to control Canada thistle (*Cirsium arvense*). Weed Sci 34:377–380

Charudattan R (1991) The mycoherbicide approach with plant pathogens. In: Te Beest DO (ed) Microbial control of weeds. Chapman & Hall, New York, pp 24–57

Chen C, Wang Z, Ye S, Feng M (2009) Synchronous production of conidial powder of several fungal biocontrol agents in series fermentation chamber system. Afr J Biotechnol 8:3649–3653

Chutia M, Mahanta JJ, Bhattacharyya N, Bhuyan M, Boruah P, Sarma TC (2007) Microbial herbicides for weed management: prospects, progress and constraints. Plant Pathol J 6:210–218

Clarke TC, Shetty KG, Jayachandran K, Norland M (2007) *Myrothecium verrucaria*—a potential biological control agent for the invasive 'old world climbing fern' (*Lygodium microphyllum*). BioControl 52:399–411

Ellison CA, Pereira JM, Thomas SE, Barreto RW, Evans HC (2006) Studies on the rust *Prospodium tuberculatum*, a new classical biocontrol agent released against the invasive alien weed *Lantana camera* in Australia. I. Life-cycle and infection parameters. Australas Plant Pathol 35:309–319

Faria M, Wraight SP (2001) Biological control of *Bemisia tabaci* with fungi. Crop Protect 20:767–778

Fiedler Z, Sosnowska D (2007) Nematophagous fungus *Paecilomyces lilacinus* (Thom.) Samson is also a biological agent for control of greenhouse insects and mite pests. BioControl 52:547–558

Fisher AJ, Smith L, Woods DM (2011) Climatic analysis to determine where to collect and release *Puccinia jaceae* var *solstitialis* for biological control of yellow starthistle. Biocontrol Sci Technol 21:333–351

Ganassi S, Grazioso P, Moretti A, Sabatini MA (2010) Effects of the fungus *Lecanicillium lecanii* on survival and reproduction of the aphid *Schizaphis graminum*. BioControl 55:299–312

Gao Y, Reitz SR, Wang J, Tamez-Guerra P, Wang E, Xu X, Lei Z (2012a) Potential use of the fungus *Beauveria bassiana* against the western flower thrips *Frankliniella occidentalis* without reducing the effectiveness of its natural predator *Oris sauteri* (Hemiptera: Anthocoridae). Biocontrol Sci Technol 22:803–812

Gao Y, Reitz SR, Wang J, Xu X, Lei Z (2012b) Potential of a strain of the entomopathogenic fungus *Beauveria bassiana* (Hypocreales: Cordycipitaceae) as a biological control agent against western flower thrips, *Frankliniella occidentalis* (Thysanoptera: Thripidae). Biocontrol Sci Technol 22:491–495

Griffiths DC, Pickett JA (1987) Novel chemicals and their formulation for aphid control. Proceedings of the 14th international symposium on controlled release of bioactive materials 14, pp 243–244

Guntli D, Burgos S, Kump I, Heeb M, Pfirter HA, Défago G (1999) Biological control of hedge bind weed (*Calystegia sepium*) with *Stagonospora convolvuli* strain LA39 in combination with competition from red clover (*Trifolium pratense*). Biol Control 15:252–258

Hernández-Torres I, Iracheta M, Gálon-Wong LJ, Hernández C, Contreras J, Jackson M, Pereyra-Alférez B (2004) A *Paecilomyces fumosoroseus* mutant over-producing chitinase displays enhanced virulence against *Bemisia tabaci*. World J Microbiol Biotechnol 20:207–210

Huang SH (1993) Application of solarization for control of soilborne fungal disease. Spl Publ Taichung Dis Agric Improv Sta 32:247–256

Ibrahim L, Hamieh A, Ganem H, Ibrahim SK (2011) Pathogenicity of entomopathogenic fungi from Lebanese soils against aphids, whitefly and non-target beneficial insects. Int J Agric Sci 3:156–164

Islam Z, Pamplone R, Atkinson AD, Azucena EJ (2010) Biological control of rice diseases. International Rice Research Institute, Los Baños

Jin S-F, Feng M-G, Chen J-Q (2008) Selection of global *Metarhizium* isolates for the control of the rice pest *Nilaparvata lugens* (Homoptera: Delphacidae). Pest Manag Sci 64:1008–1014

Kennedy DS (1986) De Vine®—the way it was developed—an industrialist's view. Weed Sci 34(suppl):15–16

Kim JJ, Goettel MS, Gillespie DR (2010) Evaluation of *Lecanicillium longisporum*, Vertalec® against cotton aphid *Aphis gossypii* and cucumber powdery mildew, *Sphaerotheca fuliginea* in a greenhouse environment. Crop Protect 29:540–544

Narayanasamy P, Duraiswamy S (2003) Plant viruses and viral diseases. New Century Book House Pvt Ltd, Chennai

Ortiz-Ribbing LM, Williams MM II (2006) Potential of *Phomopsis amaranthicola* and *Microsphaeropsis amaranthi*, as bioherbicides for several weedy *Amaranthus* species. Crop Protect 25:39–46

Ortiz-Ribbing LM, Glassman KR, Roskamp GK, Hallett SG (2011) Performance of two bioherbicide fungi of waterhemp and pigweed control in pumpkin and soybean. Plant Dis 95:469–477

Pan L, Ash GJ, Ahn B, Watson AK (2010) Development of strain specific molecular markers for the *Sclerotinia minor* bioherbicide strain IMI 344141. Biocontrol Sci Technol 20:939–959

Panyasiri C, Attathom T, Poehling H-M (2007) Pathogenicity of entomopathogenic fungi-potential candidates to control insect pests on tomato under protected cultivation in Thailand. J Plant Dis Protect 114:278–287

Pfirter HA (2000) As quoted by Muller-Scharer H, Scheepens PC, Greaves MP

Pfirter HA, Défago G (1998) The potential of *Stagonospora* sp. as a mycoherbicide for field bindweed. Biocontrol Sci Technol 8:93–101

Pilgeram AL, Sands DC (1999) Mycoherbicides. In: Hall FR, Menn JJ (eds) Methods in biotechnology, vol. 5. Biopesticides-use and delivery. Humana Press Inc, Totowa, pp 359–370

Pottinger BM, Ridgway HJ, Bourdot GW, Stewart A (2008) The potential use of ascospores as the active ingredient for a *Sclerotinia sclerotiorum*-based mycoherbicide. Australas Plant Pathol 37:443–447

Qi G, Lan N, Ma X, Yu Z, Zhao X (2011) Controlling *Myzus persicae* with recombinant endophytic fungi *Chaetomium globosum* expressing *Pinellia ternata* agglutinin. J Appl Microbiol 110:1314–1322

Roberts DW, Leger RJS (2004) *Metarhizium* spp. cosmopolitan insect pathogenic fungi: mycological aspects. Adv Appl Microbiol 54:1–70

Roditakis E, Couzin ID, Balrow K, Franks NR, Charnley AK (2000) Improving secondary pick up of insect fungal pathogen conidia by manipulating host behaviour. Ann Appl Biol 137:329–335

Rosskopf EN, Yando CB, Charudattan R (2006) Genus-specific host range of *Phomopsis amaranthicola* (Sphaeropsidales), bioherbicide agent of *Amaranthus* spp. Biocontrol Sci Technol 16:27–35

Saito T, Sugiyama K (2005) Pathogenicity of three Japanese strains of entomopathogenic fungi against the silverleaf whitefly, *Bemisia tabaci*. Appl Entomol Zool 40:169–172

Sengonca CM, Thungrabeab M, Blaeser P (2006) Potential of different isolates of entomopathogenic fungi from Thailand as biological control agent against western flower thrips *Frankliniellla occidentalis* (Pergande) (Thysanoptera-Thirpidae). J Plant Dis Protect 113:74–80

Shan L-T, Feng M-G (2010) Evaluation of the biocontrol potential of various *Metarhizium* isolates against green peach aphid *Myzus persicae* (Homoptera: Aphididae). Pest Manag Sci 66:669–675

Stapleton JJ, Elmore CL, DeVay JE (2000) Solarization and biofumigation help disinfest. Calif Agric 54:42–45

Stapleton JJ, Molinar RH, Lynn-Patterson K, McFeeters SK, Shreshta A (2005) Soil solarization provides weed control for limited-resource and organic growers in warmer climates. Calif Agric 59:84–85

Sudhakar N, Thajuddin N, Murugesan K (2011) Plant growth-promoting rhizobacterial mediated protection of tomato in the field against *Cucumber mosaic virus* and its vector *Aphis gossypii*. Biocontrol Sci Technol 21:367–386

Tang W, Zhu Y-Z, He H-Q, Qiang S, Auld BA (2011) Field evaluation of *Sclerotium rolfsii*, a biological control agent for broad leaf weeds in dry, direct-seeded rice. Crop Protect 30:1315–1320

Thomas SE, Ellison CA, Tomley AJ (2006) Studies on the rust *Prospodium tuberculatum*, a new classical biological control agent released against the invasive alien weed *Lantana camara* in Australia. 2. Host range. Australas Plant Pathol 35:321–328

Toledo AV, de Remes Lenicov AMM, Lastra CCL (2007) Pathogenicity of fungal isolates (Ascomycota: Hypocreales) against *Peregrinus maidis*, *Delphacodes kuscheli* (Hemiptera: Delphcidae) and *Dalbulus maidis* (Hemiptera: Cicadellidae), vectors of corn diseases. Mycopathologia 163:225–232

Van der Schaaf DA, Ravenberg WJ, Malais M (1991) *Verticillium lecanii* as a microbial insecticide against whitefly. IOBC/WPRS Bull 14:120–123

Van Nghiep H, Nhan NT, Hung PQ, Khang VT, Loc NT (1999) Studies on some entomogenous fungi to control brown planthopper in rice. Omonrice 7:149–157

Vestergaard S, Butt TM, Gillespie AT, Schreiter G, Eilenberg J (1995) Pathogenicity of the hyphomycete fungi *Verticillium lecanii* and *Metarhizium anioapliae* to the western flower thrips *Frankliniella occidentalis*. Biocontrol Sci Technol 5:185–192

Vidal C, Fargues J, Rougier M, Smits N (2003) Effect of air humidity on the infection potential of hyphomycetous fungi as mycoinsecticides for *Trialeurodes vaporariorum*. Biocontrol Sci Technol 13:183–198

Vu VH, Hong SI, Kim K (2007) Selection of entomopathogenic fungi for aphid control. J Biosci Bioeng 104:498–505

Vurro M, Andolfi A, Boari A, Zonno MC, Caretto S, Avolio F, Evidente A (2012) Optimization of the production of herbicidal toxins by the fungus *Aschochyta caulina*. Biol Control 60:192–198

Wakil W, Ghazanfar MU, Kwon YJ, Shamas-ul-Islam AK (2012) Toxicity of *Paecilomyces lilacinus* blended with non-conventional agents to control cotton thrips (*Thirps tabaci* Lind.) (Insecta: Thysanoptera: Thripidae). Afr J Microbiol Res 6:526–533

Wilkin JE, Shamoun SF, Ritland C, El-Kassaby YA (2005) Genetic diversity and population structure of *Valensinia heterodoxa*, a potential biocontrol agent for salal in coastal British Columbia. Can J Plant Pathol 27:559–571

Wraight SP, Carruthers RI, Jaronski ST, Bradley CA, Garza CJ, Galaini-Wraight S (2000) Evaluation of the entomopathogenic fungi *Beauveria bassiana* and *Paecilomyces fumosoroseus* for microbial control of the silverleaf whitefly, *Bemisia argentifolii*. Biol Control 17:203–217

Wraight SP, Jackson MA, de Kock SL (2001) Production, stabilization and formulation of fungal biocontrol agents—progress, problems and potential. In: Butt T, Jackson C, Magan N (eds) Fungal biocontrol agent—progress, problems and potential. CAB International, Wallingford, pp 253–287

Yandoc CB, Rosskopf EN, Pitelli RL, Charudattan R (2006) Effect of selected pesticides on conidial germination and mycelia growth of *Dactylaria higginsii*, a potential bioherbicide for purple nutsedge (*Cyperus rotundus*). Weed Technol 20:255–260

Yildiz A, Benlioğlu S, Boz Ö, Benlioğlu K (2010) Use of different plastics for soil solarization in strawberry growth and time-temperature relationships for the control of *Macrophomina phaseolina* and weeds. Phytoparasitica 38:463–473

Additional Reference for Further Reading

Muller-Scharer H, Scheepens PC, Greaves MP (2000) Biological control of weeds in European crops: recent achievements and future work. Weeds Res 40:83–98

Chapter 5
Development of Formulations and Commercialization of Biological Products

Phylogenetically diverse microorganisms have been demonstrated to be potential natural antagonists of various microbial plant pathogens. Interactions with plant pathogens such as antibiosis, competition, predation and induction of resistance in plants, result in the suppression of pathogen development. The use of microorganisms as biocontrol agents to reduce the adverse effects of plant diseases is considered as an important alternative method to minimize the application of chemicals. Because of the diversity and complexity of reactions and numerous metabolic pathways, the microorganisms form an amazing resource for the biological management of plant diseases. Although hundreds of species/strains of fungi or bacteria have been reported to possess antagonistic potential against plant pathogens, only some of them have been found to be suitable for development of formulation and commercialization as biological products or microbial pesticides for the management of crop diseases under field conditions. This situation may be due to the fact that the biocontrol systems are based on living microorganisms whose activities are dependent on different biotic and abiotic factors. Knowledge of the influence of these factors on most microorganisms that are found to have antagonistic activities, is lacking. Commercial products have to be registered as microbial pesticides, after satisfying the stringent requirements prescribed by the environmental protection agency (EPA) under the Federal Insecticide, Fungicide and Rodenticide Act (FIFRA) or the Federal Food, Drug and Cosmetics Act (FFDCA) of the Untied States of America. The European Union (EU) has established directive 91/414/CEE for harmonizing the Register of Pesticides. In the developing world, the Association of Asian Pacific Agricultural Research Institutions (AAPARI) monitors the process of development of bioproducts (Montesinos 2003; Harman et al. 2010). Various steps in the development, registration and commercialization of bioproducts and consumer concerns are discussed for facilitating development of a socially receptive environment to microbially-based biocontrol strategies for the management of crop diseases. Development of formulations and registration of bioproducts have to be carried out in different stages before the registered products are marketed for large scale application.

P. Narayanasamy, *Biological Management of Diseases of Crops*,
Progress in Biological Control 16, DOI 10.1007/978-94-007-6377-7_5,

5.1 Development of Formulations

5.1.1 Isolation and Screening of Microorganisms

The first stage in the development of formulations is the isolation of fungi, bacteria or viruses that can interfere with the biological cycles of microbial plant pathogens. Sampling using proper methods in appropriate locations or plant materials may increase the probability of obtaining effective microorganisms. Samples taken in suppressive soils or plants in epidemic areas are likely to contain microbes antagonistic to target pathogens. Isolation of fungal and bacterial antagonists by suitable methods for growing them in synthetic media or specific media or by enrichment in appropriate media has been described in detail elsewhere (Volume 1). Frequently several species or strains may be found to be effective among hundreds of microorganisms isolated from large number of soils. The species or strains that prove to be effective in in vitro tests such as dual plate or confrontation tests have to be precisely identified by classical taxonomic methods or immunological or nucleic acid-based techniques. Screening for antagonistic activity of the isolates against the target pathogen(s) is a critical step, since the type of microorganisms selected depends on the type of method(s) used. The dual culture method can select the isolates whose mechanisms of action are based on antibiosis or toxicogenesis towards the target, but not parasitism or induction of defense responses in plants. Screening methods for parasitism include burying and retrieving propagules of pathogens to isolate antagonists. Methods for determining competition for nutrients and/or space include looking for microbes that rapidly colonize sterilized soil and have the capacity to colonize infection court. Live plants have to be used for assessing the potential of the isolate to induce resistance to the target pathogen(s). Determination of principal or additional modes of action of the candidate microorganism is important to develop proper method of developing formulations.

The importance of determining the mode(s) of action of the potential biocontrol agent (BCA) may be illustrated by the example of *Trichoderma virens*. This BCA was thought to act on the target pathogens, most likely via antibiosis, by producing the antibiotic gliotoxin which showed strong activity on a wide range of plant pathogens (Howell 1998). However, later studies carried out using number of mutants that were deficient in their ability to be mycoparasitic and/or to produce gliotoxin were able to suppress the development of fungal pathogen *Rhizoctonia solani*. No relationship could be established between the ability of the strains to produce the antibiotic or to parasitize the target pathogen with the suppression of the pathogen. In contrast, an extremely high correlation was observed between the abilities of the different strains to induce terpenoid phytoalexins involved in development of resistance in cotton seedlings against *R. solani* (Howell et al. 2000; Howell 2006). In another pathosystem involving *Pythium ultimum* that causes damping-off diseases in many plant species including *Arabidopsis*, the earlier assumption of mycoparasitism being the mechanism of biocontrol activity of *Trichoderma harzianum* strain T22 was proved to be incorrect by a later investigation. The biocontrol activity of strain T22 was lost in five mutants of *Arabidopsis* with disruptions in *NPR1* gene which is essential for resistance to be induced. The deletion had no effect on mycoparasitic events. Hence, the biocontrol activity of strain T22 was due to induced resistance than to direct effects on the pathogen by strain T22 (Shoresh et al. 2010).

Characterization of effective species or strains is of great importance to establish its genetic stability and to precisely identify the most effective strains in the species showing wide genetic diversity as in the case of *Trichoderma* spp. In the process of development of modern taxonomy for *Trichoderma* and development of broader species concept, strains with specific attributes were identified in specific species. The genus *Trichoderma* was earlier divided into nine species groups based on growth, colony and spore characteristics (Rifai 1969). As tremendous diversity could be recognized within a species of *Trichoderma*, the suggested species groupings were not meaningful in terms of ecological or physiological properties. Now the taxonomy is based on genetic sequence data. As a consequence, more than 100 species and taxa have been differentiated based on their abilities and characteristics. This convenient system of classification involves the use of standard primer sequences. Once the amplification of a specific segment is performed, the sequence is entered into a *Trichoderma* barcode database and the presumed species identification returns almost instantly (International Subcommision on Trichoderma and Hypocrea Taxonomy, ISTHT). Furthermore, GenBank contains all available data on gene sequences that have not been incorporated into the barcode database. The pooling of data has resulted in species definitions that are distinct and unambiguous. Based on the characteristics included in the modern taxonomy, it is possible to separate species that are genetically different from species that possess biocontrol potential (Harman et al. 2010). Several strains of *Trichoderma* have been formulated and are available commercially. In order to detect and quantify two strains IMI206039 (*Hypocrea parapilulifera*) and IMI 206040 (*Trichoderma atroviride*), primers for sequence characterized amplified region (SCAR) were developed. In addition, fluorophore-labeled probes were also used for simultaneous detection and differentiation of these two strains when they occurred together, in combination with SCAR primers. When two different soils were incubated, after artificial inoculation and maintained under controlled conditions, the quantification through amplification with SCAR in quantitative polymerase chain reaction (qPCR) and through colony-forming units (CFUs) from plate counting correlated well (Feng et al. 2011).

The need to have easy and inexpensive screening methods whose test conditions simulate as much as possible, the real system where the putative control agent is to be applied, has been recognized. The knowledge of the mechanism of action of the selected BCA in combination with an analysis of the sequence of the corresponding genes can provide gene targets to develop high-throughput screening procedures. Modern nucleic acid-based techniques using phenotypic or genotypic markers have been shown to improve the productivity of the screening stage. However, an association between some markers and the biocontrol potential of the putative strain has been demonstrated only for a few cases (Mathre et al. 1999; Montesinos 2003). The pathogenic fungus *Sclerotinia minor* strain IMI 344141 was developed as a bioherbicide for broadleaf weed control in turfgrass. In order to differentiate this strain from similar fungal species/strains, a strain-specific molecular marker was employed to detect and monitor the strain IMI344141. This procedure was based on polymerase chain reaction (PCR) amplification of two sequence-characterized amplified regions (SCAR) primer pairs for a first round PCR and another two sets of nested primers were employed for the second round PCR, if higher level of sensitivity was needed. *S. minor* IMI344141 was efficiently tracked both in pure cultures and environmental samples originating from bioherbicide-released field trials. DNA of the strain IMI344141 could be detected in the soil at 2 months after application, but not in samples taken at 3- and 9 months after application. When applied as a bioherbicide strain IMI344141 did not persist into the following spring in turf environments. The molecular markers developed in this study could be reliably used as a tool to monitor the presence and behavior of *S. minor* in nature, particularly in soil samples (Pan et al. 2010).

In addition to efficacy and consistency of results between bioassays, other desirable attributes of the putative BCA such as effectiveness at low doses, production of antimetabolites or toxins against the target pathogen, tolerance to fungicides or other plant protection chemicals, plant activators and adaptability to general crop management practices are also considered, while selecting the BCA isolates or strains for formulation of products. Non-pathognic *Fusarium oxysporum* strain CS-20 exhibited significant antagonistic activity against *F. oxysporum* f.sp. *lycopersici* and reduced the incidence of wilt disease in tomato. The compatibility of strain CS-20 with seven fungicides recommended for tomato in Maryland, USA was investigated. The fungicides reduced the radial growth of the strain CS-20 in agar medium amended with different concentrations of respective fungicides to varying degrees. Azoxystrobin and chlorothalonil were most toxic to strain CS-20. Mefenoxam (Ridomil Gold) and mefenoxam + copper did not affect the growth of the BCA strain. In the greenhouse experiments, tomatoes were drenched with strain CS-20 at seeding and just before transplanting into the field soil infested with the pathogen. Plants were treated with fungicides at the highest label rate. The results indicated that strain CS-20 was not compatible with the fungicides tested, since disease incidence was not reduced significantly by the combination of the BCA with the fungicides (Fravel et al. 2005). *Coniothyrium minitans*, a mycoparasite, has been shown to be an effective biocontrol agent capable of suppressing the development of *Sclertotinia sclerotiorum* causing Sclerotinia stem rot disease of oilseed rape. The possibility of combining soil application of *C. minitans* with fertilizer (N, P and K) application during cultivation of oilseed rape was explored. The compound fertilizer at concentrations of 0.1–10 % inhibited conidial germination and mycelial growth of *C. minitans* in a dosage-response manner in vitro. Simultaneous application of the BCA and the compound fertilizer at various concentrations significantly reduced the number of apothecia produced by sclerotia of *S. sclerotiorum* in both pot and field plot experiments. The compound fertilizer did not adversely affect the ability of *C. minitans* to infect sclerotia of the pathogen in vitro or to suppress carpogenic germination of sclerotia of *S. sclerotiorum*. The results indicated that *C. minitans* was compatible with compound fertilizer when applied at planting of oilseed rape. Thus application of the BCA along with compound fertilizer would save labor cost, resulting in greater monetary benefit to the farmer and enhancement of production efficiency (Yang et al. 2011).

As the effectiveness of biocontrol is generally strain-dependent subspecific characteristic, generally less than 1 % of the isolates satisfy the requirements for advancing them to the next stage. After assessing the biocontrol efficacy and mechanism(s) action of the isolates in the laboratory, growth chamber and greenhouse assays, the most effective ones are tested in pilot trials under conditions as close as possible to the natural field conditions under which the BCA is placed in practice. At this stage, it would be desirable to expose the BCA in an environment where several pathogens or pests will be available. This will enable to verify the spectrum of action and to determine the effectiveness of the BCA in different environmental conditions which can be diverse enough to guarantee a wide range of applicability. It would be possible to select a few isolates under these limiting conditions. Most of the BCA isolates are eliminated at this stage, because of their narrow spectrum of action and inconsistency in their performance in various field trials conducted during different seasons.

Selection of antagonistic microorganisms may be achieved by applying methods that are frequently used in the case of most plant pathogens. However, novel methods may be required for the selection of antagonists effective against postharvest diseases originating from latent infections by fungal pathogens. Biological control of latent infections may occur on fruits in nature and microorganisms involved, could be isolated from the fruit and used for controlling this type of infection after harvest. Direct interactions were observed between the pathogen structure involved in latent infection (appressorium) and the test microorganism can be studied in vitro and then tested in situ on fruit. Appressoria of *Monilinia fructicola* causing brown rot disease were produced on parafilm or wax membranes after depositing drops containing conidia in sucrose solutions (0.25 mM) containing 10 mM cAMP and incubating at 18 °C for 16 h, followed by addition of putative BCA strain and additional incubation at 24 °C for 72 h. Microorganisms colonizing appressoria and mycelium may be further tested for biocontrol activity on fruit with latent infection induced artificially under laboratory conditions. Several antagonists effective against brown rot originating from latent infections on stone fruits were selected by applying this approach. The antagonists that are best adapted to conditions occurring during storage and handling of fruits may be selected at the next stage and advanced for developing formulations (Janisiewicz et al. 2011).

Some of the endophytic plant symbionts have been demonstrated to be effective against microbial plant pathogens. These include plant growth-promoting rhizobacteria (PGPR), free-living rhizosphere competent fungi-like some strains of *Trichoderma* spp. and mycorrhizal fungi like *Piriformaspora indica* (Kloepper et al. 2004; Harman et al. 2008; Waller et al. 2008). The effectiveness of using molecular profile data of endophytic actinobacterial community structure to rapidly identify and select ecologically important BCAs from tomato root internals was validated. More than six genera *Streptomyces*, *Nesterenkonia*, *Arthrobacter*, *Microbacterium*, *Cellulomonas* and *Propionibacterium* were detected in four actinobacterial 16S rRNA gene clone libraries derived from tomato roots. Species similar to *S. virginiae* were detected by culture-independent and culture-dependent methods from all samples. The isolated strains Y30 and E36 similar to *S. virginiae* showed potent antagonism to *Ralstonia solanacearum* causing tomato bacterial wilt. These two strains showed significant protection to seedlings against bacterial wilt disease in greenhouse assays. The results indicated that sequence directed data could be useful for rapidly selecting biocontrol strains from diverse microbial communities (Tan et al. 2011). It is likely that these endophytic BCAs may have much longer periods of efficacy than nonendophytic organisms, because of their ability to grow along within the plants and in the environment in which plants develop. In addition, some endophytic fungal strains may colonize shoots, roots or stems have shown to increase resistance to water stress and salt or temperature tolerance. *Trichoderma harzianum* T22 enhanced expression of proteins involved in photosynthesis and starch accumulation (Harman and Shoresh 2007, 2008). *Trichoderma* strains may alleviate intrinsic stresses like loss of seed vigor and improve seed germination (Shoresh et al. 2010). The usefulness of endophytic plant symbionts has to be assessed and the effective strains have to be selected by applying suitable methods.

5.1.2 Preparation of Formulations

The strains with high level of antagonistic potential against microbial plant pathogens and acting through one or more mechanisms should be genetically stable, effective at low concentrations, easy to mass-produce in culture on inexpensive media and be effective against a wide range of microbial pathogens. Furthermore, the putative BCA should occur in an easily distributable form and it should be nontoxic or nonpathogenic to human beings and other plant species grown in the same ecosystem. Different methods have been applied to prepare formulations containing fungal and bacterial biocontrol agents. A formulated microbial product is composed of biomass of the selected biocontrol agent and ingredients to improve its survival and effectiveness of the product. Formulations of microbial mass may be of two types. Liquid formulations may be variously known as flowable or aqueous suspensions consisting of biomass suspensions in water, oils or emulsions. Dry formulation products include wettable powders, dusts or granules. Wettable powders and dusts contain dry inactive or active ingredients, while granules are free-flowing, aggregated product consisting of inactive and active ingredients (Schisler et al. 2004). The development of a shelf-stable formulated product that can retain biocontrol activity at a level which is similar to that of freshly isolated microorganism is the basic requirement for formulation and commercialization of biocontrol products. A biocontrol product for application against crop diseases should be easy to prepare and apply and produce abundant viable propagules with long shelf-life, sustained efficacy and economical. The self-life of a biocontrol product is the length of time that the propagules of the microorganism remain viable and exhibit antagonistic activity. It is expected that the product should have a shelf-life of at least

12–18 months without refrigeration to be commercially acceptable (Rhodes 1993). Various aspects such as development of suitable medium for mass multiplication, shelf-life, ability to grow and survive after application under different ecological conditions, effectiveness of disease suppression under field conditions and ease and cost of application have to be studied (Narayanasamy 2006).

5.1.2.1 Fungal Biocontrol Agents

The biocontrol agents selected for development of formulations to be advanced for commercial use should be produced at the industrial scale (fermentation), preserved, stored and formulated. Generally solid- or liquid-phase fermentation for mass scale production of the desired BCA is adopted. Many fungal BCAs are fermented in solid state, while bacteria and yeast are produced by liquid fermentation, using continuously stirred tank bioreactors (Fravel et al. 1999). Irrespective of the method used for fermentation, it should be possible to achieve highest yield of the BCA with the lowest cost of culture medium prepared from molasses, peptones or industrial grade protein hydrolysates. The BCAs are concentrated by filtration or centrifugation. When required level of BCA concentration is attained, the concentrated preparation of the BCA has to be stored under proper conditions that will increase its shelf-life which is one of the principal limiting factors affecting the prospects of commercialization. For this, the viability of the BCA has to be stabilized by refrigeration or by freezing in the presence of cryoprotectant substances, if the preparation is in liquid state or by keeping it as a dehydrated product. The methods based on dehydration may be preferable, since they allow optimum conditions for storage, handling, distribution and formulation of the BCA concerned. Though lyophilization is more effective in maintaining the viability of cells, its cost is quite high. Alternatively, encapsulation of microbial cells by mixing cells with a matrix-forming material like gelatinized polysaccharide or an emulsion in lipid material which can be diluted and spray-dried to obtain granules. At industrial level spray- or fluidized bed-drying is applied, but considerable loss of viability of cells occurs. Biocompatible additives are added to increase survival, improve application and stabilization of the final product. The additives consist of wetting and dispersal agents, nutrients and ultraviolet- or osmotic-protection agents that help microbial cells survive under field conditions (Andrews and Harris 2000; Montesinos 2003). Methods of formulating some of the improved fungal biocontrol agents effective against diseases occurring in the field and during storage after harvest are discussed below.

Formulations of Fungal Biocontrol Agents Against Field Crop Diseases

Trichoderma spp.

Trichoderma spp. are ubiquitous saprophytes and they are easily isolated from soil-decaying plant materials and organic matter. Various species of *Trichoderma* such as *T. viride*, *T. harzianum*, *T. asperellum*, *T. atroviride* and *T. stromaticum* have been reported to be effective as biocontrol agents against a wide range of microbial plant pathogens. The effectiveness of different strains within each species varies significantly and the efficient strains have been formulated for commercialization. *T. harizianum* strain T-22 was produced by protoplast fusion between strains T-95 and T-12. This strain formulated as granules named RootShield® and powder named PlantShield® by BiWorks, Geneva, NY was effective against gray mold diseases, Fusarium wilts, Pythium damping-off and Rhizoctonia root rots in many crops such as corn, soybean, potato, tomato, beans, cotton and peanut (groundnut). *T. harzianum* T-39 as TRICHODEX 20P was shown to be effective against pink rot and stem rot of tomato caused by *Phytophthora erythroseptica* and against gray mold diseases caused by *Botrytis cinerea* (Etebarian et al. 2000; Paulitz and Bélanger 2001). *Trichoderma virens* was formulated and marketed as SoilGaurd by Certis LLC (Harman et al. 2010). A method for mass multiplication and formulation for *T. viride* was developed at the Department of Plant Pathology, Tamil Nadu Agricultural University, Coimbatore, India. The BCA is multiplied in molasses-yeast medium in the fermentor for reaching a concentration of 10^8 conidia/ml and then mixed with talc powder (carrier) in the proportion of 1:2 (v/w). After air-drying, carboxymethylcellulose (CMC) is added as a sticker at 5 g/kg of BCA-talc mix. This preparation is dried to a moisture level of 20 % and packed in polyethylene bags (Narayanasamy 2002). *T. harzianum* T-1 was grown in the medium containing zeolite, molasses, brewer's yeast and wheat bran. Alginate was used to pelletize the zeolite-containing conidia of *T. harzianum* T-1. Zeolite and alginate were found to be cheap and harmless to the environment (Küçük and Kivanç 2005). As the liquid fermentation-based formulations are more vulnerable to desiccation compared with solid-state fermentation-based formulations, the effect of glycerol added as a osmoticant in the production medium, was assessed. The addition of glycerol at 3, 6 or 9 % reduced water activity in the medium. Both in shake culture and fermentor, the addition of glycerol in the production medium prolonged the shelf-life of talc formulations to 7–12 months as against 4–5 months without glycerol. In the bioeffcacy tests, formulations derived with addition of glycerol 3 or 6 % in the production medium could protect the tomato plants against Fusarium wilt disease effectively (44–50 %) even after storage for 12 months indicating the efficacy of glycerol as osmoticant (Sriram et al. 2011).

The effectiveness of an invert emulsion formulation of *Trichoderma harzianum* in suppressing the development of postharvest diseases caused by *Rhizopus stolonifer*, *Botrytis cinerea* and *Penicillium expansum* was evaluated based on the diameter of decay lesions induced by the pathogens and the period of protection provided. Treatment with *T. harzianum* conidia formulated in invert emulsion significantly reduced the mean lesion diameters induced by *R. stolonifer* on apple, pear, peach and strawberry, *B. cinerea* on grape, pear, strawberry and kiwifruit and *P. expansum* on grape, pear and kiwifruit, compared with control treatment. The longest mean duration of minimum protection period (up to 59 days) was achieved for unwounded apple fruit against infection by *R. stolonifer*. The results showed the potential of invert emulsion formulation of *T. harzianum* in protecting fruits from infection by the primary postharvest pathogens up to about 2 months with reduction of decay lesion diameters up to 86 % (Batta 2007).

A procedure for formulation of *T. harzianum* maintained in sawdust: soil: molasses (5 %) was developed. The carrier consisting of flyash:soil:molasses (5 %) was used and the material was stored at 25 °C. The shelf-life of the formulation was about 32 weeks. Seed treatment with the formulation at 5 g/kg seeds of chickpea and pigeonpea effectively reduced wilt diseases caused by *Fusarium oxysporum* f.sp. *ciceri* and *F. udum* respectively. The soil populations of these pathogens were greatly reduced. Treatment with *T. harzianum* also controlled root knot (*Meloidogyne incognita*) in these crops (Khan et al. 2011). Formulations of *Trichoderma viride*, *T. virens* and *T. harzianum* for seed treatment and soil application were evaluated for their efficacy against *Rhizoctonia bataticola*, causative agent of dry root rot disease of chickpea under greenhouse and field conditions, either singly or in combination. A combination of soil application of *T. harzianum* (PBP10G) and seed treatment with *T. harzianum* (Pusa 5SD) + carboxin was the most effective in suppressing the root rot disease development, in addition to improving the seed germination, plant growth and grain yield under field conditions (Dubey et al. 2011).

Trichoderma atroviride isolate C52 was selected as a biocontrol agent against *Sclerotium cepivorum* causing onion white rot disease, since it reduced the disease incidence by 51 and 91 % in two glasshouse-based trials, when introduced into the soil as an infested solid substrate formulation (Mc Lean and Stewart 2000). *T. atroviride* was formulated into a pellet formulation where conidia were coated onto the outside of a nutritive kernel and solid-substrate formulation where conidia were combined with grain-based components. The pellet formulation maintained the fungal conidial concentration at 10^5 CFU/g of soil, whereas solid-substrate and seed-coating formulation had a concentration of 10^4 and 10 CFU/g of soil respectively. When the isolate C52 was introduced into pathogen infested soil both as pellet and solid-substrate formulations, no statistically significant differences in disease suppression could be recorded between these treatments. On the other hand, the pellet formulation increased the percentage of the healthy plant by two times that was observed in control treatment (Mc Lean et al. 2005). *Trichoderma asperellum* strain GJS03-35 earlier considered as a strain of *T. viride* was shown to be effective against a wide spectrum of plant pathogens including *Fusarium graminearum* causing head blight disease of cereals. It was produced in submerged liquid medium for 40–60 h, concentrated and carboxymethyl cellulose (CMC) and magnesium salts added as stabilizers. This formulation remained effective for 3 months, when stored at temperatures ranging from 15 to 30 °C (Kolombet et al. 2001). In a later study, addition of starch as food base, reduction of metabolic activity by lowering the pH of the biomass paste and addition of small amounts of copper were effective in prolonging the shelf-life and enhancing the competitiveness of the formulation for about 1 year under standard conditions of storage (Kolombet et al. 2008).

Gliocladium spp.

Species of *Gliocladium* are common soilborne fungi and some of them possess biocontrol activity against fungal pathogens causing root diseases in various plant hosts. *G. catenulatum* parasitizes the pathogens *Sporidesmium sclerotiorum* and

Fusarium spp. (Punja and Utkhede 2004; Viterbo et al. 2007). *G. catenulatum* strain J1446 has been formulated as a wettable powder and marketed as Primastop® by Kemira Agro Oy, Finland. When applied as soil drench or foliar spray, it effectively reduced the incidence of damping-off disease caused by *Pythium ultimum* and root rot disease incited by *Rhizoctonia solani* under greenhouse and field conditions (Viterbo et al. 2007). *G. virens* isolate GL-21 formulated as an alginate prill and named as GlioGard® by W.R. Grace Co. and as a granular formulation sold as SoilGard® developed by the Thermo Triology Corp., Colombia reduced the seedling diseases of ornamental plants and vegetable crops (Punja and Utkhede 2004). As the investigations based on molecular techniques revealed the close relationship of *G. virens* to *Trichoderma* genus, it was suggested that this taxon might be referred to as *Trichoderma virens* (Paulitz and Bélanger 2001).

Chaetomium spp.

Chaetomium species are commonly found in soil and organic matter. They have been shown to be potential antagonists of soil- and seed-borne pathogens which are suppressed by different mechanisms such as competition for substrate and nutrition, mycoparasitism or antibiosis. *C. globosum* and *C. cupreum* could effectively suppress the development of pathogens causing root rot diseases of citrus, black pepper and strawberry and damping-off diseases of sugar beet (Soytong et al. 2001; Tomilova and Shternshis 2006). These fungal BCAs were formulated in the form of powder and pellets as Ketomium®, with broad spectrum of activity against several fungal pathogens. In addition, Ketomium® has also been registered as a biofertilizer for degrading organic matter and for inducing resistance in plants and stimulating plant growth (Soytong et al. 2001). Ketomium® was highly efficient in suppressing the development of raspberry spur blight disease caused by *Didymella applanata* and it was able to reduce the incidence of *Rhizoctonia solani* infecting potatoes and enhance the potato yield (Shternshis et al. 2005). The antagonistic activity of the BCA in the formulated product was retained at high levels even after storage for 2 years (Tomilova and Shternshis 2006).

Penicillium oxalicum

Penicillium oxalicum protects by inducing resistance in the tomato plants against *Fusarium oxysporum* f.sp. *lycopersici*, causing wilt disease. Induced resistance in tomato plants by *P. oxalicum* was related to renewed or prolonged cambial activity that led to the formation of additional secondary xylem in *P. oxalicum*-treated plants (De Cal et al. 1997, 2000). Quantum of conidia of *P. oxalicum* produced in liquid and solid fermentation was assessed. *P. oxalicum* produced 250-fold more conidia in solid than in liquid fermentation at 30 days after inoculation of the substrate. Solid fermentation was carried out in plastic bags containing 50 g of peat/vermiculite (PV, 1:1 w/w) with 40 % moisture, sealed, sterilized and then inoculated with 1 ml of a conidial suspension of *P. oxalicum* (10^5 conidia/g dry substrate), sealed again and incubated in darkness at 20–25 °C for 30 days. Amendment of PV with a meal of cereal grains (barley) or leguminous seeds (lentil) enhanced the conidial production to the maximum extent (100-fold higher) among the various amendments tested. Conidial production at 5 days after inoculation was equal to that obtained after 30 days. Optimal initial moisture in the substrate was 30–40 %. This solid-state fermentation process was found to be effective in producing large quantities of conidia of *P. oxalicum* (Larena et al. 2002). Additives of different kind were evaluated for their effect on dispersal of dried conidia of *Penicillium oxalicum* in water to improve the effectiveness of BCA application. Sugars, polyalcohols, inorganic salts and detergents were added at three points of the production-formulation process. Conidial dispersal in water improved significantly, when 1.5 % sodium alginate, 60 % sucrose, 60 % D-sorbitol, 60 % fructose, 5–20 % polyethylene glycol (PEG) 8000, or 20 % glycerol were added to conidia before drying. Dispersal of conidia was enhanced with 1 % Tween20, 1 % Tween80, 1 % TritonX-100, or 1.5 % sunflower oil. Two *P. oxalicum* formulations (with conidial suspensions being maintained with 60 % sucrose or 1.5 % sodium alginate for 10 min before drying) significantly reduced tomato Fusarium wilt disease under greenhouse conditions (Sabuquillo et al. 2005). In later studies, eight different formulations of *P. oxalicum* obtained by addition of various ingredients were evaluated for their efficacy in controlling Fusarium wilt disease of tomato. *P. oxalicum* conidial formulations could be prepared by non-vacuum or vacuum packaging that included 1.5 % alginate, 20 % glycerol, 5 % sucrose and 5 % sorbitol with <15 % moisture content. The biocontrol efficacy of the formulations was high, when wilt disease pressure was moderate under field conditions (Sabuquillo et al. 2006, 2010).

Epicoccum nigrum

Epicoccum nigrum isolate 282, a component of resident mycoflora of peach twigs and flowers effectively reduced the peach twig blight and fruit rot caused by *Monilinia* sp. in the orchards in Spain. The shelf-life of conidia of *E. nigrum* was enhanced by drying them with fluidized bed dryer twice for 20 min at the highest air flow rate and at a temperature range of 30–40 °C conidia dried by this procedure maintained 100 % viability after 90 days of storage at room temperature (Larena et al. 2003). In a later study, wettable powder formulations of *E. nigrum* conidia with a shelf-life of 12 months were developed. The moisture content of conidia after fluidized bed-drying was critical for conidial viability during storage time. Conidial samples were dried for 40 min at 40 °C in a fluidized bed-dryer to obtain moisture contents <15 %. Nontoxic stabilizers and desiccants were added at different stages of production-drying process. Shelf-life of formulations retaining the highest viability were conidia produced with 1 % KCl or 50 % polyethylene glycol (PEG) 8000, conidia dried with 2.5 % methylcellulose and conidia dried with 1 % KCl + silica powder. These formulations with greater shelf-life of conidia were able to reduce the brown rot of peaches significantly (Larena et al. 2007). In order to improve the adhesion of the formulation of *E. nigrum* to peach surfaces and to enhance its bioefficacy, stickers, desiccants and commercial adhesive (NU FILM17®) were evaluated. Adhesion and viability of adherent *E. nigrum* conidia to peach surfaces were increased, when either 1.25 % sodium alginate (sticker) or methylcellulose was added to the conidial masses after fluidized-bed drying, compared to control without additives. *E. nigrum* conidial formulations with 2.5 % methylcellulose were more effective than dried *E. nigrum* conidia without additives in reducing the infection by *Monilinia laxa* causing brown rot of peaches. When 2.5 % methylcellulose was incorporated into *E. nigrum* conidial formulation, the adhesion of BCA conidia to peach surfaces significantly increased resulting in effective suppression of brown rot disease development (Larena et al. 2010).

Coniothyrium minitans

Coniothyrium minitans is a mycoparasite of different species of sclerotial pathogens such as *Sclerotinia minor*, *S. sclerotiorum* and *Sclerotium cepivorum* infecting lettuce, oilseed rape, peanut and onion. The principal mechanism of biocontrol activity of *C. minitans* is through mycoparasitism. The BCA uses the sclerotia of *S. sclerotiorum* as the source of food for survival. The medium containing ground sclerotia of *S. sclerotiorum* was found to be a good substrate for mass production of conidia of *C. minitans* (Huang and Erickson 2007). The product formulated by Prophyta Biologischer Pflanzengchutz GmbH, Germany contained conidia of *C. minitans* and registered for use in Germany, Switzerland, Norway and USA (Partridge et al. 2006; Yang et al. 2007). The product could be applied to soil or as foliar spray. The BCA could survive in the soil for several years (Li et al. 2006; Whipps et al. 2008).

Ampelomyces

Parasitism of grapevine powdery mildew pathogen *Uncinula necator* by *Ampelomyces quisqualis* was reported by Daoust and Hofstein (1996). *A. quisqualis* isolate M-10 was formulated as AQ10 Biofungicide by Ecogen Inc., USA. The formulated product is in water-dispersible granules containing conidia of *A. quisqualis*. It was found to be effective against powdery mildew diseases of different crops like grapevine, carrot, cucumber and mango (Shishkoff and McGrath 2002; Kiss 2003).

Pythium oligandrum

Pythium oligandrum showed significant biocontrol potential against sugar beet damping-off disease caused by *Pythium ultimum*, when the seeds were treated with oospores of *P. oligandrum* (Lewis et al. 1989). The fungal BCA was formulated as a granular or powder product designated Polygangron®. *P. oligandrum* was shown to act on the pathogen by indirectly enhancing the resistance of beet root seedling to infection by the pathogen. In addition, the phosphorus uptake by plants, emerging from *P. oligandrum*-treated seeds, was also enhanced (Le Floch et al. 2003).

Clonostachys rosea

Clonostachys rosea has been reported to be effective against seed- and soil-borne diseases. Application of *C. rosea* as a single strain or mixture of strains reduced the infection of cocoa crop by *Moniliophthora roreri*, causing moniliasis disease.

Motorized mist blowers and directional hydraulic spray techniques were used for applying the formulation. The mycofungicide preparation containing *C. rosea* reduced sporulation of the pathogen resulting in restriction of pathogen spread. Overall, motorized use of mist blower was less effective than the directional hydraulic spray method (Hidalgo et al. 2003). *C. rosea* strain IK726 was more efficient in suppressing disease development (Jensen et al. 2007).

Microsphaeropsis ochracea

When the biocontrol agent *Microsphaeropsis ochracea* was applied for the control of apple scab disease caused by *Venturia inaequalis* under field conditions as an unformulated spore suspension with air blast spray using low water volumes, it was less effective compared to a ground cover spray using high water volumes. In order to improve the colonization of apple leaves by *M. ochracea*, the effect of spore suspension (10^9–10^{12} conidia/ha), water volume (250–2,500 l/ha) and 20 adjuvants were evaluated. Higher conidial concentrations facilitated greater colonization of apple leaves. Water volume was not a critical factor affecting the BCA colonization. The adjuvant glycerin improved the spore germination, while mycelial growth was stimulated by Agrimer, Ekol and Tween80 in some trials. When tested on apple trees, the adjuvant Assist provided the most consistent improvement in colonization in many trials. The results indicated that use of oils, humectants and surfactants could broaden the possibility of using this biocontrol agent for the management of apple scab disease (Bailey et al. 2007).

Formulations of Fungal Biocontrol Agents Against Postharvest Diseases

Candida spp.

A white yeast *Candida oleophila* was shown to be effective against the blue and green mold diseases affecting citrus fruit and blue and gray mold diseases affecting apples. A formulation containing *C. oleophila* was formulated and marketed as Aspire by Ecogen Inc. (Droby et al. 1998). Methods of preparation of formulation of *C. oleophila* (strain O) effective against *Botrytis cinerea* and *Penicillium expansum* were standardized by Jijakli (2000). The viability, storage stability and biocontrol potential of freeze-dried *Candida sake* strain CPA-1 against *P. expansum* infecting Golden Delicious apples were assessed. Survival of *C. sake* was the highest (85 %), when the yeast cells were protected with lactose (10 %) and rehydrated with skimmed milk. However, the yeast cells freeze-dried with lactose (10 %) + skimmed milk (10 %) as protectant and peptone (1 %) as rehydration medium reduced the incidence of blue mold to the maximum extent. In general, freeze-drying reduced the biocontrol potential significantly compared to fresh yeast cells (Abadias et al. 2001). Suspensions of eight isolates were mixed with cellulose and dried. The product was then milled to a fine powder. The efficacy of this yeast-cellulose formulation, applied as a dry powder to sporulating colonies of *Botrytis cinerea* infecting kiwifruit was assessed. This formulation significantly suppressed liberation of conidia by binding directly to the pathogen spores. By using α-cellulose prepared with *Candida pulcherrima*, a reduction of about 50 % in the number of conidia released from the lesions induced by *B. cinerea* could be achieved. Such suppression of liberation by a yeast formulation may provide an effective management tool for biological management of sporulating postharvest fungal pathogens (Cook 2002).

The survival of biocontrol agents could be improved by osmotic adaption and thermotolerance of the fungal cells. When osmotically adapted cells of *Candida sake* CPA-1 were applied to apples in the field, the BCA was able to survive for longer periods. Osmotically adapted cells were obtained by modifying the water activity of the growth medium with organic solutes to impose a mild osmotic stress. Water activity is the ratio between the vapor pressure of water in a substrate (P) and vapor pressure of pure water (Po) at the same temperature and pressure. Osmotically adapted cells were significantly better adapted to survival under field conditions than the control cells without adaptation (Teixidó et al. 1998). Increasing the thermotolerance of BCAs may also improve their survival in sub-optimal conditions, when applied in natural conditions. Cells of *C. sake* CPA-1 exposed to 30 and 33 °C at different fermentation phases became thermotolerant under in vitro conditions. Furthermore, the survival of cells heat-adapted at 33 °C during the early or mid-stationary phases of liquid fermentation was significantly higher compared with non-adapted cells after spray drying (Cañamás et al. 2008b). In a later investigation, liquid formulation of *C. sake* was improved by growing the yeast cells in molasses medium with a_w modified to 0.98 with the addition of sorbitol and preserved with an isotonic trehalose solution. The viability of this formulate was 100 %, after storage for 180 days at 4 °C. The efficacy against *Penicillium expansum* on apples was more than 95 % rot reduction. *C. sake* cells grown on unmodified molasses medium exposed to mild heat treatment at 30 or 33 °C during mid or late exponential or early or mid-stationary growth phases showed an increase of survival, when they were exposed to lethal shock at 40 °C, but only marginal improvement occurred after spray drying the formulation (Usall et al. 2010). These studies suggested that *C. sake* CPA-1 might be physiologically adapted for the environmental stresses on the phylloplane.

The effect of modifying water activity (a_w) and the addition of protective substances to the preservation medium of liquid formulations of *Candida sake* stored at 4 and 20 °C, was evaluated. The a_w of the preservation medium was altered to have a range from 0.72 to 0.95 by adding glycerol or polyethyleneglycol (PEG). The viability of yeast cells was improved by a_w levels of 0.93–0.95, but not to the desired level. When sugars such as trehalose and polyols such as glycerol or PEG were used as protectants, cells of *C. sake* maintained viabilities of greater than 60 %. Storage at 4 °C was preferable to 20 °C, because of maintenance of greater viability of yeast cells at the lower temperature. The biocontrol efficiency of liquid formulations of *C. sake* in controlling *Penicillium expansum* infection in wounds on apples was similar to fresh yeast cells. Use of 10 % sucrose as protectant in the liquid formulation of *C. sake* would be an acceptable proposition, as the cost of the formulated product would not be a limiting factor for industrial production (Torres et al. 2003). In another study, preservation of *C. sake* CPA-1 strain cells in trehalose solution

(0.96 M) which was isotonic with yeast cells yielded best results. After storage for 7 months at 4 °C, the yeast cells were grown in the sorbitol modified medium and preserved with isotonic solution of trehalose (0.96 M) maintained the viability of yeast cells at high levels. The formulated product effectively controlled the blue mold disease of apples (Abadias et al. 2003).

The effectiveness of three formulation strategies was evaluated for successful establishment, survival, persistence and efficacy of *Candida sake* CPA-1 applied to grapevines for the control of *Botrytis cinerea* causing gray mold disease. The formulation strategies were (i) liquid formulated cells versus fresh cells; (ii) induced thermotolerance in the BCA cells using heat treatments during mass production process and (iii) mixture of Fungicover® (FC), an edible coating. Different formulations of *C. sake* cells were applied at flowering, pea sized berries, veraison and before harvest. Spray application of different formulations of *C. sake* resulted in colonization of bunches under field conditions. When *C. sake* combined with FC (additive) had significantly higher survival rates (up to 50 % higher) compared with *C. sake* not coated with FC. Formulation of *C. sake* cells in an isotonic solution combined with FC resulted in effective control of *B. cinerea* at a level greater than that could be achieved with conventional fungicide program, indicating the potential of *C. sake* CPA-1. Fungicover® showed beneficial effect on the BCA, by improving the persistence of *C. sake* cells on the host plant. The efficacy of the combination of *C. sake* and FC under typical hot, dry summer climatic conditions prevailing in Catalonia (Spain) was distinctively high. The results indicated that it might be possible to widen the spectrum of effectiveness of BCAs by selecting appropriate formulation strategies (Cañamás et al. 2011).

Pichia spp.

Strains of *Pichia anomala* was reported to inhibit the development of blue and gray molds caused by *Penicillium expansum* and *Botrytis cinerea* in apple and grapevine (Jijakli and Lepoivre 1998; Masih et al. 2000). A system to assess the survival of *P. anomala* strain J121 after liquid formulation and long-term storage in different storage buffers was assessed. The cells were stored in vials containing an atmosphere with water to prevent a continuous drying process of samples. The conventional method of estimating the amount of viable cells based on CFU/ml, is time-consuming and cumbersome. Hence, an assay using flow cytometry was developed to discriminate viable and dead cells of *P. anomala* in samples. Oxonol was used to stain dead cells and unstained viable cells were determined. The CFU and flow cytometry analyses provided similar results. Flow cytometry was found to be superior to the conventional method. The survival of *P. anomala* in liquid formulation with lactose, starch and trehalose amendments was determined during prolonged storage at temperatures ranging from −20 to +30 °C. After 4 weeks incubation at 4 and 10 °C, 75–90 % of the BCA cells were viable and no significant differences between the formulations were evident. Trehalose was the most effective protectant at 20 and 30 °C. Liquid formulations have an advantage over dry formulations, since there is need to rehydrate prior to usage and rehydration may become difficult, when large quantities of the dry formulations have to be handled (Melin et al. 2006). *Pichia membranefaciens* effectively suppressed the development of Rhizopus rot disease of nectarines caused by *Rhizopus stolonifer*. The effectiveness of biocontrol activity was enhanced by calcium chloride (2 %) (Tian et al. 2002). Survival of *P. membranefaciens* in liquid formulation with sugar protectants (trehalose and galactose) and L-ascorbic acid was assessed during storage at 4 and 25 °C. The BCA maintained viability >60 % after 90 days at 4 °C, when 5 % galactose served as a protectant and its combination with L-ascorbic acid was the most effective in maintaining viability of BCA cells. The liquid formulation was more effective when stored at 4 °C than at 25 °C. In comparison with another yeast BCA, trehalose was considered to be a suitable protectant for liquid formulation of *Cryptococcus laurentii*. On the other hand, galactose was found to be better as a protectant for *P. membranefaciens*. Combining L-ascorbic acid with sugars improved the protective efficiency of both yeast BCAs (Liu et al. 2009).

Cryptococcus spp.

Cryptococcus laurentii in combination with *C. infirmo-minitans* could protect pear fruits against the blue mold pathogen *Penicillium expansum* more effectively than the fungicide thiabendazole (TBZ) at high dose (528 μg/ml) (Goyal and Spotts 1997). When galactose or trehalose was used alone as protectant, *C. laurentii* maintained relatively high viability in potassium phosphate buffer. Addition of L-ascorbic acid to trehalose improved the protective effect of trehalose. The biocontrol efficacy was at high level after formulation and storage (Liu et al. 2009). *Cryptococcus nodaensis* isolate OH182.9 was shown to effectively reduce the Fusarium head blight (FHB) disease of wheat. Frozen biomass of the isolate OH182.9 reduced FHB in 15 field experiments in the United States (Milus et al. 2001). Freeze-drying of fungal biomass has advantages such as protection from contamination or infestation during storage, long viability and ease of distribution. However, dehydration of microbial biomass may adversely affect its viability and efficacy in the process of formulation and further development for commercialization. Isolate OH182.9 was grown for 48 and 72 h in a semi-defined complete liquid (SDCL) medium with different carbon-to-nitrogen (CN) ratios. In general, cells harvested at 48 h survived freeze-drying better than those harvested at 72 h. Survival of freeze-dried cells was the greatest for cells grown for 48 h in C/N ratio 30:1 medium. Cells grown in SDCL C/N 30:1 media with 7 and 14 g/l carbon loading significantly reduced FHB disease severity. Cells harvested from SDCL C/N 9:1, 11:1 and 30:1 with 14 g/l carbon increased the100-kernel weight compared with control. The results indicated the potential of improving the product quality of the yeast BCA by managing the nutritional environment of production media without compromising the biocontrol efficacy (Zhang et al. 2005). In a later investigation, a procedure was developed for the production of dried granules of *Cryptcoccus flavescens* (earlier known as *C. nodaensis* OH182.9). Small spherical granules were dried using a fluidized-bed dryer at 30 °C. The granules dried to different moisture levels were evaluated for storage stability at 4 °C for up to 1 year. Storage stability varied significantly across the range of moisture contents (4–12 %). Moisture content of 9 % showed the best short-term stability (up to 4 months), while 4 % moisture content was more efficient for the long-term survival (1 year) (Dunlap and Schisler 2010).

Rhodotorula spp.

Rhodotorula spp. have been reported to be effective biocontrol agents against post-harvest pathogens. *R. glutinis* significantly reduced the infection of apples and pears by *Penicillium expansum* and of strawberry by *Botrytis cinerea* (Benbow and Sugar 1999; Helbig 2001). *R. minuta* efficiently suppressed the development of mango anthracnose disease caused by *Colletotrichum gloeosporioides* in a preliminary field experiment. A submerged fermentation production process for *R. minuta* was developed and scaled up to a 100-l pilot fermentor. A low cost mineral enriched medium (MEM) was used for mass multiplication of *R. minuta*. The higher viable cell count obtained at pilot scale was linked to a better oxygenation of the bench and pilot bioreactors as compared with the shake flasks. The antagonistic potential of *R. minuta* was not altered, because of the scale-up of the production process, formulation and shipping. The BCA was applied onto the mango trees starting at blossoming. As the formulation was applied at monthly intervals, the yeast BCA had enough time to colonize the phylloplane, competing with the pathogen, resulting in significant reduction in the infection of fruits. Formulating the BCA at 10^9 CFU/ml and adding glycerol (20 %) or xanthan gum (5 g/l) avoided both contamination and yeast cell sedimentation and preserved the biocontrol efficacy for 6 months at 4 °C. Field tests were conducted in commercial orchards during three harvest seasons. Yeast suspension applied to mango trees reduced the fruit anthracnose severity to levels equivalent to or better than fungicide benomyl (Patiño-Vera et al. 2005).

Metschnikowia spp.

The yeast antagonistic species of *Metschnikowia* have been demonstrated to have significant biocontrol potential that could be exploited for the management of post-harvest diseases of fruits. *M. fructicola* was evaluated for the control of *Botrytis cinerea* infecting strawberry at pre- and post-harvest stages and *Rhizopus stolonifer* causing storage rot. Application of *M. fructicola* against preharvest rots was equally effective as the fungicide fenhexamid in two growing seasons in greenhouse trials. In the field experiment the disease incidence was reduced to commercially acceptable levels. Field application of *M. fructicola* reduced fruit rot during postharvest storage by 64–72 %. The population levels of the yeast *M. fructicola* on fruits treated at weekly intervals, was about 1×10^5 CFU/fruit (Karabulut et al. 2004). A strain B10126 of *Metschnikowia pulcherrima* was selected for development and commercialization. Different complex nutrient sources with or without pH control were evaluated to optimize biomass production. Addition of yeast extract, two carbon sources d-mannitol and l-sorbose at 5 g/l each significantly increased the BCA growth reaching highest biomass. The pH range of 5.0–7.5 in combination with yeast extract, d-mannitol and l-sorbose (YEMS) exhibited synergistic effect on biomass production. In efficacy trials, YEMS reduced the incidence and severity of *Botrytis cinerea* (gray mold and *Penicillium expansum* (blue mold) infection on Gala and Golden Delicious apples (Spadaro et al. 2010a). The effect of freeze-drying using different lypoprotectants at different concentrations on the viability and biocontrol efficacy of *M. pulcherrima* was assessed. Yeast cells grown for 36 h with maltose (25 %) as protectant maintained high viability for 6 months at 40 °C after freeze-drying. The efficacy of freeze-dried cells was similar to that of fresh cells on Gala apples, but was slightly lower on Golden Delicious apples. The optimization of freeze-drying conditions may pave way for commercial development (Spadaro et al. 2010b).

5.1.2.2 Bacterial Biocontrol Agents

Among the bacterial biocontrol agents, plant growth-promoting rhizobacteria (PGPR), *Pseudomonas fluorescens*, *P. putida*, *P. aeruginosa*, *Bacillus subtilis* and other *Bacillus* spp. form an important group of bacteria with significant biocontrol potential. Many strains of these bacterial BCAs have been formulated and commercialized for large scale application under different agroecological conditions. Bioformulations containing fluorescent pseudomonad strains were found to be inconsistent in their effectiveness and failed due to lack of long-term viability, because of their inability to produce endospores. In the recent years, greater attention is bestowed for developing commercial products based on *Bacillus* spp. which produce endospores capable of resisting adverse conditions like temperature, various pH and drought stress. However, strains of fluorescent *Pseudomonas* spp. are being selected currently for their potential to induce systemic resistance and/or promote plant growth resulting in high yield levels in treated plants. Other species of bacteria with significant biocontrol potential belong to *Agrobacterium*, *Burkholderia*, *Erwinia*, *Serratia* and *Streptomyces*. Different steps in the formulation and commercialization of bioproducts containing various strains/species of bacteria are discussed.

Formulations of Bacterial Biocontrol Agents Effective Against Field Crop Diseases

Pseudomonas spp.

Strains of *Pseudomonas fluorescens* have been reported to effectively suppress the development of a wide range of microbial plant pathogens infecting various crops. It is essential to precisely identify the most effective strains either by isolation-dependent or isolation-independent methods. Cell viability during different stages of formulation and maintenance of required population levels at different periods after storage and application under field conditions where the BCA concerned is expected to act on the pathogen(s) are important factors. Isolation-dependent methods require long time and results obtained may be inconsistent and inconclusive. Hence, isolation-independent methods especially polymerase chain reaction-based techniques are preferred because of their higher level of sensitivity, rapidity, reliability and reproducibility of results. A simple protocol that uses PCR to amplify a specific portion of the 16S gene, allowing recognition of *Pseudomonas fluorescens* from other group I *Pseudomonas* was developed. The amplified DNA patterns of 16S rRNA and ITS1, from the restriction fragment length polymorphisms (RFLPs) after digestion with endonucleases *Vsp*I, *Hae*III and *Taq*I, produced band patterns characteristic of biotypes of *P. fluorescens*. Comparison of RFLP profiles and culture test for levan production formed a rapid and less expensive approach to routinely identify three biotypes A, B, C and 3 of *P. fluorescens* (Scarpellini et al. 2004). The isolates of *P. fluorescens* obtained from rhizosphere of cotton plants were subjected to dual culture test for selecting the effective ones for the suppression of development of cotton bacterial blight pathogen *Xanthomonas axonopodis* pv. *malvacearum*. The efficient isolates Pf32 and Pf93 were identified based on the sequences of 16S-23S intergenic ITS region after amplification by PCR (Salaheddin et al. 2010).

Different kinds of carriers such as talc, peat, lignite, vermiculite, kaolinite and farm yard manure (FYM) have been used for formulating *Pseudomonas fluorescens* strains P7NF and Pf1 (Caesar and Burr 1991; Vidhyasekaran and Muthamilan 1995). Although peat soil has been used as a carrier to formulate *Pseudomonas* spp., it did not find acceptability by researchers, because of variability in its quality and presence of contaminants (Bashan 1998). Talc-based powder and bentonite-based powder as mineral carriers were used for formulating *Pseudomonas fluorescens* for suppressing the development of cotton damping-off disease. The organic and inorganic carriers effectively improved the stability of the formulations and effectiveness of the BCA strain in controlling the cotton damping-off disease (Ardakani et al. 2010). The suitability of talc-based formulation of *P. fluorescens* for the control of cotton bacterial blight disease was reported by Salaheddin et al. (2010). *Pseudomonas fluorescens* strain S11:P:12 was effective in suppressing potato diseases and sprouting potato tubers prematurely. During liquid cultivation, the starch produced a polysaccharide marginalan. The bioactivity of marginalan was assessed, when applied alone or in combination with washed cells, for the suppression of dry rot and pink rot diseases and inhibition of sprouting. The impact of marginalan on cell survival during drying and storage was also determined, since the formulation and storage of dried biocontrol product is preferred for commercial use. Washed bacterial strain formulated with 0–6.6 g/l polysaccharide were either applied to Hyflo granules, then slowly dried for 24 h with airflow at 50–60 % relative humidity, or in 1 μl droplets placed in replicate wells of a microplate, then rapidly dried for 1 h in a biohazard hood. Both Hyflo and microplate dry storage results indicated that marginalan significantly reduced cell death after drying. The final stable cell density was increased by 2.5–5.0 orders respectively, when compared to control without marginalan. The polysaccharide did not exhibit significant impact on disease or sprout inhibition by the strain S11:P:12. Preservation of cell viability during drying and storage was the principal benefit of marginalan. Similar beneficial effect was observed, following formulation of selected *P. fluorescens* strains with marginalan (Slininger et al. 2010).

Pseudomonas putida isolate 06909 was effective in reducing rhizosphere populations of *Phytophthora citrophthora* causing root rot disease in citrus seedling bed trials. This isolate actively colonized the hyphae of *Phytophthora* spp., restricted the growth of the pathogen and produced an iron chelating siderophore pyoverdine. *P. putida* 06090-rif/nal in the form of lyophilized inoculum in 10 % skimmed milk was applied to the soil through irrigation water. A field fermentor was used to produce the inoculum of *P. putida* 06906 rif/nal for application in commercial orchards. Rhizosphere populations of *Phytophthora parasitica* were significantly reduced in 1 out of 3 years of experimentation (Steddom et al. 2002). Talc-based formulation of *P. putida* retained the viability up to 6 months with the population remaining at 10^8 CFU/g of the product (Bora et al. 2004). A formulation of *Pseudomonas boreopolis* combining with *Brassica* seed pomace, glycerin and sodium alginate (PBGG) was developed for controlling *Rhizoctonia solani* AG-4, causing damping-off disease of cabbage. Application of PBGG (1 %) to the soil infested with the pathogen reduced the percentage of colonization of cabbage seeds by *R. solani* and also stimulated proliferation of *Streptomyces padanus* (SS-07) and *S. xantholicus* (SS-09) with strong antagonistic activity against *R. solani*. Amendment of soil with 1 % PBGG alone or 1 % PBGG mixed with *S. padanus* or *S. xantholicus* did not affect the germination of Chinese cabbage seeds and significantly reduced the damping-off disease incidence (Chung et al. 2005). Talc-based formulations of *Pseudomonas fluorescens* Pf1 were found to reduce the incidence of several plant diseases be earlier investigations. In order to develop a new formulation of Pf1 strain with enhanced biocontrol potential, different compounds like trehalose, polyvinylpyrrolidone and glycerol were evaluated. Glycerol was the most effective in maintaining higher population levels of Pf1 up to 6 months of storage. The dose of liquid-based formulation of Pf1 was standardized for seed treatment and seedling root dip. For seed treatment an aliquot of 10 ml/kg of seeds was found to be optimum, whereas large quantity (150 ml) was required for root dip treatment of seedlings to be planted in 1 ha. Storage of liquid formulation up to 6 months did not have any negative effect on bacterial cell viability. Incidence of tomato Fusarium wilt disease was reduced to the maximum extent (4.8 %), when a combination of seed treatment, seedling root dipping and soil drenching with liquid formulation of Pf1 was applied under field conditions (Manikandan et al. 2010). It remains to be demonstrated whether the benefit-cost ratio is favorable for undertaking the seedling root dipping and soil application treatments which are quite cumbersome and expensive.

Bacillus spp.

The ubiquitous presence of *Bacillus* spp. in almost all agricultural soils and extremely large range of environments has prompted the researchers to examine their potentiality as biological control agents and plant health promoters. The US Food and Drug Administration (USDFA) has granted the "generally regarded as safe (GRAS)" status to *Bacillus subtilis* which has been recognized as non-pathogenic organism which is one of the essential characteristics for development as a biopesticide. *Bacillus* spp. have the capacity to produce spores which are extremely resistant to high temperatures, pH variations, lack of nutrients and water. The spores are produced, when the environmental conditions become unfavorable, helping the bacteria to survive in the phytosphere (Piggot and Hilbert 2004). These characteristics facilitate post-culture conditioning as bacterial suspensions can be converted into "easy-to-handle" powder formulations without significant bacterial cell mortality observed in the case of non-sporulated bacterial species (Lolloo et al. 2010). Shelf-life of bioformulations based on sporulated bacteria is generally longer and need less storage precaution compared to other products prepared from living microorganisms. Furthermore, *B. subtilis* has several characteristics that enhance its survival in the rhizosphere like the ability to live as a facultative anaerobe surviving under low oxygen availability (Nakano and Hulett 1997; Losick and Kolter 2008). *Bacillus subtilis* has been reported to suppress the development of crop diseases through different mechanisms. *B. subtilis* strain BacB reduced the severity of sugar beet Cercospora leaf spot disease caused by *Cercospora beticola* (Collins et al. 2003), while the strain CL127 reduced the infection of seedling by *Botrytis cinerea* (Leifert et al. 1995). The strain GA1 could colonize apple fruit tissues resulting in significant reduction in gray mold disease caused by *B. cinerea* (Toure et al. 2004). Since different strains of *Bacillus subtilis* act on the fungal pathogens either directly by producing antifungal compounds or indirectly by enhancing the growth and/or its resistance of plants to plant pathogens, it is necessary to identify the species/strains of *Bacillus* precisely to maintain purity of efficient strains by eliminating the less efficient or inefficient strains.

Bacillus subtilis isolate S1-2010 effective against *Botrytis cinerea* causing gray mold disease of strawberry was identified based on biochemical, morphological and cultural characteristics. Further, the identity of the isolate was confirmed with 16S rRNA sequence analysis using NCBI BLAST searching tool. The sequence showed more than 99 % homology with the authentic strain of *B. subtilis* (Hang et al. 2005). Suppressive subtractive hybridization (SSH) was employed to identify genetic markers associated with biological control activities of *Bacillus subtilis* strains. The genomes of two commercialized strains GB03 and OST713 were compared with that of strain 168 which showed no biocontrol activity, in order to have a pool of DNA fragments unique to the selected biocontrol strains. Sixteen subtracted DNA fragments shared a high degree of similarity to sequences found in multiple *B. subtilis* strains with proven biocontrol potential. Oligonucleotide primers specific to nine genes involved in biocontrol activity of *Bacillus* strains were developed. The targeted five genes involved in antibiotic synthesis and four additional genes not previously associated with biocontrol were identified. All nine markers were amplified from the commercial strains GB03, OST713 and MB1600. Strains scored positive for the amplifiable markers generally were more effective at inhibiting the growth of *Rhizoctonia solani* and *Pythium ultimum* that infected several crops, than other *Bacillus* isolates that lacked the genetic markers identified in this investigation. The diversity of marker sequences could be used as a criterion for selecting novel strains that might be more suitable for different biocontrol applications (Joshi and McSpadden Gardener 2006).

Bacillus-based formulations have been shown to have certain advantages over *Pseudomonas*-based products. The durability of *Bacillus* spores allows some processes that may be chemically, physically or environmentally more severe which are not feasible for formulating less resistant microbial biomass, as in the case of *Pseudomonas* sp. and *Bacillus* sp. that have effectively reduced the incidence and/or severity of plant diseases. They have been formulated using clay, peat, chitin, methylcellulose, vegetable oil, peptone, polyvinylpyrrolidone. The biomass viability has to be preserved during different stages of formulation like stabilization, drying and rehydration to optimize fermentation. The fermentation procedures should be designed to maximize the production of efficient endospores rather than vegetative bacterial cells. Production of endospores of *Bacillus subtilis* from distillery effluent was optimized using experimental designs the results of which could be analyzed statistically. Variables influencing spore production such as ammonium sulfate, $MgSO_4$ and corn flour were predicted using Box-Behnken design. The optimum values for the variables were obtained. Under the predicted optimum conditions, the spore concentration attained was 6.95×10^8 spores/ml (Shi and Zhu 2007). The UV light may have a devastating effect on the survival of microbial cells. The ability of a water-soluble sodium salt of lignin and the optical brightener Blankophor BBH (Sigma, Detroit, USA) as UV protectants to aid the survival of dried cells of *B. subtilis* OH131.1 exposed to sun light was assessed, using microplates. In vitro, lignin and BBH enhanced the survival of dried cells of all concentrations tested, when cells were exposed to 6 h of UV treatment (Schisler et al. 2004).

The strain of *Bacillus subtilis* used for seed treatment suppressed the development of damping-off pathogen *Pythium aphanidermatum*, infecting tomato. Formulations based on talc, lignite, lignite + flyash, bentonite and polyethyleneglycol (PEG) as wettable powder and water-dispersible tablet were evaluated for their efficacy, after different intervals of storage for 2 years. The BCA population in the formulations remained at high level for 2 years of storage at 28 °C (room temperature). There was no reduction in the antagonistic ability for 1 year in the lignite, lignite + flyash, bentonite paste, wettable powder and water-dispersible tablet formulations. Damping-off incidence in tomato was significantly reduced by seed treatment with these formulations which also enhanced the plant biomass under field conditions. *B. subtilis* strain actively colonized the rhizosphere of tomato plants growing from seeds (Jayaraj et al. 2005). Formulations of *B. subtilis* prepared using glucose, talc and peat were applied as soil and seed treatments for suppressing the development of *Fusarium oxysporum* f.sp. *lentis* (FOL), causing wilt disease of lentil. Seed treatment significantly increased the biocontrol activity against FOL, compared with unformulated bacterial suspension. The formulations decreased the severity of disease by reducing the pathogen colonization of host plant roots and by promoting the growth of the plants. Formulations based on glucose and talc were more effective than peat-based formulation (El-Hassan and Gowen 2006). Plant growth-promoting rhizobacteria (PGPR) have been marketed as inoculants or plant strengthening agents or biofertilizers because of their beneficial effects on plant growth, in addition to their biocontrol potential against plant pathogens. One of the first widely marketed PGPR was a strain of *Bacillus sutilis* produced by Gustafson

Table 5.1 Effect of wettable powder (WP) formulation of *Bacillus subtilis* S1-0210 on incidence of gray mold disease on strawberry under greenhouse conditions (Hang et al. 2005)

	Gray mold disease incidence (%)			
	21 DAI		35 DAI	
Treatments	Fruit	Leaf	Fruit	Leaf
Control	5.3 a	0.0 a	8.0 a	0.0 a
Pathogen alone	92.6 b	70.6 b	93.1 b	93.3 c
WP + pathogen	2.9 a	0.0 a	11.4 a	0.0 a
Pathogen + WP	4.0 a	0.0 a	12.1 a	3.0 a
Fungicide + pathogen	7.4 a	0.0 a	6.7 a	0.0 a
Pathogen + fungicide	4.0 a	0.0 a	7.1 a	0.0 a

Means of treatment followed by the same letter in the same column are not significantly different as per Duncan's multiple range test (DMRT) at $P=0.05$
DAI days after inoculation, WP wettable powder, *Fungicide* diethofencarb-carbendazim WP

Inc., as Kodiak. This product was applied primarily for seed treatment. These organisms (usually mixtures of strains) have the ability to promote growth of plants and root development, induce systemic resistance and other possible beneficial effects. In Europe, in order to meet the registration requirements, the liquid products of PGPR are mixed with other beneficial organisms such as *Trichoderma*-compatible strains of *B. subtilis* and mycorrhizal fungi which permit sales as a biofertilizer under European guidelines. The combination of organisms could provide benefits that single organisms are unlikely to offer. This system received widespread acceptance in many other countries like Egypt, Chile, Argentina, Brazil, Peru, Colombia and China. With the use of these bioproducts, use of chemicals was almost totally dispensed with and yields were increased appreciably with less cost to the farmers (Harman et al. 2010).

Bacillus subtilis S1-0210 was selected, because of its inhibitory effect on the mycelial growth of *Botrytis cinerea* in in vitro assays. A wettable formulation of the strain S1-0210 was prepared by mixing with inert ingredients bean flour, rice flour, glucose, $FeSO_4 \cdot 7H_2O$ and $MnCl_2 \cdot 4H_2O$. The wettable powder effectively suppressed gray mold disease development on leaves and fruits of strawberry under greenhouse conditions. At 35 days after inoculation with *Botrytis cinerea*, incidence of gray mold disease on wettable powder-treated plants was 12.1 % as against 93.1 % in untreated controls. Pretreatment with the formulation was more effective than when applied as curative treatment (Table 5.1). The wettable powder treatment containing strain S1-0210 under field conditions was able to reduce the gray mold disease incidence significantly with a control efficacy of 70 %. There was significant difference in the efficacy of biocontrol formulation and fungicide (diethofencab-carbendazim) in reducing gray mold infection on leaves and fruits. The results indicated the potential of *B. subtilis* S1-0210 in suppressing the development of gray mold disease of strawberry both under field conditions and in conventional storage conditions (Hang et al. 2005). The process of microencapsulation involves the use of biodegradable matrix based on chitosan, guar gum, arabic gum, sodium alginate and other biopolymers for formulating biocontrol agent. The microencapsulation offers protection to the BCAs against environment stress and enables slow release of the BCAs slowly. The microcapsules containing the BCA can be stored for long periods at room temperature (Bashan et al. 2002; Puoci et al. 2008). Spores were mixed with calcium chloride and sodium alginate to form microcapsules and dried at 72 °C for 72 h. The microcapsules (MIC) were dissolved in sodium bicarbonate solutions (4 %) and numbers of spores were determined under the light microscope. Application of MIC containing strains of *B. subtilis* reduced the incidence of Fusarium wilt caused by *Fusarium oxysporum* f. sp. *lycopersici* and root rot caused by *Rhizoctonia solani* in tomato and also exerted positive effects on plant growth, leaf area, biomass and tomato yield (Nihorimbere et al. 2010). In another investigation, *B. subtilis* applied as MIC for seed treatment of tomato seeds reduced the incidence and severity of diseases caused by *Fusarium oxysporum* f. sp. *lycopersici* and *R. solani*. Further, plant growth, aerial and root dry weight, leaf area and fruit yield were significantly increased due to MIC treatment compared with untreated controls. The results showed that formulation of *B. subtilis* as MIC could be a feasible approach for the management of two important diseases of tomato, in addition to enhancement of tomato yields (Hernández-Suárez et al. 2011).

Rice sheath blight disease caused by *Rhizoctonia solani* is one of the important diseases seriously affecting rice production in several countries. *Bacillus megaterium* was formulated and tested under greenhouse and field conditions against sheath blight disease. The freshly prepared bacterial suspension and formulated product were effective in suppressing the development of rice sheath blight disease under greenhouse conditions. However, the formulated product was not able to suppress the development of sheath blight disease under field conditions, possibly due to dilution and inactivation of antibiotics produced by *B. megaterium* in the aquatic environment in the rice fields and climatic conditions that were not favorable for the persistence of *B. megaterium* (Pengnoo et al. 2000). In a later investigation, *B. megaterium* was formulated as floating pellets and water-soluble granules. Their efficacy in suppressing the development of rice sheath blight disease was evaluated under both greenhouse and field conditions. The floating pellet formulation was prepared using an extrusion-spheronization process. The water-soluble granule formulation contained lactose, polvinylpyrrolidone, sodium alginate and BCA suspension. Under field conditions, floating pellet formulation was placed at the center of each rice hill on top of the inoculum of *R. solani*. Water-soluble granule formulation (10 g/100 ml) was sprayed using a knapsack sprayer. The floating pellet formulation was as effective as the fungicide iprodione in suppressing the sheath blight disease under greenhouse conditions. But when sprayed onto rice plants under field conditions, the water-soluble granule formulation was no more effective than the floating pellet formulation and suppressed the development of sheath blight disease as efficiently as iprodione based on the percentage of rice tillers affected by *R. solani* (Table 5.2). When sprayed, the granule formulation might be deposited on various rice plant tissues such as leaf sheath and leaf blade, making it more effective (Kanjanamaneesathian et al. 2007). In a later study, seed treatment with water-soluble granule formulation of *B. megaterium* (1 g/100 g seed) followed by three sprays at seedling, tillering and flowering stages (200 g/20 l) was effective against rice blast

Table 5.2 Field evaluation of *B. megaterium* formulations for their efficacy in suppressing the development of rice sheath blight disease (Kanjanamaneesathian et al. 2007)

Treatments/concentrations	Percentage of tillers with sheath blight symptoms	Lesion length (cm)
Floating pellet		
10 g	82.3 b	16.6 b
20 g	84.5 ab	16.2 b
30 g	79.9 b	15.1 c
Water-soluble granules		
100 ml	80.1 b	13.8 d
200 ml	70.6 c	13.6 d
300 ml	50.6 d	12.9 d
Iprodione	55.7 d	8.5 e
Control (pathogen alone)	92.5 a	18.7 a

Means followed by the same letter are not significantly different as per Duncan's multiple range test (DMRT) at $P = 0.05$

(*Pyricularia oryzae*) and sheath blight diseases, when recommended N-fertilizer was applied and weeds were adequately controlled. This schedule of application of the BCA has been recommended to the farmers for adoption in their fields (Kanjanamaneesathian et al. 2009).

Combinations of strains of *Bacillus subtilis* (GB03), *B. pumilus* (SE34, INR 7 or T4) and *B. amyloliquefaciens* (IN937) were formulated using chitosan as carrier. The formulation of two strains of the above mentioned *Bacillus* spp. and chitosan was named as biopreparation. Five biopreparations containing two strains each were used to treat tomato plants. When the plants were challenged with *Cucumber mosaic virus* (CMV) biopreparation-treated plants had significantly greater fresh weight and flower and fruit numbers than the untreated tomato plants. Treatment with biopreparations reduced CMV disease severity ratings significantly at 14 and 28 days post-inoculation (dpi). Accumulation of CMV in young noninoculated leaves in biopreparation-treated plants was significantly less as revealed by enzyme-linked immunosorbent assay (ELISA). The biopreparation-treated plants were phenotypically and perhaps developmentally, similar to plants that were 10 days older. The results suggested that biopreparation formulation might result in increased resistance level in treated plants against *Cucumber mosaic virus* infection (Murphy et al. 2003).

Paenibacillus spp.

Paenibacillus polymyxa strain E681, a PGPR earlier showed antagonistic properties and suppressed the development of pre- and post-emergence damping-off disease of cucumber and pepper. But seed treatment with strain E681 alone did not provide consistent protection to sesame plants. Hence, the effect of pelleting with strain E681 on disease suppression was assessed. A combination of clay and vermiculite was applied to enhance seed germination and root colonization on sesame by the strain E681. Seed pelleting with E681 significantly reduced the disease incidence in conducive soil. Under field conditions, pelleting of sesame seeds with strain E681, promoted plant growth and increased the yield levels also. The results indicated that application of *P. polymxa* strain E681 via seed pelleting offered effective solution for disease problem associated with successive years of sesame cultivation (Ryu et al. 2006). The effects of two milling methods and five surfactants were assessed on the suspensibility of a wettable formulation of *Paenibacillus polymyxa* HY96-2 (*P. polymyxa* original powder or PPOP). The suspensibility of PPOP was significantly improved by both hammer milling and jet milling respectively from 8.1 to 26.4 % and 58.3 %, while no adverse effect on spore viability could be noted. The surfactants EFWR, D-425R, sodium dodecylbenzene sulfonate, tea saponin and sodium lignosulphonate and D-425 mixed with jet-milled powder (JMP) were evaluated for their effect. Sodium lignosulphonate and D-425 were equally effective and better than the other surfactants tested in improving the suspensibility of the wettable formulation of strain HY96-2. Further, sodium lignosulphonate exhibited best biocompatibility and spore viability values of JMP mixed with this (1–10 % w/w) were over 95 %. The results suggested that jet milling and sodium lignosulphonate might synergistically improve the suspensibility of *P. polymyxa* wettable powder without any detectable adverse effect on the spore viability (Liu et al. 2011).

Serratia spp.

Among *Serratia* spp., *S. plymuthica* is classified into the risk group1 by the DSMZ (German Collection of Microorganisms and Cell Cultures), indicating that this species does not inadvertently pose a threat to human health. There appears to be no compelling evidence that *S. plymuthica* is capable of causing human infections. Different strains of *S. plymuthica* has been reported to be effective in reducing the incidence of diseases caused by *Rhizoctonia solani, Phytophthora cactorum, P. capsici, Magnaporthe grisea, Botrytis cinerea, Verticillium dahliae* and *Sclerotina sclerotiorum* (Vleesschauwer and Höfte 2007). A well-characterized strain of *S. plymuthica* HRO-C48 effectively reduced the infection of strawberry by *Verticillium dahliae*, causing wilt disease and *Phytophthora cactorum*, inducing root rot disease by 18.3 and 33.4 % respectively. The beneficial effect of *S. plymuthica* on strawberry was indicated by an increase in yield by 60 %, compared with untreated controls (Kurze et al. 2001). A commercial product containing *S. plymuthica* HRO-C48 named as RhizoStar® was developed (European patent 98124694.5) by e-nema GmbH, Raisdorf, Germany. The biocontrol activity of *Serratia plymuthica* A21-4 against *Phytophthora capsici* causing pepper (chilli) Phytophthora blight disease was enhanced by modifying the minimal medium and the buffer for multiplying the BCA. Sodium citrate buffer replaced MgSO4 in buffer. The newly formulated buffer solution for suspending the bacterial cells enhanced the biocontrol efficacy even at 100 × lower concentration. This finding was considered to be crucial for producing the commercial product of *S. plymuthica* A21-4 (Shen et al. 2005). *Serratia marcescens* was found to have high potential as a biocontrol agent against *Fusarium oxysporum* f.sp. *cubense*, causing the Panama (Fusarium) wilt of banana. Bentonite, a montmorillonite clay was selected as the carrier, non-fat skimmed milk (NFSM) and sucrose as enrichment materials and para-aminobenzoic acid (PABA) as a UV-protectant. The components of the formulation of *S. marcescens* were not particularly beneficial for storage stability, but they offered more benefits to the bacterial cells as protectants against UV light, when the formulation was applied under field conditions. Bentonite-clay was shown to have protectant effects against UV, higher temperature and desiccation. Sucrose added to the cell suspension contributed to the bacterial cell viability. The additive material PABA was found to have antimicrobial effect on the formulated cells. The results showed that bentonite and sucrose contributed significantly to protection against UV light and viability of cells of *S. marcescens* (Ting et al. 2009).

Formulations of Bacterial Biocontrol Agents Effective Against Postharvest Diseases

Pseudomonas spp.

Mass multiplication of biocontrol agents is the basic requirement in the process of formulation and commercialization of bioproducts. Combinations of two carbon (sucrose and molasses) and two nitrogen (urea and yeast extract) sources were evaluated for their efficacy in rapid biomass production of the bacterial biocontrol agent *Pseudomonas fluorescens* P-35, using nutrient broth as control. The medium containing molasses and yeast (MY) extract (1:1, w/w) stimulated the growth of the strain P-35 attaining high quantities of biomass. The biocontrol efficacy of P-35 grown in MY was higher than when grown in other combinations of carbon and nitrogen sources. The percentage of fruits showing decay caused by *Botrytis cinerea* (gray mold disease) was 38.6 %, compared to 100 % infection in control treatment, after 10 days of incubation. The strain P-35 exhibited good biocontrol efficacy after growing in MY for 20 days. The antagonist development was inhibited by the presence of urea (Peighami-Ashnaei et al. 2009). The nutritional requirements, optimum temperature, pH, and other conditions for the rapid multiplication of biocontrol agents have to be determined. The compounds that differentially stimulate growth of the antagonist with no detectable stimulatory effect on the target pathogen were identified for *Pseudomonas syringae* strain L-59-66 that was found to be effective against *Penicillium expansum* causing apple blue mold disease. The BCA occurred naturally on fruit trees and was capable of controlling postharvest diseases of pear and citrus fruits. A formulation of *P. syringae* compatible with practices followed in packing houses was developed (Janisiewicz 1994). A biocontrol product was commercialized under the name Bio-Save11, from a formulation containing *P. syringae* by EcoSience Corporation. *P. syringae* strain L-59-66 (renamed as strain ESC-11) controlled blue mold (*Penicillium expansum*), gray mold (*Botrytis cinerea*) and mucor rot (*Mucor* spp.) on apple and pear (Janisiewicz and Jeffers 1997).

Pseudomonas syringae ESC-11 sold as Biosave 110 was recommended for postharvest decays of pear and apple. *P. syringae* strain ESC-10 was effective against citrus rots and it was marketed as BioSave 100 (Janisiewicz and Jeffers 1997).

The strain ESC-11 was able to reduce the crown rot of banana caused by a complex of fungi including *Fusarium semitectum* and *F. moniliforme*. In addition, potato dry rot caused by *F. sambucinum* could also be suppressed by the strain ESC-11. Furthermore, the growth of the foodborne pathogen *Escherichia coli* 0157:H7 in apple wounds could be prevented by applying the strain ESC-11 (Kenwick and Jacobsen 1998; Janisiewicz et al. 1999; Williamson et al. 1999). The powder formulation of BioSave 11 was effective as the unformulated laboratory wet preparations in controlling blue mold and gray mold in Golden Delicious and Red Delicious apples. Formulating *P. syringae* strain ESC-11 into a powder did reduce its growth and survival in fruit wounds, but it was effective in protecting the fruits against postharvest pathogens. Hence, the commercial product has the potential to replace synthetic fungicides used for the control of fruit rots (Janisiewicz and Jeffers 1997).

Pantoea spp.

Pantoea agglomerans has been shown to be antagonistic to a many postharvest pathogens affecting pome fruits and citrus. The carbon and nitrogen sources that favored maximum biomass production of *P. agglomerans* strain CPA-2 were determined. A combination of nitrogen sources such as yeast extract (5 g/l) and dry beer yeast sources such as yeast extract (10 g/l) with inexpensive carbohydrates such as sucrose (10 g/l) and molasses (20 g/l) was found to be suitable for the mass multiplication of this bacterial BCA which was effective against *Penicillium digitatum* and *P. italicum* infecting oranges. This investigation provided a reliable basis for a scale-up of the fermentation process to an industrial project (Costa et al. 2001). The bacterial species employed as biocontrol agents have to be protected against freeze-drying injury in order to preserve the cell viability and to enhance the shelf stability of formulated products. The viability of the biocontrol agents may be enhanced by using freeze-drying protective agents and rehydration media. Maximum protection from freeze-drying injury to the bacterial antagonist *P. agglomerans* strain CPA-2 was provided by trehalose (5 %) with a survival of over 60 % of the bacterial cells. Skimmed milk (10 %) was found to be the most effective rehydration medium from which 100 % of the freeze-dried bacterial cells could be recovered. *P. agglomerans* CPA-2 reduced blue mold and gray mold diseases of pome fruits significantly (Costa et al. 2000). Further study showed that freeze-dried cells of *P. agglomerans* could be stored in glass vials or in high barrier plastic bags at 4 °C for 3 months without any loss of the biocontrol activity against *Penicillium digitatum* (Costa et al. 2002).

A talc-based formulation of *Pantoea agglomerans* strain Eh 24 effectively suppressed the development of fire blight disease in pear blossoms. The bacterial cells could survive in talc-based formulations for more than 120 days at 4 °C. Addition of glycerol to the BCA cell suspension appeared to increase long-term survival of the strain Eh 24. The moisture content of the formulation was maintained at 15 % after drying. The talc-based formulation of Eh 24 at a concentration of 10^8 CFU/ml, significantly reduced the blossom blight of pear under field conditions. The population of Eh applied on pear blossoms increased from 2×10^4 CFU/blossom to

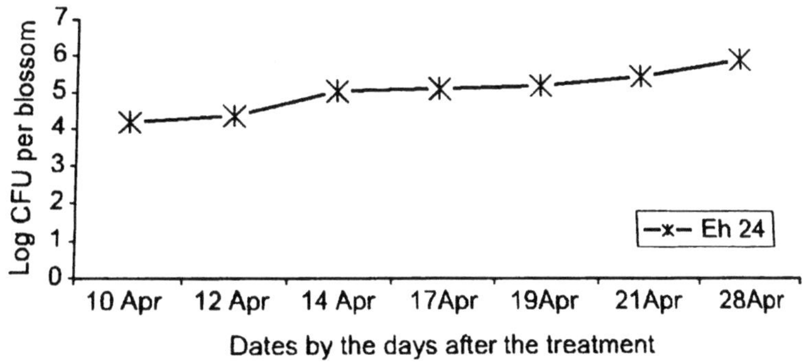

Fig. 5.1 Assessment of population sizes of *Pantoea agglomerans* in pear blossoms at different dates after treatment based on isolation using sucrose nutrient agar (SNA) medium supplemented with streptomycin (Courtesy of Özaktan and Bora 2004 and with kind permission of the Brazilian Journal of Microbiology, Brazil)

1.3×10^6 CFU/blossom over a period of 18 days (Fig. 5.1) (Özaktan and Bora 2004). *P. agglomerans* strain E325 forms the active ingredient of a commercial product effective against *Erwinia amylovora*, causing fire blight disease of apple and pear. Osmoadaptation involving the combination of saline osmotic stress and osmolyte amendment to growth media was investigated for improving the epiphytic colonization of E325 on apple flowers, particularly in dry climates. E325 was osmoadapted in nutrient yeast dextrose broth and the commercial fermentation medium, amended with NaCl and glycine betaine. The BCA was cultured and freeze-dried with cryoprotectants at an ARS-USDA laboratory and commercial facility prior to treating apple flowers in an orchard where relative humidity averaged <50 %. Osmoadaptation generally did not affect colonization of E325 on flower stigmas or near-rich hypanthia on orchard flowers and on detached crab apple flowers. The exception was the significant advantage of osmoadapted E325 on hypanthia of detached flowers at 70 % RH, resulting in osmotic conditions marginally conducive for bacteria. Osmoadaptation proved most beneficial for enhancing E325 survival during freeze-drying and storage prior to orchard application. Osmoadaptation complemented cryoprotection, by improving overall stability of freeze-dried preparations of *P. agglomerans* strain E325 (Pusey and Wend 2012).

The survival and effectiveness of different formulations of *Pantoea agglomerans* against *Penicillium* spp. infecting citrus was assessed using different preharvest treatments. High sensitivity of non-adapted and osmotic-adapted *P. agglomerans* cells to environmental conditions existing in the field was observed. Dry conditions and solar radiation were found to significantly affect population of BCA cells. Different formulation strategies were evaluated for their efficacy in improving the resistance of BCA cells to unfavorable environmental conditions. Osmotic-adapted *P. agglomerans* cells in the presence of NaCl (25 g/l) in the production medium or at water activities (a_w) of 0.98 had higher survival rates than non-adapted cells, when these cells were sprayed on oranges and stored hermetically sealed chambers at a low RH (43 %). Of the seven additives tested, Fungicover (5 %) added to the bacterial suspension enhanced the adherence and persistence of bacterial cells on oranges exposed to unfavorable conditions. Lyophilization of cells also improved the resistance of BCA cells to unfavorable conditions compared with fresh cells. The results indicated that the formulation strategies that could improve the resistance levels of BCA to unfavorable conditions have to be applied for effective management of diseases occurring at postharvest stages (Cañamás et al. 2008a). In a later investigation, by modifying growth media or temperature during growth period, the cell viability of *P. agglomerans* in spray-dried formulations could be improved. *P. agglomerans* grown during 48 h in NaCl 0.97 a_w-modified medium could increase their viability after spray-drying the formulation from 6 % in unmodified medium to nearly 30 % without affecting their biocontrol potential. The effect of combination of thermal and osmotic stress was assessed to improve effectiveness of fluidized bed drying formulations of *P. agglomerans*. By using NaCl to adjust a_w to 0.988 in the growth medium and increasing the temperature to 35 °C during 1 h in the early stationary phase, it was possible to obtain an effective formulate with only 0.5 log reductions during fluidized bed drying process (Usall et al. 2010).

Bacillus subtilis

Bacillus subtilis strain CPA-8 is effective in suppressing the development of postharvest pathogen *Monilinia fructicola* infecting peaches. A low cost medium based on commercial products and by-products was developed to maximize the growth of the strain CPA-8 and to maintain its biocontrol efficacy. Low cost media containing economical nitrogen and carbon sources such as yeast extract, peptone, soy products, sucrose, maltose and molasses were evaluated. Maximum cell growth ($>3\times10^9$ CFU/ml) was obtained in defatted soy flour (44 %) combined with sucrose or molasses media. CPA-8 production was scaled up in a 5-1 fermentor and the bacterial BCA population on dynamics, pH and oxygen consumption in optimum medium was recorded. Bacterial cell suspensions and cell-free supernatants from CPA-8 grown in optimized medium maintained their biocontrol efficacy against *M. fructicola* on peaches, resulting in reduction of disease incidence up to 95 %. Population of the BCA survived in wounds on inoculated peaches, regardless of the culture media used. It is possible to mass multiply *B. subtilis* CPA-8 in a low cost medium combining inexpensive nitrogen and carbon sources in shake flasks and a laboratory fermentor (Yánez-Mendizábal et al. 2012).

5.1.3 Combination of Bacterial and Fungal Biocontrol Agents

Rice sheath blight disease caused by *Rhizoctonia solani* is one of the major diseases in several countries. The efficacy of the formulated product based on *Bacillus megaterium*, *B. subtilis* and *Aspergillus niger* was assessed. Application of talc-based formulation of individual antagonists and mixtures of the bacterial and fungal antagonists as spray or soil application were effective in reducing the disease incidence up to 45 % at 27 days after inoculation of *R. solani* and the grain yield due to application of BCA formulations was also increased. Optimum sporulation conditions, antagonists for preparation of spore-based formulations and their commercially desirable features such as the ability to maintain spore viability in storage were determined. Culturing in the synthetic replacement sporulation medium (SRSM-2) for 72 h was the most effective for sporulation of *B. megaterium* and *B. subtilis*, whereas culturing *A. niger* in PDA for 7 days was effective for high sporulation. The talc-based formulations of all antagonists either in refrigerated storage (4 °C) or at room temperature (28 ± 2 °C) retained greater spore viability over a longer period (>6 months) than spore suspension. The bacterial spore-based formulations are more desirable than those based on vegetative cells, because of their longer survivability under natural field conditions (Soe and De Costa 2012).

5.2 Delivery Systems for Formulated Products

Biotic biological control agents (BCAs) after assessing their potential against target pathogen(s), are formulated by applying various procedures. These formulated products are applied to soils, seeds or propagative materials such as tubers, bulbs, rhizomes and setts, and plants either as protective or curative treatments. Their efficacy in reducing the disease incidence and/or severity is compared with the unformulated fresh fungal or bacterial species to determine the extent of loss of biocontrol activity, if any, due to the process of formulation and storage for various duration. Commercially available products are evaluated by state-owned research institutions, before they are recommended for use by farmers.

5.2.1 Seed Treatment

Seeds are treated with liquid, powder or granular formulations containing single or mixture of strains of fungal or bacterial biocontrol agents. Seed treatment with biocontrol products has been the most common method of application, since it is easier and more economical especially, when the target pathogens are soilborne. Seed treatment not only tackles the pathogens present on or in the seeds, but also protects the emerging young seedlings against the pathogens surviving in the soils. Treatment of seeds with plant growth-promoting rhizobacteria (PGPR) has been carried out for suppressing the development of pathogens causing diseases that affect different plant parts such as roots, stem, foliage and inflorescence. The PGPRs are effective against different kinds of pathogens because of their multiple modes of action such as antagonism, competition for nutrients and space and induction of resistance in plants growing from treated seeds. Both fungal and bacterial biocontrol agents have been shown to be effective in reducing the incidence and/or severity of diseases following seed treatment. *Trichoderma viride, T. harzianum, T. (Gliocladium) virens, Gliocladium catenulatum, Pseudomonas fluorescens, Bacillus subtilis, B. pumilus* and *Streptomyces griseoviridis* have been reported to be effective against soilborne pathogens causing seed rot, root rot and wilt diseases. Parasitism and antagonism by producing different antimicrobial compounds may be the more common mechanism of biocontrol activity of fungal agents. On the other hand, production of antimicrobial substances that have direct action on the pathogens and induction of resistance in plant hosts against pathogens are primarily associated with bacterial agents.

The biocontrol agents (BCAs) are applied as spore suspensions or dry powder to coat the seeds. The seeds are soaked in the spore suspension and air-dried before sowing. Coating or pelleting of seeds with BCAs may ensure the presence of the antagonists on or near the seed for longer periods after sowing and irrigation. Various kinds of stickers such as carboxymethyl-cellulose (CMC) (1 %), methocel (2 %), polysulf (0.8 %) and polyvinyl alcohol (20 %) have been used for efficient adherence of the BCA to seed surfaces. Several biocontrol agents like *Chaetomium, Gliocladium, Penicillium, Pseudomonas* and *Trichoderma* have been reported to be effective, following seed treatment for the control of several destructive crop diseases. Treatment of cotton seeds with *Trichoderma (Gliocladium) virens* or *Bacillus subtilis* reduced colonization of roots by *Fusarium oxysporum* f.sp. *vasinfectum*, resulting in lower levels of disease incidence and severity (Zhang et al. 1996). The corn damping-off disease caused by *Pythium ultimum, P. arrhenomanes* and *Fusarium graminearum* was more effectively controlled by coating the seeds with *T. virens* isolate Gl.3, resulting in greater seedling stand, plant height and reduction in root rot severity, compared to the fungicide captan or other antagonists tested (Mao et al. 1997). The biocontrol potential of *Bacillus* sp. strain L324-94 for the control of Pythium root rot caused by *P. irregulare* and *P. ultimum*, Rhizoctonia root rot caused by *R. solani* and wheat take-all disease caused by *Gaeumannomyces graminis* var. *tritici* was assessed. The plants growing from seeds coated with BCA were effectively protected from all three diseases, when the treated seeds were directly drilled into the soil in both winter and spring seasons. The strain L324-92 could survive and proliferate at 4 °C in winter and later in the season also (Kim et al. 1997). Seed treatment with BCAs may offer protection against foliage pathogens in certain cases. *Pseudomonas fluorescens* in coated rice seeds as a bacterial suspension or talc-based powder formulation spread to roots, stems and leaves of plants growing from the treated seeds and protected the plants growing from the treated seeds and protected the rice plants against *Rhizoctonia solani* causing sheath blight disease and *Magnaporthe grisea* causing rice leaf blast disease. This bacterial BCA may act on the pathogens by multiple modes of action such as inhibition of pathogen growth by producing antibiotics, siderophores and/or by inducing systemic resistance in rice plants against the pathogens. In addition to suppression of disease development, higher yield could also be obtained by the seed treatment with *P. fluorescens* (Mew and Rosales 1986; Vidhyasekaran et al. 1997).

Trichoderma virens was earlier considered to suppress the development of *Rhizoctonia solani* infecting cotton seedlings by mycoparasitism and/or by producing gliotoxin. However, by generating mutants incapable of being mycoparasitic or producing the antibiotics, it was shown that *T. virens* was able to induce terpenoid phytoalexins in cotton seedlings and protect the cotton seedlings. Thus, the mechanisms proposed earlier were found to be incorrect (Howell 2006). Antibiotic-producing isolates of *Streptomyces* spp. were used to treat the seeds of sugar beet tubers, a few weeks before harvest. The isolate J-2 was the most potent in inhibiting the germination of sclerotia in the soil and also effectively reduced the damping-off disease. The antifungal compounds produced by the isolate J-2 induced conspicuous inhibition zones in the in in vitro assays. The BCA was able to multiply and survive in the rhizosphere soil from naturally infested soil for more than 3 weeks (Errakhi et al. 2007). Treatment of barley seeds with *Pseudomonas chlororaphis* MA342 was effective against net blotch disease caused by *Drechslera teres*. Immediately after inoculation of seeds, the bacteria were present mainly under the seed glume, when the seeds were sown. Irregularly distributed area of bacterial aggregation were found, indicating the epiphytic colonization of glume cells. The bacteria were not seen within the embryo. Bacterial aggregates were regularly observed in the groove of each seed formed by the base of the coleoptiles and scutellum, suggesting that the BCA could localize with the pathogen which facilitate the action of antifungal compounds produced by this BCA strain (Tombolini et al. 1999). *Pseudomonas aeruginosa* strain 7NSK2 and *P. fluorescens* strain CHA0 were compared for their efficacy in reducing the intensity of symptoms induced by cotton bacterial blight disease caused by *Xanthomonas campestris* pv. *malvacearum* (*Xcm*). Cotton seeds were treated with the biocontrol agents after acid-delinting and neutralization of seeds. *P. aeruginosa* was equally efficient in inducing resistance to *Xcm* in cotton plants and also in increasing plant biomass (fresh weight) (Fallahzadeh-Mamaghani et al. 2009). Treatment of seeds with PGPR followed by drenching of roots of seedlings growing from treated seeds with BCA suspension was shown to induce systemic resistance (ISR) against virus diseases. Treatment of tobacco seeds and later the roots of seedlings with *Bacillus* spp. significantly increased the plant height and fresh weight. Severity of infection by *Tobacco mosaic virus* (TMV) was remarkably reduced at 28 days post-inoculation (dpi). Accumulation of TMV in the young non-inoculated leaves was greatly reduced for all plants treated with *Bacillus* spp. Enhanced plant growth and reduction in virus disease severity might be due to a series of physiological and biochemical changes induced by the bacterial biocontrol agent (Wang et al. 2009).

Biopriming of seeds is used commercially as a tool to increase speed and uniformity of germination and to improve final stand under environmental stress conditions. However, if the seeds are infected, biopriming can provide conditions favorable for development of the pathogens. Application of antagonistic microorganisms during biopriming may be considered as an environmentally friendly strategy for the control of seedborne pathogens, instead of chemicals. The solid matrix priming (SMP) method is combined with biological seed treatment to enhance the level of protection offered by BCAs to plants. This method involves soaking the seeds in a suspension of spores or cells of BCA (1 ml/4 kg seeds) followed by mixing the treated seeds with organic carriers such as sphagnum moss, bituminous coal or leonardite shale, using water to attain the final moisture content varying from 60 to 90 %, depending on the nature of seeds. The mixtures are then incubated at 20 °C for 4 days, followed by planting them in the soil. This technique was employed for integrating seed treatment with *Trichoderma harzianum* and solid matrix priming (Harman and Taylor 1990). Inoculation of seeds with BCAs in combination with priming has been shown to enhance their biocontrol potential especially against soilborne pathogens (Callan et al. 1990). Isolates of *Clonostachys rosea* controlled pre- and post-emergence mortality of carrot seedlings by *Alternaria dauci* and *A. radicina* as effectively as the fungicide iprodione. Isolate IK726 of *C. rosea* was used in biopriming a seed lot with 29 % *A. radicina* and 11 % *A. dauci* (high level of infection) and a seed lot with 4 % *A. radicina* and 7 % *A. dauci* (low level of infection). The seeds were primed with water alone (hydropriming) or with addition of *C. rosea* IK 726 (biopriming). During 14 days of hydropriming the incidence of *A. radicina* and *A. dauci* was dramatically increased to two- and five-folds respectively, regardless of initial level of seed infection. In contrast, biopriming reduced the incidence of *A. radicina* to 2.3 % and that of *A. dauci* to 4.8 %, while the level of infection by both pathogens was reduced to <0.5 % on primed seeds with low level of seed infection (Jensen et al. 2004). The conditions required for seed biopriming of oilseed rape (*Brassica napus*) to be more effective were studied for the bacterial antagonists *Serratia plymuthica* and *Pseudomonas chlororaphis*. The bacterial density in the biopriming suspension should be $>10^9$ CFU/ml for *S. plymuthica* and $>10^8$ CFU/ml for *P. chlororaphis*. Priming duration of 2–4 h was found to be optimum for *S. plymuthica* and *P. chlororaphis* respectively. Addition of $MgSO_4$ supported the establishment of both BCAs and also improved seed germination. The optimum temperatures to be maintained were 28 and 22 °C respectively for *S. plymuthica* and *P. chlororaphis* respectively. The results indicated the need for standardization of conditions optimal for biopriming with different BCAs to achieve the best results (Abumasha et al. 2011).

5.2.2 Treatment of Cuttings and Transplants

Asexually propagated plants, when infected by microbial pathogens, carry the infection to the subsequent generations through infected plant material such as tubers, rhizomes, corms, bulbs, cuttings or setts. These planting materials have to be treated with biocontrol agents to reduce or entirely eliminate the incidence of the diseases in the fields. Treatment of sugarcane setts with a suspension of talc formulation of *Pseudomonas fluorescens* for 1 h, followed by incubation for 18 h prior to planting reduced the incidence of red rot disease of sugarcane caused by *Colletotrichum falcatum* under field conditions. In addition, sett treatment with the BCA resulted in enhancement of plant growth (Viswanathan and Samiyappan 2002). Potato silver scurf disease caused by *Helminthosporium solani* primarily spreads during storage. The minitubers of Red Norland potato were immersed in a suspension of fungal BCA *Acremonium strictum* for 3 min and air-dried. The treated and untreated tubers (control) were inoculated with *H. solani* by spraying the conidial suspension. *A. strictum* reduced the sporulation and spore germination of *H. solani*. The BCA was effective, when applied as a protective treatment. But *A. strictum* did not reduce silver scurf development in already infected tubers (Rivera-Varas et al. 2007).

Plant growth-promoting rhizobacteria (PGPRs) have been reported to protect the plants against diseases caused by microbial plant pathogens, following treatment of roots with the PGPRs. Dipping of the wounded roots of tomato seedlings in a suspension of cells of *Agrobacterium radiobacter* strains K84 or K1026 or the commercial product 'Nogall' entirely inhibited crown gall formation by *A. tumefaciens* (Fakhouri and Khalaif 1996). The bacterial BCAs *Pseudomonas aureofaciens* and *P. fluorescens* with wide spectrum of antagonistic activity were tested against *A. tumefaciens* and *A. vitis* infecting raspberry and grapevines respectively. The effectiveness of *P. aureofaciens* and *P. fluorescens* applied as root dip to grapevine cuttings varied depending on the grapevine cultivar. Disease incidence was reduced by 56–80 % and disease severity index (DSI) by 75–86 %. Both *P. aureofaciens* and *P. fluorescens* persisted on the root surfaces of inoculated grapevine cuttings and non-sterile soil. Treatment of rooted raspberry seedlings with *P. aureofaciens* B-4117 strain provided effective protection against *A. tumefaciens* strains infecting raspberry (Khmel et al. 1998). An increase in the amounts of phenolics in cucumber roots treated with *Pseudomonas putida* BTP1 was observed. The enhanced levels of phenolics correlated with increased protection of plants against root rot disease caused by *Pythium aphanidermatum* (Ongena et al. 1999). The root system of cucumber was soaked in the suspension of *P. putida* BTP1 for 10 min, followed by inoculation with *P. aphanidermatum*. The seedling mortality due to root rot disease was as high as 80 % in the untreated control plants, whereas the disease incidence was 45 % in plants treated with *P. putida*. Accumulation of several antifungal molecules in the root system of cucumber plants due to treatment with BCA strains BTP1 was observed. Higher concentrations of fungitoxic compounds in leaves of BTP-1-treated plants were detected. The results indicated the possible systemic elicitation of phytoalexins, as the primary mechanism of biocontrol activity of *P. putida* against *P. aphanidermatum* infecting cucumber (Ongena et al. 2000).

When the roots of grapevine seedlings were soaked in a cell suspension of the nonpathogenic *Agrobacterium vitis* strain VAR03-1 for 24 h before a 1-h soak in a cell suspension of tumorigenic *A. vitis* and subsequent planting in infested soil, incidence of tumor formation was significantly reduced by strain VAR03-1 (Kawaguchi et al. 2007). In the further studies, roots of grapevine, rose (*Rosa multiflora*) and tomato were soaked in cell suspension of *A. vitis* strainVAR03-1, before planting in soil infested with tumorigenic *A. vitis*, *A. rhizogenes* and *A. tumefaciens*. Treatment with strain VAR03-1 significantly reduced the number of plants with tumors and disease severity in the grapevine, rose and tomato. The inhibitory effects of treatment with VAR03-1 and nonpathogenic *A. rhizogenes* strain K84 on crown gall of rose and tomato were almost identical and the inhibitory effect of VAR03-1 on grapevine was greater than that of strain K84. Furthermore, under field conditions VAR03-1 was found to be more efficient in controlling grapevine crown gall caused by *A. vitis*. VAR03-1 was able to multiply and establish a population averaging 10^6 CFU/g of roots in the rhizosphere of grapevine and persisted on grapevine roots for 2 years (Kawaguchi et al. 2008).

Application of the biocontrol agents to the roots may be effective in reducing other bacterial diseases also, as in the case of tomato bacterial wilt disease caused by *Ralstonia solanacearum*. The percentage of tomato seedling survival was markedly increased, when the roots were dipped in a suspension of *Pseudomonas aeruginosa* strain ATC7700 and planted in pathogen-infested soil. The highest level of disease suppression was obtained by using a concentration of 10^{10} CFU/ml of suspension (Furuya et al. 1997). The effectiveness of *Bacillus vallismortis* strain EXTN-1 in suppressing the development of tomato bacterial wilt disease was demonstrated by root bacterization with the efficient strain EXTN-1 of *B. vallismortis*. Root treatment with this BCA strain reduced the percentage of infection to 65 % from 95 % infection in non-bacterized control plants (Park et al. 2007). The endophytic actinomycete *Streptomyces grieseorubriginosus* strain S96 was evaluated for its efficacy in suppression of the development of banana Panama (Fusarium) wilt disease caused by *Fusarium oxysporum* f.sp. *cubense* (*Foc*) race4. The roots of plantlets were immersed in a suspension of strain S96 (10^6 CFU/ml) for 1 h before planting, followed by inoculation with *Foc* after 3 weeks. The wilt disease severity was reduced significantly, while the mean fresh weight of plant was increased (Cao et al. 2005). Dipping roots of tomato plants in the cell suspension of *Achromobacter xylosoxydans* (10^8 CFU/ml) reduced the Fusarium wilt disease incidence by about 50 %. While no phytotoxin symptoms were noticed due to the treatment with the BCA, the root treatment with *A. xylosoxydans* enhanced plant growth, compared with untreated control plants (Moretti et al. 2008).

5.2.3 Soil Application

Several fungal and bacterial pathogens are able to remain viable as actively proliferating propagules or as dormant spores that can survive adverse conditions. The biocontrol agents either forming a natural component of the microflora or as introduced organisms have to compete with the pathogens and other microflora for available nutrients and niches for establishment. Although several fungal and bacterial species/strains have been found to show required level of antagonistic activity against pathogens in in vitro assays, very low percentage of them exhibit consistent effectiveness and could be employed for the biocontrol of plant pathogens under field conditions. The selected strains have to survive environmental stress and be aggressive colonizers of the rhizosphere to be successful biocontrol agents. Both fungal and bacterial BCAs have been applied to the soil primarily for the control of soilborne pathogens. Some of the PGPRs like *Pseudomonas* and *Bacillus* strains have been demonstrated to induce systemic resistance in treated plants, resulting in the reduction of foliage diseases also.

Among the fungal biocontrol agents *Trichoderma viride*, *T. harzianum* and *T. virens* have been widely employed for the control of a wide range of soilborne diseases caused by *Pythium* spp., *Phytophthora* spp. *Rhizoctonia solani*, *Fusarium oxysporum*, *Sclerotinia* spp. and *Verticillium* spp. The biocontrol agents are applied as water-dispersible granules or wettable powder. Other fungal BCAs employed for the management of soilborne diseases are nonpathogenic *Fusarium oxyporum* Fo47, *Coniothyrium minitans*, *Chaetomium* spp., *Gliocladium catenulatum*, and *Pythium oligandrum*. The bacterial agents *Pseudomonas fluorescens*, *Bacillus subtilis*, *B. licheniformis*, *Burkholderia cepacia*, *Agrobacterium radiobacter* strain 64 have been shown to significantly reduce the incidence and/or severity of diseases caused by soilborne pathogens. Soil application of biocontrol agents is less preferred, because large amounts of the product may be needed to obtain satisfactory level of disease control. Further, uniform treatment of soils may be difficult to achieve. The BCAs may be applied either as a broadcast-incorporation or as placement in the seed furrow at the time of sowing. Soil application of BCAs for the control of diseases affecting glasshouse crops may be feasible. Nonpathogenic isolates of *Fusarium oxysporum* isolated from wilt-suppressive soils were more effective in controlling wilt diseases of tomatoes and watermelon and muskmelon compared to *Burkholderia cepacia*, *Trichoderma virens*, *T. hamatum* and *Pseudomonas fluorescens* (Larkin and Fravel 1998). The possibility of using BCAs acting through different mechanisms and with a wide spectrum of activity against crop pathogens receiving acceptance for large scale field application appears to be greater. *Trichoderma koningii* suppressed the saprophytic growth of *Gaeumannomyces graminis* var. *tritici* in natural soil and consequently reduced wheat take-all disease significantly (Simon and Sivasithamparm 1989). *T. koningii* applied to the seed furrow in the field trial reduced crown root infection in wheat by 40 % and also increased the yield of spring wheat by 65 %. The combination of *T. koningii* and *Pseudomonas fluorescens* Q292-80 strain was more effectively reduced the disease incidence resulting in greater yield levels (Duffy et al. 1996).

The hypovirulent binucleate *Rhizoctonia* (HBNR) was successfully employed for suppressing the development of damping-off diseases caused by *Rhizoctonia solani* in cotton, radish, wheat and cucumber and by *Pythium ultimum* in pepper (chilli). Efficient colonization of host tissues is the crucial requirement for the protection by HBNR. The ability of HBNR to protect plants against Fusarium crown and root rot of tomato (FCRR) throughout the growing period and their effects in reducing the population of *Fusarium oxysporum* f.sp. *radicis-lycopersici* (FORL) in roots and stem under soil systems in the greenhouse and field were evaluated. In the greenhouse and soil systems, HBNR isolates significantly reduced vascular discoloration and discoloration of total root system by 90–100 % and by 73–89 % respectively. Under field conditions, HBNR isolate WI significantly reduced vascular discoloration by 71 %. Application of HBNR resulted in increases of marketable and total yields of tomatoes as much as 70 and 73 % respectively, compared with untreated plants (Muslim et al. 2003). Botrytis leaf blight of lily (*Lilium formosanum*) caused by *B. elliptica* seriously affected lily production in Taiwan. The efficacy of *Bacillus cereus* and *Pseudomonas putida* strains was assessed for control of Botrytis leaf blight disease under greenhouse and field conditions. Plants treated with *B. cereus* strain CIL effectively protected *L. formosanum* against *B. elliptica* infection. The protection could last for at least 10 days and it was consistent with the high populations of *B. cereus* on lily roots. *P. putida* was less effective in protecting the plants against *B. elliptica* both under greenhouse and field conditions. The results suggested that soil application of *B. cereus* could protect lily plants against the foliar pathogen by eliciting induced systemic resistance (ISR) (Liu et al. 2008).

Most aflatoxin biocontrol programs utilize nontoxigenic *Aspergillus* spp. to competitively exclude toxigenic fungal species. Successful reduction of aflatoxin contamination through the introduction of competitive non-aflatoxin producing strains of *A. flavus* has been demonstrated in several crops including corn (Dorner 2005). Aflatoxin contamination in corn was reduced by field application of wheat grains preinoculated with non-aflatoxigenic *Aspergillus flavus* strain NRRL30797. The reliability and efficiency of replacing wheat grains with the novel bioplastic formulation, Mater-Bi® was applied to serve as a carrier matrix to this formulated BCA. Mater-Bi® granules were inoculated with a conidial suspension of NRL 30797 strain to achieve a final cell density of approximately 10^7 conidia/granule. Incubation of 20 g-soil samples receiving a single Mater-Bi® granule for 60 days resulted in $10^{4.2}$–$10^{5.3}$ propagules of *A. flavus*/g of soil in microbiologically active (non-sterile) and sterilized soil respectively. The degree of soil colonization by the BCA was not enhanced by increasing the number of granules applied to the soil. In addition, to the maintenance of rapid vegetative growth and colonization of soil, the bioplastic formulation was also highly stable under soil environment. The results indicated that Mater-Bi® had the potential for substituting wheat grains as carrier for application of *A. flavus* NRRL30797 for reducing aflatoxin contamination in corn (Accinelli et al. 2009). In a later investigation to evaluate the feasibility of bioplastic-based formulations for delivering a non-aflatoxigenic strain of *Aspergillus flavus* was assessed for monitoring *Aspergillus* population to control aflatoxin contamination in corn. Field application of inoculated bioplastic granules showed a rapid shift in composition of soil population of *A. flavus* with a significant decrease in relative abundance of indigenous aflatoxigenic isolates. Application of bioplastic granules at 30 kg/ha was more efficient in replacing aflatoxigenic isolates than a 15 kg/ha dosage. Under field conditions, in test plots, aflatoxin contamination levels at corn maturity were 4.4 and 28.9 ng/g for 2009 and 2010 field seasons respectively. However, the bioplastic formulation was effective in reducing aflatoxin contamination in both years. Soil application of 15 and 30 kg/ha of bioplastic granules reduced aflatoxin contamination by 59 and 86 % in 2009 and 80 and 92 % in 2010 respectively (Accinelli et al. 2012).

5.2.4 Application on Aerial Plant Parts

Biocontrol agents have been applied on various plant parts such as leaves, stem and inflorescence to protect them against microbial plant pathogens that may cause localized or systemic infections. Powdery mildews of roses caused by *Sphaerotheca pannosa* var. *rosae* and of cucumber caused by *S. fuliginea* could be suppressed by the application of *Ampelomyces quisqualis* which efficiently parasitized the fungal pathogens, killing them in about 5 days (Sundheim and Krekling 1982). A formulated product AQ10 (Ecogen, Jerusalem, Israel) containing *A. quisqualis* was developed for the control of powdery mildews (Elad et al. 1996). *Verticillium lecanii* effectively suppressed the development of cucumber powdery mildew disease. In addition, *V. leccanii* was also pathogenic to the aphid *Macrosiphum euphorbiae* which can transmit *Cucumber mosaic virus* to cucurbitaceous plants. This investigation revealed for the first time that a fungal biocontrol agent could be effective against a plant pathogen and an arthropod vector of a plant virus (Askary et al. 1998). Another BCA, *Sporothrix flocculosa*, when applied on roses, was equally effective in protecting the plants against powdery mildew disease as the recommended fungicide (Bélanger et al. 1994). The biofunigicide AQ10TM containing *Ampelomyces quisqaulis* has been evaluated for its efficacy in suppressing the development of powdery mildews in several countries around the world. Early application at the commencement of disease incidence and effective coverage of the entire grapevine canopy has to be adopted for achieving efficient control by the BCA (Daoust and Hofstein 1996). *Fusarium proliferatum* G6, applied at weekly intervals, reduced development of downy mildew disease caused by *Plasmopara viticola* significantly on leaves and fruits. Reduction in disease severity on cultivar Lakemont ranged from 81 to 99 % in different years. The BCA could be preferentially employed at locations where pathogen strains resistant to metalaxyl have been detected (Falk et al. 1996). Biocontrol agents have been employed for the control of diseases affecting high value horticultural crops with a view to reducing the quantity of fungicides used so that the chances of development of fungicide-tolerant pathogen strains are minimized. Among the pathogens, *Botrytis cinerea* causing gray mold diseases on several crops has been the main target. *Trichodema harzianum* and *Gliocladium roseum* have been reported to be the most effective. The formulated product Trichodex 20P containing *T. harzianum* strain T-39 was tested under field conditions in 19 different countries (O'Neill et al. 1996). In Israel, a decision support system known as Botman (*Botrytis* manager) was developed by integrating chemical and biological control strategies involving Trichodex 20P. Based on weather records and 4-day forecast, chemical spray was recommended, if conditions are extremely favorable for disease development. During other periods, Trichodex 20P spray was carried out. Significant success in the control of *B. cinerea* infection in tomatoes and strawberries could be achieved with minimum use of chemicals (Elad et al. 1994). Another promising BCA useful for management of gray mold diseases is *Gliocladium roseum* which could effectively suppress the infection of stamens and fruits of strawberry. This BCA was equal in its effectiveness to or better than the fungicide and other BCA tested. *G. roseum* was also effective in protecting raspberry and conifer seedlings against *B. cinerea* (Yu and Sutton 1994; Sutton 1995).

The effectiveness of employing BCAs for the management of grapevine diseases has been well demonstrated. The gray mold disease caused by *Botrytis cinerea*, powdery mildew caused by *Uncinula necator* and downy mildew caused by *Plasmopara viticola* were traditionally managed by heavy application of fungicides in most countries. Appearance of strains of pathogens resistant to fungicides in many regions/countries necessitated development of alternative methods like biocontrol strategies and became a feasible approach. Spraying *T. harzianum* or Trichodex 25P was found to be effective in suppressing gray mold effectively. As few as two late applications of the BCAs provided equally effective protection against bunch rot as the fungicide applied five times throughout blossom-fruit development. Application of *T. harzianum* at bloom and early fruit development stage followed by a tank-mix application of BCA and iprodione (50 % of recommended dose) provided excellent control of bunch rot disease. Inclusion of a nutritive adhesive (Pelgel, a mixture of gum arabic and carboxymethylcellulose) enhanced the effectiveness of BCA treatment (O'Neill et al. 1996). *T. harzianum* T-39 was able to induce plant defense mechanisms against *B. cinerea* in tomato, lettuce, pepper, bean and tobacco. Application of *T. harzianum* T-39 at a site spatially separated from inoculation with *B. cinerea* resulted in a 25–100 % reduction in disease severity (De Meyer et al. 1998).

Fire blight disease of pear and apple caused by *Erwinia amylovora* is one of the devastating diseases, affecting fruit production drastically. The nonpathogenic bacterium *E. herbicola* (Eh1087) (syn: *Pantoea agglomerans*) is a natural epiphytic bacteria occurring on blossoms and it is inhibitory to *E. amylovora*. This BCA, when applied on apple blossoms, multiplies rapidly and the population reaches 400–800 folds compared with naturally occurring epiphytic population. When *E. herbicola* was applied at early and mid-blossom stages, the BCA provided 70–80 % protection against fire blight disease, while late application offered only 36 % protection (Kearns and Hale 1993). *E. herbicola* acts through preemptive colonization of blossom surfaces and the extent of control depended on the natural spread of the BCA to the blossoms, when they open after BCA application. Under wet and cool conditions, *E. herbicola* may spread rapidly aided by honey bees. The BCA could protect treated plants consistently to the level equivalent to that of streptomycin (Hickey 1996). *Pseudomonas fluorescens* strain A506 is a registered and commercialized for the control of fire blight as Blightban 506 (NuFarm, TX, USA). *P. fluorescens* A506 inhibits *E. amylovora* by competing with the pathogen for growth-limiting resources available. Strain A506 populations in open flowers at inoculation were initially about 10^4 cells/flower and increased to approximately 10^6 cells/flower in flowers that were inoculated within about 5 days of opening. Large total bacterial populations on A506-treated trees were associated with significant reductions in populations of *E. amylovora* and reduced incidence and severity of fire blight disease (Lindow and Suslow 2003).

The bacterial antagonist *Pantoea agglomerans* strain E325 of commercial interest was exceptionally effective in suppressing populations of *Erwinia amylovora* on flower stigmas of crab apple flowers (Pusey 2002). *P. agglomerans* E325 was developed commercially with the product name 'Bloomtime Biological' by North West Agricultural Products, WA, USA. A comparison of the ability of BCA strains of *P. agglomerans* E325 and *P. fluorescens* A506 to grow on stigmas of various ages revealed the failure of strain A506 to grow on late senescent stages that supported the growth of *P. agglomerans* and also the pathogen *E. amylovora*. This investigation revealed antagonist limitations that possibly affected field performance like inability of strain A506 to multiply on relatively old stigmas conducive to *E. amylovora* (Pusey and Curry 2004). In a later investigation, *P. agglomerans* strain E325 was found to have very high suppressive activity toward *E. amylovora* on stigmatic surfaces and to be effective in reducing blossom blight on mature apple trees. The strain E325 suppressed *E. amylovora* not only by competing for nutrients on stigma, but also by producing antibiotics specific to the pathogen (Pusey et al. 2008).

Sclerotinia sclerotiorum causes stem rot disease in canola (*Brassica napus*), infection commencing from petals generally. Four bacterial strains *Pseudomonas chlororaphis* PA-23, *Pseudomonas* spp. (DF41) and *Bacillus amyloliquefaciens* BS6 and E16 were evaluated for their efficacy in suppressing the development of *S. sclerotiorum* by spraying the bacterial suspension at 30 and 50 % bloom stages, followed by inoculation with ascospore suspension of the pathogen. *P. chlororaphis* strain PA-23 and *B. amyloliquefaciens* strain BS6 significantly reduced stem rot disease. Ascospore germination was inhibited and plant defense enzymes were triggered by strain PA-23. The effectiveness of strains PA-23 and strain BS6 was comparable to the level of control that could be obtained with the fungicide Rovral Flo® (iprodione). No significant difference in the disease incidence in plots receiving single- or double-spray application of the BCA strains (Fernando et al. 2007).

5.2.5 Dissemination of Biological Control Agents Through Pollinators

A novel method was developed for spreading biocontrol agents through honey bees which visit flowers for their nectar and incidentally spreading the biocontrol agents as well as plant pathogens. *Pseudomonas fluorescens* effective against *Erwinia amylovora* causing fire blight disease of apple. The honey bee species *Apis mellifera* could deposit the bacterial BCA on floral parts, when the flowers open soon after application, because of their foraging habits (Thomson et al. 1992). In another investigation, an efficient procedure of spreading the fungal BCA *Gliocladium roseum* formulated with talc and corn meal was developed. The formulated material was placed inside the bee-hive in a dispenser. As the honey bees moved inside the hive, the conidia of *G. roseum* adhered to their legs and bodies. The BCA propagules were then deposited on the floral parts, when the honey bees visited the strawberry flowers for nectar later (Peng et al. 1992). *Monilinia vaccinii-corymbosi* causes the mummy berry disease in blueberry and the commercial product 'Serenade' containing *Bacillus subtilis* showed antagonistic activity against *M. vaccinii-corymbosi* causing flower infection. The efficacy of *B. subtilis* was assessed by using bee-hives equipped with dispensers containing the biocontrol product for reducing the disease incidence. On caged rabbit eye blueberry bushes in the field, population densities of *B. subtilis* vectored by honey bee reached a carrying capacity of $<10^3$ CFU per flower stigma within 2 days of exposure. Serenade treatment significantly affected the incidence of fruit mummification on caged bushes, whereby increasing bee density increased disease incidence and application of Serenade reduced the disease incidence. The results suggested that use of hive-dispersed biocontrol product like Serenade as a supplement during pollination could reduce the risk of mummy berry disease (Dedej et al. 2004). The efficacy of two pollinators *Apis mellifera* and the mason bee *Osmia cornuata* as carriers of *Bacillus subtilis* strain BD170 (present commercial product Biopro®) was assessed for the control of fire blight disease caused by *Erwinia amylovora*. Under field conditions, the efficiency of pollinating bee species in carrying *B. subtilis* from sprayed flowers to the stigmas of newly opened ones at different intervals was determined by detecting the BCA strain using PCR assays. *O. cornuata* was found to be more efficient in carrying the BCA from flower to flower. The percentages of PCR-positive flower samples were higher in the internal treated areas of the field compared with external treated areas (Maccagnani et al. 2009).

5.2.6 Electrostatic Application of Biocontrol Agents

The usefulness of electrostatic force to deliver the biocontrol agents onto specific plant structures precisely was investigated using *Bacillus subtilis* an efficient biocontrol agent capable of suppressing the development of several phytopathogens. Incorporation of electrostatic force was attempted to increase the mass transfer of the viable bacterial cells of *B. subtilis* onto stigmatic surfaces of blueberry infecting pathogens like *Monilinia* sp. The population density of 1.52×10^5 CFU of biocontrol agent electrostatically deposited per ~0.7 mm diameter stigma exceeded by 4.5-fold that was deposited by conventional full rate application of *B. subtilis*. The results revealed the potential of electrostatic spraying of *B. subtilis* for the protection of floral infection by fungal pathogens (Law and Scherm 2005).

5.2.7 Application on Postharvest Produce

Postharvest pathogens may infect fruits and vegetables either prior to harvest and/or during transport and storage causing heavy losses. The produce has to be protected against postharvest pathogens in such a way that the consumers do not lose any of the desirable attributes of the fruits and vegetables, because of the negative effects associated with the use of chemicals. Biocontrol approach has been demonstrated as an acceptable alternative for restricting the incidence and severity of postharvest diseases. Biocontrol agents have been applied on fruits and vegetables by spraying or immersing in the suspension of conidia/spores prior to storage at required temperature. The biocontrol agents have been able to restrict the pathogen development through one or more mechanisms such as antibiosis, competition for nutrients and/or space or induction of resistance in treated fruits against the pathogens. Different species of *Candida*, *Cryptococcus*, *Bacillus* and *Pseudomonas* have been effective to varying degrees, based on the strains of the BCA and pathogens.

Muscodor albus, an endophytic fungus has a different mechanism of biocontrol activity on the postharvest pathogens. *M. albus* produces a large number of volatile compounds which together can inhibit and kill several microbial pathogens causing field crop and postharvest diseases. Fumigation of apples for 7 days with a culture of *M. albus* provided complete control of apple blue mold (*Penicillium expansum*) and gray mold (*Botrytis cinerea*) diseases in wound-inoculated fruits. Direct contact between the pathogen and BCA was prevented. Since *M. albus* has sterile mycelium without any kind of spores that can be disseminated and does not require direct contact with the plant or pathogen, it could be employed as an effective biological fumigant for controlling postharvest diseases (Mercier and Jiménez 2004). An endophytic bacterial species *Pseudomonas putida* strain MGY2 reduced the incidence of anthracnose disease caused by *Colletotrichum gloeosporioides* in papaya fruits, when the papaya fruits were immersed in bacterial suspension for 10 min, followed by air-drying. The disease index and lesion diameter were also significantly reduced following treatment with the strains MGY2. Treatment with the BCA activated defense mechanisms operating normally in papaya fruits, resulting in enhanced activities of defense-related enzymes and accumulation of phenolic compounds. Treatment with the strain MGY2 accelerated the synthesis of host enzymes and compounds, but not chemicals harmful to the human beings, making the use of BCAs that induce resistance to pathogens, acceptable for large scale application (Shi et al. 2011).

5.2.8 Application of Biological Control Agents Compatible with Agricultural Inputs

Cultivation of crops has been successfully achieved by providing required in puts such as fertilizers and irrigation to sustain better plant growth and consequent higher yields. Biocontrol agents contribute to the yield increase by reducing ill-effects of microbial plant pathogens, causative agents of numerous economically important diseases. Attempts have been made to determine the effect of combining application of BCAs with fertilizer application and irrigation systems. Simultaneous application of *Coniothyrium minitans*, a mycoparasite, effective against *Sclerotinia scleorotiorum* causing Sclerotinia stem rot disease oilseed rape and the compound fertilizer (N-P-K) at various concentrations significantly reduced ($P < 0.05$), the number of apothecia produced by sclerotia of *S. sclerotiorum* in field experiments. The compound fertilizer did not adversely affect the ability of *C. minitans* to infect the sclerotia of the pathogen or to suppress the carpogenic germination of sclerotia of the pathogen. *C. minitans* was found to be compatible with the compound fertilizer when applied at planting of oilseed rape. Application of the BCA along with the fertilizer could suppress development of stem rot disease and also save labor cost, increasing production efficiency of oilseed rape (Yang et al. 2011).

Compatibility of biocontrol agents with current crop management practices is an important attribute that can widen its acceptability for the management of crop diseases. The biofungicide AQ10TM containing *Ampelomyces quisqaulis* has been evaluated for its efficacy in suppressing the development of powdery mildews in several countries around the world. Early application at the commencement of disease incidence and effective coverage of the entire grapevine canopy has to be adopted for achieving efficient control by the BCA (Daoust and Hofstein 1996). *Fusarium proliferatum* G6, applied at weekly intervals, reduced development of downy mildew disease caused by *Plasmopara viticola* significantly on leaves and fruits. Reduction in disease severity on cultivar Lakemont ranged from 81 to 99 % in different years. The BCA could be preferentially employed at locations where pathogen strains resistant to metalaxyl have been detected (Falk et al. 1996). The compatibility of *Trichoderma atroviride* strain C52, effective against *Sclerotium cepivorum*, infecting onion (*Allium* pp.) crops with management practices was assessed. Addition of two blended pellet products containing poultry manure and other organic nutrients or humic acid plus organic matter to the sandy soil stimulated the proliferation of *T. atroviride*. The BCA strain C52 was less sensitive to the field rate application of urea in field soils. However, increase in the rate of urea applied, reduced the growth of the BCA. Populations of C52 were not adversely affected when exposed to any of the fungicide soil treatments tested. Diallyl disulphide (DADS) applied to soil at 4, 6 and 8 weeks before treatment with the strain C52, did not have any detectable negative effects on the BCA. The results indicated that effects of production practices followed in a location have to be assessed before choosing the BCA for field application (Mc Lean et al. 2012).

5.3 Registration and Commercialization of Biological Control Agents

Biocontrol agents have been demonstrated to have the potential to become alternatives for synthetic chemicals applied for the management of diseases affecting a wide range of crops in different ecosystems. The biocontrol agents that have provided consistent protection have been registered and commercialized. Biocontrol products are regulated by systems originally designed for chemical pesticides that have created market entry barriers by imposing burdensome costs on the biopesticide industry. It is necessary as a first step, to ensure legal protection for the new formulation based on biocontrol organisms, as a biotechnological invention by obtaining a patent. A temporary privilege is provided by a patent for industrial or commercial exploitation given by the administration to the owner for 20 years after the application date. Biopesticide patents are considered biotechnological inventions, because they include microbial products and processes. Filing of patents for microbial pesticides is regulated by a legal series of national and international treaties.

5.3.1 Regulatory Requirements and Barriers for Registration

For registration of new formulations, different sets of regulations have to be followed depending on the country where the product is to be registered. The Food, Insecticide, Fungicide and Rodenticide Act (FIFRA) of United States of America, requires any material making pesticide claims must be registered with the U. S. Environmental Protection Agency (EPA). Registration of biopesticides in European Union Countries and Canada appears to be more difficult than in the USA. When the data package is submitted, the regulatory frame work in EU requires not only toxicological and environmental testing, but also efficacy evaluation. Efficacy tests have to be performed for almost every crop-pathogen combination, resulting in the limitation of registration on different pathogens. Costs are enormous especially for entry level products. In the USA, registration of bioproducts is authorized by the Office of Pesticides Programs in EPA. In the EU, the register is maintained by the Directorate of the Consumer Health Protection and it is regulated by Directive 911414/CEE. Some of the biopesticides commercially produced based on fungal and bacterial biocontrol agents are presented in Tables 5.3 and 5.4 respectively. Further details are available in the websites: www.ipc.orst.edu/biocontrol/biopesticides and www.epa.gov/pesticides/biopesticides (Montesinos 2003; Harman et al. 2010).

According to the definition of FIFRA, the bioagents that are not themselves toxic or parasitic on target pathogens, but capable of inducing resistance in plants against pathogen(s), are also "pesticides". The microbial pesticide regulations in FIFRA and similar regulatory frame works, are primarily designed for specific, defined active ingredients. Hence, the complex microbial systems such as microbial mixtures, composts, compost teas and other complex mixtures are not well suited to these regulations. Companies may be interested to develop biocontrol products only if the profit margin is favorable. Likewise, the decision for the farmer whether or not to adopt a new technology depends on economic terms as a cost-benefit comparison of profits to be made from applying the new versus the existing technology. Another point to be considered is that conventional chemicals are being used widely and the fixed costs associated with them are spread over large number of users. Hence, it may represent a small percentage of the total cost of pest control. On the other hand, potential adopters of bioproducts, being limited in number, have to bear high fixed costs of the adoption that can decrease only when the new products attract many more users. The inconsistency in the performance of the biocontrol agents under various environmental conditions poses considerable risk for the manufacturers and farmers who have to be fully convinced of the effectiveness of the newly developed bioproducts.

The guidance of Organization for Economic Cooperation and Development (OECD) for microbial biopesticides is that "the microorganism and its metabolites pose no concerns of pathogenicity or toxicity to mammals and other non-target organisms which will likely be exposed to the microbial product; the microorganism does not produce a known genotoxin; all additives in the microbial manufacturing product and in end-use formulations are of low toxicity and suggest little potential

Table 5.3 Fungal biocontrol agents marketed as commercial products for the control of crop diseases

Commercial product	Fungal agent	Disease/pathogen	Formulation	Producer
AQ Biofungicide	*Ampelomyces quisqualis*	Powdery mildews of apple, grapes, cucurbits, strawberry, tomato and ornamentals	Water dispersible granules	Ecogen, Inc., PA, USA
Aspire	*Candida oleophila*	*Botrytis cinerea*, *Penicillium* spp.	Water dispersible granules	Ecogen, Inc., PA, USA
Binab T WG Bineb T Pellet	*Trichoderma harzianum* ATCC 20476, *T. polysporum* ATCC 20475	Fungi causing wilt, and root rot diseases; wood decay	Wettable powder/granules	Bio-Innovation, Sweden
Bioderma Bioderma-H	*Trichoderma viride*, *T. harzianum*	*Sclerotinia*, *Rhizoctonia*, *Phytophthora*, *Pythium*, *Fusarium*, *Cercospora*, *Colletotrichum*, *Ralstonia*	Wettable powder	Biotech International, India
Biofox C	*Fusarium oxysporum* (non-pathogenic)	*Fusarium oxysporum*, *F. moniliforme*		SIAPA, Italy
Biofungus	*Trichoderma* spp.	*Sclerotinia*, *Rhizoctonia*, *Fusarium*, *Pythium*, *Verticillium*	Wettable powder, granule	Grondortsmettingen de Cuester, Belgium
Contans WG Intercept WG	*Coniothyrium minitans* strain CON/M9108	*Sclertotinia sclerotiorum*, *S. minor*	Water dispersible granule	Sylvan Bioproducts USA; Prophyta, Germany
Fungi-Killer	*Trichoderma harzianum*	*Phytophthora*, *Fusarium oxysporum*	Powder	Bangkok Organic Compost Ltd, Thailand
Fusaclean	*Fusarium oxysporum* Fo47 (non-pathogenic)	*Fusarium oxysporum*	Spores, microgranules	Natural Plant Protection, France

Ketocin	*Chaetomium cupreum*	*Fusarium oxysporum* f.sp. *lycopersici*	Powder	Neoworld Ltd, Thailand
Ketomium	*Chaetomium globosum*, *C. cupreum*	*Phytophthora* spp., *Colletotrichum* sp., *F. oxysporum* f.sp. *lycopersici*, *Pyricularia grisea*, *Sclerotium rolfsii*, *Drechslera maydis*	Pellets, powder	Gungxi Guilin Green Harvest Fertilizer Factory, China; Nova Science, Thailand
Koni	*Coniothyrium minitans*	*Sclerotinia* spp.		BIOVED Ltd, Hungary
Novacide	*Chaetomium cupreum*	*F. oxysporum* f.sp. *lycopersici*	Powder	Nova Science, Thailand
Polygandron	*Pythium oligandrum*	*Pythium ultimum*	Granules or powder	Plant Protection Institute, Slovak Republic
Polyversum				
Prestop	*Gliocladium catenulatum* Strain JI446	*Pythium* spp., *R. solani*, *Botrytis* spp., *Didymella* spp.	Wettable powder	Kemira Agro Oy, Finland; AgBio Development Inc., USA
Primastop				
Promote	*T. harzianum*, *T. viride*	*Pythium* spp., *R. solani*, *F. oxysporum*		JH Bioteh Inc., Ventura, CA, USA
RootShield/ Plant Shield	*T. harzianum* Rifai Strain T-22	*Pythium*, *Rhizoctonia*, *Fusarium*, *Sclerotinia*	Granules, Wettable powder	Bioworks, Inc., NY, USA
SoilGard (GlioGard)	*Gliocladium virens* GL21	*R. solani*, *Pythium* spp.	Granules, alginate prill	Thermo Triology, USA; Certis Columbia, USA
Trichodex	*T. harizianum* T-39	*Botrytis cinerea*, *Colletotrichum*, *Plasmopara viticola*	Wettable powder	Makhteshim-Agan De Ceuster, Belgium
Vinevax™	*Trichoderma* spp.	Grapevine wood-infecting fungi, ornamental trees	Wettable powder	Agrimms Technology Ltd. www.vinewax.com

Table 5.4 Bacterial biocontrol agents marketed as commercial products for the control of crop diseases

Commercial product	Bacterial agent	Diseases/pathogens	Formulation	Producer
Actinovate	*Streptomyces lydicus*	Soilborne disease of greenhouse crops and nursery seedlings	Water-dispersible granules	Natural Industries, Houston, USA
Avogreen	*Bacillus subtilis*	*Colletotrichum gloeosporioides*, *Cercospora* spp. (Avocado)	Water-dispersible granules	Ocean Agriculture, South Africa
Ballad	*Bacillius pumilus* QST2808	Asian soybean rust	Liquid	AgraQuest, Inc., CA, USA
BioJect Spot-less	*Pseudomonas aureofaciens*	Anthracnose, damping-off of turf and other crops	Liquid	Eco Soil Systems, Inc., CA, USA
Bios-Save COLP110	*Pseudomonas syringae*	Postharvest pathogens – *Botrytis cinerea*, *Penicillium* spp., *Mucor pyriformis*, *Geotrichum candidum*	Wettable powder	JET Harvest Solutions, Longwood, USA
BlightBan A506	*Pseudomonas fluorescens* A506	Fire blight of apple, pear, peach, strawberry	Wettable powder	NuFarm Inc., www.nufarm.com
Cedomon	*Pseudomonas chlororaphis*	Leaf stripe, net blotch, leaf spot diseases of barley and oat	Seed treatment	BioAgri AB, Uppsala, Sweden
Companion	*Bacillus subtilis* GB03, *B. licheniformis*, *B. megaterium*	Root rot, damping-off, blight diseases of greenhouse crops and nursery seedlings	Liquid	Growth Products, NY, USA
Deny	*Burkholderia cepacia* type Wisconsin	*Rhizoctonia*, *Pythium*, *Fusarium* infecting alfalfa, barley, beans, cotton, peas and wheat	Peat-based dried biomass from solid fermentation	Stine Microbial Products, Memphis, USA
EcoGuard	*Bacillus licheniformis*	Turf dollar spot	Liquid spore concentrate	Novozymes, Bagsvaerd, Denmark
Galltrol	*Agrobacterium radiobacter* strain 84	Crown gall disease of fruit and nut trees	Bacterial mass suspended in chlorinated water	AgBioChem, Inc., CA, USA
GB34 Biological Fungicide	*Bacillus pumilus* GB34	*Rhizoctonia*, *Fusarium* infecting soybean seeds	Water slurry applied to seeds	Gustafson LLC, TX, USA

HiStick N/T	*Bacillus subtilis* MBI600	*Fusarium*, *Rhizoctonia*, *Aspergillus* infecting seeds of soybean, peanut, alfalfa	Slurry for damp and dry inoculation of seeds	Becker Underwood, IA, USA
Intercept	*Burkholderia cepacia*	*R. solani*, *Fusarium* spp., *Pythium* spp., infecting maize, cotton and vegetables		Soil Technologies Corp, IA, USA
Kodiak (several formulations)	*Bacillus subtilis* GB03	*R. solani*, *Fusarium* spp., *Alternaria* spp., *Aspergillus* spp., infecting cotton, legumes	Dry powder	Gustafson Inc., TX, USA
Mycostop	*Streptomyces griseo-viridis* strain K61	*Fusarium*, *Alternaria*, *Phomopsis*, *Pythium*, *Phytophthora* causing seed rot stem rot and wilt diseases	Powder	Kemira Agro, Helsinki, Finland
Nogall	*Agrobacterium radiobacter* K1026	*Agrobacterium tumefaciens* causing crown gall disease	Pure culture grown in petriplates on agar	Bio-Care Technologies, Australia; New BioProducts, OR, USA
Serenade Sonata	*Bacillus subtilis* QST716 *Bacillus pumilus*	Powdery mildew, downy mildew, early and late blight, brown rot, fire blight, Cercosopora leaf spot in cucurbits, grapes, vegetables, peanuts and fruit crops	Wettable powder	AgraQuest Inc., Davis, CA, US
Subtilex	*Bacillus subtilis* MBI600	*Fusarium*, *Rhizoctonia*, *Pythium*, causing seed and root rots in vegetable crops	Powder	Becker Underwood, IA, USA
YieldShield	*Bacillus pumilus* GB34	Fungal pathogens causing root diseases in soybean	Dry powder for seed treatment	Gustafson, Inc., TX, USA

for human health or environmental hazard" (OECD 2003). The information to be provided include mode of action, toxicological and eco-toxicological evaluations and host range testing. It is expensive for the companies to produce and it can deter them from commercialization of biopesticides which are usually niche market products. The decision whether or not to authorize a biopesticide product is made on the basis of expert opinion available within the regulatory authority. Hence, lack of expertise with biospesticides may lead to delay in making the required decision (Chandler et al. 2011).

Microorganisms labeled as 'inoculants', 'plant strengthening agents' and 'biofertilizers' are considered as organisms for improving the plant performance. Microorganisms such as plant growth-promoting rhizobacteria (PGPRs), mycorrhizal fungi and *Trichoderma* spp. provide several beneficial effects on plants. They may be produced and sold in various agricultural systems without specific pesticide registration, if pesticidal claims are not made. Sales of biocontrol agents in this manner may exceed those of formally registered organisms. A strain of *Bacillus subtilis* was sold by Gustafson, Inc., as Kodiak for treating the seeds as a bioinoculant for increasing yields of crops. Likewise, *Trichoderma harzianum* T 22 is sold as a registered pesticide product in the USA, but it is marketed (as Trianum) in Europe as a plant strengthening agent, although its biocontrol activity against several plant pathogens is known. Development of these useful products is much rapid and less expensive compared to the full pesticide registration marketing process. In certain countries, local production systems are in operation. In Tamil Nadu, India, unskilled persons are trained by interactive and learning-by-doing techniques. Biological agents such as *Trichoderma*, *Pseudomonas*, *Azospirillum* and mycorrhizal fungi are produced by different groups with a production capacity of about 1,000 kg/month. The local production systems are likely to be much less expensive and effective distribution of the bioproducts may be achieved more easily compared with sophisticated production systems in developed countries (Selvamukian et al. 2006; Harman et al. 2010).

References

Abadias M, Teixidó N, Usall J, Benabarre A, Viñas I (2001) Viability, efficacy and storage stability of freeze-dried biocontrol agent *Candida sake* using different protective and rehydration media. J Food Protect 64:856–861

Abadias M, Teixidó N, Usall J, Viñas I (2003) Liquid formulation of the postharvest control agent *Candida sake* CPA-1 in isotonic solutions. Phytopathology 93:436–442

Abumasha R, Salman M, Ehlers R (2011) Improvement of seed bio-priming of oilseed rape (*Brassica napus* ssp. *oleifera*) with *Serratia plymuthica* and *Pseudomonas chlororaphis*. Biocontrol Sci Technol 21:199–213

Accinelli C, Saccá ML, Abbas HK, Zablotowicz RM, Wilkinson JR (2009) Use of granular bioplastic formulation for carrying conidia of a non-aflatoxigenic strain of *Aspergillus flavus*. Bioresour Technol 100:3997–4004

Accinelli C, Mencarelli M, Saccá ML, Vicari A, Abbas HK (2012) Managing and monitoring of *Aspergillus flavus* in corn using bioplastic-based formulations. Crop Prot 32:30–35

Andrews JH, Harris RF (2000) The ecology and biogeography of microorganisms on plant surfaces. Annu Rev Phytopathol 38:145–180

Ardakani SS, Heydari A, Khorasani N, Arjmandi R (2010) Development of new bioformulation of *Pseudomonas fluorescens* and evaluation of these products against damping-off of cotton seedlings. J Plant Pathol 92:83–88

Askary H, Carriére Y, Bélanger RR, Bordeur J (1998) Pathogenicity of the fungus *Verticillium leccanii* to aphids and powdery mildew. Biocontrol Sci Technol 8:23–32

Bailey KL, Carisse O, Leggett M, Holloway G, Leggett F, Wolf TM, Shivpuri A, Derby J, Caldwell B, Geissler HO (2007) Effect of spraying adjuvants with the biocontrol fungus *Microsphaeropsis ochracea* at different water volumes on the colonization of apple leaves. Biocontrol Sci Technol 17:1021–1036

Bashan Y (1998) Inoculants of plant growth-promoting rhizobacterial for use in agriculture. Biotechnol Adv 16:729–770

Bashan Y, Hernandez JP, Leyva LA, Bacilio M (2002) Alginate microbeads as inoculants carrier for plant growth-promoting bacteria. Biol Fertil Soils 35:359–368

Batta YA (2007) Control of postharvest diseases of fruit with an invert emulsion formulation of *Trichoderma harzianum* Rifai. Postharvest Biol Technol 43:143–150

Bélanger RR, Labbé C, Jarvis WR (1994) Commercial scale control of rose powdery mildew with fungal antagonists. Plant Dis 78:420–424

Benbow JM, Sugar D (1999) Fruit surface colonization and biological control of postharvest diseases of pear by preharvest yeast applications. Plant Dis 83:839–844

Bora T, Ozaktan H, Gore E, Aslan E (2004) Biological control of *Fusarium oxysporum* f.sp. *melonis* by wettable powder formulations of the two strains of *Pseudomonas putida*. J Phytopathol 152:471–475

Caesar AJ, Burr TJ (1991) Effect of conditioning, betaine and sucrose on survival of rhizobacteria in powder formulations. Appl Environ Microbiol 57:168–172

Callan NW, Mathre DE, Miller JB (1990) Bio-priming seed treatment for biological control of *Pythium ultimum* preemergence damping-off in *sh2* sweet corn. Plant Dis 74:368–372

Cañamás TP, Viñas I, Usall J, Casals C, Solsona C, Teixidó N (2008a) Control of postharvest diseases on citrus fruit by preharvest application of the biocontrol agent *Pantoea agglomerans* CPA-2: part I. Study of different formulation strategies to improve survival of cells in unfavorable environmental conditions. Postharvest Biol Technol 49:86–95

Cañamás TP, Viñas I, Usall J, Magan N, Solsona C, Teixidó N (2008b) Impact of mild heat treatments on induction of thermotolerance in the biocontrol yeast of *Candida sake* CPA-1 and viability after spray drying. J Appl Microbiol 104:767–775

Cañamás TP, Viñas I, Torres R, Usal J, Solsona C (2011) Field applications of improved formulations of *Candida sake* CPA-1 for control of *Botrytis cinerea*. Biol Control 56:150–158

Cao L, Qui Z, You J, Tan H, Zhou S (2005) Isolation and characterization of endophytic streptomycete anatagonists of Fusarium wilt pathogen from surface-sterilized banana roots. FEMS Microbiol Lett 247:147–152

Chandler D, Bailey AS, Tatchell GM, Davidson G, Greaves J, Grant WP (2011) The development, regulation and use of biopesticides for integrated pest management. Phil Trans R Soc B 366:1987–1998

Chung WC, Huang JW, Huang HC (2005) Formulation of a soil biofungicide for control of damping-off of Chinese cabbage (*Brassica chinensis*) caused by *Rhizoctonia solani*. Biol Control 32:287–294

Collins DP, Jacobsen BJ, Maxwell B (2003) Spatial and temporal population dynamics of a phyllosphere colonizing *Bacillus subtilis* biological control agent of sugar beet Cercospora leaf spot. Biol Control 26:224–232

Cook DM (2002) Effect of formulated yeast in suppressing the liberation of *Botrytis cinerea* conidia. Plant Dis 86:1265–1270

Costa E, Usall J, Teixidó N, Garcia N, Viñas I (2000) Effect of protective agents, rehydration media and initial cell concentration on viability of *Pantoea agglomerans* strain CPA-2 subjected to freeze drying. J Appl Microbiol 89:793–800

Costa E, Teixidó N, Usall J, Atares E, Viñas I (2001) Production of the biocontrol agent *Pantoea agglomerans* strain CPA-2 using commercial products and byproducts. Appl Microbiol Biotechnol 56:367–371

Costa E, Usall J, Teixidó N, Torres R, Viñas I (2002) Effect of package and storage conditions on viability and efficacy of freeze-dried biocontrol agent, *Pantoea agglomerans* strain CPA-2. J Appl Microbiol 92:873–878

Daoust RA, Hofstein R (1996) *Ampelomyces quisqualis*, a new biofungicide to control powdery mildew in grapes. Brighton Crop Protect Conf 1:33–40

De Cal A, Pascual S, Melgarejo P (1997) Involvement of resistance induction by *Penicillium oxalicum* in the biocontrol of tomato wilt. Plant Pathol 46:72–79

De Cal A, García-Lepe R, Melgarejo P (2000) Induced resistance by *Penicillium oxalicum* against *Fusarium oxysporum* f.sp. *lycopersici*: histological studies of infected and induced tomato stems. Phytopathology 90:260–268

De Meyer G, Bigirimana J, Elad Y, Hyette M (1998) Induced systemic resistance in *Trichoderma harzianum* T39 biocontrol of *Botrytis cinerea*. Eur J Plant Pathol 104:279–286

Dedej S, Delaplane KS, Schern H (2004) Effectiveness of honey bees in delivering the biocontrol agent *Bacillus subtilis* to blueberry flowers to suppress mummy berry disease. Biol Control 31:422–427

Dorner JW (2005) Biological control of alfatoxin crop contamination. In: Abbas HK (ed) Aflatoxin and food safety. CRC Press, Taylor & Francis Group, Boca Raton, pp 333–352

Droby S, Cohen L, Daus A, Weiss B, Horev B, Chalutz E, Kalz H, Keren-Tzun M, Shachnai A (1998) Commercial testing of aspire: a yeast preparation for the biological control of postharvest decays of citrus. Biol Control 12:97–101

Dubey SC, Tripathi A, Bhavani R, Singh B (2011) Evaluation of seed dressing and soil application formulations of *Trichoderma* species for integrated management of dry root rot of chickpea. Biocontrol Sci Technol 21:93–100

Duffy BK, Simon A, Weller DM (1996) Combination of *Trichoderma koningii* with fluorescent *Pseudomonas* for control of take-all on wheat. Phytopathology 86:188–194

Dunlap CA, Schisler DA (2010) Fluidized-bed drying and storage stability of *Cryptococcus flavescens* OH 182.9, a biocontrol agent of Fusarium head blight. Biocontrol Sci Technol 20:465–474

Elad Y, Shtienberg D, Niv A (1994) *Trichoderma harzianum* T-39 integrated with fungicides improved biocontrol of grey mould. Brighton Crop Protect Conf 3:1109–1114

Elad Y, Malathrakis NE, Dik AJ (1996) Biological control of *Botrytis*-incited diseases and powdery mildews in greenhouse crops. Crop Prot 15:229–240

El-Hassan SA, Gowen SR (2006) Formulation and delivery of the bacterial antagonist *Bacillus subtilis* for management of lentil vascular wilt caused by *Fusarium oxysporum* f.sp. *lentis*. J Phytopathol 154:148–155

Errakhi R, Bouteau F, Lebrihi A, Barakate M (2007) Evidences of biological control capacities of *Streptomyces* spp. against *Sclerotium rolfsii* responsible for damping-off disease in sugar beet (*Beta vulgaris* L). World J Microbiol Biotechnol 23:1503–1509

Etebarian HR, Scott ES, Wicks TJ (2000) *Trichoderma harzianum* T-39 and *T. virens* DAR74290 as potential biocontrol agents for *Phytophthora erythroseptica*. Eur J Plant Pathol 106:329–337

Fakhouri WD, Khalaif H (1996) Biocontrol of crown gall disease in Jordan. Disrat Ser B, Pure Appl Sci 23:17–22

Falk SP, Pearson RC, Gadoury DM, Seem RC, Sztejnberg A (1996) *Fusarium proliferatum* as a biocontrol agent against grape downy mildew. Phytopathology 86:1010–1017

Fallahzadeh-Mamaghani V, Ahmadzadeh M, Sharifi R (2009) Screening systemic resistance-inducing fluorescent pseudomonads for control of bacterial blight of cotton caused by *Xanthomonas campestris* pv. *malvacearum*. J Plant Pathol 91:663–670

Feng XM, Holmberg AJ, Sundh I, Ricard T, Melin P (2011) Specific SCAR markers and multiplex real-time PCR for quantification of two *Trichoderma* biocontrol strains in environmental samples. BioControl 56:903–913

Fernando WGD, Nakkeeran S, Zhang Y, Savchuk S (2007) Biological control of *Sclerotinia sclerotiorum* (Lib.) de Bary by *Pseudomonas* and *Bacillus* species on canola petals. Crop Prot 26:100–107

Fravel DR, Rhodes DJ, Larkin RP (1999) Production and commercialization of biocontrol products. In: Albajes R, Gullino LM, van Lenteren JC, Elad Y (eds) Integrated pest and disease management in greenhouse crops. Kluwer, Dordrecht, pp 365–376

Fravel DR, Deahl KL, Stommel JR (2005) Compatibility of the biocontrol fungus *Fusarium oxysporum* strain CS-20 with selected fungicides. Biol Control 34:165–169

Furuya N, Yamasaki S, Nishioka M, Shiraishi I, Iiyama K, Matsuyama N (1997) Antimicrobial activities of pseudomonads against plant pathogenic organisms and efficacy of *Pseudomonas aeruginosa* ATCC7700 against bacterial wilt of tomato. Ann Phytopathol Soc Jpn 63:417–424

Goyal TC, Spotts RA (1997) Biological control of postharvest diseases of apple and pear under semi-commercial conditions using three saprophytic yeasts. Bull OILB/SROP 21:81–89

Hang NTT, Oh S-O, Kim GH, Hur J-S, Koh YJ (2005) *Bacillus subtilis* S1-0210 as a biocontrol agent against *Botrytis cinerea* in strawberries. Plant Pathol J 21:59–63

Harman GE, Shoresh M (2007) The mechanisms and applications of opportunistic plant symbionts. In: Vurro M, Gressel J (eds) Novel biotechnologies for biocontrol agent enhancement and management. Springer, Amsterdam, pp 131–153

Harman GE, Taylor AG (1990) Development of an effective biological seed treatment system. In: Hornby D (ed) Biological control of soilborne plant pathogens. CAB International, Wallingford, pp 415–426

Harman GE, Bjorkman T, Ondik KL, Shoresh M (2008) Changing paradigms on the mode of action and uses of *Trichoderma* spp. for biocontrol. Outlooks Pest Manag 19:24–29

Harman GE, Obregón MA, Samuels GJ, Lorito M (2010) Changing models for commercialization and implementation of biocontrol in the developing and the developed world. Plant Dis 94:928–939

Helbig J (2001) Field and laboratory investigations into the effectiveness of *Rhodotorula glutinis* (isolate 10391) against *Botrytis cinerea* Pers. ex Fr. in strawberry. J Plant Dis Protect 108:356–368

Hernández-Suárez M, Hernández-Castillo D, Gallego-Morales G, Lira-Saldivar RH, Rodríguez-Herrera R, Aguilar CN (2011) Biocontrol of soil fungi in tomato with microencapsulates containing *Bacillus subtilis*. Am J Agric Biol Sci 6:189–195

Hickey KD (1996) Antagonistic bacteria for fire blight control. Pennsyl Fruit News 76:25–26

Hidalgo E, Bateman R, Krauss U, Hoopen M, Martinez A (2003) A field investigation into delivery systems for agents to control *Monolithophthora roreri*. Eur J Plant Pathol 109:953–961

Howell CR (1998) The role of antibiosis in biocontrol. In: Harman GE, Kubicek CP (eds) *Trichoderma* and *Gliocladium*, vol 4. Taylor & Francis, London, pp 173–184

Howell CR (2006) Understanding the mechanisms employed by *Trichoderma virens* to effect biological control of cotton diseases. Phytopathology 96:178–180

Howell CR, Hanson LE, Sipanovic RD, Puckhaber LS (2000) Induction of terpenoid synthesis in cotton and control of *Rhizoctonia solani* by seed treatment with *Trichoderma virens*. Phytopathology 90:248–252

Huang H-C, Erickson RS (2007) Use of sclerotia of *Sclerotinia sclerotiorum* for efficient production of conidia of *Coniothyrium minitans* in liquid culture. Phytoparasitica 35:140–145

Janisiewicz WJ (1994) Enhancement of biocontrol of blue mold with nutrient analog 2-deoxy-D-glucose on apples and pears. Appl Environ Microbiol 68:2671–2676

Janisiewicz WJ, Jeffers SN (1997) Efficacy of commercial formulation of two biofungicides for control of blue mold and gray mold of apples in cold storage. Crop Prot 16:629–633

Janisiewicz WJ, Conway WS, Leverentz B (1999) Biological control of postharvest decays of apple can prevent growth of *Escherichia coli* 0157:H7 in apple wounds. J Food Protect 62:1372–1375

Janisiewicz WJ, Pimenta RS, Jurick WM II (2011) A novel method for selecting antagonists against postharvest fruit decays originating from latent infections. Biol Control 59:384–389

Jayaraj J, Radhakrishnan NV, Kannan R, Sakthivel K, Suganya D, Venkatesan S, Velazhahan R (2005) Development of new formulations of *Bacillus subtilis* for management of tomato damping-off caused by *Pythium aphanidermatum*. Biocontrol Sci Technol 15:55–65

Jensen B, Knudsen IMB, Madsen M, Jensen DF (2004) Biopriming of infected carrot seed with an antagonist, *Clonostachys rosea*, selected for control of seedborne *Alternaria* spp. Phytopathology 94:551–560

Jensen DF, Knudsen IMB, Lübeck M, Mamarabadi M, Hockenhull J, Jensen B (2007) Development of a biocontrol agent for plant disease control with special emphasis on the near commercial fungal antagonist *Clonostachys rosea* strain IK726. Austr Plant Pathol 36:95–101

Jijakli MH (2000) Apple: storage diseases-biological control based on two yeast strains. Arboric Fruit 539:19–23

Jijakli MH, Lepoivre P (1998) Characterization of an exo-beta-1,3-glucanase produced by *Pichia anomala* strain K, antagonist of *Botrytis cinerea* on apples. Phytopathology 88:335–343

Joshi R, McSpadden Gardener BB (2006) Identification and characterization of novel genetic markers associated with biological control activities in *Bacillus subtilis*. Phytopathology 96:145–154

Kanjanamaneesathian M, Wiwattanapatapee R, Pengnoo A, Oungbho K, Chumthong A (2007) Efficacy of novel formulations of *Bacillus megaterium* in suppressing sheath blight of rice caused by *Rhizoctonia solani*. J Plant Pathol 6:195–201

Kanjanamaneesathian M, Chumthong A, Pengnoo A, Wiwattanapatapee R (2009) *Bacillus megaterium* suppresses major Thailand rice diseases. Asian J Food Agro-Indus (Spl Issue):S154–S159

Karabulut OA, Tezcan H, Daus A, Cohen L, Weiss B, Droby S (2004) Control of preharvest and postharvest fruit rot in strawberry by *Metschnikowia fructicola*. Biocontrol Sci Technol 14:513–521

Kawaguchi A, Inoue K, Nasu H (2007) Biological control of grapevine crown gall by nonpathogenic *Agrobacterium vitis* strain VAR03-1. J Gen Plant Pathol 73:133–138

Kawaguchi A, Inoue K, Ichinose Y (2008) Biological control of crown gall of grapevine, rose and tomato by nonpathogenic *Agrobacterium vitis* strain VAR03-1. Phytopathology 98:1218–1225

Kearns LP, Hale CN (1993) Biological control of fire blight by *Erwinia herbicola*: survival of applied bacteria in orchard and glasshouse trials. Acta Hortic 338:333–339

Kenwick S, Jacobsen BJ (1998) Biological control of Fusarium dry rot on potato with antagonistic bacteria in commercial formulation. Phytopathology 88:S47 (Abst)

Khan MR, Majid S, Mohidin FA, Khan N (2011) A new bioprocess to produce low cost powder formulations of biocontrol bacteria and fungi to control fusarial wilt and root-knot nematodes. Biol Control 59:130–140

Khmel IA, Sorokino TA, Lamanova NB, Liposava VA, Metlitski OZ, Burdeinaya TV, Chernin LS (1998) Biological control of crown gall in grapevine and raspberry by two *Pseudomonas* species with a wide spectrum of antagonistic activity. Biocontrol Sci Technol 8:45–57

Kim DS, Cook RJ, Weller DM (1997) *Bacillus* sp. 324-92 for biological control of three root diseases of wheat grown with reduced tillage. Phytopathology 87:551–558

Kiss L (2003) A review of fungal antagonists of powdery mildews and their potential as biocontrol agents. Pest Manag Sci 59:475–483

Kloepper JW, Ryu C-M, Zhang S (2004) Induced systemic resistance and promotion of plant growth by *Bacillus* spp. Phytopathology 94:1259–1266

Kolombet LV, Ezhov DM, Zhigletsova SK, Bystrova EV, Kosareva NI (2001) A preparation to control plant diseases. Patent of Russian Federation on Invention No. 21

Kolombet LV, Zhigletsova SK, Kosareva NI, Bystrova EV, Derbyshev VV, Krasnova SP, Schisler D (2008) Development of an extended shelf-life, liquid formulation of the biofungicide *Trichoderma asperellum*. World J Microbiol Biotechnol 24:123–131

Küçük C, Kivanç M (2005) Effect of formulation on the viability of biocontrol agent, *Trichoderma harzianum* conidia. Afr J Biotechnol 4:483–486

Kurze S, Bahl H, Dahl R, Berg G (2001) Biological control of fungal strawberry diseases by *Serratia plymuthica* HRO-C48. Plant Dis 85:529–534

Larena I, Melgarejo P, De Cal A (2002) Production, survival and evaluation of solid-substrate inoculate of *Penicillium oxalicum*, a biocontrol agent against Fusarium wilt of tomato. Phytopathology 92:863–869

Larena I, De Cal A, Linan M, Melgarejo P (2003) Drying of *Epicoccum nigrum* conidia for obtaining a shelf-stable biological product against brown rot disease. J Appl Microbiol 94:508–514

Larena I, De Cal A, Melgarejo P (2007) Effects of stabilizers on shelf-life of *Epicoccum nigrum* formulations and their relationship with biocontrol of postharvest brown rot by *Monilinia* of peaches. J Appl Microbiol 102:570–582

Larena I, De Cal A, Melgarejo P (2010) Enhancing the adhesion of *Epicoccum nigrum* conidia to peach surfaces and its relationship to the biocontrol of brown rot caused by *Monilinia laxa*. J Appl Microbiol 109:583–593

Larkin RP, Fravel DR (1998) Efficacy of various fungal and bacterial biocontrol organisms for control of Fusarium wilt of tomato. Plant Dis 82:1022–1028

Law SE, Scherm H (2005) Electrostatic application of a plant disease biocontrol agent for prevention of fungal infection through the stigmatic surfaces of blueberry flowers. J Electrostat 63:399–408

Le Floch G, Rey P, Beniziri E, Benhamou N, Trilly Y (2003) Impact of auxin-compounds produced by the antagonistic fungus *Pythium oligandrum* or the minor pathogen *Pythium* group F on plant growth. Plant Soil 257:459–470

Leifert C, Li H, Chidburee S, Hampson S, Workman S, Sigee D, Epton HAS, Harbour A (1995) Antibiotic production and biocontrol activity by *Bacillus subtilis* CL27 and *Bacillus pumilus* CL45. J Appl Bacteriol 78:97–108

Lewis K, Whipps JM, Cooke RC (1989) Mechanisms of biological diseases control with special reference to the case study of *Pythium oligandrum* as an antagonist. In: Whipps JM, Lumsden RD (eds) Biotechnology of fungi for improving plant growth. Cambridge University Press, Cambridge, pp 191–217

Li GQ, Huang HC, Miao HJ, Erickson RS, Jiang DH, Xiao YN (2006) Biological control of Sclerotinia diseases of rapeseed by aerial applications of the mycoparasite *Coniothyrium minitans*. Eur J Plant Pathol 114:345–355

Lindow SE, Suslow TV (2003) Temporal dynamics of the biocontrol agent *Pseudomonas fluorescens* strain A506 in flowers in inoculated pear trees. Phytopathology 93:727–737

Liu Y-H, Huang CJ, Chen C-Y (2008) Evidence of induced systemic resistance against *Botrytis elliptica* in lily. Phytopathology 98:830–836

Liu J, Tian S-P, Li B-Q, Qin G-Z (2009) Enhancing viability of two biocontrol yeasts in liquid formulation by applying sugar protectant combined with antioxidant. BioControl 54:817–824

Liu Z, Wei H, Li Y, Li S, Zhang L, Chen H (2011) Effects of milling and surfactants on suspensibility and spore viability in *Paenibacillus polymyxa* powder formulations. Biocontrol Sci Technol 21:1103–1116

Lolloo R, Maharaih D, Görgens J, Gardiner N (2010) A downstream process for production of a viable and stable *Bacillus cereus* aquaculture biological agent. Appl Microbiol Biotechnol 86:499–508

Losick E, Kolter R (2008) Ecology and genomics of *Bacillus subtilis*. Trends Microbiol 16:269–275

Maccagnani B, Giacomelllo F, Fanti M, Gobbin D, Maini S, Angeli G (2009) *Apis mellifera* and *Osmia cornuata* as carriers for the secondary spread of *Bacillus subtilis* on apple flowers. BioControl 54:123–133

Manikandan R, Saravanakumar D, Rajendran L, Raguchander T, Samiyappan R (2010) Standardization of liquid formulation of *Pseudomonas fluorescens* Pf1 for its efficacy against Fusarium wilt of tomato. Biol Control 54:83–89

Mao W, Lewis JA, Hebbar PK, Lamsden RD (1997) Seed treatment with a fungal or bacterial antagonist for reducing corn damping-off caused by species of *Pythium* and *Fusarium*. Plant Dis 81:450–454

Masih EI, Alie I, Paul B (2000) Can the grey mould disease of the grapevine be controlled by yeast? FEMS Microbiol Lett 189:1563–1569

Mathre DE, Cook RJ, Callan NW (1999) From discovery to use: traversing the world of commercializing biocontrol agents for plant disease control. Plant Dis 83:972–983

Mc Lean KL, Stewart A (2000) Application strategies for control of onion white rot. NZ J Crop Hortic Sci 28:115–122

Mc Lean KL, Swaminathan J, Frampton CM, Hunt JS, Ridgway HJ, Stewart A (2005) Effect of formulation on the rhizosphere competence and biocontrol activity of *Trichoderma atroviride* C52. Plant Pathol 54:212–218

Mc Lean KL, Hunt JS, Stewart A, Wite D, Porter IJ, Villalta O (2012) Compatibility of *Trichoderma atroviride* biocontrol agent with management practices of *Allium* crops. Crop Prot 34:94–100

Melin P, Håkansson S, Eberhard TH, Schnürer J (2006) Survival of the biocontrol yeast *Pichia anomala* after long term storage in liquid formulations at different temperatures, assessed by flow cytometry. J Appl Microbiol 100:264–271

Mercier J, Jiménez JI (2004) Control of fungal decay of apples and peaches by the biofumigant fungus *Muscador albus*. Postharvest Biol Technol 31:1–8

Mew TW, Rosales AM (1986) Bacterization of rice plants for control of sheath blight caused by *Rhizoctonia solani*. Phytopathology 76:1260–1264

Montesinos E (2003) Development, registration and commercialization of microbial pesticides for plant protection. Int Microbiol 6:245–252

Moretti M, Gilardi G, Gullino ML, Garibaldi A (2008) Biological control of potential of *Achromobacter xylosoxydans* for suppressing Fusarium wilt of tomato. Int J Bot 4:369–375

Murphy JF, Reddy MS, Ryu CM, Kloepper JW, Li R (2003) Rhizobacteria-mediated growth promotion of tomato leads to protection against *Cucumber mosaic virus*. Phytopathology 93:1301–1307

Milus EA, Hershman D, McMullen M (2001) Analysis of the 2001 uniform wheat fungicide and biocontrol trials across location. In: Proceeding of the 2001 national fusarium head blight forum, Kinko's Okemos, ML, USA, pp 75–79

Muslim A, Horniouchi H, Hayakumachi M (2003) Control of Fusarium crown and root rot of tomato with hypovirulent binucleate *Rhizoctonia* in soil and rock wool systems. Plant Dis 87:739–747

Nakano M, Hulett M (1997) Adaptation of *Bacillus subtilis* to oxygen limitation. Microbiologie 157:1–7

Narayanasamy P (2002) Microbial plant pathogens and crop disease management. Science Publishers, Enfield

Narayanasamy P (2006) Postharvest pathogens and disease management. Wiley, Hobokken

Nihorimbere V, Ongena M, Cawoy H, Brostaux Y, Kakana P et al (2010) Beneficial effects of *Bacillus subtilis* on field-grown tomato in Burundi: reduction of local Fusarium disease and growth promotion. Afr J Microbiol Res 4:1135–1142

O'Neill TM, Elad Y, Shtienberg D, Cohen A (1996) Control of grapevine grey mould with *Trichoderma harzianum* T-39. Biocontrol Sci Technol 6:139–146

OECD (2003) Organization of economic co-operation and development- series on pesticides no. 18. Guidance for registration requirements for microbial pesticides. www.oecd.org/dataoecd/4/23128888446.pdf

Ongena M, Daayf F, Jacques P, Thonart P, Benhamou N, Paulitz TC, Cornélis P, Koedam N, Bélanger RR (1999) Protection of cucumber against Pythium root rot by fluorescent pseudomonads: predominant role of induced resistance over siderophores and antibiosis. Plant Pathol 48:66–76

Ongena M, Daayf F, Jacques P, Thonart P, Benhamou N, Paulitz TC, Bélanger RR (2000) Systemic induction of phytoalexins in cucumber in response to treatments with fluorescent pseudomonads. Plant Pathol 49:523–530

Özaktan H, Bora T (2004) Biological control of fire blight in pear orchards with a formulation of *Pantoea agglomerans* strain Eh24. Braz J Microbiol 35:224–229

Pan L, Ash GJ, Ahn B, Watson AK (2010) Development of strain specific molecular markers for the *Sclerotinia minor* bioherbicide strain IMI344141. Biocontrol Sci Technol 20:939–959

Park K, Paul D, Kim YK, Nam KW, Lee YK, Choi HW, Lee SY (2007) Induced systemic resistance by *Bacillus vallismortis* EXTN-1 suppressed bacterial wilt in tomato caused by *Ralstonia solanacearum*. J Plant Pathol 23:22–25

Partridge DE, Sutton TB, Jordan DL, Curtis VL, Bailey JE (2006) Management of Sclerotinia blight of peanut with biological control agent *Coniothyrium minitans*. Plant Dis 90:957–963

Patiño-Vera M, Jiménez B, Balderas K, Ortiz M, Allende R, Carrillo A, Galindo E (2005) Pilot-scale production and liquid formulation of *Rhodotorula minuta*, a potential biocontrol agent of mango anthracnose. J Appl Microbiol 99:540–550

Paulitz TC, Bélanger RR (2001) Biological control in green house system. Annu Rev Phytopathol 39:103–133

Peighami-Ashnaei S, Sharifi-Tehrani A, Ahmadzadeh M, Behboudi K (2009) Interaction of different media on production and biocontrol efficacy of *Pseuedomonas fluorescens* P-35 and *Bacillus subtilis* B-3 against grey mould of apple. J Plant Pathol 91:65–70

Peng G, Sutton JC, Kevan PG (1992) Effectiveness of honey bees for applying the biocontrol agent *Gliocladium roseum* to strawberry flowers to suppress *Botrytis cinerea*. Can J Plant Pathol 14:117–188

Pengnoo A, Kusongwiriyawong C, Nilratana L, Kanjanamaneesathian M (2000) Greenhouse and field trials of the bacterial antagonists in the pellet formulations to suppress sheath blight of rice caused by *Rhizoctonia solani*. BioControl 45:245–256

Piggot P, Hilbert D (2004) Sporulation of *Bacillus subtilis*. Curr Opin Microbiol 7:579–586

Punja ZK, Utkhede RS (2004) Biological control of fungal diseases of vegetable crops with fungi and yeasts. In: Arora DK (ed) Fungal biotechnology in agricultural food and environmental applications. Marcel Dekker, New York, pp 157–171

Puoci F, Iemma F, Spizzirri UG, Cirillo G, Curcio M, Picci N (2008) Polymer in agriculture: a review. Am J Agric Biol Sci 3:299–314

Pusey PL (2002) Biological control agents for fire blight of apple compared under conditions limiting natural dispersal. Plant Dis 86:639–644

Pusey PL, Curry EA (2004) Temperature and pomaceous flower age related to colonization by *Erwinia amylovora* and antagonists. Phytopathology 94:901–911

Pusey PL, Wend C (2012) Potential of osmoadaptation for improving *Pantoea agglomerans* E325 as biocontrol agent for fire blight of apple and pear. Biol Control 62:29–37

Pusey PL, Stockwell VO, Rudell DR (2008) Anitbiotics and acidification by *Pantoea agglomerans* strain E325 may contribute to suppression of *Erwinia amlovora*. Phytopathology 98:1136–1143

Rhodes DJ (1993) Formulation of biological control agents. In: Jones DG (ed) Exploitation of microorganisms. Chapman & Hall, London, pp 411–439

Rifai MA (1969) A revision of the genus *Trichoderma*. Mycol Pap 116:1–56

Rivera-Varas V, Freeman TA, Gudmestad NC, Secor GA (2007) Mycoparasitism of *Helminthosporium solani* by *Acremonium strictum*. Phytopathology 97:1331–1337

Ryu C-M, Kim J, Choi O, Kim SH, Park CS (2006) Improvement of biological control capacity of *Paenibacillus polymyxa* E681 by seed pelleting on sesame. Biol Control 39:282–289

Sabuquillo P, De Cal A, Melgarejo P (2005) Dispersal improvement of a powder formulation of *Penicillium oxalicum*, a biocontrol agent of tomato wilt. Plant Dis 89:1317–1323

Sabuquillo P, De Ca A, Melgarejo P (2006) Biocontrol of tomato wilt by *Penicillium oxalicum* formulations in different crop conditions. Biol Control 37:256–265

Sabuquillo P, De Cal A, Melgarejo P (2010) Development of a dried *Penicillium oxalicum* conidial formulation for use as a biological agent against Fusarium wilt of tomato: selection of optimal additives and storage conditions for maintaining conidial viability. Biol Control 54:221–229

Salaheddin K, Valluvaparidasan V, Ladhalakshmi D, Velazhahan R (2010) Management of bacterial blight of cotton using a mixture of *Pseudomonas fluorescens* and *Bacillus subtilis*. Plant Protect Sci 46:41–50

Scarpellini M, Franzetti L, Galli A (2004) Development of PCR assay to identify *Pseudomonas fluorescens* and its biotype. FEMS Microbiol Lett 236:257–260

Schisler DA, Slininger PJ, Behle RW, Jackson MA (2004) Formulations of *Bacillus* spp. for biological control of plant diseases. Phytopathology 94:1267–1271

Selvamukian B, Rengalakshmi R, Tamizoli P, Nair S (2006) Village level production and use of biocontrol agents and biofertilizers. In: Uphoff N, Ball AS, Palm C, Fernandes E, Pretty J, Herren H, Sanchez P, Husson O, Sanginaga N, Liang M, Thies J (eds) Biological approaches to sustainable soil systems. Taylor & Francis, Boca Raton, pp 647–653

Shen S-S, Park S-H, Park C-S (2005) Enhancement of biocontrol efficacy of *Serratia plymuthica* A21-4 against Phytophthora blight of pepper by improvement of inoculation buffer solution. J Plant Pathol 21:68–72

Shi F, Zhu Y (2007) Application of statistically-based experimental designs in medium optimization for spore production of *Bacillus subtilis* from distillery effluent. BioControl 52:845–853

Shi J, Liu A, Li X, Feng S, Chen W (2011) Inhibiting mechanisms induced by the endophytic bacterium MGY2 in controlling anthracnose of papaya. Biol Control 56:2–8

Shishkoff N, McGrath MT (2002) AQ10 biofungicide combined with chemical fungicides or AddQ spray adjuvant for control of cucurbit powdery mildew in detached leaf culture. Plant Dis 86:915–918

Shoresh M, Harman GE (2008) Molecular basis of maize responses to *Trichoderma harzianum* T22 inoculation: a proteomic approach. Plant Physiol 147:2147–2163

Shoresh M, Matsouri F, Harman GE (2010) Induced systemic resistance and plant responses to fungal biocontrol agents. Annu Rev Phytopathol 48:21–43

Shternshis M, Tomilova O, Shaptova T, Soytong K (2005) Evaluation of Ketomium mycofungicide on Siberian isolates of phytopathogenic fungi. J Agric Technol 1:247–253

Simon A, Sivasithamparm K (1989) Pathogen-suppression: a case study in biological suppression of *Gaeumannomyces graminis* var. *tritici* in soil. Soil Biol Biochem 21:331–337

Slininger PJ, Dunlap CA, Schisler DA (2010) Polysaccharide production benefits dry storage survival of the biocontrol agent *Pseudomonas fluorescens* S11:P:12 effective against several maladies of stored potatoes. Biocontrol Sci Technol 20:227–244

Soe KT, De Costa DM (2012) Development of spore-based formulations of microbial pesticides for control of rice sheath blight. Biocontrol Sci Technol 22:633–657

Soytong K, Kanokamadhakul S, Kukongviriyapa V, Isobe M (2001) Application of *Chaetomium* species (Ketomium®) as a new broad spectrum biological fungicide for plant disease control: a review article. Fungal Div 7:1–15

Spadaro D, Ciavorella AA, Dianpeng Z, Garibaldi A, Gullino ML (2010a) Effect of culture media and pH on the biomass production and biocontrol efficacy of a *Metschnikowia pulcherrima* strain to be used as a biofungicide for postharvest disease control. Can J Microbiol 56:128–137

Spadaro D, Ciavorella AA, Lopez-Reyes JG, Garibaldi A, Gullino ML (2010b) Effect of culture age, protectants and initial cell concentration on viability of freeze-dried cells of *Metschnikowia pulcherrima*. Can J Microbiol 56:809–815

Sriram S, Roopa KP, Savitha MJ (2011) Extended shelf-life of liquid formulation derived from talc formulations of *Trichoderma harzianum* with the addition of glycerol in the production medium. Crop Prot 30:1334–1339

Steddom K, Becker O, Menge JA (2002) Repetitive applications of the biocontrol agent of *Pseudomonas putida* 06909-rif/nal and effects on populations of *Phytophthora parasitica* in citrus orchards. Phytopathology 92:850–856

Sundheim L, Krekling T (1982) Host-parasite relationships of the hyperparasite *Ampelomyces quisqualis* and its powdery mildew host *Sphaerotheca fuliginea*. Phytopathol Z 104:202–210

Sutton JC (1995) Evaluation of microorganisms: *Botrytis cinerea* and strawberry, a case study. In: Andrews JM, Tommerup IC (eds) Advances in plant pathology, vol II, sustainability and plant pathology. Academic, London, pp 173–190

Tan H, Zhou S, Deng Z, He M, Cao L (2011) Ribosomal sequence-directed selection for endophytic streptomycete strains antagonistic to *Ralstonia solanacearum* to control tomato bacterial wilt. Biol Control 59:245–254

Teixidó N, Viñas I, Usall J, Magan N (1998) Control of blue mold of apples by preharvest application of *Candida sake* grown in media with different water activity. Phytopathology 88:960–964

Thomson SV, Hensen DR, Flint KM, van den Berg JD (1992) Dissemination of bacteria antagonistic to *Erwinia amylovora* by honey bees. Plant Dis 76:1052–1056

Tian SP, Fan Q, Jiang AL (2002) Effects of calcium on biocontrol activity of yeast antagonists against the postharvest fungal pathogen *Rhizopus stolonifer*. Plant Pathol 51:352–358

Tombolini R, van der Gaag DJ, Gerhardson B, Jansson JK (1999) Colonization pattern of the biocontrol strain *Pseudomonas chlororaphis* MA342 on barley seeds visualized by using green fluorescent protein. Appl Environ Microbiol 65:3674–3680

Tomilova OG, Shternshis MV (2006) The effect of a preparation from *Chaetomium* fungi on the growth of phytopathogenic fungi. Appl Biochem Microbiol 42:76–80

Torres R, Usall J, Teixidó M, Abadias M, Viñas I (2003) Liquid formulation of the biocontrol agent *Candida sake* by modifying water activity or adding protectants. J Appl Microbiol 94:330–339

Toure Y, Ongena M, Jacques P, Guiro A, Thonart P (2004) Role of lipopeptides produced by *Bacillus subtilis* GA1in the reduction of grey mould disease caused by *Botrytis cinerea* on apple. J Appl Microbiol 96:1151–1160

Usall J, Teixidó N, Abadias M, Torres R, Cañamás T, Viñas I (2010) Improving formulation of biocontrol agents manipulating production process. Postharvest Pathol 2:149–169

Vidhyasekaran P, Muthamilan M (1995) Development of formulations of *Pseudomonas fluorescens* for control of chickpea wilt. Plant Dis 79:782–786

Vidhyasekaran P, Rabindran R, Muthamilan M, Nayar K, Rajappan K, Subramanian N, Vasumathi K (1997) Development of powder formulation of *Pseudomonas fluorescens* for the control rice blast. Plant Pathol 46:291–297

Viswanathan R, Samiyappan R (2002) Induced systemic resistance by fluorescent pseudomonads against red rot disease of sugarcane caused by *Colletotrichum falcatum*. Crop Prot 21:1–10

Viterbo A, Inbar J, Hadar Y, Chet I (2007) Plant disease biocontrol and induced resistance via fungal mycoparasites. In: Kubicek CP, Druzhinina IS (eds) Environmental and microbial relationships, the Mycota IV, 2nd edn. Springer, Berlin, pp 127–146

Vleesschauwer DD, Höfte M (2007) Using *Serratia plymuthica* to control fungal pathogens of plants. CAB Rev Perspect Agric Veterinary Sci Nutr Nat Resour 2:1–12

Waller F, Militor A, Pfiffi S, Achatz B, Kogel K (2008) The root endophytic fungus *Piriformospora indica* accelerates host plant development and primes plants for diseases resistance. Phytopathology 98:S164 (Abst.)

Wang S, Wu H, Qiao J, Ma L, Liu J, Xia Y, Gao X (2009) Molecular mechanism of plant growth promotion and induced systemic resistance to *tobacco mosaic virus* by *Bacillus* spp. J Microbiol Biotechnol 19:1250–1258

Whipps JM, Sreenivasaprasad S, Muthumeenakshi S, Rogers CW, Challen MP (2008) Use of *Coniothyrium minitans* as a biocontrol agent and some molecular aspects of sclerotial mycoparasitism. Eur J Plant Pathol 121:323–330

Williamson SM, Guzman M, Anas O, Marin DH, Jin X, Sutton TB (1999) Evaluation of potential biocontrol agents for crown rot of banana. Phytopathology 89:S85 (Abst.)

Yánez-Mendizábal V, Viñas I, Usall J, Torres R, Solsona C, Teixidó N (2012) Production of the postharvest biocontrol agent *Bacillus subtilis* CPA-8 using low cost commercial products and by-products. Biol Control 60:280–289

Yang R, Han YC, Li GQ, Jiang DH, Huang HC (2007) Suppression of *Sclerotinia sclerotiorum* by antifungal substances produced by the mycoparasite *Coniothyrium minitans*. Eur J Plant Pathol 119:411–420

Yang L, Li G, Zhang J, Jiang D, Chen W (2011) Compatibility of *Coniothyrium minitans* with compound fertilizer in suppression of *Sclerotinia sclerotiorum*. Biol Control 59:221–227

Yu H, Sutton JC (1994) Inoculum concentration of *Botrytis cinerea* and of the biocontrol agent *Gliocladium roseum* in relation to suppression of pathogen in raspberry. Phytopathology 84:1377

Zhang JX, Howell CR, Starr JL (1996) Suppression of Fusarium colonization of cotton roots and Fusarium wilt by seed treatments with *Gliocladium virens* and *Bacillus subtilis*. Biocontrol Sci Technol 6:175–187

Zhang S, Schisler DA, Boehm MJ, Slininger PJ (2005) Carbon-to-nitrogen ratio and carbon loading of production media influence freeze-drying survival and biocontrol efficacy of *Cryptococcus nodaensis*. Phytopathology 95:626–631

Additional References for Further Reading

De Curtis F, Lima G, Vitullo D, De Cicco V (2010) Biocontrol of *Rhizoctonia solani* and *Sclerotium rolfsii* on tomato by delivering antagonistic bacteria through a drip irrigation system. Crop Prot 29:663–670

ISTHT International Subcommission on *Trichoderma* and *Hypocrea* Taxonomy. www.isth.info

Ting ASY, Fang MT, Tee CS (2009) Assessment on the effect of formulative materials on the viability and efficacy of *Serratia marcescens* – a biocontrol agent against *Fusarium oxysporum* f.sp. *cubense* race 4. Am J Agric Biol Sci 4:283–288

Chapter 6
Biological Disease Management Systems for Agricultural Crops

Disease management systems are developed by integrating effective approaches that have been directed individually or in combination for reducing the pathogenic potential of disease-causing microorganisms that have access to susceptible crop plants. Approaches for reducing the negative effects of the pathogen may be due to either direct inhibitory effect on pathogen population or indirect suppressive action by enhancing the level of resistance of the host plant to the target pathogen(s). The integration of the approaches may result in increase or decrease in the effectiveness of disease control. Hence, it is essential to select the approaches that will not have any negative influence on each other. However, only limited efforts have been made to assess the interactive effects of different disease management strategies on the incidence and/or severity of crop diseases and their contribution to enhance the yield levels. The possibility of integrating different management strategies that are individually effective has been indicated in the case of some of the destructive disease(s) infecting economically important crops and available information is discussed in this chapter.

6.1 Diseases of Cereal Crops

6.1.1 Wheat Diseases

6.1.1.1 Take-All Disease

Wheat take-all disease caused by *Gaeumannomyces graminis* var. *tritici* (*Ggt*) has a historical background for providing explanation for the phenomenon of "suppressiveness of soil" as a factor for reducing incidence of take-all disease. When wheat was planted consecutively, disease severity increased progressively and yield loss was substantial in the second, third and fourth crops. The disease severity subsequently decreased in fifth and sixth crops. This decline in take-all incidence and

P. Narayanasamy, *Biological Management of Diseases of Crops*, Progress in Biological Control 16, DOI 10.1007/978-94-007-6377-7_6,

severity was due to enhancement of populations of soil microorganisms antagonistic to *Ggt*. The take-all decline persisted only as long as wheat was grown continuously. Antagonism of root-colonizing fluorescent *Pseudomonas* spp. may be involved in the process of take-all decline. Wheat cultivars differed in their ability significantly to support resident populations of peudomonads capable of producing the antibiotic 2,4-diacetylphloroglucinol (2,4-DAPG) that suppressed the development of *Ggt* and consequently take-all disease severity (Mazzola et al. 2004). It may be advantageous to select such wheat cultivars, the rhizosphere of which might provide favorable conditions for the rapid multiplication of fluorescent *Pseudomonas* spp. with antagonistic potential against wheat take-all pathogen. The importance of nutrient management for reducing the incidence and severity of diseases incited by microbial plant pathogens has been revealed by different investigations. In the case of wheat take-all disease caused by *Gaeumannomyces gramins* var. *tritici*, nutrient deficiencies at any time during crop growth increased the severity of take-all disease. Ensuring the availability of adequate N, P, K, sulfur (S) and potassium (K) at the time of sowing was found to be a critical factor to reduce the risks of take-all incidence. Yields were generally higher with ammonium chloride than with ammonium nitrate, urea or ammonium sulfate. Wheat plants receiving ammonium nitrate were shorter and yielded less compared to plants fertilized with ammonium chloride and these plants were more vigorous (Christensen and Hart 2008). The nutritional status of the crop plant has a marked influence on the susceptibility to disease(s). Wheat take-all disease was found to be favored by phosphorus deficiency. The onset and severity of the disease was appreciably reduced by repeated application of phosphates in the long-term field experiments at Rothamsted, England (Mattingly and Slope 1977). Ammonium nitrogen was suppressive to wheat take-all disease (Colbach et al. 1997). Phosphorus was required to be applied routinely, since P-deficient plants were more susceptible and infected seedlings had poorly functioning root systems (Christensen and Hart 2008).

Incidence and severity of disease caused by soilborne pathogens is generally reduced by addition of organic amendments. Take-all severity on roots of barley and wheat caused by *Gaeumannomyces graminis* var. *tritici* was significantly lower in organically-managed than in conventionally managed soils. Fluorescent *Pseudomonas* spp. and in particular phlD$^+$ pseudomonas (producing the antibiotic 2,4-diacetylphloroglucinol, DPAG), key factors in the take-all decline were present in lower population densities in organically managed soils, compared to conventionally managed soils. In addition, organic management adversely affected the initial establishment of introduced phlD$^+$ *Pseudomonas fluorescens* strain Pf32-*gfp*, but not its survival. The efficacy of biocontrol of take-all disease by introduced strain Pf32-*gfp* was significantly stronger in conventionally managed soils than in organically managed soils. The results suggested that phlD$^+$ *Pseudomonas* spp. might not have a vital role in the suppressiveness of soils included in this investigation and other bacterial genera not investigated so far that could inhibit the growth and activity of the take-all pathogen might be involved in some soils (Hiddink et al. 2005b). The compatibility of the biocontrol agents with fungicides is a desirable attribute. The fungus *Penicillium radicum* promoted the growth of wheat plants and also inhibited the growth of several soilborne pathogens including *G. graminis* var. *tritici*. A combination of *P. radicum* with fungicide fludiconazole (Jockey) significantly reduced the incidence of take-all disease (Wakelin et al. 2006).

Crop rotation is considered as the most effective strategy for the control of take-all disease, since survival of the pathogen in the absence of the susceptible host plant is poor. A 1-year break from wheat or barley seems to be sufficient to reduce the risk to an insignificant level. The crops that could effectively break the continuous availability of susceptible host plant species are oats, corn, beans, vegetables, oilseed crops and annual legumes for seed (Christensen and Hart 2008). Take-all disease infecting wheat is suppressed by naturally occurring fungi closely related to the pathogens that infect grasses and cereals. Suppression of take-all disease development by these fungi was reinvestigated, because of the changing importance and role of grass weeds and grass cover in arable farming. Natural population of the antagonist *Gaeumannomyces cylindrosporus* was allowed to develop under rye-grass and they were found to be more effective than artificially introduced populations in suppressing the development of take-all disease in the following wheat crops. The critical requirement is that the antagonist should be present in the field soil before the wheat crop is planted. The antagonistic fungi should be introduced into the preceding monocot crop not susceptible to *Gaeumannomyces graminis* var. *tritici* (*Ggt*). Further, the plant species susceptible to *Ggt* should not be allowed to remain in the field as volunteers or weeds (Gutteridge et al. 2007).

6.1.1.2 Bare Patch Disease

Soil suppressiveness against soilborne diseases appears to be nonspecific. Most natural and agricultural soils exhibit some degree of suppressiveness to soilborne pathogens. Association of microorganisms with soil suppressiveness has been indicated in wheat take-all disease. Rhizoctonia bare patch disease of cereals caused by *Rhozoctonia solani* AG-8 is a major fungal root disease in no-till cropping systems. Rhizoctonia bare patch increased when tillage was eliminated and the disease became a major limiting factor to the adoption of no-tillage technology. Since several pulse and oilseed crops are susceptible to *R. solani*, the effect of mixed cropping and rotation with different crops was studied. Mixed cropping of triticale wheat with clover and of barley with Brussels sprouts did not enhance soil suppressiveness to *R. solani* (Hiddink et al. 2005a). A 8-year crop rotation experiment showed that crop rotation had no effect on bare patch, during the first 5 years. However, from years 6 to 8, both soft white and hard white classes of spring wheat (*Triticum aestivum* L.) grown in a 2-year rotation with spring barley had an average of only 7 % of total land area with bare patches, compared with 15 % in continuous annual soft white wheat and hard white (monoculture system). Further, average yield of both soft white and hard white wheat, during these 2 years, were greater, when grown in rotation with barley than in monoculture (Schillinger and Paulitz 2006).

6.1.1.3 Fusarium Head Blight Disease

Fusarium head blight (FHB) disease of wheat caused by *Fusarium graminearum* (*Gibberella zeae*), as well as *F. avinaceum*, *F. culmorum*, *F. poae* and *Microdochium nivale* has become a major problem in several countries. Seedling bight (damping-off disease complex) is primarily due to *F. culmorum* and *F. graminearum* resulting in reduced seed germination and emergence of seedlings. Crop residue left at soil surface is the principal source of inoculum. Conservation tillage systems involve leaving all or part of the crop residue on soil surface after harvest to reduce soil erosion. Effects of previous crop residues and tillage practices on FHB were investigated. The FHB incidence and severity were greatest, when wheat followed corn and least when wheat followed soybeans. Incidence and severity of FHB were lower in moldboard ploughed plots than in either chisel ploughed or no-till plots. Wheat yields were reduced by 15 % in plots where wheat followed corn or wheat than wheat following soybean and moldboard ploughed plots gave higher yield (10 %), compared with chisel ploughed or no-till treatments (Dil-Macky and Jones 2000). The results of a later investigation indicated that conservation soil tillage might increase the incidence of *Fusarium* in soil. The deeper the tillage, the lower was the number of *Fusarium* spp. isolated from the soil samples. Moldboard ploughing reduced the occurrence of pathogenic *Fusarium* spp. (Steinkellner and Langer 2004). The effect of crop rotation on the incidence of wheat crown disease caused by *Fusarium pseudograminearum* was assessed. The incidence of crown rot was significantly higher on wheat for the treatment that included wheat and 1 year of canola, compared with other treatments. The lowest incidence and severity of crown rot were recorded in wheat with a medic-clover mixture and in the treatment that included wheat after 3 years of rotation. The results indicated that by selecting suitable crops and duration of rotation, the incidence and severity of crown rot disease could be reduced substantially (Lamprecht et al. 2006).

The effectiveness of biocontrol agents in suppressing the development of *Fusarium graminearum* (*Gibberella zeae*) has been demonstrated under in vitro conditions. Only very few biotic agents have been tested under field conditions. The major inoculum of Fusarium head blight and Gibberella ear rot disease of wheat comes from melanized structures such as pseudothecia or perithecia produced on crop residues. The antagonistic fungus *Microsphaeropsis* sp., a saprophyte is well adapted to winter conditions. The potential of *Microsphaeropsis* sp. isolate P130A was evaluated as an antagonist of *G. zeae* in in vitro assays and under field conditions. Application of the isolate P130A significantly reduced ascospore production on wheat, when applied 2 weeks before inoculation with *G. zeae*. When applied on crop residues in the field as postharvest or preplanting treatment, *Microsphaeropsis* sp. had no effect on the pattern of perithecia maturation, but significantly reduced the number of perithecia produced on two sampling dates (May1998 and July 1999). *Microsphaeropsis* sp. had significant effect on immature and mature perithecia formed on spikelets, as the number of perithecia produced was significantly reduced ($P<0.05$) under severe epidemic conditions, when 36 % of crop residues were infected by the pathogen. The results indicated that *Microsphaeropsis* sp. had the potential of biologically reducing the initial inoculum of *G. zeae* (Fig. 6.1) (Bujold et al. 2001).

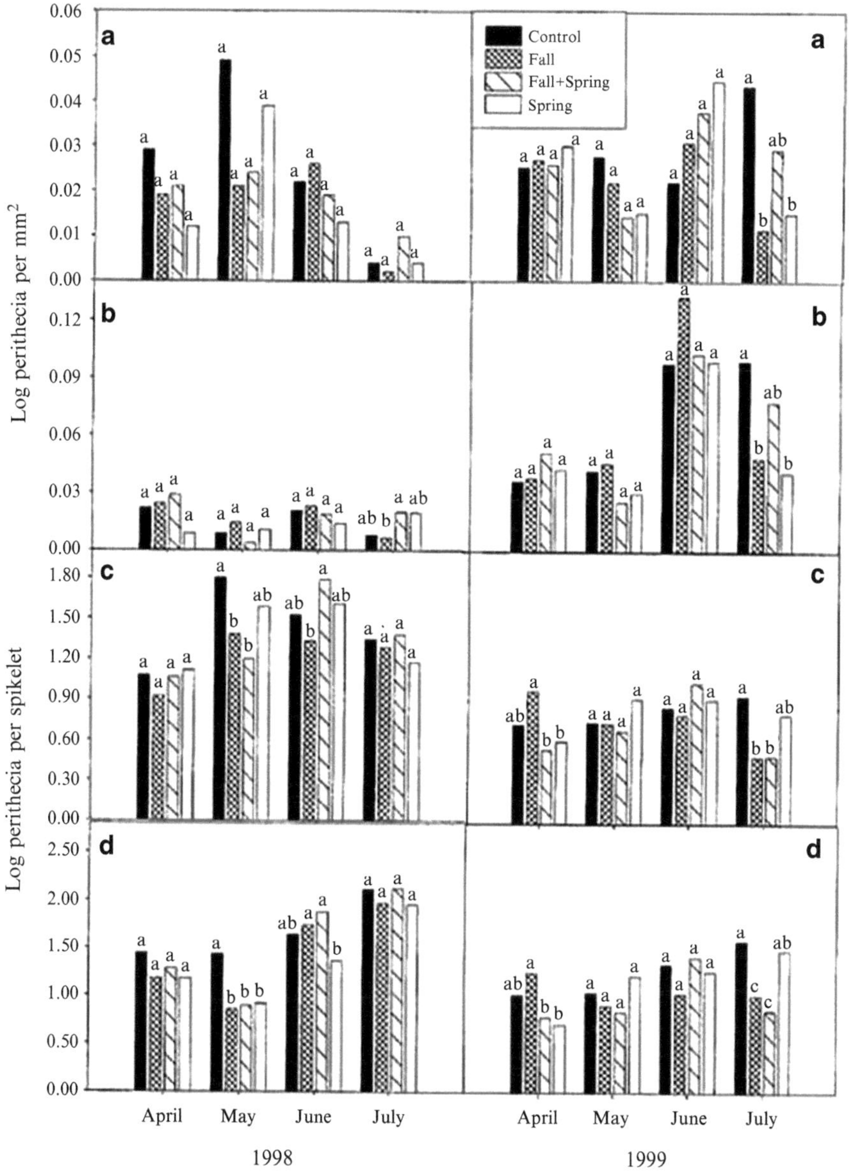

Fig. 6.1 Effect of treatment with *Microsphaeropsis* sp. on the number of perithecia produced by *Gibberella zeae* on different wheat residues in 1998 (*left*) and 1999 (*right*) (**a**) immature perithecia on straw; (**b**) mature perithecia on straw; (**c**) immature perithecia on spikelet and (**d**) mature perithecia on spikelet. *Bars with same letter* are not significantly different according to LSD test at 0.05 level of confidence (Courtesy of Bujold et al. 2001 and with kind permission of The American Phytopathological Society, MN, USA)

Of the six antagonists tested under field conditions, *Cryptococcus* sp. OH71.4 and *C. nodaensis* OH182.9 reduced the Fusarium head blight (FBH) disease severity by as much as 57 % in the trial at Peoria, IL, USA, whereas *Cryptococcus* sp. OH181.1 reduced disease severity by 59 % in the trial at Langdon, ND, USA. The yeast antagonists OH71.4, OH181.1 and OH182.9 seemed to have the highest potential for the suppression of FHB development on durum wheat under field conditions. The yeast antagonists did not influence the contents of the pathogen-produced toxin deoxynivalenol (DON) in the grains. The bacterial isolates *Bacillus subtilis* AS43.3 and AS43.4 performed better in protecting wheat plants against *G. zeae* than the yeast antagonist isolates. However, the performance of the bacterial isolates was inferior to the yeast isolates under field conditions. The application of biological control agents in the field reveals the inherently more variable nature of their biocontrol potential and challenges to be faced by them than in the greenhouse. The biological agents that can tolerate and remain viable under the harsh field environments have to be selected for large scale application (Schisler et al. 2002). The carbon/nitrogen (C/N) ratio, carbon loading of production media and cultivation time influenced the survival of *C. nodaensis* OH182.9, after freeze-drying and biocontrol efficacy of fresh cells against Fusarium head blight disease. Cells of OH182.9 harvested after 48 h from the semi-defined complete liquid (SDCL) C/N 30:1 medium with 7 and 14 g/l carbon survived better after freeze-drying than others and demonstrated levels of biocontrol efficacy comparable to cells harvested after 48 h from the standard SDCL C/N 11:1 media (Fig. 6.2). The results indicated the potential of improving the product quality of OH182.9 by managing the nutritional environment of production without compromising biocontrol efficacy (Zhang et al. 2005).

A strain of *Fusarium equiseti* G9 was selected as the most effective among 113 pre-screened microorganisms that were tested for their efficacy in suppressing the winter wheat ear blight caused by *Fusarium culmorum* and *F. graminearum* under field conditions and in reducing the production of mycotoxins particularly deoxynivalenol (DON) in wheat grains. Ears were spray inoculated during anthesis with spore suspensions, first with putative BCA strain followed by the pathogen. The strain G9 decreased DON content consistently on wheat inoculated with *F. culmorum*, compared with control, often by more than 70 %. The BCA strain was effective well under severe disease pressure (*F. culmorum* at 10^5 conidia/ml) and similarly to a standard fungicide, tebuconazole. Percentages of diseased grains and amounts of DON corresponded well. Low concentrations of nivalenol (NIV) were detected in grain samples after treatment with some *F. equiseti* strains. On wheat ears inoculated with *F. graminearum*, the BCA strain G2 decreased the percentage of diseased grains by 92 % and DON by 94 %. The strain *F. equiseti* G2 produced small quantities of NIV in pure cultures. The results indicated that *F. equiseti* strain G9 had the potential for reducing the disease incidence and mycotoxin contents in wheat grains (Dawson et al. 2004). The efficacy of seed treatment with the fungal BCA *Clonostachys rosea* (IK726) was assessed for the suppression of Fusarium head blight (FHB) disease caused by *Fusarium culmorum*, one of the *Fusarium* spp. associated with the disease. Six field experiments were conducted, in addition to growth chamber experiments. The disease severity

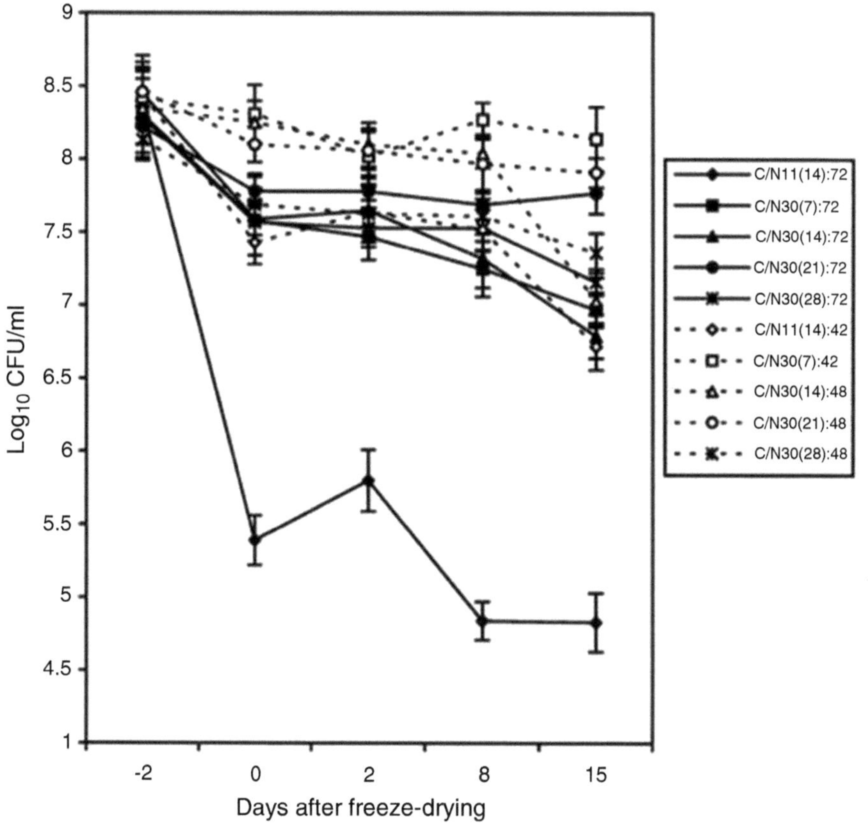

Fig. 6.2 Effect of carbon-loading in semi-defined complete liquid (SDCL) media (C/N 30:1) on survival of *Cryptococcus nodaensis* OH182.9 cells after freeze drying. Improvement of biocontrol efficacy could be achieved by appropriate management of the nutritional environment of BCA production (Courtesy of Zhang et al. 2005 and with kind permission of The American Phytopathological Society, MN, USA)

was significantly reduced. The BCA was active against the pathogen at average soil temperatures at sowing ranging from 6.2 to 12 °C. Dried and stored conidia of *C. rosea* controlled *F. culmorum* as effectively as freshly harvested conidia. A high correlation was established between disease index ratings from field experiments and corresponding growth chamber assessments of the efficacy of *C. rosea*. Addition of stickers Pelgel or Sepiret did not affect the efficacy of the BCA strain. The effective dosages of IK726 (CFU/seed) were estimated by in vitro bioassays and they were very similar for freshly harvested conidia and for dried conidia. With both types of conidia at a concentration of $>5 \times 10^3$ conidia per seed over 80 % disease control could be achieved repeatedly, indicating the stability of the two types of BCA formulations (Jensen et al. 2000). A mixture of *C. rosea* 47 and/or *Trichoderma atroviride* 312 with carboxin, thiram and triticonazole

Table 6.1 Effect of seed treatment with bacterial isolates on the yields of spring wheat cv. Curry and winter wheat cv. Tarso naturally infected by *Microdochium nivale* (Johansson et al. 2003)

	Yield (kg/ha)[a]	
Treatments	Spring wheat	Winter wheat
Untreated	4,405b	4,169c
Panoctine 400[b]	4,612a	5,366a
Isolate MF-30	4,486ab	–
Isolate MF-416	4,514ab	5,028ab
Isolate MF-588	4,466b	4,788b
Isolate MF-626	4,535ab	5,062ab
n-value	12	9

Figures with the same letter in one column are not statistically different according to Duncan's multiple range test ($P<0.05$)
[a]Average of yields in three locations
[b]Fungicide with guazatine acetate as the active ingredient

applied to wheat seeds significantly controlled *F. culmorum* in growth chambers. Under field conditions, the antagonists applied in field dose to seeds naturally infected by *F. culmorum* reduced the disease to a level comparable to triticonazole at full field dose. The results indicated the feasibility of combining the fungal BCAs with the reduced dose of fungicide for effective management of Fusarium head blight disease of wheat (Roberti et al. 2006).

6.1.1.4 Seedling Blight Disease

The bacterial isolates (164) were evaluated under field conditions for their potential in suppressing the development of wheat seedling blight disease caused by members of *Fusarium* complex, *Fusarium culmorum* and *Microdochium nivale*. The experiments were conducted for five consecutive growing seasons with spring and winter wheat. The most efficient isolates were three fluorescent pseudomonads. *Pseudomonas* sp. RNA group1 MF30 (closest to *P. veroni*), *Pseudomonas* sp. RNA group MF416 and *P. fluorescens* MF588 and one isolate of *Pantoea* sp. MF626 (closest to *P. agglomerans*). These four isolates suppressed the seedling blight disease development as effectively as the fungicide guazatine in respect of crop stand and yield. The bacterial isolates were applied as seed treatment and the disease suppressive effects were repeatable. Seed treatment with *Pantoea* sp. isolate MF626 enhanced the wheat grain yield by an average of more than 500 kg/ha in six field experiments (Table 6.1). The results indicated that seed treatment with naturally occurring and properly selected bacterial isolates could be an effective approach for efficient management of wheat seedling blight disease and for reducing the application of synthetic chemicals (Johansson et al. 2003). The biocontrol potential of *Pseudomonas fluorescens* strains MKB158 and MKB249 and *P. frederiksbergensis* strain 202 against Fusarium head blight disease caused by *Fusarium culmorum* was assessed under both greenhouse and field conditions, when the BCAs were applied 24 h prior to pathogen inoculation, the disease severity and the loss in 1,000-grain weight were significantly reduced. In the *F. culmorum*-inoculated field trials, treated with strain MKB158 or MKB 249 significantly reduced severity in wheat and barley. In addition, the levels of deoxynivalenol (DON) were also significantly reduced in wheat and barley grains by 74–78 %. The results showed that both disease severity and DON levels in grains could be reduced by the treatment of plants with *P. fluorescens* as a protective application (Khan and Doohan 2009).

Microdochium nivale causes seedling blight and snow mold diseases in winter wheat and rye. Seeds were treated with fermentate of *Pseudomonas brassicacearum* MA250 and dried before sowing or they were directly sprayed in the furrow opener at the time of sowing under field conditions. Parallel climate chamber assays were also performed to assess the effect of bacterial treatment on snow mold caused by *Microdochium nivale* and *Fusarium* sp. Significant biocontrol activity, as reflected by plant density counts, was observed both in the field and climate chamber experiments. Furrow spray application was less effective compared to seed treatment with BCA fermentate. The bacterial cell concentration of 10^9 CFU/ml was required to achieve effective suppression of the disease (Levenfors et al. 2008). In another investigation, four bacterial strains of *Pseudomonas* exerted significant effect by increasing seedling emergence and decreasing wheat seedling blight disease severity. *P. fluorescens* was the most effective in improving plant establishment and yield in infested winter wheat. The plant population increased up to 48 % and yield by 26.5 % in different field trials. Consistent and effective protection to wheat plants was provided by *P. fluorescens* for long periods during growing seasons (Amein et al. 2008). Another bacterial biocontrol agent *Lysobacter enzymogenes* strain C3 in combination with the fungicide tebuconazole provided consistent protection to wheat against FHB disease caused by *Fusarium graminearum* in all three field experiments (Jochum et al. 2006). Induction of systemic resistance by chemical compounds to different diseases affecting various crops has been reported. A foliar fertilizer containing potassium phosphate, DL-3-aminobutyric acid (BABA) and benzothiadiazole (BTH) were evaluated for their potential in suppressing the development of Fusarium head blight (FHB) disease of winter wheat caused by *Microdochium majus* (= *M. nivale* var.*majus*) under greenhouse conditions. Winter wheat plants were sprayed with aqueous solutions of the test chemicals at 7 days prior to inoculation of heads prior to inoculation of heads with the fungal pathogen. The number of bleached spikelets indicated the severity of the disease. Spraying plants with the foliar fertilizer reduced the disease severity up to 40 %. Reduced disease development was indicated by results of detached leaf bioassay. The resistance inducers BABA and BTH did not show any negative effect on the development of FHB significantly. Application of foliar fertilizer containing potassium phosphate showed significant beneficial effects that have to be confirmed by field trials (Hofgaard et al. 2010).

6.1.1.5 Foliar Diseases

The biocontrol potential of six isolates of *Trichoderma harzianum* and one isolate of *T. koningii* to reduce the incidence and severity of wheat tan spot disease caused by *Pyrenophora tritici-repentis* and leaf blotch disease caused by *Mycosphaerella graminicola* was assessed under field conditions. When the BCA isolates were either used for seed treatment or spraying on wheat leaves at different growth stages, both leaf blotch and tan spot diseases were suppressed to the maximum extent (up to 56 %). At tillering stage, six of the seven isolates reduced the severity of tan spot and leaf blotch diseases up to 39 and 12–53 % respectively compared to controls. Treatment with *Trichoderma* isolates provided some beneficial effect at head stage also against tan spot disease (Perelló et al. 2006). One of isolates of *T. harzianum* T7 was effective as the fungicide tebuconazole in suppressing the development of tan spot caused by *P. tritici-repentis* (Perelló et al. 2008). For the control of tan spot disease, no single disease management strategy could provide satisfactory long-term solution. Hence, the potential of two strains of *Trichoderma harzianum* and two synthetic compounds acibenzolar-*S*-methyl (ASM) and thiamethoxam (TM), capable of inducing systemic resistance in plants against plant diseases were evaluated for their efficacy in suppressing the development of tan spot disease, when applied alone or in combination on the wheat cultivar Klein Escorpion. The fungal BCA strains, ASM and TM were applied to field plots and the severity of tan spot was reduced in treated plots, as compared with controls. In addition, plant height, fresh and dry weights of shoots and dry weight of roots increased in comparison to the untreated controls. As preinoculation treatment ASM, TM and *T. harzianum* strain Th1 reduced necrotic lesions by more than 50 %, compared to the control. ASM alone or in combination with Th1 strain increased the dry weight to more than 60 %, whereas the effects of TM and Th1 on dry mass showed an increase ranging from 25 to 57 %. Combination of the BCA and resistance inducers resulted in a six-fold increase in dry weight over control (Perelló and Dal Bello 2011). Septoria blotch caused by *Septoria tritici* another major foliar disease of wheat could be suppressed by seed treatment with or foliar application of *T. harzianum*. Application of the BCA at later stages of crop growth did not reduce the disease severity to a satisfactory level. Two isolates T2 and T4 were equally effective as the fungicide Folicur in reducing disease incidence and severity (Perelló et al. 2009).

6.1.1.6 Common Bunt Disease

Disinfection of seeds and planting materials using heat treatments has been successfully applied against some seedborne diseases. A high precision method involving the use of warm air which is in moisture equilibrium with the grains to be treated was employed for the control of fungal seedborne infections of wheat seeds. Sanitation of wheat seeds using warm air reduced seedborne infection by *Tilletia caries* and *Microdochium nivale* to a level comparable with chemical seed treatment. No adverse effect was observed on plant development following warm air treatment. The warm air treatment can be applied for treating seeds with thick layers at short durations. As the cost of treatment is low, warm air treatment may be advantageous for large scale application (Forsberg 2001). Heat treatment of wheat and barley seeds with high level of infection by *Fusarium graminearum* (84 and 23 % respectively) was evaluated. The pathogen was eliminated from Canadian Western red spring wheat seeds treated at 60 °C for 15 days or at 70 °C for 5 days or at 80 °C for 2 days. Likewise, the barley seeds were disinfected by heat treatment at 60 °C for 21 days or at 70 °C for 9 days or at 80 °C for 5 days. Seed germination percentages were not affected in most seed samples due to heat treatment (Clear et al. 2003). The effectiveness of the biofumigant fungus *Muscodor albus* in suppressing the development of *Tilletia caries*, causing wheat common bunt disease was investigated. Dry rye grain culture of *M. albus* was ground into powder and applied (at 125 mg/g of seed) to wheat seed infested with teliospores of *T. carries*. The culture was also broken into small particles and applied in furrows at 4 g/m of row, along with teliospore-infested seed during planting. The experiments were carried out during two growing seasons and two planting dates, when the soil temperatures were optimal for disease development (5–10 °C). In the first year, treatments in the first seeding date (beginning early spring) reduced the common bunt disease to 12 and 9 % respectively in seed and in-furrow treatments, as compared to control with 44 % diseased spikes. In the planting date, the disease incidence was low (6 %) in the control and no infected head was noted in both treatments. In the second year, the incidence of common bunt disease was low even in the control plots (8 %). The disease incidence recorded for seed and in-furrow treatments were 0.5 and 0.25 % respectively in the second seeding date. The results indicated the usefulness of applying the biofumigant *M. albus* for the suppression of wheat common bunt disease (Goates and Mercier 2011).

6.1.2 Barley Basal Kernel Blight Disease

Pseudomonas syringae pv. *syringae* (*Pss*) causes barley basal kernel blight disease. The potential of strains of *Pantoea agglomerans* (syn. *Erwinia herbicola*) for the suppression of basal kernel blight disease was studied. When the BCA strains were applied to head prior to the infection by *Pss*, the disease development was suppressed. Reduction of kernel blight disease following application of BCA was from 45 to 74 %, depending on the BCA isolate and barley cultivar tested under field conditions. The efficacy of BCA strains varied due to time and rate of application. Pre-inoculation treatment with BCA strains was more effective than coinoculation treatment. A single application of *P. agglomerans* at 3 days prior to pathogen inoculation provided adequate control of the disease, since populations of the BCA were about 10^7 CFU/kernel, while the pathogen population was reduced by 100-fold to 2×10^4 CFU/kernel. Application of the BCA to the flag leaf (before heading) also reduced the kernel infection percentages significantly. Formulations of *P. agglomerans* prepared by using oil/starch/sugar encapsulation and stored at 4 °C retained viability (90–93 % survival) better than the formulations stored at 22 °C (73–79 %). Use of synthetic bactericides may be uneconomical, compared with the application of bacterial BCAs. Combination of strategies viz., application of *P. agglomerans*, selection of less susceptible barley cultivars and reduced irrigation during infection might provide effective protection against basal kernel blight disease of barley (Braun-Kiewnick et al. 2000).

6.1.3 *Corn Diseases*

6.1.3.1 Anthracnose Disease

Corn anthracnose disease caused by *Colletotrichum graminicola* occurs in two phases as leaf blight and stalk rot in most corn production areas and it may be responsible for complete crop failures in sweet corn fields. Corn residue is an important source of primary inoculum source which may increase through cultural practices such as no-tillage and continuous corn monoculture. Field experiments conducted in agricultural research stations in Arlington and Madison showed that incidence and severity of anthracnose leaf blight were higher in continuous corn crop monoculture than in soybean-corn rotations by 91 % and 24–78 % respectively. Anthracnose stalk rot was marginally increased by chisel-ploughing treatment. A positive relationship between spring residue cover and anthracnose leaf blight could be seen, but no such correlation between residue and stalk rot was evident. Severity of anthracnose leaf blight was correlated negatively with yield. But anthracnose stalk rot phase did not affect the yield in proportion of disease severity. The results showed that proper management of residue levels through crop rotation might be effective in lowering the incidence and severity of corn anthracnose disease (Jirak-Peterson and Esker 2011).

6.1.3.2 Ear Rot Disease

Fusarium verticillioides, the causative agent of maize ear rot and kernel rot disease also produces the mycotoxin fumonisins that are harmful to humans and animals, when contaminated grains are consumed. The antagonistic potential of *Pseudomonas fluorescens* against *F. verticillioides* and ability to reduce accumulation of fumonisin was assessed. *P. fluorescens* formulation was applied as seed treatment and foliar spray. The bacterial BCA was formulated using corn starch, wheat bran and talc powder. Pure culture of *P. fluorescens* and the formulations increased maize plant growth and vigor as reflected by seed germination, seedling vigor, plant height, 1,000-seed weight and yield. The incidence of disease and accumulation of fumonisin were reduced more efficiently by *P. fluorescens* than control and chemical treatment under field conditions, indicating the effectiveness of *P. fluorescens* (Nayaka et al. 2009). In a later investigation, the efficiency of *Bacillus amyloliquefaciens* and

Enterobacter hormaechei in suppressing *F. verticillioides* infection and reducing fumonisin B_1 accumulation was assessed. The antagonists were applied as seed treatment and spray on maize ears at flowering. *F. verticillioides* infection and fumonisin B_1 (FB_1) contents were determined in kernels of physiologically mature maize plants at harvest time under field conditions. Seed treatment with the bacterial BCAs reduced infection by ear rot pathogen and the FB_1 contents in both years of investigation. The number of CFU of *F. verticillioides* obtained from harvested maize kernels was positively correlated with FB_1 content. The results suggested that treatment with bacterial antagonists might improve the quality of maize grains by reducing toxin levels, although no improvement in grain yield could be possible (Pereira et al. 2010).

6.1.4 Rice Diseases

6.1.4.1 Blast Disease

Rice crops are cultivated in a wide range of environmental conditions and at different altitudes. Hence, occurrence of diseases caused by fungi, bacteria, phytoplasmas and viruses varies, depending on the levels of their resistance/susceptibility, existing environmental conditions and inputs provided to the plants. Rice blast disease, one of the major diseases is caused by *Magnaporthe grisea* which exists as numerous races that differ in their virulence (pathogenic potential). Certain cultural practices such as changing planting dates, avoiding application of excessive application of nitrogen, maintaining optimum levels of soil moisture and planting blast resistant rice varieties, have been shown to have suppressive effect on the incidence and severity of blast disease or to escape from the severe infection by the disease. Application of silicon (Si) as sodium silicate has been demonstrated to reduce the incidence and severity of blast disease. When Si was applied to Si-deficient soils, application of fungicides could be eliminated, even in areas with high endemic levels of blast as in Colombia. In addition, Si application in Si-deficient soils increased yield by 20–35 %, compared to non-amended soils. Si alone and Si combined with the fungicide edifenphos reduced leaf blast severity by 22–75 %, when compared to non-amended, untreated controls. Suppression of leaf blast by Si alone was equal to or better than the full rate tricyclazole treatment, when disease severity was low. But when disease severity was higher, a 10 % rate of tricyclazole (full rate for field) was also required, in addition to Si. Yield levels were equal for Si alone and fungicide (full rate) (Seebold et al. 2004). The mechanism of blast disease suppression has been studied. The leaf extracts from rice plants grown in soils amended with Si (Si^+) followed by inoculation with *M. grisea* contained higher concentrations of momilactone phytoalexins, compared with those from unamended control rice plants (S^-). Production of momilactones following treatment with Si was considered as a potentially important factor for reducing the blast disease severity (Rodrigues et al. 2005). *Pseudomonas fluorescens* Pf1 has been reported to provide effective

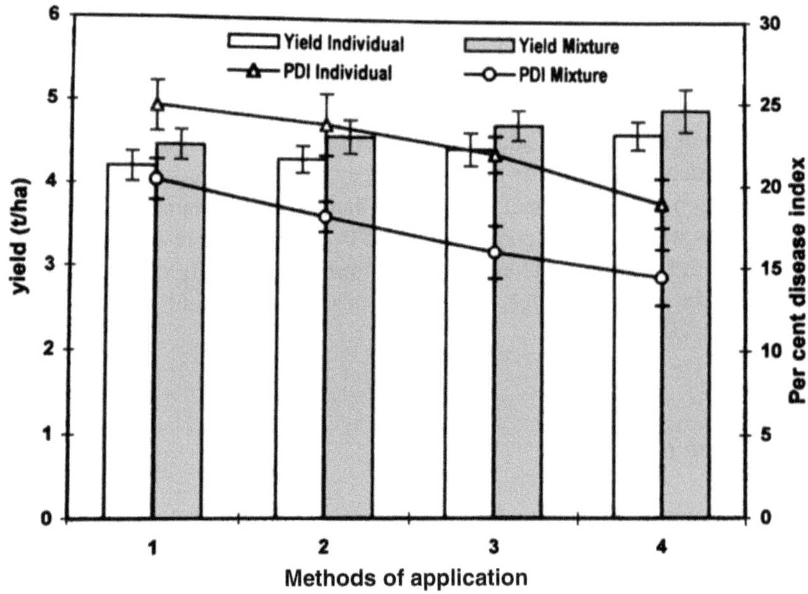

Fig. 6.3 Effect of different delivery systems on the severity of sheath blight disease and yield of rice plants treated with PF1 or FP7 strains of *Pseudomonas fluorescens* alone or in combination. Strain mixture was more effective than individual strains (Courtesy of Nandakumar et al. 2001 and with kind permission of Springer Science + B. V., Heidelberg, Germany)

protection against rice blast disease. The seed treatment followed by three foliar applications of strain Pf1 reduced the incidence and severity of blast disease (Vidhyasekaran et al. 1997). Foliar spray of *P. fluorescens* reduced the incidence of rice blast disease and also increased the grain yield (Karpagavalli et al. 2001).

6.1.4.2 Sheath Blight Disease

Rice sheath blight disease is caused by *Rhizoctonia solani* which is soilborne and has a wide host range, including many crop plants. No effective genetic source for development of rice cultivars resistant to sheath blight pathogen has been identified. Control of sheath blight disease by applying fungicides is economically not feasible. Biological management by employing fungal and bacterial biocontrol agents holds promise as an alternative approach. Talc-based formulations containing *Pseudomonas fluorescens* strain PF1 and FP7 either alone or in combination reduced the sheath blight incidence to different levels, depending on the strain or strain mixture. The extent of disease reduction was up to 29.2 % for individual strains and up to 45.1 % for strain mixture. In addition to reduction of disease incidence, treatment with PF strains enhanced the plant growth resulting in higher grain yield up to 40.4 %, when PF1 and FP7 mixture was applied (Fig. 6.3) (Nandakumar et al. 2001). Talc-based formulations of *P. fluorescens* strains PF1 and FP7 in combination with chitin amendment were more effective in reducing sheath blight disease up to 62.1 and in increasing the yield up to 21 % (Radjacommare et al. 2002). The effectiveness of suppression of mixtures of sheath blight disease by application of mixtures of fungal biocontrol agents *Trichoderma harzianum*, *T. viride* and *Gliocladium virens* under field conditions was demonstrated (Tang et al. 2002). Talc-based formulations of *P. fluorescens* and *T. viride* reduced the sheath blight incidence significantly under field conditions and their effectiveness was comparable to the level obtained with the fungicide carbendazim. Enhanced plant growth as reflected by the number of productive tillers, grains per panicle and 1,000-grain weight was observed in plants treated with *P. fluorescens* or *T. viride*. Combination of these two BCAs resulted only in a marginal increase in grain yield and reduction in disease incidence, indicating lack of additional benefit due to use of BCA combination (Mathivanan et al. 2005).

Bacillus subtilis strain NJ-18 showed antifungal activity against *Rhizoctonia solani* and *Sclerotinia sclerotiorum* causing rice sheath blight and rape Sclerotinia stem rot diseases respectively. In the field experiments, fermentation of NJ-18 strain at 5.0×10^7 CFU/ml significantly reduced sheath rot disease incidence and severity. When NJ-18 was applied alone or combined with 50 % kresoxim-methyl at 225 g a.i./ha, the sheath blight disease was controlled more effectively than 50 % kresoxim-methyl alone or Jinggangmycin at 120 g a.i./ha. Disease control obtained with NJ-18 alone was as high as 100 %, indicating the potential of the BCA strain in suppressing the development of rice sheath blight disease (Yang et al. 2009). *Bacillus megaterium* effective against the rice sheath blight pathogen *R. solani* was formulated as floating pellets and water soluble granules. The floating pellets showed good floating properties and gradually released the bacterial BCA load over time. In the greenhouse tests, even at lower dose, the floating pellet formulation was equally effective as the granular formulation in suppressing rice sheath blight disease development. The presence of endospores of *B. megaterium* on the surface of leaf sheath and on leaf blade of rice plants following application of formulated product was detected by scanning electron microscope (SEM). In the field evaluation, the population of bacterial antagonist was maintained in both formulations at equivalence. The granular formulation applied at 300 ml/rice hill was as effective as the fungicide iprodione in reducing the disease incidence and intensity and more effective than the floating pellet formulation. The floating pellet formulation might become less effective, since it might be dispersed in the relative open system, from point of application, possibly due to water movement. The effect of the antibiotic produced by *B. megaterium* might be diluted resulting in loss of effectiveness to the desired level in the floating pellet formulation (Kanajanamaneesathian et al. 2007). *B. megaterium*-based water soluble granular formulation was applied as seed treatment (1 g/100 g seed) followed by three sprays at seedling, tillering and flowering stages. Treatment with *B. megaterium* effectively reduced sheath blight and blast diseases. Yield of plants treated with the BCA was greater than the untreated controls, when weed populations was kept under check and recommended nitrogenous fertilizers were applied (Kanjanamaneesathian et al. 2009).

6.1.4.3 Bacterial Leaf Blight Disease

Bacterial leaf blight (BLB) disease of rice caused by *Xanthomonas oryzae* pv. *oryzae* (*Xoo*) causes considerable loss, when favorable environmental conditions prevail. The possibility of using antagonistic bacterial species/strains has been examined. An avirulent mutant of *Xoo* strain Du728 generated by deleting a virulence gene was evaluated for its antagonistic potential against the BLB pathogen. One single spray application of strain Du728 (10^6 CFU/ml) reduced the disease severity by 48.5 %, compared with the control. Combination of the strain Du728 and salicylic acid (10 μg/ml), the inducer of disease resistance, was more effective than the mutant strain alone in reducing the disease severity under greenhouse conditions. The BCA and salicylic acid combination applied thrice as foliar sprays under field conditions, reduced the disease by 80 % and the combined application was more effective than the chemicals tested (Liu and Wang 1998). *Lysobacter antibioticus* strain13-1 isolated from rice rhizosphere was evaluated for its biocontrol potential for the control of rice BLB disease. Whole broth culture (WBC) of the strain 13-1 was applied at maximum tillering stage of rice cultivar Huangkenuo along with the antibiotic zhongshengmycin 1 % and water (control) to assess the comparative effects under field conditions. The strain 13-1 reduced the rice BLB disease index by 78.3 and 59.1 % in 2005 and 2006 respectively relative to the control. The BCA treatment was more effective than zhongshengmycin with suppression efficiencies of 69.4 and 47.7 % in 2005 and 2006 respectively. The strain 13-1 could colonize on rice leaves as well as rice stems at 3 weeks after application. The strain 13-1 could suppress the disease development in most rice cultivars, irrespective of their susceptibility levels (Ji et al. 2008).

Harpins produced by plant pathogenic bacteria have multifunctions. They promote plant growth and induce defense responses as well. Nine fragments of the harpin protein $HpaG_{xooc}$ from *Xanthomonas oryzae* pv. *oryzicola* were characterized and these fragments possessed the properties of plant defense elicitation and plant growth stimulation relative to the intact protein. Under controlled conditions, the fragment $HpaG_{10-42}$ was more active in both properties than $HpaG_{xooc}$ and other fragments. These two fragments were evaluated under field conditions. The activity of $HpaG_{10-42}$ in rice under field conditions significantly was greater than that of $HpaG_{xooc}$. The fragment stimulated resistance to three important diseases of rice, bacterial leaf blight, blast and sheath blight and also increased the grain yields in experiments conducted in 672 experimental plots with nine rice cultivars planted at three locations. The effects of application rate, frequency and timing of application in respect of rice growth stages were also investigated. Of the concentrations (24, 12 and 6 μg/ml), number and timing of applications (1–4 stages) tested, $HpaG_{10-42}$ at 6 μg/ml applied to plants once at nursery seedling stage and three times in the field was the most effective. Incidence of bacterial leaf blight, blast and sheath blight was reduced by 61.6, 93.6 and 93.2 % and 56.4, 76.0 and 55.05 % respectively in indica and japonica cultivars relative to the untreated controls. In addition, grain yields were increased by 22–27 %, compared with control treatments. The effectiveness of application of harpin fragment $HpaG_{10-42}$ was equal to the local management practices including use of chemicals. The results showed that $HpaG_{10-42}$ protein fragment could be used effectively to control three important rice diseases and increase the yield levels of rice crops (Chen et al. 2008).

Three bacteriophages strains that lyse *Burkholderia glumae* and *B. plantarii*, causing rice seedling rot and seedling blight diseases were evaluated for their potential to suppress the development of the seedling diseases. The strain BGPP-Ar was more effective than the other two strains tested. The ratio of seedlings exhibiting disease to the total number of seedlings examined after treatment with BGPP-Ar at a concentration of 1.0×10^8 plaque-forming units (PFU)/ml was 0.0 for seedling rot and 2.0 for seedling blight, whereas the ratios were 14.3 and 15.0 respectively, after treatment with ipconazole/copper (II) hydroxide. BGPP-Ar was highly effective in suppressing seedling rot of rice, even at the low concentration of 1.0×10^5 PFU/ml. Phage treatment of seeds indoors was the most effective, since phage inactivation by UV irradiation could be avoided. Treatment with phages has the potential for effective control of seedborne bacterial diseases (Adachi et al. 2012).

6.1.5 Diseases of Sorghum and Millets

6.1.5.1 Charcoal Rot Disease

Macrophomina phaseolina, incitant of charcoal rot disease of sorghum is responsible for serious losses. Three strains of *Pseudomonas* spp. selected out of 126 isolates obtained from different sorghum growing regions in India, were evaluated for their efficacy in reducing the incidence of charcoal rot disease under field conditions. *P. chlororaphis* SRB127 proved to be the most effective in suppressing the development of charcoal rot disease. When applied as seed treatment, the incidence of charcoal rot disease was reduced by >40 %, crop-lodging by >20 % and increased the grain weight. The strain SRB127 was able to colonize the sorghum root efficiently forming microcolony-like cell aggregates in some parts of root system in the sand culture and survived in the sorghum rhizosphere without significant reduction in its population level (Das et al. 2008). Two strains of *Pseudomonas fluorescens* Pf-52 and KRU-22 suppressed the development of Setaria blast disease caused by *Magnaporthe grisea* to an extent of 86.7 and 87.8 % respectively. Plant growth was also improved by treatment with the bacterial strains significantly (Karthikeyan and Gnanamanickam 2008).

6.1.5.2 Downy Mildew Disease

The efficacy of four strains of plant growth-promoting rhizobacteria (PGPRs), *Bacillus pumilus* INR7 and SE34 and *B. subtilis* IN937b and GB03 were assessed for the control of pearl millet downy mildew disease caused by *Sclerospora graminicola* under field conditions. With fresh suspensions of bacterial cells, treatment with INR7 protected the pearl millet plants more effectively than the other strains. When applied as powder formulation also, the strain INR7 proved to be better than other strains in protecting the plants against infection by *S. graminicola*. Application of the fungicide metalaxyl (Apron) was far superior in reducing the downy mildew disease incidence (12.0 %), compared with all of the BCA strains (30–69 %) under field conditions (Raj et al. 2003). A talc-based formulation, Trichoshield containing *Trichoderma harzianum*, *T. lignorum*, *Glocladium* (= *Trichoderma*) *virens* and *Bacillus subtilis* was applied as seed treatment, slurry treatment or foliar spray for the control of pearl millet downy mildew disease. Foliar spray was found to be more effective than the other methods. Trichoshield treatment provided protection varying from 52 to 71 % under field conditions. In addition, the yield attributes were also improved by the BCA treatment. However, the fungicide metalaxyl (Apron) was more effective than the BCA formulation (Raj et al. 2005).

6.2 Diseases of Cotton

6.2.1 Verticilium Wilt Disease

The role of organic amendments in the development of suppressiveness of soil to soilborne pathogens has been investigated in respect of cotton diseases. *Verticillium dahliae* causing cotton Verticillium wilt disease exists as two distinct groups of isolates, one group causing defoliating (D) and another group inducing non-defoliating (ND) symptoms in infected cotton plant. The efficacy of inducing disease suppressiveness by organic amendments from various plant species was assessed against the microsclerotia (MS) of D and ND isolates in infested soil. The efficacy of suppressiveness depended on species of plants, incubation time, pathogen isolate and whether soil was sterilized. The ND isolate was more sensitive than D isolate to amendment treatments. Debris of *Diplotaxis virgata* was the most effective in consistently reducing the viability of MS among the plant species tested. The eradicative potential of organic amendments on MS of *V. dahliae* was confirmed by observations on disease incidence and severity of symptoms of Verticillium wilt of cotton plants grown in infested, amended soil. All organic amendments reduced the disease incidence and severity in soils infested with both D and ND isolates to different degrees. None of the cotton plants in soils infested with ND isolate and treated with amendments exhibited the symptoms of infection, indicating that organic amendments could strongly reduce the final disease incidence levels (López-Escudero et al. 2007). Several organic amendments were evaluated for their biocontrol potential against Verticillium wilt disease of cotton. Among them, crab shell (chitin) was found to be the most effective in reducing the disease severity by 72 %, whereas soybean stalk and alfalfa reduced the disease severity by 60 and 50 % respectively. Crab shell amendment stimulated the proliferation of antagonists to *Verticillium dahliae* in the rhizosphere of cotton plants. The changes in the rhizosphere populations of microorganisms by crab shells might contribute to suppression of cotton wilt disease

(Huang et al. 2006). Four strains of *Pseudomonas* spp. isolates from the rhizosphere of cotton and weeds suppressed the development of Verticillium wilt disease, when the cotton seeds bacterized with the BCA strains were sown. The cotton plants were protected, when the treated plants were challenged by stem-injection with the conidia of *V. dahliae*. The bacterial BCA *Serratia plymuthica* (HRO-C48) was also included for comparing the efficacies of *Pseudomonas* strains with the known effective BCA. The reduction of area under disease progress curve (AUDPC) by seed bacterization with four strains of *Pseudomonas* (FP22, FP23, FP30 and FP35) and *S. plymuthica* ranged from 39.2 to 50.9 % and 22.1–36.8 % in the trials conducted in 2005 and 2006 respectively. The seed cotton yield of the cultivar Sayar 314 was increased in BCA-treated plots from 13.1 to 22.3 % in the field trial in 2005. But significant increase in seed cotton yield could be obtained following treatment with BCA strains in 2006. The results showed variable effects of BCA treatment on disease suppression and seed cotton yield, depending on cotton cultivar and environmental conditions (Erdogan and Benlioğlu 2010).

Verticillium dahliae (*Vd*) and *Fusarium oxysporum* f.sp. *vasinfectum* (*Fov*) causative agents of Verticillium wilt and Fusarium wilt diseases in cotton, may cause serious losses, when they appear together in the same field, making it difficult to manage them effectively. The dry mycelium (DM) of *Penicillium chrysogenum* (PEN), a waste product of the pharmaceutical industry was evaluated for its biocontrol potential against *V. dahliae* and *Fov*. Cotton fields were amended with DM before planting or squaring or both and disease incidence was monitored from 2001 to 2004. In the experiment during 2001 and 2002, at a dose of 30 g/m^2 of DM, either as basal application alone or basal application + side dressing of DM, provided significant protection against Verticillium wilt and Fusarium wilt diseases. But side dressing alone was not effective in reducing the incidence of both diseases. In the second experiment (2003 and 2004), the efficacy of DM was found to be dependent on the application rates. Basal application plus side dressing at doses of 30, 90 and 150 g/m^2 provided protection of 20.1, 34.6 and 42.7 % against *Fov* and of 26.8, 47.8 and 49.6 % against *Vd*, compared with untreated controls. Significant increases in lint yields were also recorded in plots treated with DM. As there was no direct inhibitory effect of DM on the pathogens, disease suppression was considered, possibly due to induction of resistance in cotton against the wilt diseases (Dong et al. 2006).

6.2.2 *Root Rot Disease*

Cotton root rot complex caused by *Fusarium oxysporum* (*Fo*) and *Pythium debaryanum* (*Pd*) is a difficult disease problem to be tackled by any one approach. Mixtures of biocontrol agents *Trichoderma hamatum*, *T. harzianum* and *Paecilomyces lilacinus* and resistance inducer salicylic acid (SA) and Bion [benzo (1,2,3)-thiadiazole-7-carbolic acid-*S*-methyl ester (BTH)] were evaluated for their biocontrol potential in reducing the cotton root rot disease incidence under greenhouse and field conditions. All treatments reduced the infection by *Fo* and *Pd* under greenhouse conditions.

Disease index percentage was significantly reduced (78.8 %), while seed germination improved almost by two-fold (199 %), compared with untreated controls. Under field conditions, extent of protection achieved was the highest in the plots treated with a combination of *T. hamatum* + *P. lilacinus* + SA + BTH which reduced DI to 72.3 and 69.3 % for *Fo* and *Pd* respectively compared to control treatment. The results indicated the enhancement of control efficacy through combination of fungal BCAs and inducers of disease resistance (Abo-Elyousr et al. 2009).

6.2.3 Seedling Diseases

Cotton seedlings are infected by one or more fungal pathogens such as *Rhizoctonia solani*, *Pythium aphanidermatum*, *P. ultimum* and *Rhozopus oryzae*. The seedlings may be killed at pre-emergence or post-emergence stages, resulting in gappiness in the field. Application of organic amendments, seed treatment with chemicals and maintaining optimum soil moisture level are the strategies suggested for the management of the seedling diseases. Fungal biocontrol agents have been evaluated for their potential against one more of these fungal pathogens involved either alone or in combination in causing the disease. Q strains of *Trichoderma* (=*Gliocladium*) *virens* was demonstrated to be effective against the cotton seedling disease caused by *Rhizoctonia solani* (Howell et al. 1997). *T. virens* suppressed the pathogen development through antibiosis and/or induction of systemic resistance in cotton against *R. solani* (Lumsden et al. 1992; Howell et al. 2000). Pre-emergence damping-off disease affecting cotton caused by *Pythium* spp. and *Rhizopus oryzae* was studied for the development of effective control measures. Seed treatment with metalaxyl effectively controlled the disease incited by *Pythium aphanidermatum*, *P. ultimum* and an unidentified *Pythium* spp., but it was not effective against the disease caused by *Rhizopus oryze*. Under field conditions, seed treatment with metalaxyl was only partially effective in naturally infested soils. Hence, disease symptoms observed in 47 % of the seedlings were attributed to infection by *R. oryzae*. The efficacy of seed treatment with formulations of *Trichoderma virens* parent strains G6 and G6-5 and the hybrid strains of *T. virens* × *T. longibrachiatum* Tvl-30 and Tvl-35 were able to effectively control the disease caused by all the four pathogens (Howell 2002).

Non-pathogenic isolates of binucleate *Rhizoctonia* spp. (np-BNR) have been reported to protect the cotton seedlings against damping-off and crown and root rot diseases caused by *Pythium* spp. and *Rhizoctonia solani*. The np-BNR isolates with biocontrol potential belonged to the anastomosis groups (AGs) AG-A, AG-B(a), AG-B(o), AG-G, AG-1, AG-GK and AG-P (Sneh et al. 1991). Pretreatment of cotton seedlings with np-BNR isolates provided efficient protection against pre- and post-emergence damping-off caused by a virulent strain of *R. solani* (AG-4). Seedling stand of protected cotton was significantly higher than that of untreated control seedlings. Many of the np-BNR isolates reduced the disease severity significantly ($P<0.05$). Combined application of np-BNR and the resistance inducer BTH provided effective protection to cotton seedlings, not only against seedling rot, but also against Alternaria leaf spot disease caused by *Alternaria macrospora*. However, the degree of disease reduction was comparable to that obtained with np-BNR treatment alone. Leaf symptoms induced by *A. macrospora* on cotton cotyledons pretreated with np-BNR or sprayed with BTH were significantly reduced. The np-BNR-treatment was more effective than BTH in reducing the disease severity. The results indicated that np-BNR isolates were more efficient in protecting cotton seedlings against both root and leaf pathogens and the disease control provided by them was superior to that offered by the chemical inducer of disease resistance (Jabaji-Hare and Neate 2005).

6.2.4 Black Root Rot Disease

Thielaviopsis basicola, causal agent of cotton black root rot induces necrosis of root cortex, delayed seedling emergence and fruit set, resulting in serious yield loss in Australia. Acibenzoalr-*S*-methyl (ASM) consistently reduced disease severity on taproots by 20–30 %, when cotton seeds were soaked in ASM solution (25 or 50 μg/ml) for 3–5 h before planting under greenhouse conditions. In the field experiments, ASM was applied either as spray over seeds during sowing (in-furrow spray) or as seed dressing by soaking in the chemical solution or as a foliar spray over seedlings. Seed-soaking in ASM solution reduced the severity of black root rot disease by 33 %. The in-furrow application of ASM (25 μg/ml at 550 l/ha) reduced the severity of symptoms on tap roots by 24 % and increased the number of relatively healthy lateral roots (by 35 %) and fruit number (by 24 %). The application of ASM as foliar spray was not effective against the black root rot disease. The results showed that application of ASM could induce systemic acquired resistance against the soilborne pathogen *T. basicola* causing the black root rot disease of cotton (Mondal et al. 2005). Seed coating with bacterial BCA *Paenibacillus alvei* strain K-165 formulated in xanthan gum and talc reduced the symptoms of black root rot diseases very effectively. Plant growth was also significantly promoted by seed treatment with the strain K-165 which was found to be an efficient root and soil colonizer inhibiting extensive root colonizer inhibiting extensive root colonization by *T. basicola* (Schoina et al. 2011).

6.2.5 Alternaria Leaf Spot Disease

The effect of foliar application of potassium nitrate (KNO_3) on the development of Alternaria leaf blight of cotton caused by *Alternaria macrospora* was assessed under field conditions. Disease incidence, severity and leaf shedding were determined at the bottom (1–7 nodes), middle (8–14 nodes) and top (15 + nodes) of plants at weekly interval. Foliar application of KNO_3 at 13 kg/ha significantly reduced the mean diseases incidence, severity and leaf shedding during the experimental duration. Disease incidence, severity and leaf shedding were higher at the middle of the canopy for all treatments than those at bottom and top canopies. KNO_3 treatment significantly reduced the adverse effects of Alternaria leaf spot disease (Bhuiyan et al. 2007). The effectiveness of non-pathogenic binucleate *Rhizoctonia* spp. (np-BNR) isolates and BTH in reducing the severity was assessed. Leaf symptoms induced by *A. macrospora* on cotton cotyledons pretreated with np-BNR or sprayed with BTH were significantly reduced. The np-BNR treatment was more effective than BTH in reducing the severity of Alternaria leaf spot disease (Jabaji-Hare and Neate 2005).

6.2.6 *Bacterial Blight Disease*

The bacterial pathogen *Xanthomonas axonopodis* pv. *malvacearum*, causing bacterial blight disease infects cotton plants at different stages of crop growth and all aerial plant parts such as leaves, petioles, stem, calyx, bolls and seeds causing both quantitative and qualitative loss of lint and seeds. Crop sanitation, supply of nutrients at optimal levels, use of cultivars tolerant to bacterial blight pathogen and application of bactericides have been suggested to reduce the incidence and severity of the disease. The biocontrol potential of *Bacillus cereus* isolate L2-1 and efficacy of the resistance inducer acidbenzolar-*S*-methyl (ASM) applied either alone or in combination in suppressing the development of cotton bacterial blight disease were assessed. Treatment with ASM reduced the disease severity to the maximum extent (57.8 %), whereas the strain L2-1reduced the disease severity to 25.48 % of the untreated control. The combination of ASM and L2-1 was less effective, compared to ASM treatment alone, indicating a negative interaction between the resistance inducer and bacterial BCA (Ishida et al. 2008). A commercially developed product of seaweed *Sargassum wightii* marketed as 'Dravya' was assessed for its potential in suppressing the development of bacterial blight disease of cotton. Seed soaking with Dravya at a concentration of 1:500 followed by foliar sprays thrice at 10, 20 and 30 days after sowing, reduced the disease incidence by 66, 70 and 74 % at 40, 60, and 80 days after sowing. In addition, promotion of plant growth due to treatment with plant extract was indicated in parameters such as plant height, total number of bolls formed, and boll weight. This effect on plant growth was considered to be due to induction of systemic resistance to the bacterial blight in treated plants (Raghavendra et al. 2007).

6.3 Diseases of Pulse Crops

6.3.1 *Soybean Diseases*

6.3.1.1 Damping-Off Disease

The fungal pathogen *Rhizoctonia solani* (teleomorph: *Thanatephorus cucumeris*) has a wide host range in addition to soybean. *R. solani* infects soybean at different stages of crop growth causing damping-off, root rot, stem rot and foliar blight, accounting for heavy losses in yield. Disease incidence increased, when dry beans (*Phaseolus vulgaris*) and sugar beet (*Beta vulgaris*) were grown in soybean rotations, as they were also susceptible to *R. solani* (Nelson et al. 1996). Binucleate *Rhizoctonia* (BNR) isolates were detected in soybean and sugar beet roots. The BNRs were found to be effective against *R. solani* infecting other crops such as tomato, potato, sugar beet and snap bean (Cardoso and Echandi 1987; Herr 1988; Escande and Echandi 1991; Muslim et al. 2003). The biocontrol efficacy of BNR isolates BNR-4, BNR8-2 and BNR8-3 in suppressing the development of *R. solani* anastomosis groups AG-4 and AG2-2, infecting soybean was assessed. There was no BNR x cultivar interaction. With AG-4 of *R. solani*, BNRs significantly increased seedling emergence and survival of soybean plants and decreased disease severity. There was no negative effect of BNRs on soybean seed germination and plant height. Further, no evidence of pathogenicity of BNR isolates to soybean could be inferred. BNRs could be consistently isolated from hypocotyls and roots, indicating their ability to colonize plant tissues associated with control (Khan and Nelson 2005). Soybean foliar blight disease is caused by *Rhizoctonia solani* AG-IA, AG-IB and AG2-3 in Japan and yield loss due to this phase of disease was estimated to be up to 70 %. The resistance inducing compounds salicylic acid (SA) and acibenzolar-*S*-methyl (ASM) and fungicides were evaluated for their efficacy in suppressing the development of foliar blight disease. Application of ASM (12.5 mg a.i./l) and SA (25 mM) sprayed at 10 and 20 days before inoculation with the pathogen provided the most effective protection against the disease (Meyer et al. 2006). In a later investigation, the effect of ASM was assessed for the control of *R. solani* AG-4 causing root rot disease of soybean. At a concentration of 0.5 g/l, ASM showed its maximum antifungal activity. Seed treatment with ASM at 0.08 and 0.5 g/l significantly induced the resistance responses in the soybean hypocotyls, resulting in reduction of intensity of rotting symptoms. The reduction in disease intensity due to ASM treatment was correlated with stimulation in the activities of defense-related enzymes like chitinase. Application of ASM induced a transitional growth retardation effect on treated soybean plants. However, the treated plants recovered rapidly in optimal growth conditions (Faessel et al. 2008).

Biofumigation using organic amendments such as Brassicaceae green manure or seed meal incorporation into the soil has been demonstrated as an ecologically suitable alternative to chemical fumigation. Release of glucosinolate-derived compounds are considered to be primarily responsible for the suppression of soilborne fungal pathogens such as *Rhizoctonia solani*, *Fusarium oxysporum* and *Pythium ultimum*. The tolerance of the fungal biocontrol agents *Trichoderma* spp. to these compounds was assessed to facilitate combined application of *Brassica carinata* seed meal (BCSM) for pathogen suppression through different mechanisms. *Trichoderma* spp. were found to be generally less sensitive to BCSM than the fungal pathogens tested at the highest dose (10 μmol of sinigrin). The fungal BCAs were able to grow on BCSM and over the pathogens. The greatest inhibitory effect was observed, when BCSM was applied in combination with *Trichoderma*, irrespective of the ability of BCSM to release isothiocyanates. The results suggested that it might be possible to improve the effectiveness of pathogen suppression by the integrated approach of combining the fungal BCA with an organic amendment with biofumigant property (Galletti et al. 2008).

6.3.1.2 Sclerotinia Stem Rot Disease

Sclerotinia sclerotiorum, a soilborne pathogen causes Sclerotinia wilt or stem rot disease in soybean and it is pathogenic to many crops including lettuce and sunflower. A nonpathogenic strain of *Fusarium oxysporum* was isolated from the soil suppressive to *S. sclerotiorum*. This nonpathogenic strain S6 exhibited antagonistic activity against *S. sclerotiorum*. The strain S6 produced the antibiotic cyclosporine A which inhibited the growth of *S. sclerotiorum* and also suppressed sclerotia formation by the pathogen. A significant increase in the number of surviving soybean plants occurred, when the pathogen and strain S6 were coinoculated with the pathogen alone under the greenhouse conditions (Rodriguez et al. 2006). The biocontrol potential of the cell suspensions or cell-free filtrates of two strains of *Bacillus subtilis* SB01 and SB24 was assessed for the suppression of Sclerotinia stem rot disease under field conditions. The bacterial cell suspensions were more effective than the cell-free filtrates in suppressing the development of *S. sclerotiorum*. The cell suspensions applied on soybean leaves for up to 10 days were able to significantly reduce the disease severity by about 20–90 % at 5 days after inoculation with the pathogen. The cell suspensions could protect the plants most effectively at 3 days after application which reduced the disease severity by 40–70 % (Zhang et al. 2011).

The effectiveness of *Coniothyrium minitans* (as Contans® WG), *Streptomyces lydicus* WYEC (as Actinovate® AG), *Trichoderma harzianum* T-22 (as PlantShield® HC) and *Bacillus subtilis* QST (as Serenade® MAX) was assessed for reducing the number of sclerotia of *Sclerotinia sclerotiorum* under field conditions. At two locations, soil artificially infested with the pathogen sclerotia was treated with these BCA preparations in the top soil, before planting soybean. The fungicide Boscalid was sprayed on the foliage at the soybean growth stage R1 as positive control. The results showed that *C. minitans* was the most effective in reducing disease severity index (DST) by 68.5 % and the number of sclerotia of the pathogen in the soil by 95.3 %. *S. lydicus* and *T. harzianum* were able to reduce DSI by 43.1 and 38.5 % and sclerotia in soil by 90.6 and 70.8 % respectively. *B. subtilis* exhibited only marginal adverse effect on DSI and sclerotia in the soil. Populations of the BCAs in the soil samples were determined at different periods after application from 3 to 169 days. The populations of *Streptomyces*, *Trichoderma* spp. and *C. minitans* did not vary significantly throughout the period of sampling, indicating their persistence in the soil (Zeng et al. 2012).

6.3.1.3 Fusarium Root Rot Disease

Fusarium root rot disease caused by *F. oxysporum* and *F. graminearum* is a major disease of soybean in Ontario, Canada. *Bacillus subtilis* strains SB01, SB04, SB23, SB24, SB28 and SB33 from soybean roots and CB01 and CH22 from corn roots significantly reduced the severity of root rots due to *F. oxysporum* and *F. graminearum*, when applied as a seed or soil treatment. The strains from soybean were the most effective against both pathogens in either seed or soil treatment. When applied as seed treatments, the soybean strains reduced root rot severity by 43–63 % and increased the seedling emergence by 13–17 %, plant height by 9–18 % and root weight by 8.4–19 %. When applied as soil treatment, the soybean strains reduced the root rot severity by 68–74 % and increased seedling emergence by 14–18 %, plant height by 11–23 % and root dry weight by 16–24 %. Soil treatment with *B. subtilis* strains appeared to be a more effective method of application against both *Fusarium* pathogens, causing root rot disease of soybean than seed treatment (Zhang et al. 2009).

6.3.1.4 Rust Disease

Soybean rust disease caused by *Phakopsora pachyrhizi* is another destructive disease limiting soybean production. Silicon (Si) amendments were investigated for the ability to reduce the rust disease incidence and intensity. Silicon was applied as wollastonite ($CaSiO_2$) as soil application (at 0.96 and 1.92 t/ha of Si) or as foliar application of potassium silicate (K_2SiO_3) under greenhouse conditions. Si treatments delayed the onset of disease by about 3 days. The area under disease progress curve (AUDPC) of soybean plants receiving Si application was significantly lower than that of control treatment. In the field evaluation of the effect of Si treatment, an average of 3-day delay in the disease onset was recorded only in the case of soil application. Reduction in AUDPC values for soil and foliar treatments were up to 43 and 36 % respectively. As the rust pathogen was unable to overwinter in the major soybean production areas of the USA, the delay in the onset of disease coupled with reduction in AUDPC, Si treatments could be expected to be a useful alternative strategy for the management of soybean rust disease (Lemes et al. 2011).

6.3.1.5 Soybean Mosaic Disease

The phenomenon of cross-protection of plants against severe strains by pre-inoculating mild strain of the same virus has been successfully exploited for the control of many virus diseases like *Citrus tristeza virus* (CTV). *Tobacco mosaic virus* (TMV) and *Papaya ringspot virus* (PRSV). *Soybean mosaic virus* (SMV) is a seedborne and aphidborne potyvirus. To evaluate the effectiveness of the cross-protection strategy, black soybean seedlings were inoculated with the attenuated SMV isolate Aa15-M2. The soybean plants, at 1–3 days after protective inoculation with the SMV isolate, were transplanted in seed farms in Kyoto Prefecture, Japan. Development of severe strains of SMV was effectively suppressed in protected plants. The seeds obtained from these plants were virtually free of SMV. In the next stage, black soybean seedlings grown from these seeds were transplanted in the growers' fields. The incidence of virulent strains of SMV up to pre-flowering stage about 40 days after transplanting was 1.3 % in the first year and 0.8 % in the second year. The results revealed the effectiveness of applying cross-protection strategy for the management of soybean mosaic disease (Kosaka and Fukunishi 1994).

6.3.2 Chickpea Diseases

6.3.2.1 Fusarium Wilt Disease

Management of soilborne diseases like wilt disease caused by *Fusarium oxysporum* f.sp. *ciceris* (*Foc*) race five has been found to be difficult, since no single strategy provides effective reduction in disease incidence/severity. Crop rotation, soil solarization, use of pathogen-free seed, removal of infected plant debris and seed treatment with biocontrol agents/fungicides have been suggested for developing an integrated management system suitable for the specific location. The potential of soil solarization for the control of chickpea Fusarium wilt disease was assessed, using wilt-sick plots during 1984–1985, 1985–1986 and 1986–1987 seasons. Susceptible cultivars showed 100 % mortality due to *Fusarium oxysporum* f.sp. *ciceris* (*Foc*). Resistant cultivar JG74 was also included in the trials. Soil temperature under polyethylene sheeting exceeded 60 °C at 5 cm and reached 42 °C at 20 cm soil depth. The increase in soil temperature, following solarization, induced various biological and physicochemical changes in the soil that affected plant growth and soil microflora. The pathogen population in the experimental plots was >1,000 propagules/g of soil. Substantial reduction in pathogen population, due to solarization was observed, while in non-solarized plots pathogen population increased during this period. In non-solarized plots virtually all plants of susceptible chickpea cultivar were killed before maturity. In the case of resistant cultivar, no wilting of plants was observed both in solarized and non-solarized plots. In addition to protection against wilt disease, solarization exerted stimulatory effects on the growth of chickpea plants of susceptible and resistant cultivars. Solarization markedly increased dry matter production and seed yield of both susceptible and resistant chickpea cultivars (ICRISAT 1980).

Fusarium oxysporum f.sp. *ciceris* (Foc) occurs in severe forms causing chickpea wilt disease in most areas where chickpea is cultivated. Treatment of chickpea seeds with talc-based formulations of *Pseudomonas fluorescens* Pf1 suppressed the development of chickpea Fusarium wilt disease under field conditions and also increased the seed yield. Pf1 strain was inhibitory to the bioinoculants *Rhizobium* and *Azospirillum* and the biocontrol activity of Pf1 strain was not affected by fungicides Thiram and carbendazim used for seed treatment against seedborne pathogens (Vidhyasekaran and Muthamilan 1995). In another investigation, the efficacy of *Pseudomonas fluorescens* RGAF101, *P. fluorescens* RG26, *Bacillus megaterium* RGAF51 and *Penibacillus macerans* RGAF101 was assessed for suppressing the development of wilt disease in chickpea. Seed and soil treatments with these rhizobacterial species, especially two *P. fluorescens* strains suppressed the disease development by delaying the expression of symptoms and reducing the rate of disease increase at 20 and 30 °C, but not at 25 °C. In the absence of the bacterial BCA, higher numbers of Foc infections were necessary at 20 and 30 °C than at 25 °C for rapid development of disease severity, indicating the effect soil temperature on disease development and efficiency of the bacterial BCAs (Landa et al. 2001, 2004b). Combination of sowing date, use of partially resistant chickpea genotypes and seed and soil treatments with biocontrol agents *Bacillus megaterium* RGAF51, *Bacillus subtilis* GB03, nonpathogenic *Fusarium oxysporum* Fo90105 and *Pseudomonas fluorescens* RG26 was examined for their combined effects on chickpea wilt disease. Advancing the sowing date from early spring to winter significantly delayed disease onset, reduced final disease intensity and increased seed yield. Under conditions highly conducive for Fusarium wilt disease development, the extent of disease control depended on primarily on choice of sowing date and to a lesser extent on the level of resistance of chickpea genotypes to Foc race 5 and the biocontrol treatments. Adoption of these strategies resulted in reduction in the rate of disease intensity and an increase in emergence of chickpea seedlings (Landa et al. 2004a).

Improvement in the efficacy of biological control and overcoming inconsistencies in the performance of individual biocontrol agents may be achieved by combining biocontrol agents or combining biocontrol agents and chemical treatments. The chickpea seeds were treated with metalaxyl, because of its effectiveness against Pythium seed rot under field conditions. Metalaxyl influenced survival of the BCAs on chickpea seeds differentially. Metalaxyl had no effect on *Bacillus megaterium* RGAF51 and *B. subtilis* GB03 during storage, but reduced viability of nonpathogenic Fo90105 to some extent. Metalaxyl was more deleterious to *P. fluorescens* RG26. Although metalaxyl affected the viability of *P. fluorescens* RG26 and nonpathogenic Fo90105, treatments with these BCAs either alone or in combination were among the most effective ones, suppressing wilt disease development and increasing the seed yield. Combining nonpathogenic Fo90105 with either *B. subtilis* GB03 or *P. fluorescens* RG26 was more effective in suppressing Fusarium wilt than single application of nonpathogenic Fo90105, but was equally effective as each bacterium alone. The results indicated the enhancement of effectiveness of disease suppression by integrating the existing control practices, partially effective by themselves with other disease management strategies (Landa et al. 2004a).

6.3.2.2 Root Rot Disease

Incidence of chickpea root rot disease caused by *Macrophomina phaseolina* is favored by the association of the infestation of nematode *Meloidogyne incognita*. For the biological management of the root rot disease, the potential of *Glomus intraradices*, *Rhizoctonia* sp. and *Pseudomonas putida* and *Paenibacillus polymyxa* was assessed. Combined application of the mycorrhiza and bacterial species reduced the root rot index, nematode galling and multiplication. Plant growth and absorption of nutrients, nitrogen, phosphorus and potassium were increased by *G. intraradices*, *Pseudomonas striata* and *Rhizobium* sp. Root colonization by *G. intraradices* was enhanced in the presence of *P. putida* and *P. polymyxa* (Akhtar and Siddiqui 2007, 2008).

6.3.2.3 Damping-Off Disease

Chickpea damping-off disease caused by *Pythium ultimum* results in death of young seedlings, leading to reduction in plant stand and consequent reduction in seed yield. The efficacy of seed treatment with biological agents *Bacillus pumilus* GB34 (as YieldShield), *B. subtilis* GB03 (as Kodiak), *B. subtilis* MBI600 (as Subtilex), *Streptomyces lydicus* WYEC108 (as Actinovate), *S. grieseoviridis* K61 (as Mycostop), *Trichoderma harzianum* Rifai strain KRL-AG2 (as T-22 Planter Box) and fungicides fludioxonil (Maxim) and mefenoxam (Apron XLLS) either alone or in combination was assessed for the control of damping-off disease caused by *P. ultimum* under greenhouse and field conditions. The desi cultivar showed lower incidence of damping-off disease than the Kabuli cultivar under greenhouse and field conditions. In the field trials conducted at three locations, mefenoxam was found to be the most effective as seed treatment. Treatment with the biocontrol agents were ineffective in reducing the incidence of damping-off disease and in promoting plant growth compared with control treatment. The results indicated the superiority of fungicide treatment over all the biocontrol agents tested for the control of the chickpea damping-off and the need for comparing the efficacy of BCAs under field conditions (Leisso et al. 2009).

6.3.2.4 Ascochyta Blight Disease

Chickpea Ascochyta blight disease incited by *Ascochyta rabiei* (teleomorph—*Didymella rabiei*) causes greater losses in winter seasons. Chickpea lines CA-252 and CA-255with partial resistance to Ascochyta blight and wilt diseases might be useful in reducing the loss due to these diseases, since winter sowings could expose the chickpeas to environmental conditions highly conducive for the development of Ascochyta blight disease. Hence, an appropriate level of resistance to Ascochyta blight disease in chickpea cultivars has to be ensured as a management practice for winter crops (Navas-Cortés et al. 1998; Landa et al. 2004a, b). The infected debris is an important source of inoculum for the blight pathogen that could be carried over to the next crop. The effect of application of biological control agents on plant debris left in the field was assessed. Application of non-amended suspension of the yeast *Aureobasidium pullulans* to the postharvest chickpea debris resulted in reduction of Ascochyta blight lesions by 37.9 and 38.4 % on chickpea plants relative to the untreated controls in 2004–2005 and 2006–2007 seasons respectively. Ascospores released from the debris were identified as that of *Didymella rabiei* accounting for high proportion of airspora released from chickpea debris (Dugan et al. 2009).

6.3.3 Pigeonpea Wilt Disease

Fusarium wilt disease of pigeonpea is caused by a different species *Fusarium udum* with taxonomic characteristics different from *Fusarium oxysporum* which occurs in the form of many formae speciales. Application of organic manures to activate the resident antagonistic organisms and adoption of crop rotation may have beneficial effect on plant growth and also results in reduction of disease incidence. The fungi, *Aspergillus niger*, *Penicillium citrinum*, *Trichoderma harzianum* and *T.* (= *Gliocladium*) *virens* and the bacterial species *Bacillus licheniformis* isolated from the rhizosphere were found to have inhibitory effect on the pathogen *F. udum* in in vitro and greenhouse assays. But under field conditions all the fungi were able to reduce the incidence of wilt disease, while the bacterial species was ineffective. *T. virens* was the most effective with 50.5 % disease control followed by *A. niger* with 38.7 % disease control. *T. harzianum* was the least effective in reducing the disease incidence (Singh et al. 2002). The biocontrol potential of *Streptomyces* sp. A6 and its tolerance to fungicides that are frequently used against soilborne pathgoens, were determined. *Streptomyces* sp. A6 showed strong tolerance towards the fungicides mancozeb, sulfur and carbendazim. In addition, the strain A6 exhibited enhanced growth and mycolytic enzyme production in the presence of these fungicides. By combining the *Streptomyces* sp. and the fungicides (at EC_{50} dose), infection by *F. udum* was most effectively reduced. By this approach of integration, excessive application of fungicides could be avoided, simultaneously achieving effective control of pigeonpea wilt disease (Singh and Chhatpar 2011).

6.3.4 Mungbean Root Rot Disease

Root rot disease caused by *Macrophomina phaseolina* infects all pulse crops including mungbean (*Vigna radiata*). Application of organic amendments and maintenance of balanced soil moisture levels may be helpful in containing the disease to some extent. As the pathogen is soilborne, fungicide control of the disease would be uneconomical and often ineffective. Antagonistic microorganisms have been shown to be effective in inhibiting the growth of the pathogen in in vitro and greenhouse evaluations. Only a few have been tested under field conditions. Combined application of *Trichoderma viride* and *T. harzianum* reduced the root rot disease by 23 %, whereas another combination of *T. harzianum* and *Aspergillus versicolor* was more effective and reduced the disease incidence by 31 %. These antagonistic fungi remained at reasonably high population levels in the applied soils for about 60 days after sowing (Choudhary et al. 2010). Seed treatment or soil application of the powder formulation of the bacterial BCA *Burkholderia* sp. strain TNAU-1 reduced the root rot incidence to 16 % as against 52.6 % infection in untreated control plots. Treatment with the strain TNAU-1 was as effective as the fungicide carbendazim in protecting mungbean plants against the root disease (Satya et al. 2011).

6.4 Diseases of Oilseed Crops

6.4.1 *Peanut Diseases*

6.4.1.1 Root Rot Disease

Peanut (groundnut) root rot disease caused by *Macrophomina phaseolina* (Fig. 6.4) or *Fusarium solani* is a devastating disease in several countries like Argentina and India and the production losses could reach up to 95 % in Argentina, under conditions like drought stress. Compatibility of the biocontrol agent *Pseudomonas fluorescens* Pf1 effective against peanut root rot pathogen *Macrophomina phaseolina* and the plant growth promoter *Rhizobium* TNAU-14 was assessed under glasshouse and field conditions. Seed treatment and soil application of Pf1 was more effective than other treatments in reducing the root rot disease incidence and improving the plant growth as well (Shanmugam et al. 2002). The biocontrol potential of *Trichoderma* spp. against *Fusarium solani* causing peanut root rot was evaluated. The optimum inoculum potential of *T. harzianum* ITEM 3636 and *T. longibrachiatum* ITEM 3635 was determined under greenhouse conditions for application for field trials conducted in 2003–2004 in a commercial field with previous history of the disease. Two seed treatments—seed coated with a conidial suspension using carboxymethylcellulose (CMC) as sticker and seeds coated with the antagonistic fungal biomass on Biodac particles were carried out. *T. harzianum* in both seed treatments was more effective than *T. longibrachiatum* in decreasing the mean disease severity index (MSI), increasing the frequency of healthy plants and boosting yield. The effectiveness of *T. harzianum* in suppressing the development of root rot disease and promoting the growth of peanut plants was confirmed

Fig. 6.4 Peanut plant showing symptoms of infection by *Macrophomina phaseolina*

Table 6.2 Total colony forming units (CFUs) of potential biocontrol agents in peanut under different tillage systems and preceding crops (Gil et al. 2008b)

	log 10 CFU/g of soil of potential biocontrol agents		
Field treatments	Actinomycetes	*Trichoderma* spp.	*Gliocladium* spp.
Tillage systems			
No-tillage	7.00a	4.20a	2.90a
Disc harrow	5.90b	3.70b	1.90b
Moldboard plough	4.80c	3.10c	1.50c
Previous crop			
Maize	7.00a	5.40a	4.00a
Sorghum	6.00b	2.80b	1.40b

In the same column, figures followed by the same letter are not significantly different according to LSD test at $P < 0.05$

Actinomycetes are expressed as $\times 10^4$ and fungi as $\times 10^2$

in the trial conducted in 2004–2005 in fields artificially and naturally infested with pathogen (Rojo et al. 2006).

Cultural practices may be combined appropriately to benefit natural populations of antagonistic microorganisms that may contribute to the management of soilborne diseases in a sustainable manner within the existing ecological boundaries. The effects of rotation of crops (maize, soybean and peanut) and tillage systems (no-till, reduced and conventional tillage) on potential biocontrol agents and the incidence of root rot disease caused by *F. solani* were evaluated in two long-term field experiments in Argentina. When maize was the preceding crop and crops were under conservation tillage, populations of BCAs were higher in both trials. However, the relationship between cultural management and the incidence of root rot incidence was lower, when maize was the previous crop and peanut was under no-till, and higher under the same management conditions in the second location (Gil et al. 2008a). Soil samples were taken at sowing and harvest and root rot incidence was evaluated at harvest. There was an inverse relationship between root rot incidence in peanut and populations of potential biocontrol agents (PBAs), actinomycetes, *Trichoderma* spp. and *Gliocladium* spp. under no-till, suggesting a possible role of PBAs in the control of *F. solani*. The incidence of root rot was low under no-till and disc harrow was associated with high populations of PABs. However, no such correlation was seen, when soybean preceded peanut. The incidence of root rot was low, in spite of the presence of low populations of PBAs in the soil compared with that in maize as previous crop (Table 6.2) (Gil et al. 2008b, 2010).

6.4.1.2 Sclerotinia Blight Disease

Sclerotinia minor causes the peanut Sclerotinia blight disease which accounts for considerable crop losses. Overwintering sclerotia of *S. minor* in the soil form the primary inoculums for this soilborne pathogen. Crop rotation as a very important cultural practice followed for managing peanut diseases was examined for its usefulness for the control of Sclerotinia blight diseases. Long-term rotations with crops such as corn and cotton showed only limited effectiveness, since the soilborne sclerotia could remain viable for as long as 4 years (Melzer et al. 1997). Use of peanut cultivars with resistance to one or more diseases has been suggested to minimize the disease incidence. An advanced breeding line N92056C and cultivars Tamrun 98 and Perry showed moderate to high levels of resistance to *S. minor* and produced high yields compared with susceptible cv. NC7 (Lemay et al. 2002). In order to induce resistance in peanut plants, three applications of acibenzolar-*S*-methyl (ASM) were carried out along with the fungicide fluazinam for comparison of their effects on disease incidence and yield of treated plants. But the fungicide suppressed the disease development effectively and also increased the yield in two of three locations. Although ASM had been reported to induce resistance to many pathogens infecting different crops, application of ASM for protecting peanut plants was found to be an ineffective strategy for the management of Sclerotinia blight disease of peanut (Lemay et al. 2002).

Several microorganisms including *Coniothyrium minitans* have been demonstrated to be mycoparasites of *Sclerotinia* spp. *C. minitans* readily infected and colonized the sclerotia of *Sclerotinia* spp. in the soil and its effectiveness in controlling *S. sclerotiorum* infecting various crops has been reported (Budge et al. 1995; Gerlagh et al. 1999). The effectiveness of *C. minitans* in suppressing the development of Sclerotinia blight disease caused by *S. minor* was investigated in a field experiment conducted for 5 years and in eight short-term experiments. The commercial formulation Contans WG was repeatedly applied to the soil at 2 and 4 kg/ha. In addition, individual commercial peanut fields were treated with single dose of *C. minitans* at 4 kg/ha. The number of sclerotia of *S. minor* in the soil and the incidence of Sclerotinia blight of peanut were reduced in the 5-year field study by applying *C. minitans*. However, the reduction in disease was not observed until at least 1 year after *C. minitans* was introduced. *C. minitans* was active against the sclerotia of *S. minor* in the soil, even when the environmental conditions were favorable for the pathogen and under monoculture of peanut. Both the 2- and 4-kg/ha rates of *C. mintans* applied to the soil, decreased the number of sclerotia recovered and disease incidence. No enhanced effectiveness of *C. minitans* was observed by increasing the application rate (4 kg/ha). Sclerotinia blight incidence was less in plots receiving applications of *C. minitans* for either 1 or 3 years compared with untreated control. Integration of consecutive years of soil application of *C. minitans* at 2 kg/ha with moderately resistant cultivars and fungicide application may provide protection to the peanut plants to the required level (Partridge et al. 2006).

6.4.1.3 Collar Rot Disease

Collar rot or seedling blight disease appears to be induced by *Sclerotium rolfsii* or *Aspergillus niger* in different geographical regions. The disease due to *S. rolfsii* is also referred to as Southern blight disease. Seedborne infection by *S. rolfsii* adversely affected nodulation, leghemoglobin and nitrogenase activity of peanuts. Seed treatment with formulated *Bacillus subtilis* counteracted the negative effects of the pathogen.

In addition, the crop vigor index, total nitrogen content and survivability of *Rhizobium* spp. were also improved by the application of *B. subtilis* (Abd-Allah and El-Didamony 2007). In a later investigation, talc-based formulations of *Beauveria bassiana* (B2) and *Pseudomonas fluorescens* (TDK1 and Pf1strains) alone and their combination with and without chitin were evaluated for their efficacy in reducing the incidence of collar rot disease caused by *S. rolfsii* and the leaf miner (*Aproaerema modicella*) on peanut under greenhouse and field conditions. The combination of B2 + TDK1 + Pf1 was the most effective in reducing the incidence of both collar rot disease and the leaf miner infestation. In addition, the treatment with fungal and bacterial biocontrol agents with fungal and bacterial biocontrol agents promoted the plant growth and enhanced the yield, compared with untreated control plants (Senthilraja et al. 2010).

Bacteria isolated from six habitats of peanut were evaluated for their broad spectrum antifungal activity and suppression of collar rot caused by *Aspergillus niger* in peanut. *Pseudomonas aeruginosa* GSE18 and GSE19 strains were more effective than the other 391 isolates tested. The strain GSE18 was more efficient than strain GSE19. When applied as seed treatment, strain GSE18 reduced the pre-emergence rotting and post-emergence wilting by 60 %. Further, the strain GSE18 effectively colonized the peanut plant rhizosphere both in native and in infested soils. *P. aeruginosa* GSE18 was tolerant to thiram and in combination with the fungicide higher level of suppression of collar rot disease could be achieved, indicating the feasibility of integrating the BCA application with fungicide treatments applied against other peanut pathogens (Kishore et al. 2005). Pre- and post-harvest aflatoxin contamination of groundnut is due to *Aspergillus flavus* which causes aflroot disease in the field. Strains of *Pseudomonas* spp., *Bacillus* spp. and *Trichoderma* spp. isolated from the geocarposphere (pod zone) of peanut efficiently reduced the infection by *A. flavus* and also reduced the populations of *A. flavus* by 50 % in the geocarposphere of peanut (Anjaiah et al. 2006). In a later investigation, a strain of marine bacterial species *Bacillus megaterium* was evaluated for its biocontrol potential against postharvest decay of peanut kernels caused by *Aspergillus flavus*. The concentration of *B. megaterium* influenced its biocontrol efficacy in vivo. The washed bacterial cell suspension at a concentration of 1×10^9 CFU/ml reduced the kernel rot significantly to a level that was markedly lower than that of the control treatment, after 7 days of incubation at 28 °C. The washed and unwashed cells of *B. megaterium* were equal in their effectiveness against *A. flavus*. The *B. megaterium* strain could significantly reduce the biosynthesis of aflatoxin by *A. flavus* and the expression of the genes *aflR and aflS* was repressed by the bacterial strain resulting in reduction in production of aflatoxins (Kong et al. 2010).

6.4.2 Oilseed Rape Diseases

6.4.2.1 Sclerotinia Stem Rot Disease

Oilseed rape (*Brassica napus*) is affected by Sclerotinia stem rot disease caused by *Sclerotinia sclerotiorum* which has a wide host range that includes many crops like soybean and peanut. As the sclerotia of the pathogen remain viable for long time (>4 years), crop rotation with nonhosts has limited applicability for the effective management of this disease. The mycoparasite, *Coniothyrium minitans* has been reported to be effective against *S. scletotiorum* in other crops. A petal inoculation technique revealed that *C. minitans* was effective in inhibiting infection initiated by ascospores of *S. sclerotiorum* on flower petals by restricting the mycelial growth of the pathogen. When *C. minitans*-treated rapeseed leaves were inoculated with the mycelia of *S. sclerotiorum*, the BCA failed to prevent infection of leaves, but caused significant reduction in the number of sclerotia produced on diseased leaves. Under field conditions, *C. minitans* (10^6/ml) alone, *C. minitans* + benomyl (50 μg/ml), benomyl alone (100 μg/ml), *C. minitans* + vinclozolin (100 μg/ml) and vinclozolin (500 μg/ml) alone were evaluated during 1997–2004. No significant differences in the disease suppressive ability of the treatments were observed. All the treatments applied as aerial sprays on leaves reduced the stem rot disease incidence significantly. The sclerotia of *S. sclerotiorum* collected from the diseased plants in plots treated with *C. minitans* showed infection by *C. minitans* at a frequency ranging from 21.3 to 54.5 %, indicating the mycoparasitic ability of *C. minitans* under field conditions (Li et al. 2006).

The effectiveness of *Pseudomonas chlororaphis* PA-23 and *Bacillus amyloliquefaciens* BS6 in suppressing the development of stem rot disease of canola caused by *Sclerotinia sclerotiorum* was assessed under field conditions for 2 years (2003–2004). Application of the strains PA-23 and BS6 twice at 30 and 50 % bloom significantly reduced the percent stem rot disease incidence. The difference between single and double application of BCAs in disease reduction was not significant. The BCAs were as effective as the fungicide Rovral-Flo and both were effective in reducing petal infection by *S. sclerotiorum*. Two sprays with PA-23 and BS6 more effectively reduced petal infection than single application of the BCA strains. Application of PA-23 triggered the expression of PR-3 protein, indicating the underlying mechanism of its biocontrol activity, might be through induction of systemic resistance in canola against *S. sclerotiorum* (Fernando et al. 2007). *Bacillus subtilis* strain NJ-18 exhibited antifungal activity against two economically important fungal pathogens *Sclerotinia sclerotiorum* and *Rhizoctonia solani*, causing Sclerotinia stem rot of rapeseed and sheath blight disease of rice respectively. In the field experiments, fermentation of NJ-18 at 1.0×10^7 CFU/ml significantly reduced the disease incidence and severity with control percentage of 77.1 %. The effectiveness of disease control was equivalent to that was obtained with dimethachlon and better than carbendazim, when applied at the rate of 750 g a.i/ha. The results showed that the potential of *B. subtilis* strain NJ-18 for suppressing the development of oilseed rape Sclerotinia stem rot disease (Yang et al. 2009). *Pseudomonas fluorescens* PS1 formulated in saw dust-oil as a carrier was effective in suppressing the development of *S. sclerotiorum*, causing stem blight disease in Indian rapeseed (*Brassica campestris*). Seed bacterization enhanced seed germination, increased plant growth and reduced stem blight incidence and increased the yield (Aeron et al. 2011).

6.4.2.2 Blackleg Disease

Canola (*Brassica napus*) is affected by blackleg disease which is a complex in which two fungal species, *Leptosphaeria biglobosa* and *L. maculans* are involved. *L. biglobosa* is weakly virulent and classified as pathogenicity group1 (PG-1), while *L. maculans* is aggressive, highly virulent and classified into pathogenicity groups PG-2, PG-3 and PG-4. Under field conditions, blackleg disease severity was reduced in plants inoculated with PG-2 alone or prior to PG-1. Inoculations on canola plants with PG-1 at the six-leaf stage provided evidence for the operation of processes resulting in development of systemic acquired resistance (SAR). When plants were treated with the PG-1 isolate on lower leaves and then treated with PG-2 on upper leaves, resistance was induced. The results of field experiments suggested that the severity of blackleg disease could be reduced significantly by the weakly virulent PG-1 isolate alone at 24 h prior to PG-2 inoculation (Chen and Fernando 2006).

6.4.3 Sesamum Damping-Off Disease

Incidence of damping-off and wilt diseases increases to alarming proportions, when sesamum is cultivated in the same field for two or more successive years. *Paenibacillus polymyxa* E681, with plant growth-promoting and antagonistic properties was evaluated for its efficacy against the pathogens involved in these diseases. The unformulated strain E681 did not provide protection to the plants consistently. Hence, the bacterial BCA was formulated with a combination of clay and vermiculite and the seeds of sesamum were pelleted with this material. In the greenhouse evaluation, formulations of the strain E681 reduced the disease incidence in disease-conducive soil. Under field conditions, pelleting of seeds with strain E681 significantly reduced pre- and post-emergence damping-off compared with untreated controls. In addition, pelleting with strain E681 promoted the growth of plants growing from treated seeds. The results revealed the potential of *P. polymyxa* E681 for protecting the sesamum plants against the damping-off disease (Ryu et al. 2006).

6.4.4 Sunflower Diseases

6.4.4.1 Head Rot Disease

Head rot disease incited by *Sclerotinia sclerotiorum* attacks the grains directly, resulting in huge losses. No single disease management strategy provides efficient control of the disease. Hence, different methods have to be integrated to achieve disease reduction to the required level. Early sowing of sunflower was reported to reduce the probability of having cool and humid weather conducive for disease development, when the crop reached flowering stage. Application of chemicals to contain the disease incidence and spread is not an economically viable proposition (Mantecón and Pereyra 1997). Suppression of development of *Sclerotinia sclerotiorum* by targeting the sclerotia of the pathogen has been investigated, as an alternative approach for its effectiveness. *Trichoderma* spp. was isolated from sunflower heads and a composite mixture of *Trichoderma* spp. was found to be effective against *S. sclerotiorum*. A novel method of employing honeybees as vectors of *Trichoderma* spp. and disseminating them under field conditions, was developed by Escande et al. (1994). A mixture of six isolates including *Trichoderma koningii*, *T. aureoviride* and *T. longibrachiatum* was tested for their efficacy in reducing sunflower head rot caused by *S. sclerotiorum* under field conditions. *Trichoderma* formulation (TF) containing *Trichoderma* conidia and viable hyphal fragments, industrial talc and milled corn kernels was developed. Honeybees (*Apis mellifera*) were employed to disperse TF for 6 weeks from the onset of flowering. Two days after the first TF delivery, sunflower heads were inoculated with ascospores of the pathogen. When honeybees had dispersed 100 g of TF at 10 h/day, the incidence of head rot was significantly reduced. A delay of epidemic onset of head rot through TF dissemination by honeybees could be observed. By combining with a partially resistant genotype, reduction in disease incidence from 75 to 15 % or from 90 to 23 % could be achieved. The results indicated that the new delivery system of employing the honeybees for the dissemination of the BCA could be integrated into the disease management system (Escande et al. 2002).

6.4.4.2 Necrosis Virus Disease

Incidence of *Sunflower necrosis virus* (SNV) in several States in India has raised concern and the need to contain its spread has been emphasized. As in the case of other virus diseases, adoption of cultural practices such as alteration in planting dates, elimination of sources of virus inoculum and crop sanitation may be useful in reducing the disease incidence. Another strategy of employing biocontrol agents that can induce resistance in sunflower plants against SNV was studied, since none of the biocontrol agents showed direct antiviral effect. Strains of *Streptomyces* sp. PMS, *Streptomyces fradiae* MML1042, *Bacillus licheniformis* MML2501, *Pseudomonas aeruginosa* MML2212 and *Bacillus* sp. MML2551 were evaluated for their efficacy in reducing the incidence of SNV. The bacterial strains were employed for seed treatment and soil application either alone or in combination. Among the treatments, the combination of *B. licheniformis* + *Bacillus* sp. + *S. fradiae* was the most effective in reducing the disease incidence. Treatment with these bacterial BCAs promoted plant growth and yield attributes under field conditions, resulting in favorable benefit-cost ratio (1:8) (Srininvasan and Mathivanan 2011).

6.4.5 Palm Bud Rot Disease

Bud rot disease caused by *Thielaviopsis paradoxa* (teleomorph-*Ceratocystis paradoxa*) is a serious disease occurring on various species of palm. Application of fungicides was found to provide protection only to a limited extent and uneconomical. Screening of *Chaetomium* spp. to identify the efficient strains that could suppress the development of *T. paradoxa* was undertaken in Thailand for the last two decades. A biological formulation from *C. cupreum* CC1-10 and *C. globosum* CG1-12 was developed into pellet and powder formulations and patented for use in palms. Under field conditions, the *Chaetomium* biological product significantly reduced the disease incidence by 75 % within 30 days after application. Five year old bottle palms (*Hyophorbe lagenicaulis*) completely recovered from the disease, when the biological product was applied to the infested soil at the rate of 20 g/plant. When the antagonistic compounds produced by *C. cupreum* and *C. globosum* were sprayed, the terminal bud rot severity was reduced. The treated trees recovered within 30 days after application of biological products and new healthy leaves emerged, indicating the curative effect of the compounds produced by *Chaetomium* spp. (Fig. 6.5). The results showed the potential of the biological product developed from *Chaetomium* spp. for the effective control of bud rot disease of palms under field conditions (Soytong et al. 2005).

6.5 Postharvest Diseases

Seeds of agricultural crops infected by various microbial plant pathogens form the primary sources of infection, carrying the infection to the next generation and also to new locations where the pathogen(s) may be absent. Seed treatments based on physical, chemical and biological methods have been employed with different degrees of effectiveness for suppressing the development of seedborne pathogens. These methods have their own advantages and disadvantages. Application of biocontrol agents is considered to be more desirable and advantageous in certain pathosystems. The grains and feed materials that are contaminated by mycotoxins produced microbial pathogens are particularly suitable for treatment with biocontrol agents (BCAs). The antagonistic organisms with genetic stability and efficiency have to be selected and environments conducive for their rapid establishment and multiplication should be provided. The competition from other seed microflora for the space and nutrition has to be minimized by providing conditions that will differentially favor the development of the BCAs. The ability to remain viable and retain the biocontrol potential for long periods (about 1 year) at room temperature has to be ensured, while screening the isolates of putative BCAs (Narayanasamy 2006).

Fig. 6.5 Curative effect of *Chaetomium*-based biological fungicide on the recovery of Bottle Palm from bud rot disease caused by *Hyophorbe lagenicaulis* under field conditions (Courtesy of Soytong et al. 2005 and with kind permission of Journal of Agricultural Technology, Thailand)

6.5.1 Physical Methods

Internally seedborne fungal pathogens may be eliminated by exposing the seeds to dry heat (air) or wet heat (hot water) treatments. Hot water treatment has been shown to be effective in the case of cereal pathogens. The comparative efficacy of electrons and hot water was assessed for the control of wheat bunt disease caused by *Tilletia caries*. No adverse effect on seed germination due to electron was detectable, while hot water treatment at 52 °C for 10 min reduced seed germination (Winter et al. 1998). Pretreatment of maize seeds was performed at 18–22 °C for 4 h and at 60 °C for 5 min to eliminate *Fusarium moniliforme* (Daniels 1983). Hot water treatment at 60 °C for 10 min was the most effective in suppressing the colonization of rice seeds by *Drechslera oryzae* (Krishanamurthy et al. 2001). A new high precision method, involving the use of warm air which is in moisture equilibrium with the grain to be treated, has been developed. Sanitation of wheat grain using warm air resulted in control of seedborne fungal pathogens such as *Tilletia caries* and *Microdochium nivale*, comparable to the chemical seed treatment. No adverse effect on plant development, following warm air treatment, could be seen. This procedure is suitable for treatment of seeds with thick layers at short durations and has the potential for large scale application at low cost (Forsberg 2001). The effectiveness of heat treatment of wheat and barley seeds with high level of infection by *Fusarium graminearum* (84 and 23 % respectively) was evaluated. The pathogen was eliminated from Canada Western red spring wheat seeds treated at 60 °C for 15 days, or at 70 °C for 5 days, or at 80 °C for 2 days. Likewise, barley seeds were freed of the pathogen after treatment at 60 °C for 21 days, or at 70 °C for 9 days or at 80 °C for 5 days. Germination in most samples was not reduced. The results suggested that thermotherapy could be an effective tool for minimizing the national and international movement of the fungal pathogens through seeds/grains (Clear et al. 2003). Infrared radiation has been employed as another method of seed sanitation. Wheat seeds contaminated with microorganisms were exposed to infrared irradiation. The bacterial counts were effectively reduced from 5.7×10^4 CFU/g to 0.73×10^3 CFU/g by irradiation at 2.0 kW range for 60 s. Intermittent irradiation was more effective in maintaining the internal quality of wheat seeds. Reduction in bacterial counts was proportional to the irradiation dose (Hamanaka et al. 2000).

6.5.2 Biological Methods

Various fungal and bacterial species/strains have been evaluated for the potential to minimize the loss of grain quality used as food or feed. The efficacy of the yeast *Pichia anomala* for the control of *Penicillium roquefortii* during storage of wheat grains under airtight conditions in outdoor silos was assessed. *P. anomala* could survive long-term storage in airtight sealed tubes better at 15 °C than at –20 °C. In addition, *P. anomala* was able to protect barley and oats stored under airtight conditions (Petersson and Schnürer 1998). Wheat loose smut caused by *Ustilago segetum* var. *tritici* infects the entire earheads and converts them into a mass of spores, resulting in considerable loss. The antagonistic fungi and bacteria *Trichoderma viride*, *T. harzianum*, *Gliocladium virens* and *Pseudomonas fluorescens* were evaluated for their efficacy in reducing the infection. Seed treatment with these BCAs reduced the loose smut incidence. As these BCAs were compatible with the fungicide carboxin used for seed treatment, the concentration of the fungicide could be reduced to half of normal dose to achieve higher level of disease control (Singh and Maheshwari 2001).

The populations of seedborne fungi may be significantly reduced, leading to the production of healthy and more vigorous seedlings. Corn seeds infected by *Fusarium moniliforme* and *F. graminearum* were treated with the BCAs *Bacillus subtilis* or *Chaetomium globsum*. These BCAs effectively controlled seedling blight disease to a level equal to that could be obtained by treatment with the fungicides captan or thiram and consequently healthy seedlings were produced from the treated seeds (Mew and Kommedahl 1968). *Gibberella zeae* causes Fusarium head blight (FHB) disease inflicting heavy losses in wheat and barley. Bacterial and yeast strains reduced the severity of FHB on cultivar Renville. *Bacillus subtilis* strain AS43-4 reduced the severity of FHB by as much as 90 % in the greenhouse assays. However, under field conditions, the yeast *Cryptococcus* sp. OH71-4 and *C. nodaensis* OH182-9 could reduce the disease severity more effectively, compared with other antagonists tested. The contents of the mycotoxins deoxynivalenol (DON) of the grains did not show significant difference, due to treatment with any of the BCA evaluated (Schisler et al. 2002).

Peanut kernels infected by *Aspergillus flavus* and *A. parasiticus* contain the mycotoxins aflatoxin at high levels capable of causing ailments in humans and animals, when contaminated kernels or other products are consumed. The peanut cv. Florunner plants were treated with the conidial suspensions of nontoxigenic strains of *A. flavus* and *A. parasiticus* in field plots. After harvest, one half of the pods was also treated with the conidial suspensions of the nontoxigenic strains. Peanuts (treated and untreated control) were stored for 3–5 months under high temperature and relative humidity conditions designed to promote aflatoxin contamination. The concentrations of aflatoxins were determined in the samples from different treatments. Treatment of pods with nontoxigenic strains prior to storage did not provide any additional protection against aflatoxin contamination. Field application of nontoxigenic strain reduced aflatoxin contamination because of apparent carryover effect (Dorner and Cole 2002). Application of *Clonstachys rosea* and *Gliocladium catenulatum* on fresh petals of alfalfa at anthesis stage reduced the pod infection by *Botrytis cinerea*. Further, these BCAs suppressed the development of *B. cinerea* on pods and seeds of alfalfa, when they were applied on senescent petals at the pod development stage. In the three field trials conducted during 2000–2002, application of *C. rosea* to the upper parts of alfalfa plants significantly reduced pod rot and seed rot caused by *B. cinerea*, resulting in enhanced production of healthy seeds. The results revealed the potential of *C. rosea* as an effective alternative to the synthetic fungicides (Li et al. 2004).

References

Abd-Allah EF, El-Didamony G (2007) Effect of seed treatment of *Arachis hypogaea* with *Bacillus subtilis* on nodulation in biocontrol of southern blight (*Sclerotium rolfsii*) disease. Phytoparasitica 35:8–12

Abo-Elyousr KAM, Hashem M, Ali EH (2009) Integrated control of cotton root rot disease by mixing fungal biocontrol agents and resistance inducers. Crop Protect 28:295–301

Adachi N, Tsukamoto S, Inoue Y, Azegami K (2012) Control of bacterial seedling rot and seedling blight of rice by bacteriophages. Plant Dis 96:1033–1036

Aeron A, Dubey RC, Maheshwari DK, Pandey P, Bajpai VK, Kang SC (2011) Multifarious activity of bioformulated *Pseudomonas fluorescens* PS1 and biocontrol of *Sclerotinia sclerotiorum* in Indian rapeseed (*Brassica campestris* L.). Eur J Plant Pathol 131:81–93

Akhtar MS, Siddiqui ZA (2007) Biocontrol of root rot disease complex with *Glomus intraradices*, *Pseudomonas putida* and *Paecilomyces polymyxa*. Australas Plant Pathol 36:175–180

Akhtar MS, Siddiqui ZA (2008) Biocontrol of root rot disease complex of chickpea by *Glomus intraradices*, *Rhizobium* sp. and *Pseudomonas straita*. Crop Protect 27:410–417

Amein T, Omer Z, Welch C (2008) Application and evaluation of *Pseudomonas* strains for biocontrol of wheat seedling blight. Crop Protect 27:532–536

Anjaiah V, Thakur RP, Koedam N (2006) Evaluation of bacteria and *Trichoderma* for biocontrol of preharvest seed infection by *Aspergillus flavus* in groundnut. Biocontrol Sci Technol 16:431–436

Bhuiyan SA, Boyd MC, Dougall AJ, Martin C, Hearnden M (2007) Effect of foliar application of potassium nitrate on suppression of Alternaria leaf blight of cotton (*Gossypium hirsutum*) in northern Australia. Australas Plant Pathol 36:462–465

Braun-Kiewnick A, Jacobsen BJ, Sands DC (2000) Biological control of *Pseudomonas syringae* pv. *syringae*, the causal agent of basal kernel blight of barley, by antagonistic *Pantoea agglomerans*. Phytopathology 90:368–375

Budge SP, McQuilken MP, Fenlon JS, Whipps JM (1995) Use of *Coniothyrium minitans* and *Gliocladium virens* for biological control of *Sclerotionia sclerotiorum* in glasshouse lettuce. Biol Control 5:513–522

Bujold I, Paulitz TC, Carisse O (2001) Effect of *Microsphaeropsis* sp. on the production of perithecia and ascospores of *Gibberella zeae*. Plant Dis 85:977–984

Cardoso JE, Echandi E (1987) Biological control of Rhizoctonia root rot of snap bean with binucleate *Rhizoctonia*-like fungi. Plant Dis 71:167–170

Chen Y, Fernando WGD (2006) Induced resistance to blackleg (*Leptosphaeria maculans*) disease of canola (*Brassica napus*) caused by a weakly virulent isolate of *Leptosphaeria biglobosa*. Plant Dis 90:1059–1064

Chen L, Zhang S-J, Zhang S-S, Qu S, Ren X, Long J, Yin Q, Qian J, Sun F, Zhang C, Wang L, Wu L, Wu X, Wu T, Zhang Z, Cheng Z, Hayes M, Beer SV, Dong H (2008) A fragment of the *Xanthomonas oryzae* pv. *oryzicola* harpin $HpaG_{Xooc}$ reduces disease and increases yield of rice in extensive grower plantings. Phytopathology 98:792–802

Choudhary S, Pareek S, Saxena J (2010) Efficacy of biocontrol agents singly and in combinations against dry root rot (*Macrophomina phaseolina*) of mungbean. J Mycol Plant Pathol 40:141–144

Christensen NW, Hart JM (2008) Combating take-all of winter wheat in western Oregon. Oregon State Univ Extn Ser, pp 1–8

Clear RM, Patrick SK, Turkington TK, Wallis R (2003) Effect of heat treatment on seedborne *Fusarium graminearum* and other cereal pathogens. Can J Plant Pathol 25:489–498

Colbach N, Lucas P, Maynard JM (1997) Influence of crop management on take-all development and disease cycles on winter wheat. Phytopathology 87:26–32

Daniels BA (1983) Elimination of *Fusarium moniliforme* from corn seeds. Plant Dis 67:609

Das IK, Indira S, Annapurna A, Prabhakar, Seetharama N (2008) Biocontrol of charcoal rot in sorghum by fluorescent pseudomonads associated with the rhizosphere. Crop Prtoect 27:1407–1414

Dawson WAJM, Jestoi M, Rizzo A, Nicholson P, Bateman GL (2004) Field evaluation of fungal competitors of *Fusarium culmorum* and *F. graminearum*, causal agents of earl blight of winter wheat for the control of mycotoxin production in grain. Biocontrol Sci Technol 14:783–799

Dil-Macky R, Jones RK (2000) The effect of previous crop residues and tillage on Fusarium head blight of wheat. Plant Dis 84:71–76

Dong H, Zhang X, Choen Y, Zhou Y, Li W, Li Z (2006) Dry mycelium of *Penicillium chrysogenum* protects cotton plants against wilt diseases and increases yield under field conditions. Crop Protect 25:324–330

Dorner JW, Cole RJ (2002) Effect of application of nontoxigenic strains of *Aspergillus flavus* and *Aspergillus parasiticus* on subsequent aflatoxin contamination of peanuts in storage. J Stored Prod Res 38:329–339

Dugan FM, Akamatsu H, Lupien SL, Chen W, Chilvers ML, Peever TL (2009) Ascochyta blight of chickpea reduced 38 % by application of *Aureobasidium pullulans* (anamorphic Dothioraceae, Dothideales) to postharvest debris. Biocontrol Sci Technol 19:537–545

Erdogan O, Benlioğlu K (2010) Biological control of Verticillium wilt on cotton by the use of fluorescent *Pseudomonas* spp. under field conditions. Biol Control 53:39–45

Escande AR, Echandi E (1991) Protection of potato from Rhizoctonia canker with binucleate *Rhizoctonia* fungi. Plant Pathol 40:197–202

Escande AR, Laich F, Cuenca G, Bailez O, Pereyara V (1994) Dispersión de inóculo de *Trichoderma* spp. mediante abjas (*Apis mellifer*) pará el control de la pudrición del capitulo del girasol (*Sclerotinia sclerotiorum*). Fitopatologia 29:35

Escande AR, Laich FS, Pedraza MV (2002) Field testing of honeybee-dispersed *Trichoderma* spp. to manage sunflower head rot (*Sclerotinia sclerotiorum*). Plant Pathol 51:346–351

Faessel L, Nassr N, Lebeau T, Walter B (2008) Effect of a plant defense inducer, acibenzolar-*S*-methyl, on hypocotyls rot of soybean caused by *Rhizoctonia solani* AG-4. J Phytopathol 156:236–242

Fernando WGD, Nakkeeran S, Zhang Y, Sarchuk S (2007) Biological control of *Sclerotinia sclerotiorum* (Lib.) de Bary by *Pseudomonas* and *Bacillus* species on canola petals. Crop Protect 26:100–107

Forsberg G (2001) Heat sanitation of cereal seeds with a new efficient, cheap and environmentally friendly method. In: Biddle AJ (ed) Seed treatment: challenges and opportunities. British Crop Protection Council, Farnhamm, pp 69–72

Galletti S, Sala E, Leoni O, Burzi PL, Cerato C (2008) *Trichoderma* spp. tolerance to *Brassica carinata* seed meal for a combined use in fumigation. Biol Control 45:319–327

Gerlagh M, Goosen-van de Geijn HM, Fokkema NJ, Vereijken PFG (1999) Long-term biosanitation by application of *Coniothyrium minitans* on *Sclerotinia sclerotiorum* infected crops. Phytopathology 89:141–147

Gil SV, Haro R, Oddino C, Kearney M, Zuza M, Marinelli A, March GJ (2008a) Crop management practices in the control of peanut diseases caused by soilborne fungi. Crop Protect 27:1–9

Gil SV, Pedelini R, Oddino C, Zuza M, Marinelli A, March GJ (2008b) The role of potential biocontrol agents in the management of peanut root rot in Argentina. J Plant Pathol 90:35–41

Gil SV, Meriles JM, Haro R, Casini C, March GJ (2010) Crop rotation and tillage systems as a proactive strategy in the control of peanut fungal soilborne diseases. BioControl 53:658–698

Goates BJ, Mercier J (2011) Control of common bunt of wheat under field conditions with the biofumigant fungus *Muscodor albus*. Eur J Plant Pathol 131:403–407

Gutteridge RJ, Jenkyn JL, Bateman GL (2007) The potential of non-pathogenic *Gaeumannomyces* spp. occurring naturally or introduced into wheat crops or preceding crops for controlling take-all in wheat. Ann Appl Biol 150:53–64

Hamanaka D, Dokan S, Yasunaga E, Kuroki S, Uchino T, Akimoto K (2000) The sterilization effects of infrared ray on the agricultural products spoilage microorganisms. In: Proceedings of the ASAE Annual International Meeting, Wisconsin, pp 1–9

Herr LJ (1988) Biocontrol of Rhizoctonia crown rot of sugar beet by binucleate *Rhizoctonia* spp. and *Laetisaria arvalis*. Ann Appl Biol 113:107–118

Hiddink GA, Termorshuizen AJ, Raaijmakers JM, van Bruggen AHC (2005a) Effect of mixed and single crops on disease suppressiveness of soils. Phytopathology 95:1325–1332

Hiddink GA, van Bruggen AHC, Termorshuizen AJ, Raaijmakers JM, Semenor AV (2005b) Effect of organic management of soils on suppressiveness to *Gaeumannomyces graminis* var. *tritici* and its antagonist, *Pseudomonas fluorescens*. Eur J Plant Pathol 113:417–435

Hofgaard IS, Ergon A, Henriksen B, Trons AM (2010) The effect of potential resistance inducers on development of *Microdochium majus* and *Fusarium culmorum* in winter wheat. Eur J Plant Pathol 128:269–281

Howell CR (2002) Cotton seedling preemergence damping-off incited by *Rhizopus oryzae* and *Pythium* spp. and its biological control with *Trichoderma* spp. Phytopathology 92:177–180

Howell CR, De Vay JE, Garber RH, Baston WE (1997) Field control of cotton seedling diseases with *Trichoderma virens* in combination with fungicide seed treatments. J Cotton Sci 1:15–20

Howell CR, Hanson LE, Stipanovic RD, Puckhaber LS (2000) Induction of terpenoid synthesis in cotton roots and control of *Rhizoctonia solani* by seed treatment with *Trichoderma virens*. Phytopathology 90:248–252

Huang J, Li H, Yuan H (2006) Effect of organic amendments on Verticillium wilt of cotton. Crop Protect 25:1167–1173

ICRISAT (1980) Soil solarization. www.icrisat.org/what-we-do/learning-opportunities/lsu-pdfs/Soil Soarization/pdf

Ishida AKN, Souza RM, Resende MLV, Cavalcanti FR, Oliveira DL, Pozza EA (2008) *Rhizobium* and acibenzolar-*S*-methyl (ASM) in resistance induction against bacterial blight and expression of defense responses in cotton. Trop Plant Pathol 33:27–34

Jabaji-Hare S, Neate SM (2005) Nonpathogenic binucleate *Rhizoctonia* spp. and benzothiadiazole protect cotton seedlings against Rhizoctonia damping-off and Alternaria leaf spot in cotton. Phytopathology 95:1030–1036

Jensen B, Knudsen IMB, Jensen DF (2000) Biological seed treatment of cereals with fresh and long term stored formulations of *Clonostachys rosea*: biocontrol efficacy against *Fusarium culmorum*. Eur J Plant Pathol 106:233–242

Ji G-H, Wei F-F, He Y-Q, Wu Y-P, Bai X-H (2008) Biological control of rice bacterial blight by *Lysobacter antibioticus* strain 13-1. Biol Control 45:288–296

Jirak-Peterson JC, Esker PD (2011) Tillage, crop rotation and hybrid effects on residue and anthracnose occurrence in Wisconsin. Plant Dis 95:601–610

Jochum CC, Osborne LE, Yuen GY (2006) Fusarium head blight biological control with *Lysobacter enzymogenes* strain C3. Biol Control 39:336–344

Johansson PM, Johnsson L, Gerhardson B (2003) Suppression of wheat seedling diseases caused by *Fusarium culmorum* and *Microdochium nivale* using bacterial seed treatment. Plant Pathol 52:219–227

Kanajanamaneesathian M, Wiwattanapatapee R, Pengnoo A, Oungbho K, Chumthong A (2007) Efficacy of novel formulations of *Bacillus megaterium* in suppressing sheath blight of rice caused by *Rhizoctonia solani*. Plant Pathol J 6:195–201

Kanjanamaneesathian M, Chumthong A, Pengnoo A, Wiwattanapatapee R (2009) *Bacillus megaterium* suppresses major Thailand rice diseases. Asian J Food Agro-Indust (Spl Iss): S154–S159

Karpagavalli S, Marimuthu T, Jayaraj J, Ramabadran R (2001) An integrated approach to control rice blast through nutrients and biocontrol agent. Res Crops 2:197–202

Karthikeyan V, Gnanamanickam SS (2008) Biological control of *Setaria* blast (*Magnaporthe grisea*) with bacterial strains. Crop Protect 27:263–267

Khan MR, Doohan FM (2009) Bacterium-mediated control of Fusarium head blight disease of wheat and barley and associated mycotoxin contamination of grain. Biol Control 48:42–47

Khan FU, Nelson BD (2005) Greenhouse evaluation of binucleate *Rhizoctonia* for control of *R. solani* in soybean. Plant Dis 89:373–379

Kishore GK, Pande S, Podile AR (2005) Biological control of collar rot disease with broad-spectrum antifungal bacteria associated with groundnut. Can J Microbiol 51:123–132

Kong Q, Shan S, Liu Q, Wang X, Yu F (2010) Biocontrol of *Aspergillus flavus* on peanut kernels by use of a strain of *Bacillus megaterium*. Int J Food Microbiol 139:31–35

Kosaka Y, Fukunishi T (1994) Application of cross-protection to the control of black soybean mosaic disease. Plant Dis 78:339–341

Krishanamurthy CD, Lokesh S, Shetty HS (2001) Occurrence, transmission and remedial aspects of *Drechslera oryzae* in paddy (*Oryza sativa* L.). Seed Res 29:63–70

Lamprecht SC, Marasas WFO, Hardy MB, Calitz FJ (2006) Effect of crop protection on crown rot and the incidence of *Fusarium pseudograminearum* in wheat in the Cape, South Africa. Australas Plant Pathol 35:419–426

Landa BB, Navas-Cortés JA, Hervás A, Jiménez-Díaz RM (2001) Influence of temperature and inoculum density of *Fusarium oxysporum* f.sp. *ciceris* on suppression of Fusarium wilt of chickpea by rhizosphere bacteria. Phytopathology 91:807–816

Landa BB, Navas-Cortés JA, Jiménez-Díaz RM (2004a) Integrated management of Fusarium wilt of chickpea with sowing date, host resistance and biological control. Phytopathology 94:946–960

Landa BB, Navas-Cortés JA, Jiménes-Díaz RM (2004b) Influence of temperature on plant-rhizobacteria interactions related to biocontrol of potential for suppression of Fusarium wilt of chickpea. Plant Pathol 53:341–352

Leisso RS, Miller PR, Burrows ME (2009) The influence of biological and fungicidal seed treatments on chickpea (*Cicer arietinum*) damping-off. Can J Plant Pathol 31:38–46

Lemay AV, Bailey JE, Shew BB (2002) Resistance of peanut to Sclerotinia blight and the effect of acibenzolar-*S*-methyl and Fluazinam on disease incidence. Plant Dis 86:1315–1317

Lemes EM, Mackowiak CL, Blount A, Marois JJ, Wright DL, Coelho L, Datnoff LE (2011) Effects of silicon applications on soybean rust development under greenhouse and field conditions. Plant Dis 95:317–324

Levenfors JP, Eberhard TH, Levenfors JL, Gerhardson B, Hökeberg M (2008) Biological control of snow mould (*Microdochium nivale*) in winter cereals by *Pseudomonas brassicacearum* MA250. BioControl 53:651–665

Li GQ, Huang HC, Acharya SN, Erickson RS (2004) Biological control of blossom blight of alfalfa caused by *Botrytis cinerea* under environmentally controlled and field conditions. Plant Dis 88:1246–1251

Li GQ, Huang HC, Miao HJ, Erickson RS, Jiang DH, Xiao YN (2006) Biological control of Sclerotinia diseases of rapeseed by aerial applications of the mycoparasite *Coniothyrium minitans*. Eur J Plant Pathol 114:345–355

Liu FQ, Wang JS (1998) A preliminary study on controlling rice bacterial leaf blight with a virulence gene deleted strain DU728. Chin J Biol Control 14:115–118

López-Escudero FJ, Mwanza C, Blanco-López MA (2007) Reduction of *Verticillium dahliae* microsclerotia viability in soil by dried plant residues. Crop Protect 26:127–133

Lumsden RD, Locke JC, Adkins ST, Walter JF, Ridout CJ (1992) Isolation and localization of the antibiotic gliotoxin produced by *Gliocladium virens* from alginate prill in soil and soilless media. Phytopathology 82:230–235

Mantecón JD, Pereyra VR (1997) Integrated control methods for managing sunflower head rot in Argentina. Int J Pest Manag 43:143–144

Mathivanan N, Prabavathy VR, Vijayanandraj VR (2005) Application of talc formulations of *Pseudomonas fluorescens* Migula and *Trichoderma viride* Pers. ex S. F. Gray decrease the sheath blight disease and enhance the plant growth and yield in rice. J Phytopathol 153:697–701

Mattingly GEG, Slope DB (1977) Phosphate fertilizer and "take-all" disease of wheat and barley. J Sci Food Agric 28:658–659

Mazzola M, Funnell DL, Raaijmakers JM (2004) Wheat cultivar-specific selection of 2,4-diacetylphloroglucinol-producing fluorescent *Pseudomonas* species from resident soil populations. Microbiol Ecol 48:338–348

Melzer MS, Smith EA, Boland GJ (1997) Index of plant hosts of *Sclerotinia minor*. Can J Plant Pathol 19:272–280

Mew IC, Kommedahl T (1968) Biological control of seedling blight of corn by coating kernels with antagonistic microorganisms. Phytopathology 58:1395

Meyer MC, Bueno CJ, de Souza NL, Yorinori JT (2006) Effect of doses of fungicides and plant resistance activators on the control of Rhizoctonia foliar blight of soybean and on *Rhizoctonia solani* AG1-IA in vitro development. Crop Protect 25:848–854

Mondal AH, Nehl DB, Allen SJ (2005) Acibenzolar-*S*-methyl induces systemic resistance in cotton against black root rot caused by *Thielaviopsis basicola*. Australas Plant Pathol 34:499–507

Muslim A, Horinouchi H, Hyakumachi M (2003) Control of Fusarium crown and root rot of tomato with hypovirulent binucleate *Rhizoctonia* in soil and rock wool systems. Plant Dis 87:739–747

Nandakumar R, Babu S, Viswanathan R, Sheela J, Raguchander T, Samiyappan R (2001) A new bio-formulation containing plant growth-promoting rhizobacterial mixture for the management of sheath blight and enhanced grain yield. BioControl 46:493–510

Narayanasamy P (2006) Postharvest pathogens and disease management. Wiley, Hoboken

Navas-Cortés JA, Trapero-Casas A, Jiménez-Díaz RM (1998) Phenology of *Didymella rabiei* development on chick debris under field conditions in Spain. Phytopathology 88:983–991

Nayaka SC, Shankar ACU, Reddy MS, Niranjana SR, Prakash HS, Shetty HS, Mortensen CN (2009) Control of *Fusarium verticillioides*, cause of ear rot of maize, by *Pseudomonas fluorescens*. Pest Manag Sci 65:769–775

Nelson B, Helms T, Christenson T, Kural I (1996) Characterization and pathogenicity of *Rhizoctonia* from soybean. Plant Dis 80:74–80

Partridge DE, Sutton TB, Jordan DL, Curtis VL, Bailey JE (2006) Management of Sclerotinia blight of peanut with the biological control agent *Coniothyrium minitans*. Plant Dis 90:957–963

Pereira P, Nesci A, Castillo C, Etcheverry M (2010) Impact of bacterial biological control agents on fumonisin B_1 content and *Fusarium verticillioides* infection of field-grown maize. Biol Control 53:258–266

Perelló AE, Dal Bello GM (2011) Suppression of tan spot and plant growth promotion of wheat by synthetic and biological inducers under field conditions. Ann Appl Biol 158:267–274

Perelló AE, Mónaco CI, Moreno MV, Cordo CA, Simón MR (2006) The effect of *Trichoderma harzianum* and *T. koningii* on the control of tan spot (*Pyrenophora tritici-repentis*) and leaf blotch (*Mycosphaerella graminicola*) of wheat under field conditions in Argentina. Biocontrol Sci Technol 16:803–813

Perelló AE, Moreno V, Mónaco C, Simón MR (2008) Effect of *Trichoderma* spp. isolates for biological control of tan spot of wheat caused by *Pyrenophora tritici-repentis* under field conditions in Argentina. BioControl 53:895–904

Perelló AE, Moreno MA, Mónaco C, Simón M, Cordo C (2009) Biological control of *Septoria tritici* blotch on wheat by *Trichoderma* spp. under field conditions in Argentina. BioControl 54:113–122

Petersson S, Schnürer J (1998) *Pichia anomala* as a biocontrol agent of *Penicillium roquefortii* high-moisture wheat, rye, barley and oats stored under airtight conditions. Can J Microbiol 44:471–476

Radjacommare R, Nandakumar R, Kandan A, Suresh S, Bharathi M, Raguchander T, Samiyappan R (2002) *Pseudomonas fluorescens*-based bioformulation for the management of sheath blight disease and leaf folder insect in rice. Crop Protect 21:671–677

Raghavendra VB, Lokesh S, Prakash HS (2007) Dravya, a product of seaweed extract (*Sargassum wightii*), induces resistance in cotton against *Xanthomonas campestris* pv. *malvacerum*. Phytoparasitica 35:442–449

Raj SN, Chaluvaraju G, Amruthesh KN, Shetty HS, Reddy MS, Kloepper JW (2003) Induction of growth promotion and resistance against downy mildew on pearl millet (*Pennisetum glaucum*) by rhizobacteria. Plant Dis 87:380–384

Raj SN, Shetty NP, Shetty HS (2005) Synergistic effects of Trichoshield on enhancement of growth and resistance to downy mildew in pearl millet. BioControl 50:493–509

Roberti R, Badiali F, Pisi A, Veronesi A, Pancaldi D, Cesari A (2006) Sensitivity of *Clonostachys rosea* and *Trichoderma* spp. as potential biocontrol agents to pesticides. J Phytopathol 153:100–109

Rodrigues FA, Jurick WM II, Datnoff LE, Jones JB, Rollins JA (2005) Silicon influences cytological and molecular events in compatible and incompatible rice-*Magnaporthe grisea* interactions. Physiol Mol Plant Pathol 66:144–159

Rodriguez MA, Cabrera G, Godeas A (2006) Cyclosporine A from a nonpathogenic *Fusarium oxysporum* suppressing *Sclerotinia sclerotiorum*. J Appl Microbiol 100:575–586

Rojo FG, Reynoso MM, Ferez M, Chulze SN, Torres AM (2006) Biological control by *Trichoderma* species of *Fusarium solani* causing peanut brown root rot under field conditions. Crop Protect 26:549–555

Ryu C-M, Kim J, Choi O, Kim SH, Park CS (2006) Improvement of biological control capacity of *Paenibacillus polymyxa* E681 by seed pelleting on sesame. Biol Control 39:282–289

Satya VK, Vijayasamundeeswari A, Paranidharan V, Velazhahan R (2011) *Burkholderia* sp. strain TNAU-1 for biological control of root rot in mungbean (*Vigna radiata* L.) caused by *Macrophomina phaseolina*. J Plant Protect Res 51:273–278

Schillinger WF, Paulitz TC (2006) Reduction of Rhizoctonia bare patch in wheat with barley rotations. Plant Dis 90:302–306

Schisler DA, Khan NI, Boehm MJ, Slininger PJ (2002) Greenhouse and field evaluation of biological control of Fusarium head blight on durum wheat. Plant Dis 86:1350–1356

Schoina C, Stringlis IA, Pantelides IS, Tjamos SE, Paplomatas EJ (2011) Evaluation of application methods and biocontrol efficacy of *Paenibacillus alvei* strain K-165, against the cotton black root rot pathogen *Thielaviopsis basicola*. Biol Control 58:68–73

Seebold KW Jr, Datnoff LE, Correa-Victoria FJ, Kucharek TA, Snyder GH (2004) Effects of silicon and fungicides on the control of leaf and neck blast in upland rice. Plant Dis 88:253–258

Senthilraja G, Anand T, Durairaj C, Raguchander T, Samiyappan R (2010) Chitin-based bioformulation of *Beauveria bassiana* and *Pseudomonas fluorescens* for improved control of leaf miner and collar rot in groundnut. Crop Protect 29:1003–1010

Shanmugam V, Senthil N, Raguchander T, Ramanathan A, Samiyappan R (2002) Interactions of *Pseudomonas fluorescens* with *Rhizobium* for their effect on the management of peanut root rot. Phytoparasitica 30:169–176

Singh AK, Chhatpar HS (2011) Combined use of *Streptomyces* sp. A6 and chemical fungicides against Fusarium wilt of *Cajanus cajan* may reduce the dosage of fungicides required in the field. Crop Protect 30:770–775

Singh D, Maheshwari VK (2001) Biological seed treatment for the control of loose smut of wheat. Ind Phytopathol 54:457–460

Singh R, Sing BK, Upadhyay RS, Rai B, Lee YS (2002) Biological control of Fusarium wilt disease of pigeonpea. Plant Pathol J 18:279–283

Sneh B, Burpee L, Ogoshi A (1991) Identification of *Rhizoctonia* species. The American Phytopathological Society, St. Paul

Soytong K, Pongak W, Kasiolarn H (2005) Biological control of Thielaviopsis bud rot of *Hyophorbe lagenicaulis* in the field. J Agric Technol 1:235–245

Srininvasan K, Mathivanan N (2011) Plant growth promoting microbial consortia mediated classical biocontrol of sunflower necrosis virus disease. J Biopest 4:65–72

Steinkellner S, Langer I (2004) Impact of tillage on the incidence of *Fusarium* sp. in soil. Plant Soil 267:13–22

Tang JB, Ma BT, Li LX, Wang P, Zheng AP, Chen H (2002) Biological control of rice sheath blight with *Trichoderma* and *Trichoderma*-like species. Chin J Rice Sci 16:63–66

Vidhyasekaran P, Muthamilan M (1995) Development of formulations of *Pseudomonas fluorescens* for control of chickpea wilt. Plant Dis 79:782–786

Vidhyasekaran P, Rabindran R, Muthamilan M, Nayar K, Rajappan K, Subramanian N, Vasumathi K (1997) Development of powder formulation of *Pseudomonas fluorescens* for control of rice blast. Plant Pathol 46:291–297

Wakelin SA, Anstis ST, Warren RA, Ryder MH (2006) Role of pathogen suppression on the growth promotion of wheat by *Penicillium radicum*. Australas Plant Pathol 35:253–258

Winter W, Banziger L, Krebs H, Rüegger A, Frei P, Gindrat D (1998) Alternative methods for the control of cereal bunts and barley stripe disease. Agrarforschung 5:29–32

Yang D, Wang B, Wang J, Chen Y, Zhou M (2009) Activity and efficacy of *Bacillus subtilis* strain NJ-18 against rice sheath blight and Sclerotinia stem rot of rape. Biol Control 51:61–65

Zeng W, Kirk W, Hao J (2012) Field management of Sclertotinia stem rot of soybean using biological control agents. Biol Control 60:141–147

Zhang S, Schisler DA, Boehm MJ, Slininger PJ (2005) Carbon-to-nitrogen ratio and carbon loading of production media influence freeze-drying survival and biocontrol efficacy of *Cryptococcus nodaensis* OH182.9. Phytopathology 95:626–631

Zhang JX, Xue AG, Tambong JT (2009) Evaluation of seed and soil treatments with novel *Bacillus subtilis* strains for control of soybean root rot caused by *Fusarium oxysporum* and *F. graminearum*. Plant Dis 93:1317–1323

Zhang JX, Xue AG, Morrison MJ, Meng Y (2011) Impact of time between field application of *Bacillus subtilis* strains SB01 and SB24 and inoculation with *Sclerotinia sclerotiorum* on the suppression of Sclerotinia stem rot in soybean. Eur J Plant Pathol 131:95–102

Additional Reference for Further Reading

Petersson S, Johnsson N, Schnürer J (1999) *Pichia anomala* as a biocontrol agent during storage of high-moisture feed grain under airtight conditions. Postharvest Biol Technol 15:175–184

Chapter 7
Biological Disease Management Systems for Horticultural Crops

Horticultural crops are cultivated in varied environments from temperate to tropical conditions in different ecosystems. They may be grown under natural conditions or controlled conditions in greenhouses where temperature, humidity and nutrient supply can be controlled. As the plants are exposed to different environment and nutrient regimes, disease problems are likely to vary widely. Consequently, the extent of losses due to diseases caused by microbial pathogens may depend on the susceptibility/resistance levels of cultivars and available inoculum potential. Vegetable and fruit crops are high value crops and hence, greater attention is bestowed to protect them against diseases caused by microbial pathogens, compared to diseases affecting agricultural crops. Generally, number and frequency of application of chemicals are high, leading to the possibility of development of resistance in pathogens to chemicals. Intensive research efforts have been made to integrate various strategies showing effectiveness to the required level for the development of suitable disease management systems for the horticultural crops, in order to minimize the incidence and severity of diseases both under field conditions and during storage. Disease management systems for diseases affecting major vegetable, fruit and plantation crops are discussed in this chapter.

7.1 Diseases of Vegetable Crops

7.1.1 Diseases of Tomato

7.1.1.1 Fusarium Wilt Disease

Application of soil amendments to differentially encourage the antagonistic microorganisms has been followed for the reduction of incidence of soilborne diseases. Chitosan has been applied as a soil amendment to suppress the development of *Fusarium oxysporum* f.sp. *radicis-lycopersici* (FORL) in soilless tomato system.

P. Narayanasamy, *Biological Management of Diseases of Crops*,
Progress in Biological Control 16, DOI 10.1007/978-94-007-6377-7_7,

The toxicity of chitosan to FORL was assessed. The tomato wilt pathogen was sensitive to treatment with chitosan. Ultrastructural studies using transmission electron microscope (TEM) revealed marked alterations in the sensitive fungal pathogen cells. The results suggested that chitosan application could be combined with biotic biological control agents to enhance the effectiveness of disease suppression (Lafontaine and Benhamou 1996; Palma-Guerrero et al. 2008).

Soil solarization has been demonstrated to have marked negative effects on soilborne plant pathogens like *Fusarium* spp. Soil solarization using photo-selective low density polyethylene film for a period of 32–92 days prior to planting tomatoes significantly reduced the population density of *F. oxysporum* f.sp. *lycopersici* and *F. oxysporum* f.sp. *radicis-lycopersici* in the upper 5 cm of soil. The increases in temperatures in solarized soils at depths of 5, 15, and 25 cm, were 5.7, 7.1 and 5.0 °C respectively over nonsolarized soils. The sensitivity of the pathogens to enhanced temperature was an important factor in their elimination (Chellemi et al. 1994). In another investigation, biodegradable plastics and plastic films were evaluated for solarization of soil for their comparative efficiency in suppressing the development of tomato wilt disease pathogen *F. oxysporum* f.sp. *lycopersici*. Plastic cover was more effective and consistent in, increasing soil temperature, compared to biodegradable plastics. Solarization with plastic films significantly suppressed weed development, while biodegradable plastic offered limited effect on weed growth. The results indicated that the biodegradable solarizing materials have to be considered for the control of soilborne diseases, because of the difficulties associated with the disposal of used plastic materials (Bonanomi et al. 2008).

The effectiveness of the biocontrol activity of the fungi depends on the type of fermentation employed for their mass multiplication. Both solid and liquid fermentation systems have been adopted for mass production of BCAs. *Penicillium oxalicum* applied as conidial suspension by watering the tomato seedlings in seedbeds 7 days before transplanting, provided the most effective protection against tomato Fusarium wilt disease caused by *F. oxysporum* f.sp. *lycopersici* (FOL) (De Cal et al. 1999). The volumetric yield of *P. oxalicum* conidia was 250-fold higher in solid fermentation than in liquid fermentation at 30 days after inoculation. Solid fermentation has been shown to offer several advantages over conventional submerged culture such as lower capital and operating costs, simpler equipment and media and easier downstream processing. Conidia of *P. oxalicum* in solid fermentation had a long shelf-life of >180 days, if stored at −20 °C (Larena et al. 2003). The efficacy of strains of nonpathogenic *Fusarium oxysporum* isolated from the wilt-suppressive soil, in controlling the Fusarium wilt disease of tomato was demonstrated (Larkin and Fravel 1998). The ability of the BCA strain CS-20 to suppress the development of Fusarium wilt disease in different soil type, pathogen isolate and race, environmental and cropping conditions and tomato cultivars was studied. The strain CS-20 reduced the Fusarium wilt disease at all temperature regimes and four different field soils varying in texture and organic matter content. This strain was also effective against all three races of FOL and reduced the disease incidence by 48–66 %. The results indicated the need for assessing the efficacy of the BCA strains under varied conditions to avoid the inconsistencies in the performance of the selected BCA strain(s) under field conditions (Larkin and Fravel 2002). The efficacy of *Streptomyces griseoviridis* strain K61 formulated and marketed as Mycostop® alone or combined with soil solarization was assessed for the control of tomato Fusarium wilt disease under field conditions for over 2 years (2001 and 2002). The biofungicide was very effective against *Fusarium oxysporum* f.sp. *lycopersici* (FOL) in artificially infested soils. Soil spraying was more effective than soil irrigation for the control of tomato wilt disease. *S. griseoviridis* was not effective against Fusarium crown and root rot disease caused by *F. oxysporum* f.sp. *radicis-lycopersici* (FORL), when applied alone. Soil solarization provided good control of FOL, but it was slightly less effective, when combined with *S. griseoviridis*. A significant increase in fruit mass and a higher yield was achieved, when solarization and the BCA were applied together in 1 year, indicating a possible additive effect of the BCA and solarization (Minuto et al. 2006).

Presence of several fungal biocontrol agents, effective against *Fusarium oxysporum* f.sp. *lycopersici* (FOL), has been reported in different countries. *Penicillium oxalicum* was evaluated for its efficacy by applying one to four times to the substrate prior to infestation with the pathogen. Repeated application of *P. oxalicum* prolonged the duration of protection against FOL, especially when the disease incidence was high. Reduction in disease incidence was not related with decrease in FOL population density, irrespective of the number of applications of BCA (De Cal and Melgarejo 2001). Formulations of *P. oxalicum* were tested under glasshouse and field conditions. All formulations (FOR1 to FOR8) were applied to seedlings in seedbeds at 7 days before transplanting. The initial conidial viability of formulations just after fluid bed-drying was approximately 80 % for FOR2, FOR3, FOR5, FOR6 and FOR8 and 30 % for the remaining ones. The percent disease reduction varied from 22 to 64 % with all formulations. Disease incidence in untreated plants was negatively correlated ($r=0.54$) with percentage of disease control (Sabuquillo et al. 2006). The effectiveness of the biocontrol agents is influenced by the nature of additive and storage conditions of varying durations and temperatures. *Pseudomonas fluorescens* Pf1 suppressed the development of tomato wilt pathogen *Fusarium oxysporum* f.sp. *lycopersici* (FOL). Glycerol, among different amendments tested, more effectively maintained the population level of the strain Pf1 for a period of 6 months. The optimum doses of liquid-based formulation of Pf1 for seed treatment and seedling dip were standardized. The Pf1 formulation at 10 ml/kg seeds and 150 ml/ha were found to be optimum for seed treatment and seedling root dip treatments to achieve effective control of tomato wilt disease (Manikandan et al. 2010). In another investigation, the optimal conditions for the biocontrol activity of *Penicillium oxalicum* against FOL were determined. A *P. oxalicum* conidial formulation for maintaining high biocontrol activity and a long shelf-life was developed with a non-vacuum-packed or vacuum-packed formulation that contained 1.5 % sodium alginate, 20 % glycerol, 5 % sucrose and 5 % sorbitol and had <15 % moisture content (Sabuquillo et al. 2010).

In another study, the effect of glycerol, an osmoticant in the production medium on the shelf-life of *Trichoderma harzianum* was assessed. Addition of glycerol in the production medium, reduced the water activity in the medium and extended the shelf-life of talc formulation. The bioefficacy assays showed that addition of glycerol (3 or 6 %) in the production medium reduced the tomato Fusarium wilt disease by 44–50 % (Sriram et al. 2011). The efficacy of consortium of bioagents in suppressing the development of Fusarium wilt disease and enhancing the yield of tomato was investigated under greenhouse and field conditions. *Pseudomonas* spp. and *Trichoderma harzianum* were applied by biopriming and the mycorrhiza *Glomus intraradices* by soil treatment. The combination of the bacterial and fungal and mycorrhizal bioagents protected the tomato plants in a highly effective manner, resulting in a reduction of disease incidence by 74 and 67 % respectively under greenhouse and field conditions. Yield of tomato fruits due to treatment was also increased by 20 %. Addition of cowdung compost still further reduced disease incidence and improved the yield in all treated plots (Srivastava et al. 2010).

Compatibility of fungal bioagents with fungicides can be expected to widen its applicability to crops for which application of fungicides might have become unavoidable. The nonpathogenic strain of *Fusarium oxysporum* strain CS-20 reduced the incidence of Fusarium wilt disease of tomato significantly. The compatibility of *F. oxysporum* CS-20 with the fungicides recommended for tomato was investigated. Mefenoxam (Ridomil Gold) and mefenoxam + copper did not affect the growth of the strain CS-20. Other fungicides azoxystrobin, chlorothalonil, mancozeb, mancozeb + copper inhibited the growth of the strain CS-20 in the in vitro assays, but they were less toxic in greenhouse tests. The results suggested that the compatibility of the bioagents with the fungicides generally applied to the crop concerned will have to be determined before advancing the BCAs to the next stage in commercialization (Fravel et al. 2005). The antagonistic potential and compatibility with fungicides of *Bacillus megaterium* (strain 96) and *Burkholderia cepacia* (strain 91) were investigated for the control of tomato crown and root rot caused by *F. oxysporum* f.sp. *radicis-lycopersici* (FORL). Both bacterial strains were highly tolerant to carbendazim. Application of carbendazim at a low concentration (1 μg/ml) in combination with *B. cepacia* or *B. megaterium* reduced the disease symptoms by 46 and 84 % respectively, compared with carbendazim alone (77 %) and untreated control. The combination of bacterial BCA slightly outperformed the treatment with carbendazim at 100 μl/ml, indicating the possibility of reducing the fungicide use, by combining it with the bacterial strains for the control of Fusarium crown and root rot disease of tomato (Omar et al. 2006).

7.1.1.2 Damping-Off Diseases

Damping-off disease may affect the tomato seedlings in the nursery bed at pre-emergence and post-emergence stages of development. *Pythium* spp., *Rhizoctonia solani* and *Sclerotium rolfsii* are the causative agents of the diseases. Bacterial bioagents have been reported to suppress the development of the damping-off disease. In the rockwool system, *Pseudomonas fluorescens*, *P. putida*, *P. marginalis*, *P. corrugata* and *P. viridiflava* reduced the incidence of damping-off caused by both

P. aphanidermatum and *P. ultimum* (Gravel et al. 2005). Treatment of tomato seeds with *Bacillus subtilis* AUBS-1 formulations in lignite, lignite+flyash, bentonite paste resulted in effective suppression of the damping-off disease caused by *P. aphanidermatum* and enhancement of plant biomass under glasshouse and field conditions (Jayaraj et al. 2005). Soil application of *Bacillus subtilis* RB14-C protected the tomato seedlings against the damping-off disease incited by *Rhizoctonia solani* (Szezech and Shoda 2006). An integrated management system for tomato damping-off disease caused by *Pythium ultimum* was developed. *Pseudomonas fluorescens* isolate CW2 in combination with fungicides azoxystrobin, metalaxyl-M or pyraclosporin was assessed for their efficacy. The fungicides were fungitoxic to *P. ultimum*, but did not inhibit the development of *P. fluorescens* in in vitro assays. Under greenhouse conditions, two concentrations (5 and 10 μg/ml) of fungicides were applied. The degree of control varied significantly depending on the fungicide applied. On the other hand, combined seed treatment with *P. fluorescens* and fungicides resulted in significant improvement in disease control and improved plant growth too, as indicated by shoot and root dry weights. Metalaxyl-M alone or in combination with *P. fluorescens* more effectively protected tomato seedlings than other fungicides tested. Strobilurin alone and in combination with the bacterial BCA stimulated plant growth to a greater extent with moderate level of disease control (Salman and Abuamsha 2012).

Rhizoctonia solani and *Sclerotium rolfsii* cause damping-off disease in the nursery beds and induce crown and stem rot disease in transplanted main fields. Many biocontrol agents and commercial formulations, when sprayed on the plants could not reduce the disease incidence significantly, because the suspension containing the antagonists may not reach the crown region in sufficient concentration covered by dense foliage. In order to overcome this obstacle, two bacterial isolates *Burkholderia cepacia* T1A-2B and *Pseudomonas* sp. T4B-2A were compared with two commercial biofungicides based on *Bacillus subtilis* (BSF4) and *Trichoderma asperellum* (Tv1). These BCAs were applied to the soil, proximal to the plant crowns and main roots by means of an effective and specific system of drip irrigation under field conditions for 2 years. Four synthetic fungicides tolchlofos-methyl, azoxystrobin, fosetyl-Al and fosetyl-Al + propamocarb were also applied in the same manner for comparing their effect on disease incidence. In all experiments, the bacterial BCAs significantly reduced the disease incidence as well as the disease severity induced by *R. solani* or *S. rolfsii*. The level of effectiveness of disease control by the strains T1A-2B and T4B-2A was comparable to Tv1, better than BSF4 and comparable with synthetic fungicides, except tolchlofos-methyl which was the most effective among all the treatments tested. The strain T1A-2B reduced the disease severity caused by *S. rolfsii* and *R. solani* up to 58.33 % and up to 63.8 % respectively. On the other hand, the strain T4B-2A reduced the disease severity induced by *S. rolfsii* and *R. solani* up to 73.2 and 62.7 % respectively. The results indicated the importance of delivery system in providing effective protection to crop plants, although the BCAs have the potential for controlling the disease in question (De Curtis et al. 2010).

Table 7.1 Effect of treatment of tomato with PGPR strain alone or mixtures on the severity of southern blight disease and fruit yield in rainy and winter seasons (Jetiyanon et al. 2003)

	Reduction of disease (%)		Total fruit weight (kg/plot)	
Treatments	Rainy season	Winter season	Rainy season	Winter season
Control	70.8a	72.7a	1.4b	2.5a
IN937a alone	16.7b	48.1ab	2.0ab	2.9a
IN937a + IN937b	23.9b	38.7b	2.4a	3.7a
IN937b + SE34	58.0ab	46.9b	1.6ab	3.4a
IN937b + SE49	34.7ab	53.6ab	1.7ab	2.8a
T4 + INR7	57.0ab	41.4b	1.6ab	3.4a
LSD (P=0.05)	44.4	25.2	0.8	1.9

Means followed by the same letter in one column are not significantly different as per the least significant difference (LSD) test at P=0.05

7.1.1.3 Southern Blight Disease

Southern blight disease of tomato caused by *Sclerotium rolfsii* is responsible for serious losses in Thailand. The biocontrol potential of the plant growth-promoting rhizobacteria (PGPR), *Bacillus amyloliquefaciens* strain IN937a and *B. pumilus* strains IN937b, SE34, SE49, T4 and INR7 was evaluated for the suppression of southern blight disease under field conditions. Application of INR937a alone and the mixture of IN937a + IN937b reduced the disease severity significantly by 75 and 65 % respectively, compared with untreated control. Further, the PGPR mixture IN937a + IN937b increased the yield of marketable fruit by 1.5 times over that of the control in the rainy season. On the other hand, in the winter season, the mixtures IN937a + In937b, T4 + INR7 and IN937b + SE34 were also able to provide protection to tomato plants to a level that was significantly greater than the untreated control treatment (Table 7.1) (Jetiyanon et al. 2003).

7.1.1.4 Late Blight Disease

Management of late blight disease of tomato caused by *Phytophthora infestans* in the greenhouse, has been difficult, because of the limited possibility of manipulating the environmental conditions of prolonged periods of leaf wetness and temperatures favoring the development of this disease. The late blight pathogen infects all plant parts including fruits, when favorable conditions prevail (Fig. 7.1) (Douglas 2010). Polyethylene mulch alters the environmental conditions in the soil and also above the mulch. The effects of using the polyethylene mulch to cover the soil before planting tomato plants and application of fungicides commonly used by organic farmers on late blight disease incidence were assessed. Application of fungicide Kocide 2000 plus Neemguard resulted in inconsistent and insufficient suppression of late blight disease (34.5 ± 14.3 %). On the other hand, the polyethylene mulch provided consistent, effective and highly significant suppression, the disease control efficacy being 83.6 ± 5.6 %. The combined effect of polyethylene mulch and

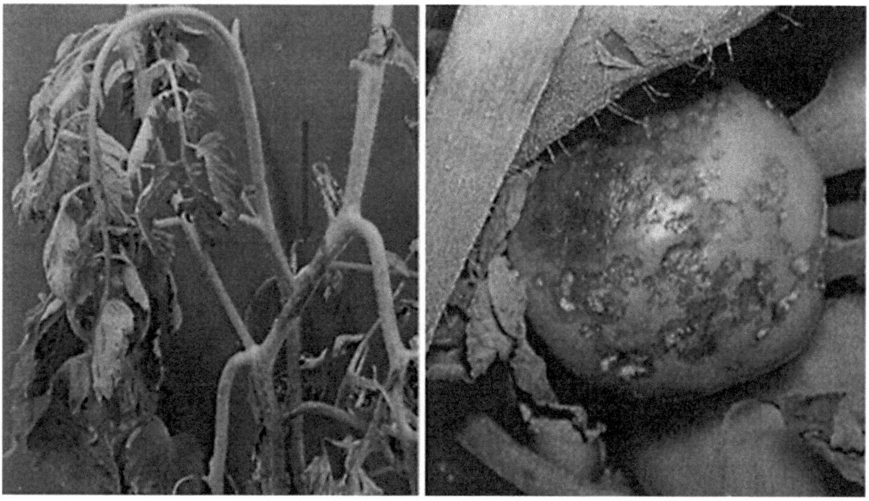

Fig. 7.1 Symptoms of late blight disease on aerial plant parts of tomato (Courtesy of Douglas 2010 and with kind permission of the Connecticut Agricultural Experiment Station, CT, USA)

fungicide was additive. As the next step, the effect of type of polyethylene mulch on disease suppression was studied, by employing bicolor aluminized, clear or black sheets. The type of polyethylene mulch did not affect the efficacy of late blight disease suppression, control efficacy of late blight disease suppression varying from 60.1 to 95.8 %. The disease suppressive effect of the polyethylene mulch appeared to be due to a reduction in leaf wetness duration, since both frequency of nights when dew formed and the number of dew hours/night when it formed, were reduced following polyethylene mulching in the greenhouse. In addition, mulching also possibly resulted in reduced sporulation. Additional benefits of prevention of weed infestation and saving up to 30 % of irrigation water also have been realized by tomato growers in Israel (Shtienberg et al. 2010).

The effectiveness of different kinds of polyethylene mulches was assessed for suppressing or preventing the development of potato tuber blight caused by *Phytophthora infestans* under field conditions. A water-permeable agricultural textile treated with copper hydroxide was also tested for its efficacy. Two barriers that covered the potato hills, black polyethylene film and copper hydroxide-treated agricultural textile reduced the incidence of tuber blight, relative to appropriate controls. Other barriers (mulches) had little effect on the tuber blight incidence. The effectiveness of black polyethylene film might be attributed to blocking of water, resulting in prevention of infiltration of inoculum through potato hill and alteration of the soil environment within the hill. Hills covered with black plastic sheets were drier and slightly warmer than uncovered hills, thus markedly affecting inoculum movement and infection potential (Glass et al. 2001). Induction of systemic resistance, as a disease management strategy, has been investigated for protecting the tomato plants against the late blight pathogen *Phytophthora infestans* by inoculating lower leaves of tomato with the pathogen and assessing the induction of resistance to zoospore germination on the non-inoculated upper leaves (Heller and Gessler 1986). *Bacillus pumilus* SE34 *and Pseudomonas fluorescens* 89B61 were employed to elicit systemic resistance in tomato against late blight disease. The disease severity was reduced by a level equivalent to systemic acquired resistance (SAR) induced by the pathogen or induced local resistance by ß-aminobutyric acid (BABA) (Yan et al. 2002).

7.1.1.5 Anthracnose Disease

Compost amendments have been shown to improve biological, chemical and physical properties of amended soils. In addition, they stimulated the proliferation of microorganisms antagonistic to microbial plant pathogens. Compost incorporation into soil was reported to reduce the severity of diseases caused by foliar pathogens (Tränkner 1992). The effect of compost amendments on the development of tomato anthracnose disease was assessed under field conditions for 2 years. The incidence of anthracnose fruit rot was reduced in organic tomato plots amended with high rate of composted cannery wastes compared with the disease incidence in non-amended control plots, when the disease pressure was high. Marketable yield was increased by 33 % in compost-amended organic plots. In addition, plots amended with high compost rate produced more ripe fruits than the non-amended control plots. The beneficial effects of compost amendments which influence plant growth and physiology, can be realized, only if the composts have been stabilized adequately before application in the soil. It is possible that the effects of compost on fruit infection by *Colletotrichum coccodes* might be due to systemic induced resistance (Abbasi et al. 2002).

7.1.1.6 Stem Canker Disease

Botrytis cinerea causes stem canker disease in greenhouse tomatoes. Bioagents and chemical treatments were evaluated for their effectiveness against the stem canker disease. Commercial bioagent Prestop and the fungicide Decree applied as preventive or curative sprays, PlantShield applied as curative spray and S33 (*Rhodosporidium diobovatum* and Quadra 136 applied as preventive or preventive plus one spray to wounded surface, protected the treated tomato plants against *B. cinerea* infection for the entire growing season. Treatments with bioagents enhanced the tomato plant growth resulting in increased fruit yield, in addition to the protection against the stem canker disease (Utkhede and Mathur 2006). The possibility of inducing systemic resistance in tomato plants to *B. cinerea* by applying defense activator acibenzolar-*S*-methyl (ASM) and O-hydroxyethylorutin (HER) was examined. Pretreatment of tomato plants with ASM or HER at 24 h prior to inoculation suppressed the development of *B. cinerea* (Malolepsza 2006).

7.1.1.7 Bacterial Wilt Disease

Bacterial wilt disease caused by *Ralstonia solanacearum* occurs in most areas where tomatoes are grown. Various disease management strategies have been evaluated for their efficacy, when applied alone or in combination. The potential of soil solarization for the suppression of tomato bacterial wilt disease was assessed. Soil solarization using transparent plastic sheets as mulch during summer months was found to effectively reduce the incidence of bacterial wilt disease. The bacterial BCA *Pseudomonas fluorescens* was incorporated uniformly in soils infested with *R. solanacearum* followed by solarization for 8–10 weeks at two locations in Himachal Pradesh, India. The soil temperature was increased in solarized soils by 8.9 and 10 °C in the two locations. The bacterial antagonist proliferated in the solarized soil in the first year and the population of the pathogen declined subsequently. The incidence of bacterial wilt disease was reduced by 43–63 % in the solarized soil. A progressive decrease was recorded in the terminal wilt incidence with a corresponding increase in duration of solarization. A 10-week solarization period was the most effective in reducing the terminal wilt phase of the bacterial wilt disease (Ambadar and Sood 2010).

Different plant growth-promoting rhizobacteria (PGPRs) have been evaluated for their potential biocontrol activity against tomato bacterial wilt pathogen *Ralstonia solanacearum* (race1, biovar1). In addition, acibenzolar-*S*-methyl (ASM) (Actigard) and a soil amendment with S-H mixture containing agricultural and industrial wastes such as bagasse, rice husk, oyster shell powder, urea, potassium nitrate, calcium superphosphate and mineral ash were also included for assessing their comparative effect. *Pseudomonas putida* 89B61 significantly reduced bacterial wilt disease incidence, when applied to the transplants at the time of seeding and 1 week prior to inoculation with *R. solanacearum*. BioYield, containing two *Bacillus* strains, decreased the incidence of bacterial wilt significantly. Equity containing more than 40 different microbial strains did not reduce the disease incidence significantly, compared with controls. Incorporation of S-H mixture into infested soils at 2 weeks prior to transplanting reduced bacterial wilt incidence in one experiment, while combination of Actigard and S-H mixture significantly reduced bacterial wilt incidence in tomato in both experiments (Amith et al. 2004). The effect of application of *Paenibacillus polymyxa* strains or chitosan on the development of tomato bacterial wilt disease caused by *Ralstonia solanacearum* was studied under greenhouse conditions. Chitosan and *P. polymyxa* showed strong antibacterial activity required for suppression of development of bacterial wilt disease. Chitosan applied as seed treatment at a concentration of 10 mg/ml for 2 h, followed by overnight drying or as soil drench at a concentration of 10 mg/ml significantly reduced wilt disease incidence by 48 and 72 % respectively. *P. polymyxa* MB02-1007 as seed treatment with bacterial suspension for 8 h, followed by drying overnight or soil drench with bacterial suspensions (1×10^9 CFU/ml) was effective in reducing the disease incidence by 88 and 82 % respectively. *P. polymyxa* was more effective than chitosan in reducing bacterial wilt disease incidence. Chitosan was more effective as a soil drench, whereas *P. polymyxa* activity was not affected by the method of application. Irrespective of application methods, both chitosan and *P. polymyxa* promoted plant growth, resulting in greater height, fresh weight and dry weight of plants (Algam et al. 2010).

The effects of application of acibenzolar-*S*-methyl (ASM) alone or in combination with *Pseudomonas fluorescens* strain Pf2 on the suppression of tomato bacterial wilt disease caused by *Ralstonia solanacearum* were investigated. Treatment of tomato seedlings with either Pf2 or ASM significantly reduced severity of bacterial wilt disease by 58 and 56 % respectively. Highest disease suppression (72 %) was achieved by combining both Pf2 and ASM. Application of ASM alone increased seedling biomass, relative to infected control with 64.3 %. Changes in the activities of polyphenol oxidase (PPO), ß-glucosidase (ß-GL) and peroxidase (PO) in tomato, following application of ASM and Pf2 and inoculation with *R. solanacearum* were investigated. Significant changes ($P<0.05$) in the activities of PPO, ß-GL and PO were also observed. The results of field trials indicated that application of ASM and *P. fluorescens* as foliar sprays and soil drench could be integrated with disease management program for tomato bacterial wilt disease (Abo-Elyousr et al. 2012). The effectiveness of the filamentous bacteriophages ØRSM3 in suppressing the development of tomato bacterial wilt disease was demonstrated. Inoculation of ØRSM3-infected pathogen cells into tomato plants did not cause bacterial wilt disease. Pretreatment of tomato plants with ØRSM3 infected cells effectively protected them from infection by virulent *R. solanacearum* strains. The effective dose of ØRSM3-infected cells for prevention of disease incidence was 10^5 CFU/ml. ØRSM3 infected pathogen cells could grow and continue to produce infectious phage particles under appropriate conditions. The use of ØRSM3 for the control of tomato bacterial wilt seems to be a desirable approach (Addy et al. 2012).

The effectiveness of acibenzolar-*S*-methyl (ASM) in suppressing the development of tomato bacterial wilt pathogen *Ralstonia solanacearum* was investigated using susceptible and moderately resistant tomato cultivars under greenhouse and field conditions. ASM was ineffective in reducing bacterial wilt disease incidence in susceptible cultivar Equinox and FL47. However, ASM significantly enhanced the level of resistance in cultivars with moderate resistance like Neptune and BHN466. The results suggested that ASM-mediated resistance was partially due to prevention of internal spread of the bacterial pathogen toward upper stem tissues of inoculated tomato plants. The results of field experiments were consistent with those of greenhouse assessments of the effects of ASM on bacterial wilt disease incidence. ASM-treated BHN466, FL7514 and Neptune with moderate resistance to bacterial wilt disease showed significantly lower level of disease incidence, compared with untreated controls. Further, the cultivars BHN466 and FL7514 treated with ASM, yielded more than the control treatment. The results indicated the need for combining two different disease management strategies of using genetic resistance and resistance inducers to achieve higher level of disease control (Pradhanang et al. 2005).

Thymol (from *Thumus* spp.) and palmarosa (from *Cymbopogon martinii*) are volatile plant essential oils with antimicrobial properties. Thymol and palmarosa were evaluated for their efficacy in suppressing the development of tomato bacterial wilt disease under field conditions. Two hours after infestation of field soils with *Ralstonia solanacearum*, the essential oils were applied at a concentration of 0.7 % and the plots were sealed with plastic mulch for 3 or 6 days. Tomato seedlings were planted after 7 days. Thymol was more effective than palmarosa oil with final disease incidence of 33.1 and 48.1 % respectively. The untreated control plots showed 92.5 % wilted plants. Soil treatment with thymol or palmarosa increased the yield of tomato, compared with untreated control plots. In the second year, thymol alone was tested and its application significantly reduced bacterial wilt incidence on the susceptible cultivar SolarSet, with 12 % infection, as against 65.5 % infection in untreated controls. The yield of tomato fruit was also increased significantly by treatment with thymol. The results indicated that thymol had the potential for use as soil fumigant as an alternative to chemicals conventionally applied for soil fumigation (Ji et al. 2005).

With a view to enhancing the effectiveness of soil application of thymol for the control of tomato bacterial wilt (*Ralstonia solanacearum*) and root knot nematode (*Meloidogyne arenaria*), acibenzolar-*S*-methyl (ASM) was applied in conjunction with thymol. This essential oil was applied as preplant fumigant, after infestation with the bacterial pathogen and nematode, through drip irrigation lines under polythene mulch at the rate of 73 kg/ha during 2004 and 2005. ASM was applied primarily as foliar spray at 25 mg/l. Thymol application significantly reduced incidence of bacterial wilt disease in both years of field trials, the percent disease incidence being 26 and 22.6 % respectively in 2004 and 2005, as against >95 % in untreated control plots. In plots treated with both thymol and ASM, the number of root knot juveniles was significantly reduced both in 2004 and 2005 trials. In thymol-ASM-treated plots, the yield of tomatoes was significantly increased, compared with single treatment of thymol or ASM. The development of drip irrigation system for thymol application has the potential for wider use for delivery of biocontrol agents for disease management (Ji et al. 2007). In a later field experiment, the effect of combined application of thymol and ASM on the incidence of tomato bacterial wilt disease was investigated. Thymol (9.43 kg/ha) was applied after soil infestation with *R. solanacearum*, followed by foliar application of ASM (3.59–8.98 l/ha) on tomato seedlings once in the greenhouse and for five times on tomato plants in the field after transplanting. The combined application of thymol and ASM significantly reduced disease incidence in tolerant tomato genotype 7514, compared with thymol alone treatment. The yield was also increased by the combined treatment, whereas ASM or thymol alone did not provide beneficial effect either on diseases incidence or enhancement of yield of tomato (Hong et al. 2011).

7.1.1.8 Bacterial Spot Disease

Xanthomonas axonopodis pv. *vesicatoria* (*Xcv*), causing tomato bacterial spot disease is primarily transmitted through infected seeds and becomes endemic posing a challenge in commercial tomato production areas, necessitating utilization of different strategies for effective management of the disease. In the field trials conducted in Alabama and Florida, *Pseudomonas syringae* strain Cit7 was found to be the most effective in suppressing bacterial spot disease in two of three trials. *Bacillus pumilus* SE34 suppressed the development of bacterial leaf spot in two field trials. Combined application of these two strains was effective against bacterial spot and bacterial speck of *Pseudomonas syringae* pv. *tomato* in all the trials. Both bacterial strains appear to enhance the level of resistance in treated tomato plants (Ji et al. 2006). The mutants generated from the pathogen *Xcv*, 75-3S *hrpG*, *hrpX*, *hrpF* and *hrpE*1 were evaluated for their biocontrol potential against bacterial spot pathogen *Xcv* under greenhouse conditions. The mutant 75-3S *hrpG* was the most effective and consistently suppressed the disease development with a mean reduction in disease severity by ~58 %. These mutants were further evaluated under field conditions along with the previously reported BCA *Pseudomonas syringae* Cit7 and *Pseudomonas putida* B56 in Alabama (AL) and Florida (FL). Two mutants 75-3S *hrpG* and 75-3 *hrpF* provided significant reductions in disease severity, compared to pathogen-only control treatment in all three field trials. Averaged across all three field trials, the 75-3S *hrpG* mutant provided a mean reduction in foliar disease severity of ~76 %, compared to the mean of ~29 % for *P. syringae* Cit7 averaged across all previous trials. The results revealed the effectiveness of the mutants 75-3S *hrpG* and its superior performance over other BCAs reported earlier (Moss et al. 2007).

Use of bacterial viruses or bacteriophages for the control of plant bacterial diseases was evaluated as an alternative approach. In the field experiments conducted in 1997 and 1998. Spray application of four phages specific to *Xanthomonas campestris* pv. *vesicatoria* (*Xcv*) in the early morning twice a week reduced the disease severity of bacterial spot disease on tomato plants by an average of 17 %. On the other hand, the copper-mancozeb application (standard treatment) could reduce the disease severity by 11 %. In addition, phage treatment promoted plant growth resulting in the production of more extra large-sized fruits (Flaherty et al. 2000). In a later investigation, the efficacy of phage was found to be greatly reduced due to its short residual activity on plant foliage. Hence, three formulations (Agriphage, OmniLytics Inc., USA) containing six to eight phage mixtures with enhanced longevity on plant surface were evaluated under greenhouse and field conditions. In the greenhouse tests, the nonformulated pregelatinized corn flour (PCF), Casecrete- and skim milk-formulated phage mixtures reduced the disease severity by 1, 30, 51 and 62 % respectively, compared with untreated controls. In the field experiments PCF- and Casecrete-formulated phage mixtures reduced the disease severity significantly, compared with unformulated phage mixture and controls. PCF-formulated phage was more effective, when applied in the evening than in the morning (Balogh et al. 2003).

Bioactive products, referred to as plant activators, have been used to induce systemic acquired resistance (SAR) in plants to limit the process of initiation of infection, leading to development of symptoms characteristic of the disease. Acibenzolar-*S*-methyl (ASM, CGA245704, Actigard 50WG and Bion 50WG has been reported to induce systemic resistance in several plant species against various microbial plant pathogens. Plant activators have no direct antimicrobial activity, but elicit in plants preinfectional biochemical processes that confer resistance against the same spectrum of pathogens as biological elicitors. ASM was evaluated for its effectiveness against tomato bacterial leaf spot pathogen *Xanthomonas axonopodis* pv. *vesicatoria* (*Xcv*) under field conditions. Application of ASM at 35 g a.i./ha reduced severity of bacterial spot in 14 of 15 experiments. Disease control was similar or superior to that obtained with standard bactericide copper. The results indicated that ASM could be integrated as a viable alternative to copper-based bactericides for field management of bacterial spot disease (Louws et al. 2001). *Pseudomonas fluorescens* and benzothiadiazole (BTH) were applied as seed treatment or foliar spray for the control of bacterial spot disease under field conditions. All treatments effectively reduced the severity of bacterial spot disease, compared with untreated control plants. Foliar application of *P. fluorescens* was the most effective treatment in reducing disease severity. The combined application of *P. fluorescens* and BTH reduced the pathogen population effective and also promoted plant growth (Abo-Elyousr and El-Hendawy 2008).

Phosphorus acid salts (PASs) were evaluated for their ability to suppress the development of tomato bacterial spot disease under field conditions for a period of 3 years. The treatments were a weekly schedule of PAS alone, PAS combined with standard copper bactericide at full rate or half-rate, PAS alternated with a standard copper-bactericide and PAS every week plus biweekly applications of acibenzolar-*S*-methyl (ASM). The effectiveness of PAS combined with standard copper bactericide full rate, PAS alternated with standard copper bactericide and PAS every week plus biweekly applications of ASM was on par with that of standard copper-bactericide program. None of the treatments improved the tomato yield in any of the field experiments conducted. Phytotoxicity due to PAS treatment was observed under greenhouse conditions. The usefulness of phosphorus salts has not been demonstrated in clear terms by this investigation (Wen et al. 2009).

7.1.1.9 Bacterial Speck Disease

Pseudomonas syringae pv. *tomato*, causative agent of tomato bacterial speck disease, occurs worldwide and it is responsible for considerable quantitative and qualitative losses, since fruit infection reduces the market value of the fruits considerably. Nonpathogenic strains of *P. syringae* Cit7 most effectively reduced the disease intensity in the greenhouse conditions, when *P. syringae* strains TLP2, *Pseudomonas fluorescens* strain A506 and *P. syringae* pv. *syringae* DC3000 *hrp* mutants were also tested for their efficacy. The strain Cit7 provided a mean level of disease reduction of 78 % and hence, this strain was tested under field conditions at different locations in Alabama and Florida, USA and Ontario, Canada. *P. syringae* Cit7 was the most effective in reducing disease severity. The mean level of disease reduction was 28 % over ten different field experiments. *P. fluorescens* A506 available commercially as BlightBan provided a mean level of disease reduction of 18 % over nine different field experiments. As both strains Cit7 and A506 were sensitive to copper, they cannot be integrated with the standard copper bactericides (Wilson et al. 2002). Application of acibenzolar-*S*-methyl (ASM) at 35 g a.i./ha reduced the severity of tomato bacterial speck disease in all seven field experiments conducted in the USA (four states) and Canada (one state). The severity values of foliar bacterial speck ranged from 0 (complete absence) to 22.6 and fruit with bacterial speck ranged from 1.4 to 9.2 for ASM, whereas the untreated control plots had severity values ranging from 1.3 to 45.0 for foliage infection and from 9.2 to 22.6 for fruit infection. Tomato yield, however, was not improved by the application of ASM in the field (Louws et al. 2001). Commercially available plant activators like benzothiadiazole (BTH) (inducer of SAR) and plant growth-promoting rhizobacteria (PGPRs) (inducers of induced systemic resistance, ISR) have been shown to be effective, when applied individually. Use of these products together as an integrated control strategy was investigated for the control of tomato bacterial speck disease caused by *Pseudomonas syringae* pv. *tomato*. BTH and ISR-inducer BioYield concentrate (containing the endospores of *Bacillus subtilis* GB03 and *B. amyloliquefaciens* IN937a). Application of BTH on greenhouse-grown tomatoes effectively reduced bacterial speck disease incidence and severity, both alone and in combination with the ISR-inducing bacterial product. No antagonism between BTH and the bacterial product was indicated, since the extent of disease control was similar to or better than BTH. The results indicated that incorporation of BTH into greenhouse transplant production, in place of copper bactericide, might be useful for bacterial speck disease control (Herman et al. 2008).

7.1.1.10 Bacterial Canker Disease

The effect of soil solarization on the tomato bacterial canker disease caused by *Clavibacter michiganensis* subsp. *michiganensis* (*Cmm*) was investigated. Transparent polyethylene sheets were used to cover the soil in commercial tomato production areas. Soil mulching with the plastic film was applied for 6 weeks in plastic houses. Disease severity induced by *Cmm* was drastically reduced by soil solarization/mulching throughout the tomato growing season. The bacterial species *Pseudomonas*, *Bacillus* and the actinomycete *Streptomyces* antagonistic to *Cmm* present in the tomato rhizosphere survived the increased temperatures following soil solarization. The fluorescent pseudomonads isolated from the tomato rhizosphere soil, could induce resistance to *Cmm* in plants growing from seeds treated with peudomonads. Tomato growers in Greece appeared to be convinced about the beneficial effects of soil solarization using polyethylene films (Antoniou et al. 1995). The effectiveness of *Bacillus subtilis* (Quadra 136 and 137), *Trichoderma harzianum* (RootShield) and *Rhodosporidium diobovatum* (S33) in suppressing the development of tomato bacterial canker disease was indicated by Utkhede and Koch (2004).

Acibenzolar-*S*-methyl (ASM) was evaluated for its ability to induce resistance in tomato plants against *Clavibacter michiganensis* subsp. *michiganensis* (*Cmm*). Pretreatment of tomato plants with ASM reduced the severity of the canker disease. Development of resistance to canker disease required an interval of 1–7 days between inducer application and challenge inoculation with *Cmm*. Highest level of protection could be obtained, when plants were inoculated at 3 days after ASM application

(Soylu et al. 2003; Baysal et al. 2003). DL-ß-aminobutyric acid (BABA) has been demonstrated to induce resistance in tomato plants against canker pathogen *Cmm*. Foliar sprays at 500 μg/ml of BABA suppressed canker disease development up to 54 % by 14 days after inoculation. Bacterial populations were reduced by 84 % in planta by treatment with BABA (Baysal et al. 2005). In another investigation, the effect of BABA application alone, or in combination with *Pseudomonas fluorescens* isolate CW2 in suppressing the development of tomato canker disease was assessed. Soil treatment with BABA or isolate CW2 significantly reduced the incidence of bacterial canker disease. Combined sequential treatments with BABA and isolate CW2 were found to be more effective in reducing the disease severity, compared to treatment with either BABA or isolate CW2. The combined application was effective not only for protecting the tomato plants against canker disease, but also for promoting the growth of tomato plants (Hassan and Buchenauer 2008).

7.1.2 Tomato Spotted Wilt Disease

Tomato spotted wilt virus (TSWV) infects tomato and it has a wide host range that includes several crops and weeds growing in and adjacent to tomato fields. Several species of thrips belonging to the genera *Frankliniella* and *Thrips* transmit TSWV in a persistent manner. The virus and the thrips have to be managed by integrating all possible strategies of disease management. The ultraviolet (UV)-reflective mulches have been shown to reduce thrips colonization onto tomato and mulch painted silver and UV-reflective silver mulch have been reported to reduce TSWV incidence in tomato (Riley and Pappu 2000; Stavisky et al. 2002). Hence, the effects of UV-reflective mulch, acibenzolar-*S*-methyl (ASM) and insecticides for reducing the incidence of virus and the vector species *F. occidentalis* were investigated under field conditions. The UV-reflective mulch was more effective than the black polyethylene in reducing the colonization of thrips and consequently the primary infections by TSWV was reduced, since the thrips migrating from other plants/areas are responsible for the primary infections on tomato plants. During the TSWV epidemic in 2000, the primary spread of the virus reduced early in the season in tomatoes grown on UV-reflective mulch versus black mulch. Application of ASM further reduced TSWV incidence in 2000 and 2002, when the disease pressure was quite high. Reproduction of thrips on the tomato plants in the experimental plots was poor. Insecticide application (methamidophos/spinosad) reduced the thrips population and prevented secondary spread in 2000 and 2002. The combination of UV-reflective mulch, acibenzolar-*S*-methyl and insecticides was very effective in reducing tomato spotted wilt incidence in tomato (Momol et al. 2004).

Lecanicilliium lecanii, *Metarhizium anisopliae* and *Beauveria bassiana* were reported to be pathogenic to *F. occidentalis* (Vestergaard et al. 1995; Sengonca et al. 2006). A strain of *L. lecanii* originally isolated from glasshouse whitefly and later commercially produced in the Netherlands was also effective against thrip species *Frankliniella occidentalis* (van der Schaaf et al. 1991), the vector of *Tomato spotted wilt virus*. Isolates of *Paecillomyces fumosoroseus* were reported to be pathogenic to *F. occidentalis* and an isolate of *P. fumosoroseus* was developed as a commercial product to be used against whiteflies and thrips (Saito and Sugiyama 2005). *Paecilomyces lilacinus*, a soil inhabiting nematophagous fungus produces chitinases and proteases capable of breaking down egg shell, facilitating penetration into insect body. This mechanism was demonstrated to operate effectively against the western flower thrip *F. occidentalis* (Fiedler and Sosnowska 2007). In another investigation, five strains of *Beauveria bassiana* were evaluated for their efficacy against *F. occidentalis*. The strain RSB of *B. bassiana* was the most virulent causing 69–96 % mortality at concentrations of 1×10^4 to 1×10^7 conidia/ml, at 10 days after inoculation of first instar larvae. In greenhouse evaluation, RSB strain applied to broccoli foliage significantly reduced adult and larval populations of *F. occidentalis* (Gao et al. 2012b). The insecticidal property of *Beauveria bassiana* on *Orius sauteri*, an important predator of western flower thrips *Frankliniella occidentalis* was assessed under in vitro conditions. *B. bassiana* strain RSB (Bb-RSB) was not insecticidal against *O. sauteri*, irrespective of the concentration applied to the first instars. The developmental rate of *O. sauteri* was also not affected by direct treatment with Bb-RSB. However, significant differences in the development rates and adult longevity were noted between *O. sauteri* that fed on Bpb-RSB infected *F. occidentalis* cadavers and those fed on untreated thrips. Developmental time (from first instar to adult) increased from 0.3 to 0.7 days for predators-fed thrips treated with low and high concentration of strain Bb-RSB respectively, compared with predators fed on untreated thrips. But these differences were only 3–13 % of mean values for the controls. The results suggested that the observed adverse effects on *O. sauteri* were relatively minor. The usefulness of combination of *B. bassiana* and *O. sauteri* for the control of *F. occid*entalis has to be demonstrated under field conditions (Gao et al. 2012a).

7.1.3 Diseases of Potato

7.1.3.1 Verticillium Wilt Disease

Among the various cultural practices, addition of green manures is practiced with a view to improving soil fertility and structure, in addition to promoting the activities of antagonistic organisms in the soil. The efficacy of sweet corn varieties (Jubilee Sweetcorn and Jubilee Supersweet corn) as green manure was assessed for suppressing the Verticillium wilt disease of potato caused by *Verticillium dahliae*. The sweet corn varieties suppressed the disease incidence by 60–70 %. These treatments did not influence the pathogen populations directly, but the colonization of *V. dahliae* on potato feeder roots and in potato tissues of stem pieces were reduced. Feeder-root colonization was positively correlated with Verticillium wilt disease incidence ($P<0.05$) and negatively corrected with yield. In addition, corn green manures increased the populations of several soil fungi such as *Ulocladium* and *Fusarium equiseti*. When potato was grown consecutively for 2 years, the beneficial effects of sweet corn green manures was almost entirely lost. But following two consecutive years of potato, a single sweet corn crop was enough to restore the original benefit of disease suppression and enhanced yields, although the pathogen populations had increased by four-fold. The results indicated the effectiveness of growing green manure crops that could reduce disease incidence and increase the yield as well (Davis et al. 2010a). Austrian peas, Sudan grass, rape, oats and rye also exerted similar beneficial effects by reducing disease incidence and enhancing potato yields (Davis et al. 2010b).

Bacterial isolates (5) and extracts of plant species (4) were tested for their efficacy in suppressing the development of Verticillium wilt disease caused by *Verticillium dahliae* in two potato cultivars Russet Burbank (moderately susceptible) and Kennebec (highly susceptible). *Pseudomonas fluorescens* biotype F isolate DF37 was more effective in reducing the disease incidence, severity and vascular discoloration due to infection in both cultivars over two seasons in growth-room experiments. *Bacillus pumilus* isolate M1 reduced the wilt parameters only on Kennebec. Among the plant extracts, Canada milkvetch extract (MVE) was the most effective in reducing the wilt incidence by 55–84 %, compared with untreated controls. In the field experiments, the isolate DF37 and plant extract MVE were effective in reducing Verticillium wilt on Russet Burbank and Kennebec respectively. Reductions in percent infection and vascular discoloration by DF37 were 26 and 67 %, whereas treatment with MVE reduced the parameters by 45 and 55 % respectively during the first field trial. During the second year trial, bacterial isolates DF37 and MI and the plant extract MVE reduced all wilt parameters by percentages ranging from 19 to 31 % and increased the yield by 18 % on cultivar Kennebec. Bacterial isolate DF37 could reduce the Verticillium wilt disease by 29–43 % and increased the yield of the cultivar Russet Burbank by 24 % (Uppal et al. 2008).

7.1.3.2 Stem Rot Disease

Sclerotinia sclerotiorum, causing potato stem rot disease, infects several other crops also, accounting for appreciable losses in production. Sixteen isolates belonging to 11 species of *Trichoderma* were evaluated for their potential to suppress the development of stem rot disease. In addition, one isolate of *Talaromyces flavus* was also included in the evaluation. Spore suspensions of these fungi were sprayed on the foliage in the greenhouse assays. *T. koningii*, *T. virens*, *T. ceramicum* and *T. viridescens* were more effective in reducing disease severity, while *T. flavus* was the least effective against the stem rot pathogen. Under field conditions, the spore suspensions of the BCAs were applied to the soil and foliage five times. *T. viridescens* followed by *T. ceramicum* provided the best protection to the potato plants against *S. sclerotiorum*. The results indicated the need for confirming the biocontrol efficacy of the microorganisms that perform well in the greenhouse assay, under the field conditions also (Ojaghian 2011).

7.1.3.3 Late Blight and Tuber Blight Disease

Phytophthora infestans infects the foliage of the plants destroying the photosynthetic tissues seriously. Shortly after late blight is established on the potato foliage, the inoculum consisting of sporangia and zoospores produced on the foliage is deposited onto the surface of the potato hill. Then the water from irrigation or rain can carry the inoculum into the soil. Infection of potato foliage by *P. infestans* results in late blight phase of the disease and later tubers are also infected leading to tuber blight disease. The late blight disease is a challenging plant health problem defying human efforts to achieve complete control. Different disease management strategies have been applied alone or in combination to restrict the disease incidence and spread of the disease through forecasting models, genetic resistance, cultural practices and chemical application and varying degrees of success have been obtained under different environmental conditions. It is necessary to combine different approaches that are suitable for the geographical localities concerned. Use of disease-free certified seed tubers is the basic step to effectively eliminate initial sources of inoculum of *P. infestans*. Production of disease-free tubers depends on the prevention of tuber infection in the previous crop to a great extent. Crop sanitation has to be strictly adopted. All volunteer plants and solanaceous weeds have to be thoroughly eliminated and properly disposed off, since the pathogen can overwinter in the plant debris by forming oospores that can be a potential source of inoculum for the subsequent crops of potatoes. During the growing period, incidence of the disease has to be carefully monitored and the infected plants have to be eliminated immediately. Observations on the pattern of incidence of late blight disease in a give location may be useful to select the date of planting, when the natural incidence of the disease is less. This approach was followed for reducing the disease incidence in West Bengal State, India (Basu 2009). The effects of strip-cropping of potatoes with cereals or a grass-clover mix on the incidence of late blight disease were assessed during 2001–2002. In the plots with strip-cropped potatoes, incidence of the disease was significantly reduced by 4–20 %, depending on the plot size. The smaller plots (6×18 m) contracted less disease, compared to larger plots (6×36 m). Grass-clove mix in and perpendicular to the main wind direction also contributed to reduction in late blight disease incidence. The results indicated strip intercropping might be a useful component in an overall management strategy to reduce incoming late blight inoculum (Bouws and Finckh 2008).

The effectiveness of different kinds of polyethylene mulches was assessed for suppressing or preventing the development of potato tuber blight caused by *Phytophthora infestans* under field conditions. A water-permeable agricultural textile treated with copper hydroxide was also tested for its efficacy. Two barriers that covered the potato hills, black polyethylene film and coppery hydroxide-treated agricultural textile reduced the incidence of tuber blight relative to the appropriate controls. Other barriers (mulches) had little effect on the tuber blight incidence. The effectiveness of black polyethylene film might be attributed to blocking of water through potato hill and alteration of the soil environment within the hill. The hills covered with black plastic sheets were drier and slightly warmer than uncovered hills, thus markedly affecting the inoculum movement and infection. The effect of mulches and hill sizes as barriers on development of potato tuber blight was studied. The polyurethane spray foam in an 8-cm diameter area immediately surrounding the plant stem and black polyethylene film over the entire hill except near the stem alone or in combination were tested. Water permeable agricultural textile treated with copper hydroxide was also used to cover the hill as in black polyethylene. Black film and textile mulches reduced tuber infections significantly, indicating that inoculum of *Phyophthora infestans* could move from foliage to tubers through soil and that inoculum movement was not limited to large channels in the hills such as those created by potato stems. Mulching the stem area with spray foam did not reduce incidence of tuber blight incidence, compared with control. Comparison of blight incidence in tubers in different hill sizes, showed insignificant differences among the hill size treatments, although tubers covered with more than 15 cm soil had lower blight incidence (1–14 %) than tubers with less soil cover (13–59 %). The mulch treatments could provide only partial protection to the tubers, limiting its practical use for wider application by growers. Suppression of tuber infection might be better achieved by preventing/reducing foliar epidemics or through host resistance (Glass et al. 2001).

Several fungal and bacterial species with known antagonistic activity against other microbial plant pathogens have been reported. Different species of *Trichoderma*, *Pseudomonas* and *Bacillus* were evaluated for their efficacy in suppressing the development of late blight pathogen. The fungal BCAs *Penicillium aurantiogriseum* and *Stachybotrys atra* suppressed the lesion development following inoculation with *P. infestans* in greenhouse-grown potato plants. The disease severity was reduced by 93 and 84 % respectively. The efficacy of these BCAs was not, however, tested under field conditions (Jindal et al. 1988). An isolate of *Pseudomonas putida* had no direct inhibitory effect on *P. infestans* in in vitro assays, but it induced systemic resistance in potato plants against the pathogen. On the other hand, *Serratia plymuthica* inhibited the growth of *P. infestans* through antibiosis, but it did not induce systemic resistance in potato plants. Both *P. putida* and *S. plymuthica* effectively suppressed the development of late blight disease through different mechanisms of action on the pathogen *P. infestans* U8-clonal lineage which is a highly aggressive isolate (Daayf et al. 2003). Tuber infection could be reduced significantly by treatment of tubers with strain mixtures of *P. fluorescens* (Slininger et al. 2007). *Trichoderma harzianum*, among the six antagonists tested under field conditions, was found to be the most effective in reducing the disease severity percentage, followed by *Pseudomonas fluorescens* compared with untreated control treatment. Seed treatment with *T. harzianum* (5 g/kg) and *P. fluorescens* (2.5 g/kg) provided effective protection to plants and also increased the tuber yield to 19.5 and 19.0 t/ha respectively, compared with control (13.8 t/ha). These BCAs are being used widely by farmers in the West Bengal State, India (Basu 2009). Late blight forecast systems have been adopted in many countries and adapting them to suit different conditions existing in various locations for the application of biological control agents may become feasible in future.

Late blight management has been heavily dependent on application of fungicides and in many countries fungicide use has increased over the years, due to the appearance of new, more aggressive (virulent) genotypes of *Phytophthora infestans*. Even in organic production of potato application of copper-based fungicides for the control of late blight is on the increase, despite the existing ceiling of 6 kg of elemental copper/ha (Kato et al. 1997; Ghorbani et al. 2004). Hence, there is enormous pressure to develop suitable and effective alternative non-chemical strategy to protect potatoes against late blight and other economically important crop diseases. Numerous plant-derived products especially plant extracts and essential oils have been shown to be effective in in vitro assays and greenhouse tests. However, none of them had the stability to be used under the field conditions for effective reduction of incidence and/or severity of late blight disease (Dinz et al. 2006; Soylu et al. 2006). Many organic compounds have been demonstrated to induce resistance in plants against diseases caused by microbial pathogens. The protective effect of ß-aminobutyric acid (BABA) against the potato late blight pathogen *Phytophthora infestans* was demonstrated using two potato cultivars Bintje and Pampeana with different levels of horizontal resistance. Foliar pre-treatment at 30 days after emergence provided a 60 % protection in cv. Pampeana against *P. infestans*. BABA treatment stimulated the expression of molecules associated with development of disease resistance (Altamiranda et al. 2008). Plants receiving four applications of BABA throughout the crop cycle produced tubers with greater resistance to both *P. infestans* and another potato pathogen *Fusarium solani*, than those from non-treated control plants. Treatment with BABA improved the growth of potato plants, in addition to the protection provided against the fungal pathogen. The degree of enhancement of resistance by BABA depended on the natural level of genetic resistance of the potato cultivars to different races of *P. infestans* (Olivieri et al. 2009). Combination of BABA and fungicides appeared to have an additive effect both under greenhouse and field conditions. With increase in the concentration of BABA, the protective effect became stronger proportionally. The partially resistant potato cultivars Ovatio and Superb reacted to lower concentrations of BABA, whereas no significant effect of treatment with BABA was discernible under field conditions. BABA treatment was able to compensate a 20–25 % reduction in the dose of fungicide Shirlan, as the combination was effective as the full dose of Shirlan alone. The results indicated that BABA could be applied as a component of disease management system directed towards an important disease of potato (Liljeroth et al. 2010).

Phosphites (Phi) are alkalimetal salts of phosphorus acid with the potential for protecting plants against plant pathogens. The effects of treatment of seed tubers and foliage of potato cv. Shepody and Kennebec with Phi on the induction of resistance to *Phytophthora infestans*, *Fusarium solani* and *Rhizoctonia solani* were assessed. The degree of protection provided by Phi was high against *P. infestans*, intermediate against *F. solani* and low against *R. solani*. In addition, seed tubers treated with calcium or potassium phosphate (CaPhi or KPhi) at 1 % commercial product emerged earlier than untreated ones. When Phi sprays were applied on foliage two to four times at different doses, high levels of protection against *P. infestans* were observed on both cultivars. Kennebec was more responsive to Phi application than Shepody, as reflected by the degree of resistance developed in treated plants. Application of Phi did not cause any detectable adverse effect on plant growth. Leaves in treated plants were darker green in color than those in untreated plants, in addition to a delay in setting of leaf senescence in Phi-treated plants (Lobato et al. 2008).

7.1.3.4 Black Scurf Disease

Potato black scurf pathogen *Rhizoctonia solani* can infect a wide range of plant species including several crop plants on which damping-off and root rot diseases inflict serious damages. Different disease management tactics have been applied with varying degrees of success. The effects of 2- and 3-year crop rotations with conventional and minimum tillage treatments on the severity of soilborne diseases of potato, canker and black scurf disease (*Rhizoctonia solani*), dry rot (*Fusarium* spp.) and silver scurf (*Helminthosporuin solani*) and root zone bacteria were investigated. The 2-year rotation included spring barley, while the 3-year rotation had barley and red clover, in addition to potato. The disease incidence/severity was reduced to a greater extent in the 3-year rotational soils, compared with the 2-year rotational soils. The soil agroecosystem can be modified through crop rotation and conservation tillage practices to improve disease suppression by enhancing antagonistic activities of endophytic and root zone bacteria, as revealed by the analysis of root zone bacteria using in vitro assays (Peters et al. 2003, 2005). A cropping system was designed specifically for suppressing soilborne diseases affecting potato. The disease suppressive (DS) system included diverse crops with known disease-suppressive potential, such as *Brassica* and Sudangrass, green manure crops and fall cover crops. High crop diversity resulted in greatest reduction in stem and stolon canker and black scurf caused by *Rhizoctonia solani* and common scab due to *Streptomyces scabiei* under both irrigated and non-irrigated conditions, compared to other systems for soil management. The DS system also caused significant shifts in soil microbial community characteristics different from all other rotations. Biofumigation was considered to be the mechanism of action for the crops included in the DS system (Larkin et al. 2011).

Verticillium biguttatum an effective biocontrol agent against potato black scurf pathogen *Rhizoctonia solani* was examined for its compatibility with broad-spectrum fungicides azoxystrobin, chlorothalonil and thiabendazole. These fungicides were found to be fungitoxic to *V. biguttatum* in the in vitro bioassays. On the other hand, the oomycete-specific fungicides cymoxanil and propamocarb did not affect the growth of *V. biguttatum*. In contrast, *Rhizoctonia*-specific fungicides pencycuron and flutalonil, when co-applied with *V. biguttatum*, showed additive effects and improved the effectiveness of black scurf biocontrol. When combinations of *V. biguttatum* and cymoxanil or propamocarb were applied to immature potato tubers at green crop lifting, a reduction in both black scurf and *Pythium*- or *Phytophthora*-incited tuber rot was observed at harvest. The results indicated that *V. biguttatum* could be combined with selected chemicals that had specific activity against *R. solani* or *Pythium* spp. or *Phytophthora* spp. infecting potato tubers (van den Boogert and Luttikholt 2004).

7.1.3.5 Dry Rot Disease

Potato dry rot disease caused by *Fusarium sambucinum* has significant economic importance, because of its potential to induce serious quantitative and qualitative losses of tubers. Isolates of *Bacillus licheniformis* and *B. cereus* effectively reduced the development of *Fusarium roseum* var. *sambucinum* causing potato dry rot disease and also stimulated emergence of plants from treated tubers. *B. licheniformis* 132 and *B. cereus* X16 were more efficient in reducing disease severity. Field evaluation of the efficacy of the efficient isolates 132 and X16 showed that they were effective in controlling Fusarium rot on seed tubers and increased yield parameters. The seed tuber colonization by bacteria increased until 61 days after planting with a rate significantly higher for *B. cereus* X16. Furthermore, during conventional storage conditions for 6 and 8 months, dry rot disease incidence was significantly lower on potato boxes treated with each isolate of the antagonist alone or by the mixture (consisting of X16 + *Bacillus thuringiensis* var. *galleriae* and *Trichoderma viride* 55T), as compared with pathogen inoculated control boxes and to those treated with the fungicide carbendazim (Sadfi et al. 2002).

The biocontrol potential of *Pseudomonas fluorescens* and *Enterobacter cloacae* in suppressing the development of the pathogen, when applied as seed treatment under field conditions for 2 years was assessed. The fungicide fludioxonil and mustard meal were included in the field trial for comparison of effectiveness of disease suppression. After the harvest of tubers, severity of dry rot and also other diseases including silver scurf (*Helminthosporium solani*) and common scab (*Streptomyces scabiei*) was recorded. Significant reduction in dry rot severity was observed in all treatments, compared with non-treated control inoculated with *F. sambucinum*. The descending order of effectiveness of dry rot disease suppression was noted in fludioxonil (55.7 %), mustard meal (47.5 %), *P. fluorescens* (35 %) and *E. cloacae* (26.5 %). In addition, all treatments significantly reduced the severity of common scab and silver scurf diseases, compared with non-treated controls. On an average, seed tubers treated with *P. fluorescens* and *E. cloacae* produced higher total number of tubers. *E. cloacae* was more effective than *P. fluorescens* in increasing both total and marketable tuber yields than non-treated inoculated controls. The results indicated that *P. fluorescens* and *E. cloacae* and mustard meal could be considered as viable options for control of potato tuber diseases along with the fungicide (Al-Mughrabi 2010).

7.1.3.6 Pink Rot Disease

Phytophthora erythroseptica, a soilborne fungus causes the pink rot disease of potato tubers. The frequency and composition of the crops in a potato rotation influence the incidence and development of soilborne diseases affecting potato crops. *P. erythroseptica* can survive in the soil for several years as oospores and propagules of the pathogen, making it an endemic disease in most cultivated soils where potatoes are grown repeatedly. Crop rotation with at least one alternate crop was found to reduce the pink rot disease incidence significantly (Lambert and Salas 2001). In a later investigation, the effect of 2- and 3-year crop rotations and conservation tillage practices on the severity of pink rot disease was assessed. Barley was included in the 2-year rotation, while red clover was also included additionally in the 3-year rotation. The development of pink rot disease was significantly less pronounced in potatoes from the 3-year rotational soils than from the 2-year rotational soils. Potato plants grown in the greenhouse, using soils from field plots, had less disease in 3-year rotation than in 2-year rotation (P=0.05). The results suggested that potatoes grown in soils managed under a 3-year rotation appeared to be intrinsically more resistant to infection by *P. erythroseptica* than those managed under a 2-year rotation (Peters et al. 2005).

Fungal and bacterial antagonistic species have been evaluated for their efficacy in suppressing the development of pink rot disease. *Trichoderma virens* DAR74290 and Trichodex, a commercial formulation containing *T. harzianum* T-39 were tested alone or in combination. These bioagents, when applied alone or in combination reduced the disease severity in shoots and roots of potatoes at 10 weeks after the pathogen inoculation. In addition to disease suppression, the fungal BCA increased the tuber yields also compared with untreated control plots (Etebarian et al. 2000). The zoospores or germinating encysted zoospores of *Phytophthora erythroseptica* could induce new infections via potato tuber eyes, lenticels and cracks and cuts that result during harvesting. The possibility of preventing pathogen entry through these infection courts, *Enterobacter cloacae*, *Enterobacter* spp. and *Pseudomonas* spp. were used as suspension of cells for tuber treatment. *E. cloacae* strain S11:T:07 was more efficient in reducing the lesion size than other two bacterial species that also reduced the lesion development significantly. Treatment with the bacterial BCAs provide protection to both potato cultivars Russet Norkotah and Russet Burbank, indicating that cultivars did not influence the performance of antagonists (Schisler et al. 2009).

7.1.3.7 Scab Diseases

Potato scab and common scab diseases are due to *Streptomyces scabies* and *S. turgidiscabies* respectively. They are capable of surviving in the soil for a long time. Cultivation and incorporation of specific green manure crop(s) have been suggested as a management tactic for soilborne diseases. The effects of green manures (buck wheat and canola) and crop sequences on potato scab disease (*Streptomyces scabies*) and Verticillium wilt disease (*V. dahliae*) were assessed in a 2-year field trial. Indigenous streptomycete densities and in vitro pathogen inhibitory activity were also determined. Potatoes grown in soil planted to corn or alfalfa in the previous year had lower potato scab and Verticillium wilt ratings significantly, as well as higher yields than potatoes grown in soil previously planted with potato. Streptomycetes from soil collected from green manure (GM)-treated plots tended to have greater in vitro pathogen inhibitory activity than those from fallow-treated plots. The streptomycete communities in GM-treated soils had consistently greater proportion of pathogen antagonists than communities in fallow-treated plots in the second season of the investigation. GM crops may selectively enrich the abundance or activity of antibiotic producers within the soil microbial community (Wiggins and Kinkel 2005).

The effect of green manure and fallow on the incidence of potato common scab diseases caused by *Streptomyces turgidiscabies* was determined under field conditions. The number of diseased tubers harvested from soil incorporated with lopsided oat or woolly pod vetch was significantly fewer, compared with those from oat and continuous potato cultivation in a planter experiment. In the field experiments conducted during 1999–2000, treatment with lopsided oat, followed by lopsided oat or woolly pod vetch was significantly more effective at suppressing the disease severity than oat and continuous potato cultivation ($P<0.001$). An increase in marketable tuber ratio was also more significant than for oat and continuous potato cultivation ($P<0.001$). In another field experiment (2000–2001), lopsided oat cultivation alone and with the application of Ferosand (Fe, mainly $FeSO_4$ to decrease soil pH) at 1.8 t/ha or resistant potato cultivar treatment were significantly more effective in suppressing the disease severity and incidence than sugar beet cultivation ($P<0.001$), even under high disease pressure in the field. However, the potato yield tended to decrease, after lopsided oat treatment with an application of Ferosand (1.8 t/ha), compared with lopsided oat alone or the application of Ferosand at 600 kg/ha, due to low pH conditions. In the third field experiment (2001–2002), the lowest disease severity and incidence of common scab disease were observed in soil treated with lopsided oat than with other treatments. The results suggested that lopsided oat could be a useful option as fallow green manure for reducing common scab disease severity and increasing the tuber yield (Sakuma et al. 2011).

7.1.4 Pepper Diseases

7.1.4.1 Phytophthora Blight Disease

Several investigations have indicated the contribution of cultural practices to restriction of incidence and severity of diseases affecting various crops. In the presence of the stubble from a fall-sown, no-till wheat cover crop, the dispersal of soil inoculum of *Phytophthora capsici*, causing Phytophthora blight disease of bell pepper was markedly reduced, resulting in considerable reduction in disease incidence. The final disease incidences ranged from 2.5 to 43 % in no-till plots, as against 71–72 % incidence in pepper planted in a bare field, where all dispersal mechanisms operated without hindrance (Ristaino et al. 1997). In order to retain crop residues and satisfactory level of disease suppression, reduced or no-till practice may be coupled with proper crop sequence. Rotation of pepper (chilli) (*Capsicum annuum*) with peanut or sesame resulted in the reduction of Phytophthora blight disease by 39 and 11 % respectively. Nonhosts included in the rotation and/fallowing reduced the selection pressure for a given soilborne pathogen and 'starve' it out, preventing the buildup of large populations of pathogen (Kim 1989). The type of irrigation may become an important factor in terms of disease risk. *P. capsici* produces zoospores that initiate infection of roots primarily. Conditions of wet and dry cycles in soil are required for completion of different stages in its life cycle. Rainfall and periodic furrow irrigation provide the wet-dry cycle in the soil favoring production of sporangia during dry period and zoospores release during wet period. The role of irrigation methods (drip, basin and furrow) in the incidence of pepper crown rot disease caused by *P. capsici* was investigated. Disease percentage was higher in the drip irrigation system (78.39 %) than in basin (56.27 %) and furrow (55.92 %) irrigation systems (Sağir et al. 2005).

Antagonistic bacterial species have been evaluated for their potential against *Phytophthora capsici*, causing Phyophthora blight disease of pepper which can infect the different plant parts and at different growth stages. Biocontrol efficiency of *Bacillus* preparation (BB11 and FH17 strains and a mixture of both strains(BF at a 1:1 ratio by concentration) and different application methods was assessed. In the greenhouse assays, BF showed greater biocontrol efficacy (60.3 %) and yield increase (200 %) than that could be obtained with BB11 (55.8 and 80.6 % respectively) or FH (37.1 and 50.0 % respectively). In the field trials, the most effective dosage of BF mixture (10^{10} CFU/ml) was 7.5 l/ha, when BF was mixed with rape seed residue and made compost before application. When preparations were applied at the best dosage in the same field, the BF mixture provided superior biocontrol efficiency and greater yield increase with this treatment. Addition of BF mixture to rape seed residue followed by composting and applying the compost provided better disease control and higher yield than other methods of application of the bacterial BCA *Bacillus* sp. (Jiang et al. 2006). Phytophthora blight disease incited by *P. capsici* causes devastating damages to the crop, when it occurs along with *Rhizoctonia solani*, *Fusarium oxysporum* and *F. solani*. In order to develop an effective biocontrol approach, three bacterial chitinolytic BCAs, *Serratia plymuthica* strain C-1, strongly antagonistic to *P. capsici*, *Chromobacterium* sp. strain C-61 strongly antagonistic to *R. solani* and *Lysobacter enzymogenes* strain C-3 antagonistic to *R. solani* and *Fusarium* spp. were evaluated under greenhouse conditions. Application of all three bacterial strains suppressed Phytophthora blight more effectively than individual strains. The formulation derived from simple medium containing chitin was tested under field conditions. The efficacy of the formulated product depended on the dose and timing of application. The undiluted product suppressed Phytophthora blight under all field conditions. A ten-fold diluted product was effective in solar-sterilized greenhouses and in fields with crop rotation (Kim et al. 2008). The availability of phosphorus, an important nutrient for plant growth is a vital factor, since it is generally found in insoluble form in the soil. The possibility of controlling pepper crown blight caused by *P. capsici* by employing phosphate-solubilizing bacteria like plant growth-promoting rhizobacteria (PGPR) was examined under growth room and field conditions. Three phosphate solubilizing strains of *Bacillus megaterium* were effective, when applied alone or in combination in reducing disease severity under field conditions, although their efficacy was lower under field conditions than in the growth room conditions. Treatment with strain M1-3 reduced crown blight severity to half, having an efficacy of 50.4 % higher than all other treatments. Increase in yield due to treatment with M1-3 was also observed (Akgül and Mirik 2008).

Acibenzolar-*S*-methyl (ASM), a known inducer of disease resistance was evaluated for its potential to protect pepper plants against root rot phase of the disease induced by *Phytophthora capsici*. ASM was applied four times on greenhouse-grown pepper plants and inoculated with *P. capsici* or in soil naturally infested with the pathogen. Inhibition of stem canker development on pepper cvs. Bell Tower and AZ9 after four treatments with ASM (75 μg/ml) was significantly greater than on plants receiving single application of ASM. Survival of pepper plants grown in field soil naturally infested with *P. capsici* was significantly increased by three foliar applications of ASM (75 μg/ml), compared with untreated plants. The length of survival among pepper plants treated with 25, 50, or 75 μg/ml of ASM and grown in soil infested with *P. capsici* was similar. ASM could be employed in combination or in alternation with mefenoxam to manage the crown and root rot disease. This will be useful to reduce the development of resistance to mefenoxam by *P. capsici* (Matheron and Porchas 2002). The suppressive ability of silicon applied as calcium silicate (+Si) or calcium carbonate (−Si) on the development of Phytophthora blight disease in bell pepper was studied by incorporating them in the soil and by planting seedlings were planted after 6 weeks. The soil was infested with the pathogen *Phytophthora capsici* isolates Cp30 and Cp32. The disease development was assessed based on the area under disease progress curve (AUDPC) and area under wilting progress curve. An increase of 40 % in the Si concentration in the roots, but not in the stem was observed in the treated plots. The rate of disease progress was significantly reduced in +Si treatment, as compared to −Si treatment. The dry root weight and stem weight were increased due to Si treatment, compared with control, indicating the growth promotion effect of Si on pepper plants, in addition to protection against Phytophthora blight disease (French-Monar et al. 2010).

7.1.4.2 Verticillium Wilt Disease

Verticillium wilt disease caused by *Verticillium dahliae* infects several crop plants especially those belonging to the family Solanaceae including pepper. The efficacy of Polyversum® containing *Pythium oligandrum*, the bioagent and the conventional fungicides benomyl and propamocarb-hydrochloride in suppressing the development of Verticillium wilt disease was assessed. Benomyl was the most efficient in suppressing the development of the disease to the extent of 94.6 and 88.2 %, as pre- and post-inoculation treatments respectively. Polyversum was more effective (66.6 %), when applied prior to inoculation and its effectiveness was equal to that of propamocarb-hydrochloride, but less efficient, when compared with benomyl. However, Polyversum-treated plots had significantly less disease, compared with untreated control plots (Rekanovic et al. 2007). The bioagent *Trichoderma asperellum* and the chemical dazomet either alone or in combination were evaluated for their potential to suppress the development of Verticillium wilt disease in bell pepper (*Capsicum annuum*). Six field trials in three consecutive years were carried out in commercial farms in Poland. In five of six trials, no significant differences in the area under disease progress curve

(AUDPC) values could be observed between methyl bromide (MB, used for comparison of effectiveness) and dazomet (DZ) alone or DZ combined with *T. asperellum* B35. The most consistent positive effect on yield and reduction in root rot complex and Verticillium wilt was obtained in the treatment with DZ + *T. asperellum*. The mean yield increase due to the combination was 40.1 % which was equal to that of MB treatment. Further, DZ + *T. asperellum* combination provided the highest net marginal return and a higher return on investment than MB. *T. asperellum* alone provided highly variable protection which did not merit the recommendation for wider adoption by growers (Slusarski and Pietr 2009).

7.1.4.3 Anthracnose Disease

Anthracnose disease incited by *Colletotrichum acutatum* infects all aerial plant parts including fruits. Myxobacteria produce a large number of bioactive secondary metabolites that have suppressive effect on microbial plant pathogens. The biocontrol potential of three strains of *Myxococcus* spp. KYC1126, 1136 and 2001 was determined for the control of anthracnose disease under greenhouse and field conditions in 2009 and 2010. *Myxococcus* KYC 1126 was the most effective as shown by in vitro assays and greenhouse tests. The culture filtrate of KYC1126 was able to reduce the incidence of anthracnose disease on seedlings to 29 %, whereas the disease incidence in the untreated control plots was 74 %. This strain was more effective with a control value of 60 %, compared with the fungicide dithianon (42 %). However, the biocontrol activity was not maintained for long time in the field experiments (Kim and Yun 2011).

7.1.4.4 Fusarium Wilt and Rhizoctonia Damping-Off Diseases

Pepper and tomato crops are infected by Fusarium wilt and Rhizoctonia damping-off diseases which occur commonly in most of the areas where these crops are cultivated. Three PGPRs *Bacillus licheniformis* CECT5106, *Pseudomonas fluorescens* CECT5398 and *Chrysobacterium balustinum* CECT5399 and the bioproduct LS213 derived from *Bacillus subtilis* strain GB03, *B. amyloliquefaciens* strain IN937 and chitosan were evaluated for their potential to protect pepper and tomato plants against Fusarium wilt and Rhizoctonia damping-off diseases. The treatments that combine LS213 and each of the three PGPRs provided significantly higher level of protection as reflected by higher percentages of healthy pepper and tomato plants than that can be obtained with LS213 alone. Similar synergistic effect of LS213 and the individual bacterial species on plant growth promotion was observed. Overall, the positive effects of treatment with the BCAs and the bioproduct LS213 were greater on pepper than on tomato. The results revealed that the combinations of microorganisms might be more effective probably because of their ability to act on the pathogens through different mechanisms (Domenech et al. 2006). The compatibility of endophytic bacterial strains *Bacillus subtilis* EPCO16 and EPC5 and *Pseudomonas fluorescens* Pf1 for the suppression of *Fusarium solani* incitant of pepper wilt disease was investigated. The bacterial strains applied alone or in combination effectively reduced the incidence of wilt disease. Combinations of EPCO16 + EPC5 + Pf1 were more effective than the individual strains. The results indicated that the bacterial strains suppressed the disease development by inducing systemic resistance in pepper plants against Fusarium wilt disease (Sundaramoorthy et al. 2012).

7.1.4.5 Sclerotium Root and Stem Rot and Ralstonia Wilt Diseases

Occurrence of two and more diseases simultaneously results in disease complex, making their management more difficult and complicated. The biocontrol potential of strains of *Streptomyces* spp. was assessed for the suppression of development of root and stem rot disease caused by *Sclerotium rolfsii* and bacterial (Ralstonia) wilt disease caused by *Ralstonia solanacearum* infecting pepper (chilli) under greenhouse conditions. The efficacy of *Streptomyces philanthi* RL-1-178 in suppressing Sclerotium root and stem rot of pepper was approximately equal to that of *Trichoderma harzianum* NR-1-52 or that of the fungicide carboxin. *S. mycarofaciens* SS-2-243 and *S. philanthi* RL-1-178 suppressed the development of Ralstonia wilt disease and its antibacterial activity was similar to that of streptomycin sulfate treatment. *T. harzianum* NR-1-52 had no adverse effect on the bacterial pathogen. Under field conditions where the soil was infested with the pathogens, *S. philanthi* RL-1-178 was able to protect the pepper plants against both *S. rolfsii* and *R. solanacearum* more effectively than *S. mycarofaciens* SS-2-243 or *T. harzianum* NR-1-52. Treatment with *S. philanthi* RL-1-178 resulted in 58.75 % survival of pepper plants and its effectiveness was equivalent to that of carboxin and streptomycin sulfate treatment (Boukaew et al. 2011).

7.1.5 Diseases of Cucurbitacous Crops

7.1.5.1 Pythium Root Diseases

Different species of *Pythium* can cause frequently the root rot disease of greenhouse-grown cucumbers. Physical and biological management strategies have been shown to be effective in reducing damping-off or root rot disease. Trials were conducted in commercial greenhouses to assess the effect of solarization and biofumigation on the incidence of cucumber damping-off disease caused by *Pythium aphanidermatum*. For biofumigation, cabbage residue (*Brassica oleracea*) was incorporated into top 20-cm soil at the rate of 5 kg/m^2. Clear polythene (0.2 mm thick) was used to cover the soil. Solarized and biofumigated plots had significantly reduced disease incidence, compared with untreated controls. Damping-off disease incidence was significantly lower in biofumigated than in solarized plots at the final assessment at 35 days after planting. Biofumigation and solarization treatments enhanced plant growth as additional benefit

(Deadman et al. 2006). *Pythium ultimum* is another fungal pathogen involved in damping-off disease. Compost amendments have been reported to have disease suppressive properties. Municipal biosolid compost was suppressive to *P. ultimum* and disease incidence on cucumber was significantly reduced. Suppression of damping-off disease incidence in cucumber was shown to be due to microbial consortia colonizing cucumber seeds within 8 h of sowing (Chen and Nelson 2008). The bacterial strains *Bacillus subtilis, Pseudomonas fluorescens* and *P. corrugata* and two fungal strains *Trichoderma viride* and *T. (Gliocladium) virens* were evaluated for their ability to suppress the damping-off disease of cucumber caused by *P. ultimum.* Among the antagonists, *Pseudomonas* spp. were superior to *Bacillus subtilis* in reducing the incidence of damping-off disease in cucumber. Combination of antagonists did not show any additive effect. The effectiveness of disease suppression was greater, when the bacterial antagonists were applied by drenching or by coating the cucumber seeds with bacteria in a peat carrier (Georgakopoulos et al. 2002).

7.1.5.2 Fusarium Wilt Diseases

Fusarium wilt diseases are caused by *Fusarium oxysporum* f.sp. *radicis-cucumerinum* (FORC) in cucumber and by *F. oxysporum* f.sp. *melonis* (FOM) in melon. The effect of soil amendments on the incidence of Fusarium root rot and stem rot caused by FORL was studied. The composts containing greenhouse waste material (GC), windrow composted dairy solids (WDS) and vermicomposted dairy solids (VMC) were evaluated for their efficacy in suppressing the development of FORC in infecting cucumber. Greenhouse compost significantly reduced cucumber plant mortality due to FORC to 13 % as against 63 % in untreated control. The effectiveness of GC was equal to that of the fungicide benomyl (Punja et al. 2002). The composts consisting of dairy and greenhouse wastes significantly reduced the severity of FORC. The number of total culturable microbes in the composts showed a positive relationship with disease suppressive potential of the compost. The strains of *Pseudomonas aeruginosa* isolated from the composts exhibited the greatest antagonistic activity against FORC. Further, internal stem colonization of FORC was significantly reduced by *P. aeruginosa* (Bradley and Punja 2010).

Disease suppressiveness of composts is influenced by several factors. Effects of the raw materials, aeration conditions during composting process and compost maturity were investigated to determine their influence on suppressiveness to melon Fusarium wilt disease caused by *Fusarium oxysporum* f.sp. *melonis* (FOM). Application of compost reduced the incidence of Fusarium wilt disease of melon. *Aspergillus* spp. isolated from the compost was the most effective among the large number of microorganisms isolated from the compost. The suppressiveness of the compost was associated with the population of *Aspergillus* spp. (Suárez-Estrella et al. 2007). The efficacy of two citrus composts – C1 composed of 40 % citrus wastes, 20 % sludge obtained from citrus industry waste-water treatment facility and 40 % green residues and C2 composed of 60 % citrus wastes amended with *Trichoderma harzianum* T-78 – was evaluated for the control of *Fusarium oxysporum* f.sp. *melonis* (FOM). Incidence of Fusarium wilt disease on melon and growth promotion effect of the treatments was recorded. Disease incidence was significantly reduced in C2Th (extract amended with T-78), while C1Th was not effective. Population of T-78 significantly decreased at first sampling time compared to the initial level, but later recovered over time. The results indicated that combination of citrus compost and *T. harzianum* T-78 could become a viable alternative to peat and also adoption of this strategy could minimize the chemical use for the management of Fusarium wilt disease in greenhouse nurseries for melon seedling production (Lopez-Mondejar et al. 2010). Residues of various plant species were evaluated for their efficacy in suppressing the development of cucumber crown and root rot disease caused by *Fusarium oxysporum* f.sp. *radicis-cucumerinum* (FORC). Disease incidence and severity of the disease in cucumber plant inoculated with the pathogenic FORC were reduced by 20–80 %, when seedling were planted in soils incorporated with residues of *Diplotaxis tenuifolia* (Wild rocket, WR), *Artemisia dracunculus*, *Salvia officinalis* and *Brassica oleracea* var. *italica* (broccoli). Effective soil suppressiveness persisted after repeated inoculations and plantings in the same soil without additional treatment between inoculations. Furthermore, residues of WR induced soil suppressiveness in two additional tested soils differing in their physical and chemical properties. Soil suppressiveness to Fusarium crown and root rot disease was induced, when cucumber seeds were sown in soils which were initially amended with WR residues and later infested with FORC chlamydospores. The results indicated that the possibility of containing the incidence of soilborne disease caused by *Fusarium oxysporum* by incorporating suitable plant residues that contribute to development of soil suppressiveness (Klein et al. 2011).

Disease suppressiveness of composts is influenced by several factors. Effects of the raw materials, aeration conditions during composting process and compost maturity were investigated to determine their influence on suppressiveness to melon Fusarium wilt disease caused by *Fusarium oxysporum* f.sp. *melonis* (FOM). Application of compost reduced the incidence of Fusarium wilt disease of melon. *Aspergillus* spp. isolated from the compost was the most effective among the large number of microorganisms isolated from the compost. The suppressiveness of the compost was associated with the population of *Aspergillus* spp. (Suárez-Estrella et al. 2007). The efficacy of two citrus composts – C1 composed of 40 % citrus wastes, 20 % sludge obtained from citrus industry waste-water treatment facility and 40 % green residues and C2 composed of 60 % citrus wastes amended with *Trichoderma harzianum* T-78 – was evaluated for the control of *Fusarium oxysporum* f.sp. *melonis* (FOM). Incidence of Fusarium wilt disease on melon and growth promotion effect of the treatments was recorded. Disease incidence was significantly reduced in C2Th (extract amended with T-78), while C1Th was not effective. Population of T-78 significantly decreased at first sampling time compared to the initial level, but later recovered over time. The results indicated that combination of citrus compost and *T. harzianum* T-78 could become a viable alternative to peat and also adoption of this strategy could minimize the chemical use for the management of Fusarium wilt disease in greenhouse nurseries for melon seedling production (Lopez-Mondejar et al. 2010).

The effect of soil solarization and semi-solarization were investigated during five consecutive years alone or in combination with calcium cyanamide soil amendment, in order to develop alternative and ecologically compatible methods for the management of Fusarium wilt diseases of melons caused by *Fusarium oxysporum* f.sp. *melonis* (FOM). Semi-solarization and solarization increased the mean soil temperature at 25 cm depth by 8.6–12.6 °C and 12.6–16.3 °C degrees respectively. Solarization reduced the population of FOM in sterile soil samples and also reduced the Fusarium wilt disease incidence by 82–90 % in three out of five trials. The reduction in disease incidence (%) was proportional to the period, when the soil temperatures remained at or above 40 °C or at or above 42 °C at 25 cm depth. Calcium cyanamide amendment at 80 g/m^2 had no effect on the efficiency of soil solarization (Tamietti and Valentino 2006). The nonpathogenic mutant strain 4/4 and 15/15 of FOM generated through UV mutagenesis were employed to protect the muskmelon plants. No symptoms of infection by the wild-type strain of FOM race 1, 2 or any detrimental effect could be observed, following inoculation of muskmelon plants with strain 4/4. In contrast, strain 15/15 caused mortality of susceptible muskmelon cultivars to a lesser extent compared with the wild-type strain. The mutant 4/4 colonized the roots to a great extent (100 %) and between 30 and 70 % of the lower stem tissues in 7 days after planting. The mutant 4/4 was able to significantly reduce mortality of watermelon seedling caused by *F. oxysporum* f.sp. *niveum* race 2. The results indicated the wide spectrum of activity of the nonpathogenic mutant FOM and its potential to protect other cucurbitaceous crop plants also (Freeman et al. 2002).

Hairy vetch (*Vicia vilosa*), a legume, can be planted as a winter crop that can reduce soil erosion, decrease nutrient run-off and maintain soil productivity. Hairy vetch as a soil amendment was assessed for its ability to suppress Fusarium wilt disease of watermelon caused by *Fusarium oxysporum* f.sp. *niveum* (FON) under greenhouse, microplot and field conditions. Soil amended with hairy vetch at 0.25 or 0.5 % (w/w) in microplots resulted in 54–69 % decrease in wilt disease incidence and 100–220 % increase in watermelon biomass. Hairy vetch winter cover crop incorporated into field plots under black plastic sheets reduced wilt disease incidence by 42–48 % which was comparable to the level that could be obtained with fumigants methyl bromide or 1,3-dichloropropene + 35 % chloropicrin. Hairy vetch amendment significantly reduced propagules of the pathogen FON and this might be due to increased temperatures under field conditions. Hariy vetch soil amendment significantly reduced propagules of FON and this might be due to increased temperatures under field conditions. Hairy vetch amendment, along with increased temperatures under plastic film provided another benefit of suppressing the growth of some annual grass and broad leaf weeds. Hairy vetch soil amendment has little human and environmental risks and it is compatible with current crop production systems in some parts of the United States (Zhou and Everts 2004).

Another disease management tactic that has provided promising results is the use of organic compounds that may aid in reducing disease incidence/severity in treated plants. The immobile phytohormone 24-epibrassinolide (EBL) with antistress activity was evaluated for its efficacy in reducing the incidence of Fusarium wilt disease of cucumber caused by *Fusarium oxysporum* f.sp. *cucumerinum* (FOC) by treating the roots and foliage of test plants. Pretreatment with EBL to either roots or shoots of cucumber plants significantly reduced the disease severity and also improved the plant growth, irrespective of the method of application. Population of FOC on root surfaces was significantly reduced in treated plants, while population of antagonistic fungi and bacteria on root surfaces increased. In addition, EBL application nullified the effects of the pathogen on the bacterial community in the rhizosphere of treated cucumber plants. The results indicated that EBL treatment of cucumber plants, might improve the plant growth, encourage the multiplication of antagonistic organisms and reduce the incidence of Fusarium wilt disease (Ding et al. 2009). The bioorganic fertilizer containing an organic fertilizer and *Paenibacillus polymyxa* (3×10^7 CFU/g) and *Trichoderma harzianum* (5×10^7 CFU/g) was evaluated for its efficacy in suppressing the development of watermelon Fusarium wilt disease caused by *F. oxysporum* f.sp. *niveum*. The incidence of Fusarium wilt disease was reduced by 84.9 and 75.0 % at 27 and 63 days after treatment with bioorganic fertilizer (0.5 %) under greenhouse conditions (Wu et al. 2009). In the further study conducted under field conditions, best control was obtained by applying the bioorganic fertilizer product B10 into soil during the nursery phase of watermelon seedling, followed by a second application to Fusarium-infested soil at the time of transplantation of watermelon seedling. The incidence of Fusarium wilt disease was reduced by 60–100 % in the greenhouse and by 59–73 % under field conditions. Nursery application of B10 reduced the pathogen population in the soil significantly. The bacterial BCA *Paenibacillus polymyxa* present in the product, effectively colonized the rhizosphere of watermelon and prolifierated along the extending plant roots. The results indicated the potential of B10 to be considered as an alternative for the management of watermelon wilt disease (Ling et al. 2010).

7.1.5.3 Phytophthora Blight Disease

Phytophthora capsici causes the Phytophthora blight disease of squash and it is responsible for appreciable loss in production. The possibility of suppressing the development of Phytophthora blight of squash by applying acibenzolar-*S*-methyl (ASM), an inducer of systemic acquired resistance (SAR), under field conditions was examined. ASM was applied as foliar sprays before and after transplanting at rates of 17.5, 8.8 and 4.4 a.i./ha. Application of ASM did not reduce the final Phytophthora blight incidence to the desired level. However, the values of area under disease progress curve (AUDPC) were reduced significantly by ASM in all the three experiments. The combined application of ASM with mefenoxam (Ridomil Gold), copper hydroxide and mandipropamid (Revus), reduced both AUDPC values and final disease incidence consistently and more effectively than the fungicides alone. ASM treatment enhanced the squash yield, when applied alone or in combination with fungicides. The results suggested that ASM might induce the resistance of squash plants under field conditions and the combination of ASM with fungicides might enhance the consistency of the level of protection offered to treated squash plants against Phytophthora blight diseases (Ji et al. 2011).

7.1.5.4 Anthracnose Disease

Anthracnose disease caused by *Colletotrichum orbiculare* (Teleomorph: *Glomerealla cingulata* var. *orbiculare*) infects leaves and fruits of cucumber and other cucurbits. Plant growth-promoting rhizobacteria (PGPR) strains capable of inducing resistance in cucumber against naturally occurring cucumber diseases were evaluated under field conditions with and without methyl bromide (MB) fumigation in 1996 and 1997. The strains of *Bacillus pumilus* strain INR7, *Bacillus subtilis* GB03 and *Curtobacterium fluccumfaciens* strain ME1 either alone or in combination of strains reduced the severity of anthracnose disease caused by *Colletotrichum orbiculare* and bacterial angular leaf spot caused by *Pseudomonas syrringae* pv. *lachrymans.* The mixtures of PGPR strains provided higher level of protection to the foliar diseases in both years with and without methyl bromide. The results indicated that PGPRs might help compensate for the reduced plant growth attributed to fumigation with MB which was to be phased out of vegetable integrated pest management programs (Raupach and Kloepper 2000). *Serratia marcescens* suppressed the development of anthracnose disease caused by *Colletotrichum orbiculare* in cucumber through induced systemic resistance (ISR). The level of ISR mediated by *S. marcescens* strain 90-166 varied with iron concentrations of the potting mix in which cucumber plants were grown. ISR by the strain 90-166 was affected by external iron concentration, indicating that iron-regulated production of components such as siderophores might be required for the ability of this strain to induce ISR. When the iron concentration of the planting mix was reduced by addition of an iron chelator, suppression of cucumber anthracnose by the strain 90-166 was significantly improved. The mutant 90-166-2882, lacking catechol siderophore production had a transposon insertion in an *entA* homolog. The *entA* mutant 90-166-2882 could not induce resistance in cucumber to *C. orbiculare*, but it retained the ability to produce salicylic acid. Severity of anthracnose on cucumber plants treated with the *entA* mutant was similar to that in control plants, whereas plants treated with the wild type 90-166 showed highly significant reduction in the incidence of disease ($P=0.05$). Total (internal and external) population sizes of 90-166 and the *entA* mutant on roots did not differ significantly ($P=0.05$) at any sampling time, whereas internal population size of the *entA* mutant were significantly lower than those of the wild-type strain at two sampling times (Fig. 7.2). The loss of ISR capacity by the *entA* mutant could not be attributed to lack of rhizosphere colonization. The inability of the mutant to produce the catechol siderophore which was produced by the wild-type strain under iron-limiting conditions in vivo might contribute to induction of ISR in cucumber against anthracnose pathogen (Press et al. 2001). *Bacillus mycoides* isolate BmJ and *B. mojavensis* isolate 203-7 were tested for their ability to induce systemic resistance in treated plants against the anthracnose disease. The isolates BmJ and 203-7 delayed disease onset and reduced total (43 and 56 %) and live spore production (38 and 49 %) per mm^2 of lesion area by inducing systemic acquired resistance (SAR) in cucumber. Field experiments (2004 and 2005) were conducted to evaluate the efficacy of applications of BmJ and fungicides for the control of anthracnose in cucumber cv. General Lee and cantaloupe. BmJ applied 7 days before inoculation with the pathogen reduced disease severity by 41 % in

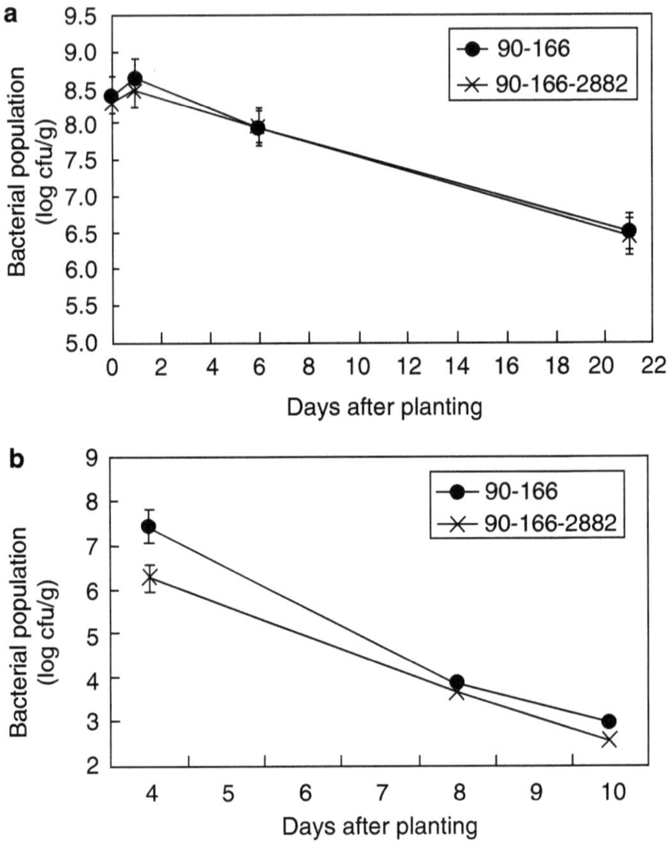

Fig. 7.2 Population sizes of *Serratia marcescens* 90-166 and the *entA*⁻ mutant 90-166-2882 on the roots of cucumber at different days after planting. The mutant was less effective than the parent strain (Courtesy of Press et al. 2001 and with kind permission of The American Phytopathological Society, MN, USA)

cucumber in 2004 and by 21–14 % in cantaloupe for both years, compared with water controls which were in equivalence to fungicides azoxystrobin and chlorothalonil. BmJ applied 1 week prior to inoculation with the pathogen significantly reduced the AUDPC values (P=0.05) in cucumber, compared with control plots (Neher et al. 2009).

Resistance to cucumber anthracnose disease has been induced by different inorganic and organic chemicals. Spraying inexpensive phosphate solution on the foliage induced resistance to anthracnose disease. Hyaluronic acid (HA) isolated from *Streptococcus* sp. strain KL01888 was evaluated for its efficacy in suppressing the development of cucumber anthracnose disease. Application of HA, even at the lowest concentration (0.1 ppm) significantly suppressed development of anthracnose disease. Drenching HA was more effective than injection or spray application. Drenching with 10 ppm of HA provided about 71.9 % disease suppression, compared to control.

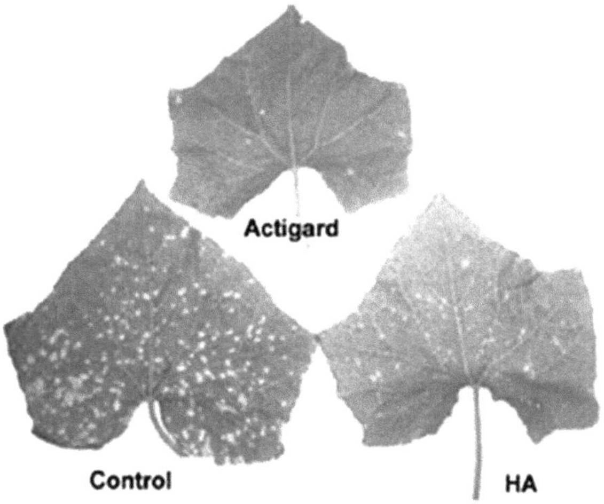

Fig. 7.3 Effect of treatment of cucumber plants with hyaluronic acid (HA) on the development of anthracnose disease in comparison to Actigard (BTH) and untreated control (Courtesy of Park et al. 2008 and with kind permission of Springer Science and Business Media B.V., Heidelberg, Germany)

On the other hand, Actigard (benzothiadiazole, BTH) showed 97.14 % protection over control (Fig. 7.3) (Park et al. 2008). Phosphate-mediated resistance in cucumber was shown to be associated with localized cell death and increases in the activities of defense-related enzymes (Orober et al. 2002). Acibenzolar-*S*-methyl (ASM) was evaluated for its ability to induce systemic resistance in cucumber against the anthracnose pathogen *Colletotrichum orbiculare*. Application of ASM significantly reduced the disease severity. The studies on the mechanism of biocontrol activity of ASM indicated that antifungal compounds capable of inhibiting the development of *C. orbiculare* were formed in ASM-treated plants, resulting in the suppression of anthracnose disease development (Lin et al. 2009).

The possibility of developing a sustainable nonchemical strategy was explored for the management of watermelon anthracnose disease caused by *Colletotrichum orbiculare* and gummy stem blight caused by *Didymella bryoniae* which require multiple applications of fungicides. A split-plot field trial was conducted over 3 years to evaluate the effects of a no-till-hairy vetch (*Vicia villosa*) cover crop on disease severity, plant growth and fruit yield, compared with two conventional bedding systems and fungicide application. Hairy vetch mulch provided more than 65 % reduction in area under disease progress curve (AUDPC) of anthracnose and gummy stem blight diseases and greater than an 88 % decrease in diseased fruits compared with bare ground or polyethylene mulch. Extent of reduction was comparable to that was achieved by application of fungicides. Watermelon vine length in plots with hairy vetch was similar to or greater than those in plots with polyethylene or bare ground that were treated with fungicides. Marketable fruit yield was also increased in plots with hairy vetch and the increase was similar to the level obtained with fungicide application. Combination of fungicide application and hairy vetch provided additive effect on fruit yield. The results demonstrated the feasibility of adopting no-till-hairy vetch as an effective strategy for managing anthracnose and gummy stem blight diseases of watermelon (Zhou and Everts 2012).

7.1.5.5 Powdery Mildew Diseases

Powdery mildew disease of cucumber caused by *Sphaerotheca fuliginea* occurs in severe forms frequently, resulting in withering of shoots and drastic reduction in fruit production. The biocontrol potential of fungal biocontrol agents has been demonstrated. The hyperparasite, *Ampelomyces quisqualis* was able to effectively parasitize *S. fuliginea*, killing the pathogen in about 5 days (Sundheim and Krekling 1982). The conidia of *A. quisqualis* may be spread by wind currents to untreated plants (Elad et al. 1996). *A. quisqualis* was found to be compatible with fungicides and pesticides applied to cucumber crops. The fruit yield of cucumber was increased by 50 %, when *A. quisqualis* and triforine were alternated (Sundheim and Amundsen 1982). A formulated product AQ10 (Ecogen, Jerusalem, Israel) was developed for use against powdery mildews (Elad et al. 1996). *Verticillium* (= *Lecanicillium*) *lecanii*, another fungal biocontrol agent was also reported to effectively suppress the cucumber powdery mildew development. In addition, *V. lecanii* was also pathogenic to the aphid *Macrosiphum euphorbiae* which is an efficient vector of *Cucumber mosaic virus* infecting cucumber and other cucurbitaceous crops. The effectiveness of *V. lecanii* for the control of a fungal pathogen and the aphid vector of an important plant virus has opened up new avenues for the development of integrated disease management systems (Verhaar and Hijwegen 1994; Askary et al. 1998). The biocontrol efficacy of antagonistic microorganisms depends on a combination of factors such as the characteristics of the antagonistic microorganism, the epidemiology of the target pathogen and the environmental conditions in which the relationship between the pathogen and the antagonist(s) is taking place (Paulitz and Bélanger 2001). The biocontrol efficacy of two mycoparasite-based products AQ10® containing *Ampelomyces quisqualis* and Mycotal® containing *Lecanicillium lecanii*, as well as three strains of *Bacillus subtilis* was assessed for the control of the melon powdery mildew disease caused by *Podosphaera fusca*. The mycoparasites were more effective, when the relative humidity values in the greenhouse were 90–95 %. The effectiveness of the mycoparasites *A. quisqualis* and *L. lecanii* was absolutely dependent on mineral oil. The mycoparasites were most effective only in combination with the mineral oil ADDIT, showing a disease reduction of 80–95 %. On the other hand, the strains of *Bacillus subtilis* without any complementary additive were very effective in providing disease reduction similar to mycoparasites with mineral oil or the fungicide azoxystrobin. The results revealed the effectiveness of these BCAs for suppressing the development of powdery mildew disease of greenhouse melons/cucurbits either as single products or as a component of integrated control programs (Romero et al. 2007).

Powdery mildew disease of pumpkin (*Cucurbita pepo*) is incited by three fungal species *Podosphaera xanthii*, *Golovinomyces cucurbitacearum* and *G. orontii*. Three commercially available products Actinovate (*Streptomyces lydicus* WYEC108), Companion (*Bacillus subtilis* GB03) and Sonata ASO (*Bacillus pumilus* QRD2808) were evaluated along with the fungicide quinoxyfen for their efficacy in suppressing the development of powdery mildew in pumpkin under field conditions where the disease pressure was high. Penetration of foliar sprays of the bacterial BCA strains to the lower surfaces appeared to be restricted, resulting in higher incidence and severity of the disease in the lower surface of leaves. Disease incidence was lower in *B. subtilis*-treated foliage than on leaves treated with *S. lydicus* or *B. pumilus*. Disease incidence on the upper surface of leaves treated with three species in rotation, was similar to levels in *B. subtilis*-treated plots. The fungicide quinoxyfen was the most effective in reducing the disease severity and outperformed all three bacterial BCAs tested (Janousek et al. 2009).

Silicon is considered to have significant effect on plant growth and development. Addition of silicon (100 mg/l) to nutrient solution in hydroponic system adopted for cucumber production reduced the number and area of colony of the powdery mildew pathogen, *Sphaerotheca fuliginea* (Bélanger et al. 1995). In a later investigation, the effect of temperature on suppression of powdery mildew by silicon was investigated. The greatest effect of temperature on powdery mildew suppression was observed at 20 °C, when cucumber plants irrigated with silicon at 100 mg/l exhibited significant reduction in the number of powdery mildew colonies per leaf, indicating the crucial role of temperature in disease suppression (Schuerger and Hammer 2003). The effect of foliar application of silicon (Si) on the resistance of cucumber to infection by *Podosphaera xanthii* (syn. *Sphaerotheca fuliginea*) was assessed. Two cucumber cultivars differing in the level of resistance to powdery mildew, Ningfeng No.3 (susceptible) and Jinchun No. 4 (resistant) were treated with Si by foliar and root application. Root-applied Si significantly suppressed powdery mildew development, the disease index being lower in Si-supplied than in control (Si-deprived) plants. The resistant cultivar showed a more consistent lower disease index than the susceptible cultivar, irrespective of Si treatment. On the other hand, foliar-applied Si had no influence on either the suppression of subsequent infection by the powdery mildew pathogen, or on the activity of defense-related enzymes, irrespective of inoculation. The results indicated that foliar application of Si might suppress powdery mildew development through the physical barrier of Si deposited on leaf surfaces, whereas root-applied Si might induce systemic resistance response to infection by the foliar pathogen *P. xanthii* infecting cucumber (Liang et al. 2005).

7.1.5.6 Virus Diseases

The phenomenon of cross-protection between plant viruses and their strains has been exploited for large scale application to protect susceptible plants against infection by severe strains, employing mild strains. Multiple inoculations of cucumber seedlings with attenuated isolates of *Cucumber mosaic virus* (CMV), *Zucchini yellow mosaic virus* (ZYMV) and *Watermelon mosaic virus-2* (WMV-2) reduced yield loss due to mixed infection by virulent strains of these viruses under field conditions, when severe epidemic prevailed in 1994 and 1995. This approach of multiple inoculations of viruses was found to be an effective virus disease management strategy for the summer-early autumn production of cucumber, when economic losses due to the concurrent incidence of CMV, WMV-2 and ZYMV were greater than the loss in yield due to the attenuated strains of these viruses

(Kosaka and Fukunishi 1997).In a further study, an attenuated isolate of ZYMV was obtained by cold treatment. The attenuated strain of ZYMV-2002 was obtained from a virulent isolate, after repeated low temperature treatment at 12.5–15 °C, followed by five cycles of single plant transfer. The cucumber plants protected with ZYMV-2002 had very similar fruit productivity to healthy plant under field conditions. The protected plants exhibited very mild or no symptoms after inoculation with ZYMV-2002 under field conditions. Protected cucumber plants exhibited significant suppression of infection by ZYMV, progression of disease severity and reduction of fruit yield and quality. The results revealed the potential of the attenuated ZYMV-2002 strain for protecting the cucumber plants against severe strain of ZYMV under field conditions (Kosaka et al. 2006).

Reflective, metalized plastic mulch formed by adhering a thin coat of aluminum ions has been demonstrated to reduce the incidence of aphid-borne viruses infecting cucurbitaceous crops (Summers et al. 2004). The effect of integration of UV-reflective plastic mulch and commercially available resistance-inducing compound BioYield™ on the incidence of *Watermelon mosaic virus* (WMV) infecting summer squash (*Cucurbita pepo* var *melopepo*). Incidence of WMV was significantly reduced in plants grown on silver-on-black (UV-reflective mulch), compared with plants grown on black mulch (control) in the spring trial. Absorbance values obtained with enzyme-linked immunosorbent assay (ELISA) tests, reflecting the concentration of WMV in treated plants, were less than those in control. In the fall trial, highly UV-reflective silver mulch was also included for testing. WMV incidence and ELISA values were significantly lower for squash plants in the silver mulch treatment, compared with silver-on-black and black mulch treatments. However, yields of squash plants were significantly greater for plants in the silver-on-black mulch treatment than for those in the silver or black mulch treatments. Treatments with BioYield™ neither reduced WMV incidence, nor increased the squash fruit yields in both spring and fall trials (Murphy et al. 2008).

7.1.6 Diseases of Allium Crops

7.1.6.1 White Rot Disease

Allium white rot disease caused by *Sclerotium cepivorum* is a major disease of Allium crops in several countries around the world. White rot disease infects garlic causing considerable loss in production. Soil solarization was highly effective in suppressing the development of garlic white rot disease by bringing down the soil populations of *S. cepivorum* to negligible levels. Soil solarization provided the beneficial effect for the second consecutive crops of garlic with significant improvement in yield. Spraying tebuconazole to the stem bases of garlic plants was also effective as solarization. When soil solarization and application of tebuconazole were combined, the yield of garlic bulbs was almost doubled, compared with untreated control plots. The bulb quality was also enhanced by the combined treatment, even under high disease pressure conditions. In contrast, selected isolates of *Trichoderma harzianum* and *Bacillus subtilis* were used for treating the cloves and the soil, the levels of disease control provided by them were significantly lower than that could be obtained by using soil solarization and the fungicide (Melero-Vara et al. 2000).

Application of different kinds of organic amendments to the soil is one of the disease management strategies followed for soilborne diseases. The efficacy of mixtures of wet vegetables (*Brassica*, carrot or onion) and dry onion wastes composted at 50 °C for 7 days was assessed for the suppression of white rot disease. The viability of sclerotia was reduced by incorporation of raw or composted vegetable mixtures into sandy loam, silt and peat soils in the glasshouse assays. Onion waste was the most effective in reducing the viability of sclerotia in all the three types of soils tested. The efficiency of onion waste in reducing the sclerotial viability was less under field conditions, compared with that under glasshouse conditions. The most consistent control by onion waste was observed in the peat soil, indicating the effect of soil type on the effectiveness of the organic amendments against onion white rot disease (Coventry et al. 2005). As the efficiency of onion waste compost (OWC) was variable, the fungal BCA *Trichoderma viride* S17A was used as an OWC amendment. The effects of OWC and spent mushroom waste (SMC) with and without *T. viride* on sclerotial viability of *S. cepivorum* were studied. Incorporation of OWC into the soil reduced the viability of sclerotia and the incidence of Allium white rot disease on onion plants in the glasshouse assays. On the other hand, SMC or *T. viride* could reduce the incidence of the disease only. In the field experiments, OWC reduced sclerotial viability and also decreased disease incidence as effectively as the fungicide. Addition of *T. viride* to SMC facilitated proliferation of*T. viride* in the soil and increased healthy onion bulb yield. The disease suppressive activity of OWC was attributed to the presence of sulfur compounds in the compost (Coventry et al. 2006).

Trichoderma viride isolates L4 and S17A were selected based on their high level of efficacy in degrading the sclerotia of *Sclerotium cepivorum*, the causative agent of Allium white rot disease. Field experiments were conducted in 2000 and 2001 for testing these two isolates for their potential to suppress disease development. The isolates L4 and S17A consistently reduced the white rot symptoms in both years. These two isolates fluid-drilled in guar gum with bulb onion seed reduced the white rot symptoms more effectively, while stem base treatment, applied at mid-season was ineffective in reducing the disease (Clarkson et al. 2002). In a later investigation, clear relationships between the efficacy of *T. viride* isolates L4 and S17A to degrade *S. cepivorum* sclerotia and soil water potential or temperature were observed. Degradation of *S. cepivorum* sclerotia by both *T. viride* isolates increased with temperatures from 5 to 25 °C and soil water potential of −0.022 MPa, with most degradation occurring at 10 °C. The results indicated that *T. viride* L4 and S17A could effectively degrade sclerotia of *S. cepivorum*, but their efficacy and performance for biological control is modified by pathogen isolate, soil type, temperature and soil water potential (Clarkson et al. 2004). Commercially available formulations of the vesicular, arbuscular mycorrhiza (VAM) *Glomus intraradices* were evaluated for the efficacy in controlling Allium white rot (AWR) disease caused by *Sclerotium cepivorum* on onions (*Allium cepa*) in organic soils in comparison with the fungicide Folicur 3.6F (430 g a.i./l, tebuconazole) under field conditions during 2000 and 2001. The VAM product MIKRO-VAM applied in transplanted onions, reduced the disease incidence by about 50 %, compared with the untreated control. The effectiveness of VAM product was comparable with that of the fungicide applied at label recommendations. Consistent difference in the incidence of white rot was observed between cultivars Hoopla and Fortress. The cv. Hoopla was found to be more susceptible than cv. Fortress in 10 of 13 field trials. Significant negative correlation between disease incidence and VAM colonization was observed, indicating the importance of root colonization by the VAM for effective suppression of white root rot disease development. This investigation has demonstrated, under field conditions, the potential of the VAM as a biocontrol agent against a soilborne fungal pathogen (Jaime et al. 2008).

An integrated disease management strategy was developed for suppressing the Allium white rot (AWR) caused by *Sclerotium cepivorum*. Two isolates of *Trichoderma viride* L4 and S17A were evaluated for their efficacy with different onion accessions and cultivars alone or in combination with tebuconazole-based seed treatment or composted onion waste. All the 23 bulb onion accessions did not show variation in their susceptibility to white rot disease. But when combined with S17A isolate, disease incidence was reduced by up to two-thirds over all accessions. L4 and S17A isolates, tecbuconazole or composted onion waste controlled AWR and at least halved the proportion of diseased plants. Combination of treatments of *T. viride* with either tebuconazole or compost enhanced disease control and in some treatments, disease incidence was almost eliminated. Control of AWR by the isolate S17A was significant in 17 out of 18 field trials using individual or mixed bulb-onion accessions, with disease reduced overall by more than half. Combining the isolate S17A and tebuconazole resulted in same level of control of AWR as the fungicide alone. The results indicated that use of *T. viride*, tebuconazole seed treatment and composted onion waste as soil amendment had potential for suppressing the development of AWR. Integrating these approaches might provide protection to bulb onion crops for the entire season (Clarkson et al. 2006). Compatibility of *Trichoderma atroviride* Karsten strain C52 with organic amendments, nitrogen fertilizer, fungicides and diallyl disulphide (DADS – a stimulant of germination of sclerotia of *Sclerotium cepivorum*, causative agent of white rot disease) was examined. Addition of two blended pellet products containing poultry manure and other organic matter to the sandy soil enabled *T. atroviride* populations to proliferate well. Urea at twice field rate application reduced the mycelial growth of *T. atroviride*, as well as the spore germination and elongation of germ tube. However, *T. atroviride* was less sensitive to field rate of urea in the field soil. *T. atroviride* C52 populations were not adversely affected by fungicides tested. Volatiles of DADS reduced *T. atroviride* C52 mycelial growth in vitro at high rates, but not at half field rate. However, when DADS was applied to soil at 4, 6 and 8 weeks, before application of *T. atroviride* C52, populations were not affected. An integrated white rot disease management strategy could be proposed based on the results obtained in this investigation (McLean et al. 2012).

7.1.6.2 Leaf Blight Disease

Onion leaf blight disease caused by *Botrytis squamosa* is a widespread disease found in most onion growing countries. Conidia formed from overwintering sclerotia form the principal source of inoculum for newly planted onion crops, whereas the conidia from lesions formed on onion leaves contribute to the secondary inoculum buildup. The biocontrol potential of *Microsphaeropsis ochracea* was investigated for suppressing the pathogen development from sclerotia-borne inoculum, by colonizing onion leaves and reducing production of conidia under field conditions. Colonization of *M. ochracea* of onion leaves was monitored at different growth stages of onion crop and its effect on *B. squamosa* sporulation on necrotic leaves was evaluated. Field plots for different treatments were applied with either Dithane® or with *M. ochraceae* at 7–10 days interval and according to inoculum production index (IPI). The number of conidia produced per sclerotium treated with *M. ochracea*, was reduced by 75.5 %. In the field, *M. ochracea* colonized only senescent or necrotic leaves and reduced the production of conidia on the infected leaves by an average of 82 %, as compared with untreated leaves. The fungicide Dithane® was more effective in reducing the leaf blight disease. During the period of 3 years of investigation, no significant differences in the concentrations of airborne conidia in plots treated with Dithane® or *M. ochracea* at 7–10 day-interval were observed. Application of *M. ochracea* in the fall could be used as a sanitation practice to reduce initial inoculum or as a part of an integrated disease management program for onion crops (Carisse et al. 2006).

7.1.6.3 Black Mold Disease

Black mold disease of onion incited by *Aspergillus niger* is responsible for both quantitative and qualitative yield losses of onion bulbs. Application of composts prepared from alfalfa and sunflower stalks reduced the incidence of black mold disease. The leachates from alfalfa and sunflower stalks were used at 10 and 20 % concentrations for treating onion seeds. The leachates reduced the disease incidence in sets, but not disease severity in onion seedlings. No significant difference in the antifungal activity of the leachates against *A. niger* was observed. The presence of fluorescent pseudomonads and *Pantoea agglomerans* was detected in the leachates of alfalfa and sunflower composts. The sunflower compost leachate had higher populations of *P. agglomerans*, compared with that of alfalfa leachate. It is possible that the bacterial species with antagonistic activity might have a role in the suppression of development of black mold pathogen by producing antifungal compounds effective against this pathogen (Özer and Köycii 2006).

7.1.6.4 Basal Rot Disease

The soilborne pathogen, *Fusarium oxysporum* f.sp. *cepae* (FOC) causes the basal rot disease of onions. The biocontrol of *Trichoderma harzianum* (*Th*) to suppress the development of the disease was assessed under greenhouse and field conditions.

Seeds were treated with *T. harzianum* at 10 g/kg and sown in naturally infested soil. Seed treatment with *Th* reduced the disease incidence to a level comparable with that of the imidazole fungicide prochloraz under both greenhouse and field conditions. Extracts from onion sets grown from the *Th*-treated seed showed high antifungal activity against FOC, indicating the mechanism of biocontrol activity of FOC was through antibiosis (Coşkuntuna and Özer 2008).

7.1.6.5 Xanthomonas Leaf Blight Disease

Xanthomonas axonopodis pv. *allii* (*Xaa*) causes the bacterial leaf blight disease which limits the onion yields considerably in the western United States. Use of disease-free seeds, adopting crop rotation including nonhosts of *Xaa* and growing resistant onion cultivars have been suggested for minimizing the disease incidence. Copper bactericides amended with an ethylenebisdithiocarbamate (EBDC) applied preventively and regularly increase the production cost. In addition, these fungicides are considered as class B2 carcinogens and their registrations are likely to be cancelled in the future (Houeto et al. 1995). The potential of acibenzolar-*S*-methyl (ASM) and the bacterial biocontrol agents (BCAs) and the bacterial biocontrol agents (BCAs) *Pantoea agglomerans* strain C9-1 and *Pseudomonas fluorescens* A506 was assessed for the control of the onion Xanthomonas leaf blight disease. Application of ASM reduced in planta and epiphytic populations of *Xaa* as effectively as copper hydroxide-mancozeb in growth chamber assays. Under field conditions, four weekly applications of ASM reduced the severity of Xanthomonas leaf blight disease as or more effectively than 9–12 weekly sprays of copper hydroxide or copper hydroxide-mancozeb. When ASM was applied ten times at weekly intervals, yield reduction of 22–27 % was recorded in the absence of disease. The commercial formulation of the mixture of strain C9-1 and strain A506 reduced the disease severity, under field conditions, to the level comparable with that of copper hydroxide-mancozeb. This investigation indicated an effective and economical means of reducing the use of copper bactericides and EBDC fungicides for suppressing the Xanthomonas leaf blight disease of onion by combining ASM and bacterial biocontrol agents (Gent and Schwartz 2005).

7.1.7 *Sugar Beet Diseases*

7.1.7.1 Rhizoctonia Crown and Root Rot Disease

Rhizoctonia solani AG2-2 causes the crown and root rot disease which is one of the most damaging diseases of sugar beet. In addition, this pathogen may also cause post-emergence damping-off of sugar beet seedlings. *Pseudomonas fluorescens* strain 96.578 applied as seed treatment significantly reduced the incidence of crown and root rot disease. The biocontrol activity of the strain was associated with its ability to produce a cyclic lipopeptide tensin (Nielssen et al. 2000). Three yeast species *Candida valida*, *Rhodotorula glutinis* and *Trichosporon asahii*, when applied individually or in combination significantly reduced damping-off, crown and root rot diseases of sugar beet. In addition, they promoted the plant growth also (El-Tarabily 2004). The efficacy of antagonistic *Bacillus* sp. strain MSU-127 and fungicides in suppressing the development of crown and root rot disease of sugar beet was assessed under field conditions. The fungicides azoxystrobin and tebuconazole were applied as in-furrow sprays at planting or as band sprays directed at the crown at the 4-leaf stage, 4- plus 8-leaf stage, whereas the bacterial BCA was applied at the 4-leaf stage only. The strain MSU-127 provided long-term protection to sugar beet plants equal to the low rate of azoxystrobin (76 g a.i./ha) or tebuconazole in both years (1996 and 1997). The efficacy was increased by the combined application of MSU-127 with low rate of azoxystrobin. The combination of BCA and fungicide resulted in best disease reduction and greatest increase in root and sugar yield (15.9 %). The sugar yield was further increased to 23.0 %, when the BCA-fungicide combination was applied twice during the growing season at the 4- and 8-leaf stages (Kiewnick et al. 2001). The effectiveness of two *Streptomyces* sp. was evaluated under field conditions for 3 years (2005–2007). Seed treatments with BCA reduced the seedling mortality by Rhizoctonia damping-off in naturally and artificially infested soils. The BCAs were more effective than the fungicide Vitavax and there was no difference in the efficacy of the two *Streptomyces* spp. and the root yield of beet was significantly increased (Sadeghi et al. 2009).

7.1.7.2 Damping-Off Disease

Sugar beet damping-off disease caused by *Pythium ultimum* is responsible for high rate of mortality of seedlings. Bacterial strains *Bacillus subtilis, Pseudomonas fluorescens* and *P. corrugata* and fungal strains *Trichoderma viride* and *Gliocladium (Trichoderma) virens* were evaluated for their efficacy in suppressing the development of damping-off disease due to *P. ultimum*. Bacterial seed treatments and fungal compost treatments were effective in controlling the disease. *Pseudomonas* spp. were superior to *Bacillus subtilis* in reducing disease incidence (Georgakopoulos et al. 2002).

7.1.8 Bean Diseases

7.1.8.1 White Mold Disease

White mold disease of bean caused by *Sclerotinia sclerotiorum* is a major disease of several vegetable crops, including bean. The disease management strategies applied for Sclerotinia diseases such as application of organic amendments and soil solarization may be effective. The effectiveness of *Ulocladium atrum* in suppressing the development of white mold disease was assessed under field conditions for 3 years. Foliar application of conidial suspensions of *U. atrum* (300 ml/m^2 of 10^6 spores/ml) significantly reduced incidence and severity of white mold disease. In addition, application of *U. atrum* conidial suspension enhanced seed yield and reduced contamination of bean seed with the sclerotia of *S. sclerotiorum*. The BCA *U. atrum* and the mycoparasite *Coniothyrium minitans* were equally efficient in suppressing the development of white mold disease of bean. Application of fungicide Ronilan (vinclozolin) at the rate of 1,200 g/ha or Lance (boscalid) at the rate of 750 g/ha were more efficient in protecting the bean plants against white mold disease than both fungal BCAs (Huang and Erickson 2007).

7.1.8.2 Root Rot Disease

Rhizoctonia solani causes root rot disease in many leguminous crops. Application of different organic amendments, adoption of crop rotation and soil solarization have been followed for minimizing the incidence of root rot disease. The biocontrol potential of fungal and bacterial bioagent has been assessed. The long-term activity of some antagonistic fungal and bacterial agents against the faba bean root rot pathogen was evaluated, when applied as bio-priming seed treatment. The antagonistic activity of *Trichoderma viride, T. harzianum, Bacillus subtilis* and *Pseudomonas fluorescens*, was enhanced significantly under greenhouse conditions, when applied as biopriming seed treatment. All the tested and 2-months-stored bioprimed faba bean seeds showed significant reduction in disease incidence at both pre- and post-emergence stages, compared with untreated control treatment. Seeds stored for 4 and 6 months showed less protective effect against the disease. After storage for 3 months under field conditions, the biopriming treatment could protect seeds against infection by *R. solani*. Priming the seeds only with the adhesive had no influence on disease incidence. The results indicated primed seeds might be considered as a safe, cheap and easily applied biocontrol tactic, providing effective protection to plants against soilborne pathogens like *R. solani* (El-Mougy and Abdel-Kader 2008).

7.1.9 Lettuce Diseases

7.1.9.1 Fusarium Wilt Disease

Fusarium oxysporum f.sp. *lactucae* (FOL) causes wilt disease of lettuce, causing heavy losses in several countries. The effects of soil solarization on the incidence of Fusarium wilt disease and growth of lettuce plants were determined. In four trials under field conditions where soils were infested with FOL, soil solarization reduced Fusarium wilt disease incidence to 42 %, compared with the control plots with 91 % infection. No significant extra benefit was obtained due to a 1-month over a 2-month solarization period under natural field conditions, where the mean soil temperature at a depth of 5 cm during 1-month solarization period was 47 and 49 °C respectively in 2005 and 2006. The growth of lettuce plants in solarized plots was consistently greater than the plants in nonsolarized plots. Retention of soil moisture by the plastic film during solarization process was considered as an important factor contributing to the reduction in the pathogen population, since microorganisms are much more resistant to heat under dry conditions. Soil solarization offered extended benefit to the subsequent lettuce crops. Solarization significantly reduced the number of plants affected by Fusarium wilt disease in a subsequent planting of susceptible lettuce cultivar, indicating the long-term benefit of solarization (Matheron and Porchas 2010). The effectiveness of flooding in suppressing the development of Fuasrium wilt disease was assessed. Flooding treatment was applied by completely saturating the soil with water and maintaining a 2-cm layer of water on the soil surface by adding water daily to replace water that was lost due to evaporation in experiments conducted in microplots. Growth of lettuce in flooded soil containing the pathogen occasionally was significantly higher than the plants in non-flooded soil. However, the effect on plant growth was not as consistent as that recorded for soil solarization with which flooding treatment was compared (Matheron and Porchas 2010).

The biocontrol potential of nonpathogenic isolates of *Fusarium oxysporum* was assessed for the control of lettuce Fusarium wilt disease caused by *F. oxysporum* f.sp. *lactucae*, along with commercially available bioproducts containing *Trichoderma harzianum* T-22 (RootShield) and *T. viride* (Remedier) under field conditions. With high disease pressure, the best results in terms of disease control as well as increased growth promotion were provided by *T. harzianum* T-22 applied at 3 g/l of substrate. *F. oxysporum* IF23 gave effective disease control in two of the five trials. Another nonpathogenic strain *F. oxysporum* MSA25 at 3 g/l of substrate and *T. viride* TV1 provided significant, but less consistent disease control. The bioproduct Mycostop containing *Streptomyces griseoviridis* was less effective in reducing the disease incidence, compared with the other BCAs tested in this investigation which indicated the feasibility of using the biocontrol agents for minimizing the incidence of Fusarium wilt disease of lettuce (Gilardi et al. 2007).

7.1.9.2 Lettuce Drop Disease

Lettuce drop disease is caused by two species of *Sclerotinia*, *S. sclerotiorum* and *S. minor*. Different disease management strategies have been evaluated for their efficacy in suppressing the development of lettuce drop disease. The effects of cultural practices in the incidence of lettuce drop disease have been studied. Deep ploughing reduced the population of *S. minor* sclerotia in soil, but the increased post-deep ploughing operations to prepare land redistributed the inoculum from an aggregated pattern to less aggregated patterns. The tillage operations, thus, transferred sclerotia to areas within the field that were devoid of them. Hence, the incidence of lettuce drop disease increased significantly, after deep ploughing (Subbarao et al. 1996). The investigation to assess the effect of irrigation types on the incidence of lettuce drop showed that surface drip irrigation reduced both the number of soilborne sclerotia and incidence of the disease (Subbarao et al. 1997). As the initial investment costs are quite high and the subsurface irrigation system requires intensive management, practical application for wider use has been very much limited. The fungal pathogens causing lettuce drop disease have wide host range that includes several crop plant species affected by stem rot disease. The possibility of using crop rotation as a disease management strategy was examined. The efficacy of broccoli as a rotation crop was assessed for reducing the number and viability of sclerotia of *Sclerotinia minor* and the incidence of lettuce drop disease in the field. Continuous lettuce did not increase lettuce drop incidence for at least 2 years, although an increase in soilborne sclerotia was observed annually. Rotation with broccoli resulted in small reductions in disease incidence only in the first year. In another location, rotation with broccoli significantly reduced both sclerotia of the pathogen and lettuce drop incidence. The number of broccoli crops, rather than the sequence of lettuce rotations with broccoli was critical for reducing the number of sclerotia in the soil. Fallowing after lettuce crop resulted in a marginal reduction in sclerotial number and disease incidence. Incorporation of broccoli residue reduced the number of sclerotia, possibly because of volatiles formed during the breakdown of glucosinolates present in broccoli. The results indicated that broccoli, as a nonhost of *S. minor* and glucosinolates present in its tissues could be a desirable candidate for use as a component of crop rotation for the control of lettuce drop disease (Hao et al. 2003). The efficacy of soil solarization in suppressing the development of two soilborne diseases of lettuce, lettuce drop disease caused by *S. minor* and lettuce bottom rot incited by *Rhizoctonia solani* AG1-1 was assessed. Soil solarization was applied for 60 days during the summer in lettuce fields. Incidence of both diseases was very low in the first year after solarization and the crop cycle was shortened by 10 days, as compared to nonsolarized plots. In the nonsolarized plots, a loss of 25–40 % of lettuce heads was recorded. Soil solarization, in addition to providing protection against lettuce drop and bottom rot diseases, offered additional advantages such as preservation of the population of fluorescent pseudomonads and increased nutrient availability to lettuce plants (Patricio et al. 2006).

Use of *Coniothyrium minitans*, the sclerotial mycoparasite of *Sclerotinia sclerotiorum*, has been demonstrated to be an effective approach for the control of lettuce drop disease. The efficiency of *C. minitans* decreased, when disease pressure increased under field conditions. Compatibility of *C. minitans* with fungicides and other chemicals, one of the important attributes that decide the suitability of the BCA for further advancement in the process of development and commercialization was studied. An iprodione tolerant *C. minitans* isolate was selected. Disease caused by *S. sclerotiorum* was significantly reduced by *C. minitans* and its effectiveness was enhanced by a single application of iprodione. The combination of *C. minitans* and a single application of iprodione was found to be as effective as the prophylactic sprays with iprodione every 2 weeks. The results showed that the integrated control of *S. sclerotiorum* with soil applications of *C. minitans* and reduced foliar applications of iprodione was feasible (Budge and Whipps 2001). *C. minitans* significantly increased the percentage of sclerotia infected and reduced the percentage viability of sclerotia, when applied to lettuce plant showing the earliest symptoms of lettuce drop caused by *S. minor* in commercial crops. The viability of untreated sclerotia of *S. minor* in the plant debris on the surface of soil declined greatly, providing evidence that prolonging the exposure of pathogen sclerotia on the soil surface before cultivation for the next crop may substantially reduce sclerotial inoculum. Infected mature plants have to be cut at an early stage of symptom development and removed from the field and disposed off properly (Isnaini and Keane 2007).

7.1.9.3 Bottom Rot Disease

Rhizoctonia solani, incitant of lettuce bottom rot disease has a wide host range, including many crop plants which suffer from root rot diseases induced by this pathogen. The biocontrol potential of three bacterial strains of *Pseudomonas fluorescens* B1, *P. fluorescens* B2 and *Serratia plymuthica* B4 was assessed for suppressing the development of *R. solani* under growth chamber and field conditions. All three BCAs entirely or significantly limited the dry mass (DM) losses on lettuce and disease severity (DS) caused by *R. solani* on potato sprouts (assay host). Under field conditions, the disease severity on lettuce and potato plants which were bacterized, decreased significantly and the biomass losses on lettuce decreased as well. The strain B1 provided best protection and disease suppressive effect among the three bacterial strains tested under high disease pressure conditions. As *P. fluorescens* B1 proved to be the most effective against *R. solani*, this strain has the potential for further development and commercialization (Grosch et al. 2005).

7.1.10 Cabbage Diseases

7.1.10.1 Cabbage Yellows Disease

Yellows disease of cabbage is associated with infection by *Verticillium longisporum* and *Fusarium oxysporum* f.sp. *conglutinans* (FOC). The fungal BCA *Heteroconium chaetospira* could colonize the roots of cabbage plants and suppress the disease under field conditions. *H. chaetospira*-treated plants displayed slight yellowing on leaves and reduced the disease by 60 % under high disease pressure conditions (Narisawa et al. 2000). In order to enhance the effectiveness of biocontrol, fungal endophytes were screened and two isolates of *Phialocepahalia fortinii* and a dark, septate endophyte (DSE) effective against *V. longisporum* were identified. Chinese cabbage (*Brassica campestris*) preinoculated with *P. fortinii* DE isolate were transplanted into the field infested with the pathogen. *H. chaetospira* was also included in the trial for comparison. The DSE isolate was more effective than *H. chaetospira* and suppressed the development of Verticillium yellows disease, with reductions in disease percentages of external and internal symptoms of 84 and 88 % respectively. The BCA-treated plants produced heads with better marketable quality (Narisawa et al. 2004). The development of FOC causing cabbage yellows disease was effectively suppressed by *Pseudomonas fluorescens* LRB3W1 under glasshouse conditions and the bacterial BCA survived in the roots and rhizosphere for about 4 weeks, after initial application. Under field conditions, combined application of the strain LRB3W1 and a low dosage of benomyl was more effective than the bacterial BCA alone. The survival of LRB3W1 was not affected by the fungicide. The combination of the BCA and fungicide appeared to be a desirable option for effective management of cabbage yellows disease caused by *Fusarium oxysporum* f.sp. *conglutinans* (Someya et al. 2007).

7.1.10.2 Damping-Off Disease

Damping-off disease of Chinese cabbage is caused by *Rhizoctonia solani* which infects many vegetable crops, inciting pre- and post-emergence damping-off diseases. Addition of organic amendments and soil solarization may be effective in reducing pathogen populations and consequent reduction in disease incidence. A granular formulation, PBGG was prepared by combining *Pseudomonas boreopolis*, *Brassica* seed pomace, glycerin and sodium alginate. Application of PBGG (1 %) to the soil infested with *R. solani* AG-4 significantly reduced the percentage of colonization of cabbage seeds by the pathogen and also stimulated the development of *Streptomyces padanus* strain SS-07 and *S. xantholiticus* strain SS-09 which exhibited high antagonistic activity against *R. solani.* Soil treatment with *S. padanus* or *S. xantholiticus* alone or in combination with 1 % PBGG significantly reduced the percentage of colonization of cabbage seeds by *R. solani*, compared with untreated control. *S. padanus* was the most effective in suppressing the development of damping-off disease. Under field conditions, soil amendments with PBGG (1 %) alone or in combination with *S. padanus* or *S. xantholiticus* effectively reduced the incidence of damping-off disease caused by *R. solani* and no adverse effect due to treatment with PBGG or actinomycetes was observed. The results of the investigation indicated the effectiveness of the combination of PBGG and actinomycetes applied before sowing the seeds for suppressing the development of damping-off disease of Chinese cabbage (Chung et al. 2005).

7.1.11 Radish Diseases

7.1.11.1 Fusarium Wilt Disease

Fusarium oxysporum f.sp. *raphani* (FOR) causes the widespread Fusarium wilt disease of radish, being responsible for considerable losses. Use of different antagonistic microorganisms has been shown to be a promising alternative approach for the control of the disease. The additive or inhibitory effects of biocontrol agents, when applied in combination have been investigated. *Pseudomonas putida* strain RE8 inhibited the growth of *P. fluorescens* strain RS111 in in vitro assays. The combination of these two strains did not improve the effectiveness of control of Fusarium wilt of radish. On the other hand, the combination of RE8 with RS111-a, a spontaneous mutant of RS111 did not have any inhibitory effect and the combination enhanced the suppression of pathogen development (De Boer et al. 1999). In a later investigation, the effect of combining two BCA strains *P. putida* WCS358 and RE8 on the development of Fusarium wilt disease of radish. The strain WCS358 suppressed the wilt pathogen FOR by competing with it for iron through the production of the siderophore pseudobactin. The strain RE8 induced systemic resistance against Fusarium wilt disease. When both strains WCS358 and RE8 were mixed through the soil together, disease suppression was enhanced to approximately 50 % as compared to the 30 % reduction for the single strain treatment. Moreover, when one strain failed to suppress the disease development in the single strain application, the combination still provided higher level of protection to radish plants. The results indicated that consistency of biocontrol performance could be ensured by combining biocontrol strains that may have different disease suppressive mechanisms, leading to enhancement of effectiveness of biocontrol of crop diseases (De Boer et al. 2003).

7.1.11.2 Damping-Off Disease

Radish damping-off disease, due to infection by *Pythium aphanidermatum* and *Rhizoctonia solani*, causes high mortality of seedling under favorable soil moisture conditions. Different kinds of organic amendments have been evaluated for their efficacy in suppressing the development of damping-off disease. The biocontrol potential of fish emulsion in suppressing the damping-off disease of radish was assessed. Fish emulsion (1–4 %) or equivalent inorganic fertilizer (N-P-K) was incorporated in pathogen-infested peat mix. After 7 days, seeds were planted. In the peat mix amended with 4 % fish emulsion, 70–80 % of the seedlings remained disease-free. The inorganic fertilizer treatment did not have any effect on disease incidence. Hence, it was considered that disease suppression was not due to increased plant nutrition supplied by fish emulsion. Fish emulsion incorporarted into the soil at 5 days before planting protected the radish seedlings effectively against damping-off disease. Fish emulsion (2 and 4 %) applied on naturally infested muck soil effectively and consistently suppressed damping-off disease of cucumber seedlings. In addition, the plant growth was promoted in soil treated with fish emulsion. Fish emulsion might prove to be a desirable organic amendment for use in organic or conventional transplant production (Abbasi et al. 2004).

7.1.12 Asparagus Crown and Root Rot Disease

Infection by *Fusarium oxysporum* f.sp. *asparagi* and *F. proliferatum* causes the Fusarium crown and root rot disease of asparagus which is economically important in many countries, including the United States. As in the case of other soilborne diseases of various crops, different disease management tactics such as addition of organic amendments, use of mulches and establishing general soil anaerobic conditions have been evaluated for their efficacy in minimizing disease incidence. However, asparagus is a deep-rooted perennial, making application of chemical or biological control methods difficult, once the crop is established. Infection of seedlings in the nursery predisposes the crop to severe decline, making the seedling stage a critical time to apply management strategies for Fusarium crown and root rot disease. Under field conditions, aerobic conditions develop, when oxygen consumption mostly by the soil microflora, exceeds resupply of oxygen by diffusion from the atmosphere. A general anerobic condition was established by increasing microbial respiration through incorporation of readily decomposable organic amendments into moist soil and by reducing the resupply of oxygen by covering the soil with ensilage plastic with low oxygen-permeability characteristics. Field plots were amended with fresh broccoli or grass (3.4–4.0 kg fresh weight/m^2) and covered with plastic sheeting. In these plots, aerobic and strongly reducing conditions developed rapidly, as indicated by rapid depletion of oxygen and a decrease in redox potential values to as low as −200 mV. After 15 weeks, the survival of *F. oxysporum* f.sp. *asparagi* and other soilborne pathogens *Rhiozoctonia solani* and *Verticillium dahliae* in inoculum samples buried 15 cm was strongly reduced in amended, covered plots in two field experiments. No adverse effects on the fungal pathogens were observed in amended, non-covered or non-amended, covered plots. The results indicated the potential of this approach to become an alternative to chemical disinfestations for high value crops under conditions where other alternatives such as soil solarization or soil flooding are not effective or feasible (Blok et al. 2000).

The ability of fungal biocontrol agent *Trichoderma harzianum* strain 2-2 and nonpathogenic isolates of *Fusarium oxysporum* and fungicides to suppress the development of Fusarium crown and root rot disease of asparagus caused by *F. oxysporum* f.sp. *asparagi* (FOA) was assessed. *T. harzianum* T-22, benomyl and fludioxonil treatments increased the root weight and decreased the disease incidence in asparagus seedlings, compared with the infested control, when a low level of FOA and *F. proliferatum* (FP) was used. Fludioxonil treatment limited plant mortality caused by FOA and FP at high inoculum levels, whereas strain T-22 was not effective (Reid et al. 2002). Nonpathogenic strains of *Fusarium oxysporum* (NPFO) alone and in combination with sodium chloride were evaluated for their efficacy in suppressing the development of crown root rot disease of asparagus. Crowns of asparagus were treated with NPFO at planting and NaCl was broadcast at 280 and 560 kg/ha in later years. The NPFO CWB318 reduced disease incidence at two locations tested and increased yield by 29 % in one location over controls, but did not increase the yield in another location. Application of NaCl did not affect the biocontrol activity of NPFO strains (Elmer 2004).

7.1.13 *Postharvest Diseases of Vegetables*

Vegetables may be infected in the field prior to harvest and during harvest, handling, transit and storage, by various kinds of microbial pathogens. It is necessary to ensure that close coordination, constant surveillance and efficient technical support for rapid detection and precise identification of microbial pathogen(s) associated with the disease(s) concerned and feedback on the effectiveness of corrective measures applied to restrict the incidence and subsequent spread of diseases gathered and made available for development of effective systems of disease management. Various disease management strategies applied at preharvest stage may lead to reduction in pathogen inoculum, such as use of disease-free seeds and planting materials, modification of cultural practices and application of organic amendments and mulches, soil solarization, eradication of alternative and alternate hosts and proper disposal of infected plant debris have to be applied. These ecofriendly methods will not pollute the environment and also, there will not be any accumulation of toxic materials in the vegetables, as in the case of treatments involving use of chemicals. Furthermore, the chances of causing wounds to the vegetables during harvest, transit and storage should be considered and necessary precaution should be taken to avoid wounds which form the primary entry points for the pathogens. Various physical, chemical and biological methods have to be integrated for effectively minimizing or avoiding incidence and severity of postharvest diseases.

7.1.13.1 Tomato Diseases

Physical, chemical and biological agents have been evaluated for their efficacy in suppressing the development of postharvest pathogens infecting tomatoes. Application of ultra-violet (UV)-C (254 nm) hormesis in tomato fruits stimulated beneficial responses resulting in reduction of incidence of soft rot disease caused by *Rhizopus stolonifer* (Stevens et al. 1998). Mature green tomato fruits irradiated with UV-C light (24–36 kJm^2) showed delayed ethylene production and reduced respiration rate. Infection by *Alternaria alternata* was reduced, in addition to retarded ripening and color development in tomato fruits (Rong and Feng 2001). Treatment of tomato fruit with UV-C irradiation suppressed the development of gray mold disease caused by *Botrytis cinerea* by enhancing the level of resistance to diseases. Induction and maintenance of resistance to gray mold rot disease in tomato fruit during storage by hormic dose of UV-C (3–7 kJ/m^2) was investigated. Treated fruits became more susceptible to disease soon after treatment with UV-C, but thereafter, they became gradually resistant and the resistance was maintained until the end of the storage period of 35 days. There was a significant correlation between the phytoalexin, rishitin accumulation in UV-treated fruit both before and after inoculation and disease resistance (Charles et al. 2008). A short pre-storage hot water rinsing and brushing (HWRB) method was shown to be effective for reducing infection by *Botrytis cinerea* in tomatoes. Fruit of the pink tomato cv. 189 were placed at 5 or 12 °C for 15 days and plus 3 days at 22 °C. Application of HWRB treatment at 52 °C for 15 s, or dipping the fruit at 52 °C for one min (hot water dip, HWD) significantly reduced decay development. Further, chilling injury symptoms were entirely arrested after storage. In addition, HWRB treatment enhanced the resistance of tomato fruits artificially inoculated with *B. cinerea.* Further, their storability was also extended to over 3 weeks at 5 °C by minimizing chilling injury (Fallik et al. 2002).

Tomato fruits may come in contact with the soil and the pathogens causing decay may spread and during the transit and storage. Gray mold (*Botrytis cinerea*) and Alternaria rot (*Alternaria alternata*) could be suppressed effectively by employing the yeast *Pichia guillermondii* to wounds of tomatoes after harvest (Chalutz and Droby 1998). Fruit rot (soil rot) caused by *Rhizoctonia solani* may infect tomatoes through seeds and soilborne inocula. *Trichoderma harzianum*, when added to soil or applied as a coating on tomato fruits, provided significant control (Strashnov et al. 1985). Formulations of *T. harzianum* KRL-AG2, *Trichoderma* spp. and *Pseudomonas cepacia* available commercially, reduced the soil rot caused by *R. solani* effectively (Janisiewicz and Korsten 2002a, b). Strains of the yeast species *Rhodotorula rubra* and *Candida pelliculosa* effectively arrested the development of *Botrytis cinerea*, causing gray mold disease of tomato which seriously reduced the shelf-life of harvested tomatoes. *R. rubra* strain 231 was the most effective in reducing the lesion diameter by 60 %, compared to untreated control fruits. The results indicated that the application of antagonistic yeasts could be a promising and environment-friendly alternative to the fungicide treatments for the control of postharvest gray mold disease of tomatoes (Bello et al. 2008). The marine yeast *Rhodosporidium paludigenum* effectively controlled the gray mold disease in cherry tomatoes at 15 °C (Wang et al. 2010a, b).

7.1.13.2 Potato Diseases

Several fungal pathogens infect the potato tubers prior to harvest in the field and they are carried, during transit and storage, whereas others infect the tubers at postharvest stage. *Rhizoctonia solani* (sclerotial stage), *Macrophomina phaseolina* (conidial stage) and *Thanatephorus cucumeris* (teleomorph) may infect potato plants in the field causing charcoal rot disease of tubers. The percentage of infection by *M. phaseolina* at harvest could be reduced by treating the tubers with the bacterial antagonist *Bacillus subtilis* (Thirumalachar and O'Brien 1977). Potato black scurf disease caused by *R. solani* is initiated by both seed- and soilborne inocula. Pathogen sclerotia were parasitized by *Verticillium biguttatum.* The extent of reduction in black scurf symptoms on tubers by the antagonist was comparable to, and in some cases greater than, that could be obtained by using chemical disinfectants of soil (Velvis and Jager 1983). The bacterial species *Azotobacter chroococcum* in combination with *V. biguttatum* provided more effective protection to potato plants, sprouts and stolons. Production of sclerotia on new tubers was significantly reduced by inoculating the seed tubers with three isolates of *V. biguttatum* separately or in mixtures (Meshram 1984; Jager and Velvis 1986). Application of *Trichoderma viride* and *T.* (= *Gliocladium*) *virens* as dusts to seed tubers inoculated with *R. solani* reduced the disease incidence and severity, since both soil- and tuber-borne inocula were significantly reduced (Beagle-Ristaino and Papavizas 1985). The biocontrol products formulated from *Trichoderma* spp., *Trichoderma* strain KRL-AG2 and *Streptomyces griseoviridis* strain K61 were also found to be effective against *R. solani* (Janisiewicz and Korsten 2002a, b).

Phytophthora infestans causes the highly destructive late blight disease of potato crops. All aerial plant parts are infected by the pathogen. Tubers in the soil are also infected by this pathogen causing tuber blight. Bacterial strains (18) patented as biological control agents of both sprouting and Fusarium dry rot were evaluated for their efficacy in suppressing the development of *P. infestans* causing tuber blight. *Pseudomoas fluorescens* S22:T:04 consistently suppressed the pathogen development to the maximum extent. Small scale pilot testing of three strains of *P. fluorescens* and *Enterobacter cloacae* S11:T:07 was carried out under conditions simulating a commercial application. Tubers (30 tubers per box) after inoculation with the pathogen were randomized in storage and maintained for 4 weeks at 7.2 °C and 95 % RH. All BCA treatments reduced the disease significantly. The extent of disease suppression ranged from 35 % up to 86 % in the first year and from 35 to 91 % in the second year. Four strain mixtures were the most effective in suppressing the development of tuber blight disease. The BCA strains and strain mixtures were earlier found to be effective against Fusarium dry rot and sprouting. These results suggested that the bacterial strain with broad spectrum of biocontrol activity and sprout suppression could be an effective alternative to the synthetic chemicals (Slininger et al. 2007). The polysaccharide marginalan isolated from *P. fluorescens* S11:P:12, when added to the bacterial cells protected them against drying and storage survival. The effects of marginalan were more subtle and dry rot suppression was not impacted (Slininger et al. 2010).

Potato dry rot disease appears to be due to infection by one or more species of *Fusarium* such as *F. solani* var. *coeruleum*, *F. sulphureum* (syn. *F. sambucinum*), *F. thricothecioides*, *F. equiseti*, *F. sporortrichoides* and *F. avenaceum*. Dry rot disease is a serious worldwide storage disease caused by different *Fusarium* spp. individually or in combination. Strains of *Pseudomonas fluorescens* and *Enterobacter cloacae* that were effective in suppressing the development of Fusarium dry rot disease of stored potato tubers and also amenable to production in liquid culture were selected for bin trials conducted at commercial storage houses. In the first year of pilot trials, *P. fluorescens* S22:T:04 decreased dry rot caused by *Fusarium sambucinum* (teleomorph: *Gibbrella pulicaris*) by 19 %, when coinoculated with the pathogen, compared to controls and to the fungicide thiabendazole (TBZ) applied at label rates. In the second year, *P. fluorescens* P22:Y:05 and *E. cloacae* S11:T:07 reduced the disease severity by 25 and 17 % respectively, when the antagonists were applied after pathogen inoculation. In the commercial storage bin trials, *E. cloacae* reduced naturally occurring levels of dry rot by an average of 21 %, compared to

14 % for TBZ, indicating the suitability of the bacterial BCA for application to potato tubers for suppression of Fusarium dry rot disease, as an alternative to the fungicide (Schisler et al. 2000).

In another investigation, *Bacillus* spp. were evaluated for their efficacy in suppressing the development of dry rot disease. *B. cereus* (X16) was the most effective for the control of the dry rot on seed tubers and increased the yield parameters as well. During the traditional cold storage for 6–8 months, dry rot incidence was reduced significantly in potato boxes treated with the antagonist, compared to the treatment with fungicide carbendazim (Sadfi et al. 2002). The biocontrol potential of *Pseudomonas fluorescens* was assessed for the control of dry rot disease. In addition, their ability to inhibit tuber sprouting during storage was also investigated. *P. fluorescens* bv.VS11:P:12 and two strains of *Enterobacter* sp. S11:T:07 and S11:P:08 exhibited maximum effectiveness in controlling sprouting of tubers, in addition to the superior level of protection to tubers against the dry rot disease. The BCA strains showed the potential for use as an alternative to the synthetic chemical inhibitor, 1-methylethyl-3-chlorophenyl carbamate (Burkhead et al. 2003). *Heliminthosporium solani* causes the silver scurf disease of potato tubers. The presence of thiabendazole-resistant strains of *H. solani* is widespread, underscoring the need for alternative strategy for minimizing the incidence and spread of silver scurf disease. Application of *Pseudomonas corrugata*, as a postharvest treatment, reduced the disease severity and secondary transfer of the pathogen to daughter tubers under glasshouse conditions (Chun and Shetty 1994). It is essential to determine the in vitro heat tolerance of different kinds of spore forms or infection structures of postharvest pathogens of concern to develop effective heat treatment systems. The growth and proliferation of undisturbed colonies of *H. solani* could be eliminated in vitro by radiant heat treatment above 50 °C and lasting 5 min. In contrast, detached conidia were not affected by a 3-min treatment at 70 °C. These conidia can pose problems, if tubers are heat-treated. It is suggested that removal of conidia by washing the tubers in hot water, followed by heat treatment lethal to the remaining mycelium may be effective in reducing fungal inoculum on seed tubers (Johnson et al. 2003).

7.1.13.3 Carrot Diseases

Heat treatment has been applied for the control of postharvest diseases in different forms. The effectiveness of steam treatment for the reduction of decay of carrot caused by *Alternaria* spp. and *Sclerotinia sclerotiorum* was demonstrated. Winter carrots were exposed to steam for three seconds prior to packaging. After 60 days of storage at 0.5 °C plus an additional week at shelf conditions (20 °C), treated carrots exhibited only 2.0 % decay as against 23 % decay in untreated carrots. Artificial inoculation with *Alternaria alternata*, *A. radicina* and *Sclerotinia sclerotiorum* followed by steam treatment resulted in 5 % decay, whereas the untreated carrots showed 65 % decay (Afek et al. 1999). Brushing the carrots before storage enhanced the infection by black root rot disease caused by *Thielaviopsis basicola*. The efficacy of combining physical, low residue chemical and bioagents was evaluated, as an alternative to the conventional chemical control method. The technology for the precise application of steam and combined application with stabilized hydrogen peroxide (Tsunami®100) or the yeast commercial product, Shemer™ were tested. Both steam and Tsunami were highly effective in reducing decay, when applied alone, but they were phytotoxic to the carrots. Application of combined treatments of sublethal steam, followed by a sublethal dosage of Tsunami or Shemer improved the efficacy and disease control by 80 and 86 % respectively. These combinations showed a synergistic effect as compared with the effect of individual treatment. The results indicated that the disease-suppressive agents could be used for a short period, then washed off, if needed and efficiently followed by application of a bioagent (Eshel et al. 2009).

7.1.13.4 Melon Diseases

Postharvest diseases of melon are caused by fungal pathogens such as *Alternaria alternata*, *Fusarium semitectum*, *Rhizopus stolonifer* and *Trichothecium roseum* (Yang et al. 2006). Among the various kinds of bioagents employed for the suppression of postharvest diseases of vegetables, *Bacillus* spp. offers some advantages. *Bacillus* spp. including *B. subtilis* produce spores that are resistant to various physical and chemical treatments such as desiccation, heat, UV-irradiation and organic solvents and they produce different antibiotics and cell wall-degrading enzymes that adversely affect the development of fungal pathogens of postharvest diseases. *B. subtilis* EXWB1 was effective in suppressing the development of *Alternaria alternata* infecting melon (*Cucumis melo*). Treatment of melon fruits with *B. subtilis* limited the development of *A. alternata* around the inoculated sites on the fruit surface and the necrotic lesions induced by the pathogen were significantly reduced by 77.2 %. In addition, the treated fruits retained high levels of sugar and vitamin C and low levels of organic acids and maintained water content and turgidity at room temperature. The results indicated that the strain EXWB1 could restrict the pathogen development and maintain fruit quality parameters to the required levels (Wang et al. 2010a, b). Silicon (Si) application has been shown to induce resistance in treated vegetables against postharvest pathogens. Application of Si (100 mM) was the most effective in suppressing the development of *Alternaria alternata*, *Fusarium semitectum* and *Trichothecium roseum* infecting melon fruits. Si at 100 mM was the most effective concentration in reducing decay by these three pathogens, whereas higher concentration (200 mM) was phytotoxic which led to lack of general acceptance. Si showed direct inhibitory effect on the fungal pathogens. Disease suppression was correlated with induction of peroxidase and chitinase activities, indicating the possible induction of resistance to the postharvest diseases as an alternative mode of action of Si (Bi et al. 2006).

7.2 Management of Diseases of Fruits

7.2.1 *Apple Diseases*

7.2.1.1 Root Rot Disease

Rhizoctonia root rot of apple caused by *R. solani* AG-5 becomes severe after a few years of orchard establishment. A soil suppressive to Rhizoctonia root rot of apple was identified in Washington where the soil had been in wheat monoculture, prior to planting apple at that site. However, the soil became conducive within 3 years and the transition of the soil from a suppressive to conducive state was associated with significant changes in the microbial community structure including a decrease in recovery of actinomycetes and *Burkholderia cepacia* from apple. Some isolates of *Pseudomonas putida* were able to provide protection against *R. solani* AG-5 (Mazzola 1998, 1999). The role of host genotype in elicitation of the essential transformations in soil microbial community structure that led to disease suppressiveness was investigated. Apple orchard soils were planted with three successive 28-day cycles of specific wheat cultivars prior to infestation with *R. solani* AG-5. Wheat cultivars that induced disease suppression enhanced populations of specific pseudomonad genotypes with antagonistic activity toward *R. solani* AG-5 and AG-8. Wheat cultivar-specific response in terms of the transformation of the fluorescent pseudomonad community and subsequent suppression of Rhizoctonia root rot of apple was observed in three different orchard soils. The transformation was characterized by a change in the dominant fluorescent pseudomonad from *P. putida* to *P. fluorescens* bv. III. When the same soil was cultivated to wheat prior to planting apple, infection of apple roots by the resident population of *R. solani* was reduced and associated with an increase in the relative proportion of *P. putida* in the fluorescent population recovered from the apple rhizosphere (Mazzola and Gu 2000, 2002).

7.2.1.2 Scab Disease

Apple scab disease caused by *Venturia inaequalis* is considered to be the single most important disease of apples in several countries, as it is responsible for both quantitative and qualitative losses. The strategy of apple scab control relies mainly on multiple applications of fungicides. The fungal bioagents *Microsphaeropsis* sp., *M. arundinis*, *Ophiostoma* sp., *Diplodia* sp. and *Trichoderma* sp. were evaluated for their efficacy in suppressing the development of apple scab disease. Another antagonist *Athelia bombacina* and urea were also tested. In the autumn, the antagonistic fungi were applied to leaf disks artificially inoculated with *V. inaequalis* and to scabbed leaves of apple (*Malus domestica*) incubated under controlled and natural conditions. In addition, large-scale trials were also conducted with

Microsphaeropsis sp. applied as foliar spray after harvest of apples or as a ground application at 90 % leaf fall. All treatments except *Ophiostoma* sp. resulted in significant reduction in *V. inaequalis* ascospore production on the leaf disks incubated under controlled conditions or in the orchard. Leaves with apple scab lesions treated with urea or *Microsphaeropsis* sp. produced significantly fewer ascopspores of *V. inaequalis* than did nontreated leaves, with a reduction of 73.0 and 76.3 % respectively. *Microsphaeropsis* sp. was the most effective in restricting ascospores production and it reduced the total amount of airborne ascospores trapped by 70.7 and 79.8 %, as compared with non-treated plots in 1997 and 1998 respectively (Carisse et al. 2000). In a further study, field and in vitro trials were conducted to establish the influence of the BCA *Microsphaeropsis ochracea* on the ejection pattern of ascospores by *Venturia inaequalis* and on apple scab development and to establish the best timing of application. Fall application of *M. ochracea* combined with delayed-fungicide program was evaluated in orchards with intermediate and high scab risk. Ascospores production was reduced by 97–100 % on leaf disks inoculated with *M. ochracea* less than 6 weeks after inoculation with *V. inaequalis*. But ascospores production increased with increasing interval between inoculation and BCA application. In the orchard, the greatest reduction in production of ascospores (94–96 % in 2000 and 99 % in 2001) occurred on leaves sprayed with *M. ochracea* in August (Fig. 7.4). The results showed that *M. ochracea* should be applied in August or September and ascospores maturation models and delayed-fungicide program could be adopted in orchards treated with *M. ochracea* (Carisse and Rolland 2004).

Another disease management strategy applied for the control of scab diseases of apple (*Venturia inaequalis*) and pear (*V. pirina*) was induction of resistance using commercially available systemic resistance inducers, Messenger (a.i. harpin protein), Phoenix (a.i. potassium phosphate) and Rigel (a.i. salicylic acid derivative). These products were applied at four different growth stages of tree development and the fungicide penconazole was also included in the field evaluation for comparison. When the resistance inducers were applied only at two growth stages, they did not have any influence on plant responses. However, when they were applied at three or four growth stages, they were able to induce resistance in treated trees and protected them against scab disease. The fungicide was the most effective in reducing scab disease incidence in both years (2006 and 2007) of field evaluation. Among the three resistance inducers, no difference in their efficiency of protecting apple and pear trees against scab disease could be observed. The results indicated the usefulness of the strategy of using resistance inducers for the management of scab disease of apple and pear (Percival et al. 2009).

7.2.1.3 Powdery Mildew Disease

Powdery mildew disease of apple caused by *Podosphaera leucotricha* occurs in severe forms during cool and dry weather conditions. The mycoparasite *Ampelomyces* sp. parasitizes powdery mildew pathogens infecting many crops, including apple.

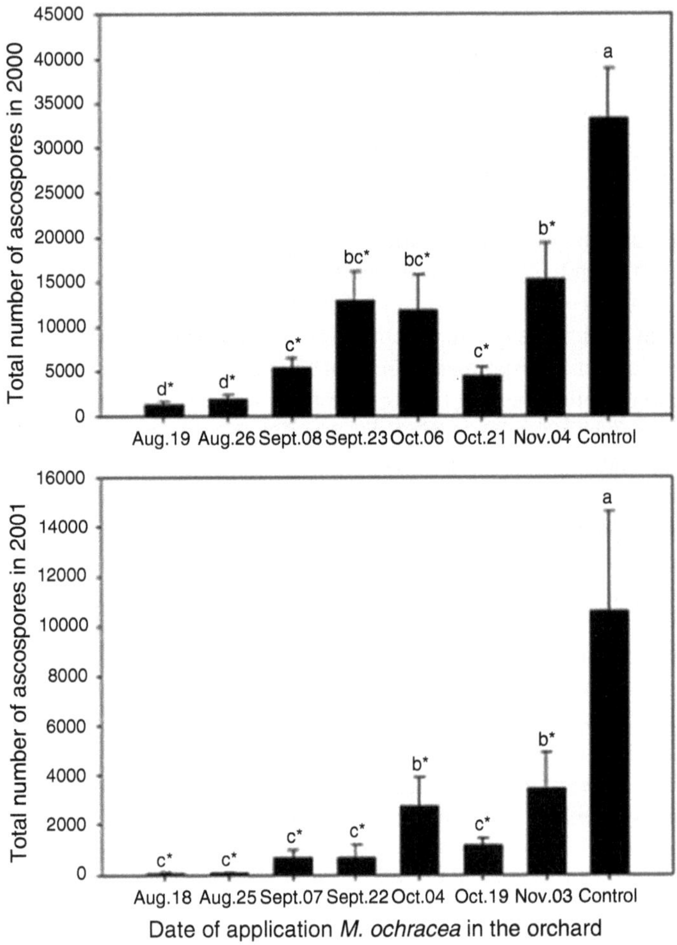

Fig. 7.4 Effect of spraying *Microsphaeropsis ochracea* at different dates in the fall on production of ascospores of apple scab pathogen *Venturia inaequalis* (Courtesy of Carisse and Rolland 2004 and with kind permission of The American Phytopathological Society, MN, USA)

Ampelomyces sp. infecting apple powdery mildew pathogen, produces pycnidia and hibernates in the buds of apple trees infected with *P. leucotricha*. In the investigation conducted over a period of 6 years, apples heavily infected with powdery mildews were widely parasitized by *Ampelomyces* every year. *Ampelomyces* mycoparasites survived the winter in two ways: as pycnidia, produced in the ascomata and in the conidiophores of powdery mildews and as thick-walled resting hyphae in the dried powdery mildew mycelia. Conidia of *Ampelomyces* released in the spring from the overwintered ascomata, germinated and penetrated into the freshly formed mycelia of *P. leucotricha*. The results of this investigation showed that *Ampelomyces* sp. could survive the winter in the field as pycnidia and as resting hyphae in the dried mycelia of their fungal hosts (Szentivânyi and Kiss 2003). The yeast Y16 effectively suppressed the development of apple powdery mildew disease. The effects of application of the yeast BCA on the nontarget fungal pathogen *Venturia inaequalis*, causing apple scab disease were assessed. The yeast did not affect the conidial germination or penetration of the leaf tissue, but suppressed the scab disease development. In addition, the quality parameters of apple fruit were not affected by the application of the yeast BCA (Alaphilippe et al. 2008).

7.2.1.4 Fire Blight Disease

Fire blight disease caused by *Erwinia amylovora* is known to be a destructive disease of apple and pear for over two centuries (Baker 1971). Fire blight disease management using chemicals has been difficult, because of the availability of limited number of registered chemicals. Copper derivatives are phytotoxic and the antibiotic streptomycin has been shown to induce resistance in the pathogen and to have increased risk of resistance in the pathogen and to have increased the risk of resistance of human pathogens as well. As an alternative to chemicals, the potential of epiphytic bacteria have been assessed for the suppression of *E. amylovora*. The strains of *Pantoea agglomerans* (syn. *Erwinia herbicola*) and *Pseudomonas fluo*rescens have been evaluated for their efficacy. *P. fluorescens* A506, the active ingredient of the commercial product BlightBan A506 and *P. agglomerans* were tested under field conditions for their ability to protect apple blossoms from infection by *E. amylovora*. The strains E325 and C9-1 of *P. agglomerans* and *P. fluorescens* A506 were applied to open blossoms on enclosed or non-enclosed trees, in order to minimize or allow honey bee-dispersal of the bacteria. Dispersal of antagonistic bacterial strains was less frequent on blossoming apple trees surrounded by plastic enclosures. All BCA strains were able to colonize flower stigmas and multiplied. The BCA strains A506 and C9-1 were equally effective and they were more efficient than other strains (Fig. 7.5) (Pusey 2002). In another study, the comparative ability of two strains of *Pantoea agglomerans* Eh252 (capable of producing antibiotic) and a near isogenic antibiotic-deficient derivative strain 10:12 was assessed. The BCA strains were applied to apple or pear trees at 30 and 70 % bloom. Aqueous suspensions of freeze-dried cells of *Erwinia amylovora* (*Ea*) were applied at bloom. Both Eh252 and 10:12 strains reduced the growth of *Ea* on blossoms, compared with control treated with water. In three of the seven field trials, Eh252 was more effective in decreasing the fire blight incidence than the mutant strain. Overall, Eh252 reduced the disease incidence by 55 ± 8 % and strain 10:12 by 30 ± 6 % (Stockwell et al. 2002). The effect of *Pantoea agglomerans* E325 (Pa) and *Erwinia amylovora* (Ea) on pH of Gala apple stigmas was investigated in field inoculations in 2006 and 2007. The mean pH values for samples increased over time and in a narrow range. The pH values were only slightly lower than controls. Under low phosphate and low pH conditions, an antibacterial product

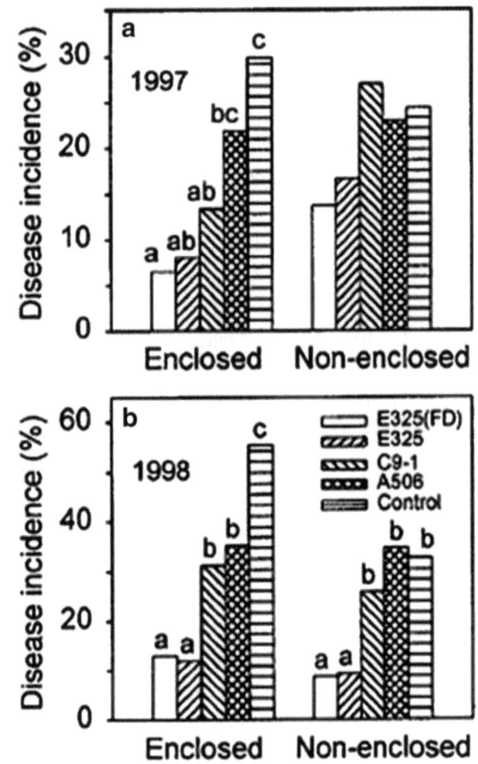

Fig. 7.5 Suppressive effects of bacterial biocontrol agents *Pantoea agglomerans* E325, *P. agglomerans* C9-1, *Pseudomonas fluorescens* A506 prepared from fresh cultures and strain E325 from a freeze-dried (FD) culture on the incidence of fire blight disease on enclosed or non-enclosed trees of (**a**) Jonagold apple in 1997 and (**b**) Gala apple in 1998. *Bars with different letters* are significantly different as per LSD test ($P \leq 0.005$) (Courtesy of Pusey 2002 and with kind permission of The American Phytopathological Society, MN, USA)

of E325 specifically inhibitory to *E. amylovora* was produced, restricting the development of the pathogen (Fig. 7.6) (Pusey et al. 2008).

A strain of *Pantoea agglomerans*, P10c isolated from pear blossoms reduced the fire blight incidence, when applied prior to inoculation with *Erwinia amylovora* on apple and pear flowers (Vanneste et al. 2002). The strain P10c was selected for its ability to colonize apple and pear flowers rapidly which is an essential requirement for the bacterial BCA to be a successful antagonist of *E. amylovora*. The number of healthy and diseased clusters of flowers was recorded. In all experiments, the level of control obtained with P10c was similar to that obtained with streptomycin (Vanneste et al. 2002). A talc-based formulation of *Pantoea agglomerans* strain Eh24 was applied at 30 and 100 % bloom on two pear orchards in the Aegean Region in Turkey during 1999–2000. The BCA formulation was sprayed on pear trees naturally infected with *Erwinia amylovora*. Application of the strain Eh24 reduced the percentage of blighted blossoms by 63–76 % in different orchards. The standard bactericides copper oxychloride + maneb was less effective in reducing the incidence of blossom blight phase of the disease, compared with the talc-based bioformulation. Experiments to study the colonization and population dynamics using the spontaneous streptomycin-resistant mutant of *P. agglomerans* ($Eh24^{strR+}$) showed that the population of Eh24 on pear blossom increased from 2×10^4 to 1.3×10^6 CFU/blossom

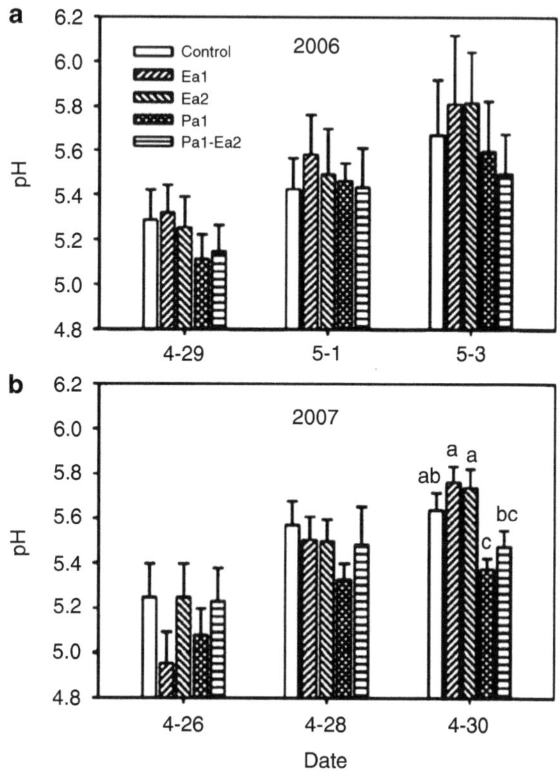

Fig. 7.6 Effect of *Pantoea agglomerans* E325 (Pa) and *Erwinia amylovora* Ea153 (Ea) of Gala apple stigmas in field inoculants in 2006 (**a**) and 2007 (**b**), as determined by pH of extracted exudates. *Vertical lines* on *bars* represent standard error; *bars with different letters* are significantly different as per LSD ($P \leq 0.05$) (Courtesy of Pusey et al. 2008 and with kind permission of The American Phytopathological Society, MN, USA)

over 18 days (Özaktan and Bora 2004). The bacterial strains *Pseudomonas fluorescens* A506 and *Pantoea agglomerans* strains C9-1 and Eh 325 and preparations of *Bacillus subtilis* QST713 containing bacterial endospores and lipopeptide metabolites were evaluated for their biocontrol potential against fire blight disease of apple under field conditions in three states in the United States. The bacterial strains in individual assessments were not consistent in the biocontrol activity as reflected by reduction in blossom infection which ranged from 9.1 to 36.1 % in the treated trees. Colonization of large proportions of flower stigma and rapid growth to reach a large population size of 10^5–10^6 CFU/blossom are important requirements, among other attributes for the bacterial antagonists to be successful. The bioproduct Serenade containing lipopeptides produced by *B. subtilis* QST713 was slightly more efficient in controlling blossom blight than other antagonists and more consistent from year to year and between locations. However, Serenade was less efficient, compared with streptomycin. Integration of biological control preparations with streptomycin might be a desirable approach to reduce the use of streptomycin, while achieving better control of apple fire blight disease (Sundin et al. 2009).

Plant activators that stimulate plant growth, in addition to protection offered against diseases caused by several microbial pathogens, have been investigated for their ability to suppress the development of fire blight disease of apple. Acibenzolar-*S*-methyl (ASM) was able to protect Golden Delicious apple seedlings, scions and trees effectively against fire blight disease, when applied before inoculation of *Erwinia amylovora*. ASM provided protection to apple seedlings to the level comparable to the standard antibiotic plantomycin (100 mg/l, streptomycin sulfate) applied just before inoculation with the pathogen. Protection of apple seedlings by ASM was consistently associated with the activation of defense-related enzymes. ASM (as Actigard, BION) provided significant level of fire blight disease control under very favorable conditions for fire blight infection in Michigan in 1999 and 2000. Application of ASM, which stimulates the apple trees' natural defense mechanism through systemic acquired resistance (SAR) pathway, has to be carried out before infection occurs to allow sufficient time for the induction of resistance (Brisset et al. 2000). In most cases, the effectiveness of the resistance induced against *E. amylovora* did not appear to persist in apple. The best control of blossom blight and terminal infections was obtained, when sprays with ASM were repeated at weekly rather biweekly schedules. The effectiveness of ASM for shoot blight control increased with increasing concentration from 75 to 100 mg a.i/l. Although multiple applications of ASM reduced the incidence and severity of fire blight, it was not nearly as effective as streptomycin (Maxson-Stein et al. 2002; Norelli et al. 2003). The potential of acibenzolar-*S*-methyl (ASM) and DL-3-aminobutyric acid (BABA) for suppressing the development of apple fire blight disease was assessed. ASM or BABA alone or in sequence were applied as foliar sprays on apple seedlings at different days before inoculation (dbi). ASM was slightly more effective than BABA in suppressing the development of fire blight disease. Spray treatments of seedlings with BABA (4 dbi) and ASM (4 dbi) in sequence drastically reduced the fire blight symptoms, while single treatment separately with ASM or BABA was less effective. Further, BABA/ASM treatment induced long-lasting resistance in apple seedlings, as indicated by the absence of disease symptoms on treated apple seedlings for more than 2 months (Hassan and Buchenaeur 2007).

The organic compound trinexapac-ethyl (Moddus, Palisade), a growth retardant, when applied reduced the shoot growth of apple trees and also suppressed the secondary spread of fire blight disease in apple orchards (Jones 2001). The ability of another commercially available product prohexadione-calcium (Apogee), when applied on apple, controlled vegetative growth and also reduced the incidence and severity of fire blight shoot infection (Yoder et al. 1999). Prohexadione-calcium did not show any direct antibacterial effect on the pathogen *Erwinia amylovora*, increased host resistance by reducing plant vigor and altered the phenylpropanoid biosynthesis pathways that might result in increase in the level of resistance of pretreated plants to the fire blight disease. The potential of prohexadione-calcium (PC) to suppress the development of fire blight under severe natural infection conditions was assessed. Prohexadione-calcium was applied at petal fall, either as a single spray or as two sprays with the total amount of chemical applied per hectare being the same. The level of disease control under the severe infection conditions was 60–75 %. Both single and split applications of prohexadione-calcium significantly reduced the severity of blossom and shoot blight in 1999. A single application of the

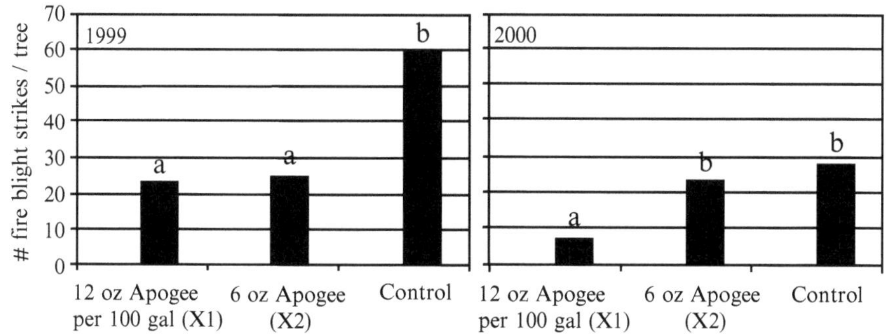

Fig. 7.7 Effect of spraying prohexadione-calcium (PC) once (X1) or twice (X2) on the incidence of apple fire blight disease under natural conditions. *Letters* indicate statistical significance (P=0.05) according to LSD test (Courtesy of Norelli et al. 2003 and with kind permission of The American Phytopathological Society, MN, USA)

compound was more effective than the split application in 2000 (Fig. 7.7). The effectiveness of prohexadione-calcium (PC) was demonstrated in ten orchards in Virginia State. Application of PC suppressed the shoot blight development. The incidence of shoot blight was reduced by 88–96 % in four orchards where fire blight disease occurred (Evans et al. 1999; Norelli et al. 2003).

7.2.2 *Grapevine Diseases*

7.2.2.1 Root and Stem Diseases

Armillaria root disease of grapevine is caused by *Armillaria mellea* which decays the woody roots of grapevine and many common forest trees. Infection of grapevine occurs in locations where forests are converted to vineyards, as the mycelium surviving in partially decayed tree roots can infect grapevines up to several years after planting (Baumgartner and Rizzo 2001). Management of the Armillaria root disease has been primarily through the use of soil fumigants to kill the pathogen inoculum in the decayed roots. In order to replace the cumbersome and endless cycles of removing dead vines, fumigating and replanting, the possibility of employing antagonistic fungi was examined for suppressing the development of *A. mellea*. A soil inoculant commercially available, Vesta (Biologically Integrated Organics, Inc., USA), prepared by a proprietary compost fermentation process was tested. The product contains viable populations of bacteria that serve as antagonists of *A. mellea*. The bacterial species *Bacillus subtilis, B. lentimorbus, Comamonas testosteroni, Pseudomonas aeruginosa* and *P. mendocina* were isolated from the bioproduct Vesta. The efficacy of Vesta for postinfection control of Armillaria root disease was assessed in an Armillaria disease-infested vineyard in

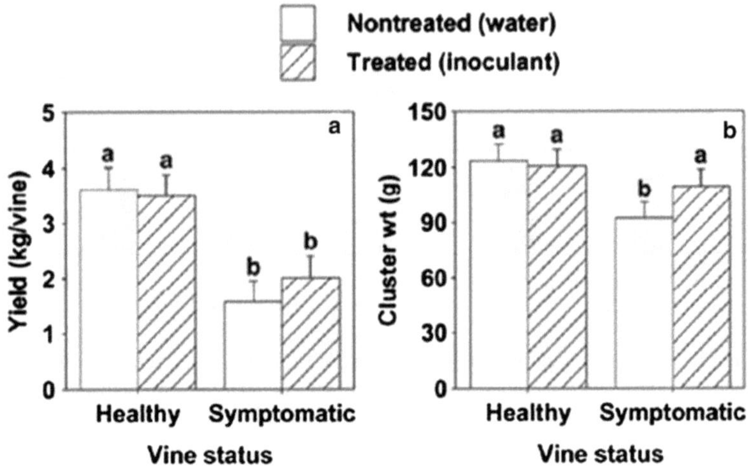

Fig. 7.8 Biocontrol potential of bioinoculants containing mixture of bacterial species against Armillaria root rot disease of grapevine. (**a**) yield of grapes (kg/vine); (**b**) cluster weight of healthy and symptomatic grapevines. *Bars with same letter* are not significantly different at P≤0.05 as per Turkey's test (Courtesy of Baumgartner and Warnock 2006 and with kind permission of The American Phytopathological Society, MN, USA)

northern California. The bioproduct was applied via drip-irrigation to vine rows in replicated blocks in 2003 and 2004. Treatment with the bioproduct significantly increased cluster weights of symptomatic vines, relative to the comparable symptomatic non-treated vines, to levels comparable to those of healthy vines (Fig. 7.8). However, the bioproduct did not reduce the rate of symptom development or mortality of treated vines. The results suggested that the inoculants may not prevent root disease caused by *A. mellea*, but it could provide therapeutic benefit by improving the productivity of infected vines (Baumgartner and Warnock 2006).

Wound protection during all stages of grapevine propagation is of utmost importance to prevent infection of propagation materials by decline and die-back pathogens. *Trichoderma harzianum* applied to grapevine wounds in a spore suspension and in the commercial formulations of Trichoseal, Trichoseal spray and Vinevax pruning wound dressing reduced recovery of the pathogen *Eutypa lata* in the glasshouse and field experiments. Recovery of *E. lata* was significantly reduced (P<0.001), when fresh wounds were treated with viable *T. harzianum* at 2 or 7 days before inoculation with ascospores of the pathogen in the glasshouse. Under field conditions, recovery of *E. lata* was significantly reduced (P<0.001), when fresh wounds were treated with spores of *T. harzianum*, *Fusarium lateritium* or Vinevax at 1 or 14 days before ascospores were applied. In general, a dely of 14 days between wounding and inoculation with ascospores of *E. lata* reduced recovery of the pathogen, compared with inoculation on the day after wounding (John et al. 2005). The efficacy of chemical and biological protection of grapevine propagation material against trunk disease pathogens was assessed in semi-commercial nursery trials.

Grapevine rootstock and scion cuttings were soaked in chemical or biological sanitation products or water (for control) prior to cold storage, prior to grafting (machine- or hand-grafting) and prior to planting in field nurseries. Levels of natural infection in basal ends and graft unions of uprooted nursery grapevines were evaluated at 8 months after planting. Untreated control plants showed pathogen infection up to 30 % in basal ends and 13 % in graft unions. The fungal pathogens associated with infection were *Phaeomoniella chlamydospora*, being the most common, followed by *Phaeoacremonium*, *Cylindrocarpon* + *Campylocarpon*, *Botryosphaeria* sp. and *Phomopsis* spp. Machine-grafted unions generally had lower pathogen incidence, compared with hand-grafted graft unions. Repeated-soak treatments of propagation materials in the different products, generally were more effective in reducing incidence of the pathogens in the nursery grapevines. But products containing *Trichoderma harzianum* (Trichoflow-T), hydrogen peroxide (Bio-sterilizer) and 8-hydroxy-quinoline sulfate (Chinosol) gave inconsistent results. On the other hand, Broncide (a blend of halogenated alcohols and water) was the effective sterilizing agent, but it reduced the certifiable plant yield significantly. Benomyl (1 g/l), Sporekill (a patented didecyldimethyl ammonium chloride formulation) at 150 ml/100 ml) and Captan (10 ml/l) were the most effective in reducing pathogen incidence without negatively impairing growth parameters of the treated grapevines (Fourie and Halleen 2006).

7.2.2.2 Powdery Mildew Disease

Powdery mildew disease caused by biotrophic pathogen *Uncinula necator* infects leaves and the berries. As the infected berries split, the loss is quite high due to the disease. In addition, it may have negative effects on the wine production in terms of quantity and quality. Development of resistance in the pathogen to selective fungicides and emphasis on the production of grapes with minimal fungicide residues have led to the active consideration of other alternative approaches of powdery mildew disease suppression. The basidiomycetous yeast *Pseudozyma flocculosa* (syn. *Sporothrix flocculosa*) was found to be an efficient natural antagonist of powdery mildews (Bélagner and Labbe 2002). Extracts from the plant *Reynoutria sachalinensis* (formulated extract under the name Milsana®) were found to be effective against powdery mildew diseases infecting vegetables and ornamentals. The efficacy of Sporodex®L (based on *P. flocculosa*) and two formulations of Milsana® was assessed for the control of powdery mildew disease of grapevines. Six field trials were conducted in Greece. In three trials, Milsana® was tested alone. Both formulations VP1999 and 2001 significantly reduced disease severity on berries and its efficacy was moderate to low, but with the range of sulfur treatment alone. Significant increase in yield was recorded in one trial. Sporodex®L tested in two field trials was effective, when the disease pressure was moderate to high on grape bunches. But its efficacy declined under extremely high disease pressure. The Sporodex®L was either similar or inferior to sulfur alone treatment in its efficacy. Alternated applications of Milsana® and Sporodex®L assessed in one trial, improved the efficacy of

Fig. 7.9 Symptoms of infection on grape bunch by downy mildew pathogen *Plasmopara viticola*

Milsana®, but not that of Sporodex®L. There was no influence of alternation of application of Milsana® or Sporodex®L with sulfur. The results indicated that Milsana® or Sporodex®L could be used in organic farming or in IPM systems to reduce the use of chemicals (Konstantinidou-Doltsinis et al. 2007).

7.2.2.3 Downy Mildew Disease

Downy mildew disease of grapes is of historical and economic importance (Fig. 7.9). The well known Bordeaux mixture was formulated by preparing a mixture of copper sulfate and lime in the nineteenth century. Since then, its effectiveness against several other fungal diseases of various crops has been reported. The need for preparing the Bordeaux mixture afresh every time was found to be a serious limitation. Many organic compounds like ß-aminobutyric acid (BABA) was investigated for its ability to induce resistance against the downy mildew pathogen *Plasmopara viticola.* The efficacy of DL-3-amino-n-butanoic acid,

DL-ß-aiminobutyric acid (BABA) or a mixture of BABA and different fungicides at reduced rates in reducing the incidence of grape downy mildew disease was compared. Two foliar sprays of BABA or the mixture of BABA reduced the disease effectively by >90 % in the foliage of field-grown grapevines. In five field trials, BABA sprays reduced significantly the infected leaf area, fungal sporulation and necrosis of oilspots. In four trials, spray application of BABA was as effective as metalaxyl-Cu or Acrobat Plus (dimethomorph + mancozeb). Two-way tank-mixtures of BABA + fosetyl-Al, BABA + folpet or BABA + Bion (benzothiadiazole) each at half recommended rate, provided an additive effect against the downy mildew disease. These treatments were in equivalence in their effectiveness to the full rate of each fungicide alone. BABA did not induce phytotoxic symptoms and also did not affect the pH, total titrable acids or Brix of the juice, as determined by commercial fungicidal standards. The results indicated that the foliar application of BABA could be advantageously integrated into a downy mildew control program in vineyards (Reuveni et al. 2001). Application of BABA resulted in strong reduction in mycelial growth and sporulation of *P. viticola* in the susceptible cv. Chasselas. Best protection was provided by BABA, followed by jasmonic acid (JA). On the other hand, acibenzolar-*S*-methyl (ASM), salicylic acid (SA) and abscisic acid (ABA) had no influence on the level of resistance in treated grapevine plants (Md. Hamiduzzaman et al. 2005). The effect of Pen, an aqueous extract of the dry mycelium of *Penicillium chrysogenum* was investigated on the suppression of grapevine downy mildew and grapevine powdery mildew diseases under greenhouse and field conditions. Pen extract had no direct inhibitory effect on the grapevine pathogens. Hence, it is considered that Pen might act indirectly by inducing resistance in treated plants. Under certain conditions, Pen caused phytotoxic side effect producing small necrotic spots. Particularly purified fraction of Pen was much less toxic than the crude preparation. However, the level of protection provided by crude and partially purified preparations was similar (Thuerig et al. 2006).

7.2.2.4 Gray Mold Disease

Biological control methods for the suppression of the gray mold pathogen *Botrytis cinerea* may be successful only with adequate knowledge of the ecology and epidemiology of *B. cinerea* in the vineyards. In spring, conidia from saprophytically surviving mycelium and germinating sclerotia infect senescing floral tissues, tender shoots and leaves of grapevine plants (Seyb 2004). An isolate of *Ulocladium oudemansii* effectively suppressed the development of *B. cinerea* on necrotic grape tissues. An aggressive colonization of necrotic bunch trash tissues by *U. oudemansii* early in the season is the primary mode of action of the BCA (Elmer et al. 2003; Shorten et al. 2003). A commercial formulation of *U. oudemansii* (BOTRY-Zen®) was evaluated under field conditions. The formulation was found to be as effective as the fungicide program, as suggested by Hawke's Bay Wineries (Elmer et al. 2005). The efficacy of the chemical elicitor 5-chloro-salicylic acid (5CSA) and the fungal antagonist *U. oudemansii* in suppressing the development of *Botrytis cinerea* infecting Chardonnay grapevines was compared, when applied alone and in combination under greenhouse conditions. Treatment with 5CSA reduced the lesion diameter significantly, when the plants were treated at 7 days before challenge inoculation with *B. cinerea*. Application of 5CSA or *U. oudemansii* and in combination every 10–14 days from flowering until 1 week prior to harvest significantly reduced Botrytis bunch rot severity in the greenhouse. Levels of disease suppression by the treatments did not significantly differ, but they were significantly more effective in reducing bunch rot, compared with untreated controls. Assessment of Botrytis disease severity after 14 days of incubation postharvest showed that bunches from plants treated with the combination of 5CSA + BCA showed significant low incidence of disease (<15.2 %), compared with those from untreated controls (83 %). The results indicated the effectiveness of the combination of the BCA and elicitor of disease resistance in suppressing the bunch rot disease (Reglinski et al. 2005).

Biosuppression of gray mold disease pathogen *Botrytis cinerea* by bioagents has been inconsistent, because of the variability of environmental conditions in the field. The effectiveness of combining different formulations strategies on the establishment, survival, persistence and efficacy of *Candida sake* CPA-1 applied in the field in order to control *Botrytis cinerea* in grape was assessed. The formulations, liquid formulated cells and fresh cells of the strain CPA-1 were applied at flowering, pea-sized berries, veraison and before harvest. In addition, the compound Fungicover® (FC), an edible coating was evaluated as a potential additive for CPA-1 treatments. Spray applications of different formulations of *C. sake* resulted in colonization of bunches under field conditions and when combined with FC, the BCA had significantly higher survival rates (up to 50 % higher), compared with the CPA-1 without the additive FC. All treatments with CPA-1 significantly reduced the incidence of *B. cinerea* to between 36 and 40 %, while the disease incidence in the untreated control was 64 %. No significant differences were observed between CPA-1 treatments and the conventional fungicide program recommended. Addition of FC significantly increased the populations of *C. sake* CPA-1 on grape berries throughout the growing season and there was significantly less Botrytis bunch rot, as a result of *C. sake* CPA-1 applied with FC. However, FC did not have any direct inhibitory effect on *B. cinerea*. The results indicated that good consistent Botrytis control could be achieved by combining *C. sake* CPA-1 and FC which provided more effective protection than the conventional fungicide program (Cañamás et al. 2011).

7.2.2.5 Pierce's Disease

The bacterial pathogen *Xylella fastidiosa* (*Xf*), causing the grapevine Pierce's disease, can also infect several economically important crops such as citrus, peach, plum and coffee, in addition to grapevines. The bacterial pathogen is transmitted by the members of the leafhopper subfamily Cicadelinae (sharpshooters) and of the spittlebug subfamily Cercopidae (Purcell 1990). Naturally occurring weakly virulent or avirulent strains of *Xf* were identified and their biocontrol potential against virulent strains of *Xf* was investigated. The weakly virulent or avirulent strains were inoculated into the lower nodes of grapevine cv. Carignane plants in the greenhouse and challenged with pathogenic strains 2 weeks later. Several strains provided some reduction in symptom development. A strain Syc86-1 from sycamore and two strains from elderberry were the most effective in reducing the symptom severity. In a 2-year field evaluation, the strain Syc86-1was effective in limiting the development of Pierce's disease in the cv. Himrod under vineyard conditions. In new vineyards with cvs. Flame Seedless and Cabernet Sauvignon, the strain EB92-1 (from elderberry) provided effective protection to both grapevine cultivars against the disease. Strain Syc86-1 was not effective in these vineyard evaluations. Two strains from grapevine were less effective, but delayed the symptom development. The results indicated that use of benign strains of *Xf*, especially strain EB92-1 appeared to have potential for use in commercial vineyards for the protection of susceptible vine cultivars against the Pierce's disease (Hopkins 2005).

7.2.2.6 Crown Gall Disease

Crown gall disease caused by *Agrobacterium vitis*, one of the most destructive diseases of grapevines, occurs in several countries. The usefulness of biocontrol approach using antagonistic microorganisms to contain the crown gall disease of grapevine was investigated. *Agrobacterium rhizogenes* strain K84 was effective against *A. tumefaciens* infecting peach, but not against *A. vitis* (Kerr 1980). *Rahnella aquatilis* isolated from vineyard soils in Beijing, was evaluated for its biocontrol potential against *A. vitis. R. aquatilis* strain HX2 exhibited significant suppressive effect on the development of crown galls in grapevines. Under field conditions, immersion of the basal ends of grape cuttings with HX2 cell suspensions inhibited or completely prevented crown gall formation caused by *A. vitis* K308 in the roots of the plants growing from the cuttings. The 3-year average disease incidence in grapevine plants treated with HX2 was 30.8 %, compared with 93.5 % in untreated control plants. Treatment with HX2 strain reduced the proportion of plants with visible symptoms of crown gall disease and the severity of the disease as well. The culture supernatant of HX2 exhibited a stronger suppressive effect on crown gal disease management than did the cell suspension. The strain HX2 could be detected in the grape rhizosphere grown under field conditions for up to 90 days after inoculation. The supernatant and cell suspension of HX2 had no adverse effect on the total cell density of the microbial community consisting of fungi, bacteria and actinomycetes (Chen et al. 2007). Nonpathogenic strains of *Agrobacterium vitis* ARK-1, ARK-2 and ARK-3 were evaluated for their efficacy in suppressing crown gall development in grapevine seedlings. Stems of grapevine seedlings were inoculated with cell suspension of seven mixed strains of *A. vitis* (Ti) as pathogen and one of the nonpathogenic strains or *A. vitis* strain VAR03-1 (reported earlier as effective suppressor of crown gall), as competitors to determine the effectiveness of suppression of tumor formation. The nonpathogenic strains ARK-1, ARK-2 and ARK-3 reduced the tumor incidence, when a 1:1 cell ratio of pathogen/nonpathogenic strain was applied. Strain ARK-1 was the most effective in inhibiting tumor formation and it established high populations on roots of grapevine tree rootstock and persisted on roots for 1 year. The dead cells or culture filtrate of ARK-1 did not have inhibitory effect on tumor formation. The strain ARK-1 showed potential for wider application (Kawaguchi and Inoue 2012).

7.2.3 *Citrus Diseases*

7.2.3.1 Phytophthora Root Rot Disease

Phytophthora root rot disease is caused by *P. parasitica* (= *P. nicotianae*) and *P. citrophthora*. The disease induces a slow decline that reduces plant vigor, and size and number of fruits produced. No biological control approach appeared to have been made in the commercial fields. Of the large number of rhizosphere bacterial isolates (1,600) tested, only 26 % of the bacterial isolates were effective in reducing the rhizosphere populations of *Phytophthora citrophthora* or *P. parasitica* in greenhouse experiments. *Pseudomonas putida* 06909 was effective in seedling bed trials. The population of *P. citrophthora* was reduced by 45–73 % (Turney et al. 1994). The bacterial strain actively adhered to the hyphae of *Phytophthora*. In the field trials, application of *Pseudomonas putida* 06909-rif/nal increased the populations of the BCA with every irrigation, as well as its biocontrol potential against *P. para*sitica over that of a single annual applications at the start of the irrigation season (Steddom and Menge 1998, 1999). A commercial field fermentor as a means of culturing and delivering the BCA *P. putida* 06909-rif/nal through irrigation was developed. The fermentor produced 120 l of inoculum at 5×10^8 CFU/ml after 12-h fermentation, allowing two applications per day (Steddom and Menge 2001). By employing the field fermentor, *P. putida* 06909-rif/nal was applied repetitively during the irrigation season in two citrus orchards over 3 years (1997–1999). With weekly applications of the BCA, the pathogen populations of *P. parasitica* were significantly reduced in 1999, but not in 1997 or 1998. Yearly application of the BCA did not reduce the rhizosphere populations of *P. parasitica*. Repeated applications of *P. putida* strain resulted in significant changes in soil microbial community (Steddom et al. 2002a, b).

Viroids are subviral infectious agents capable of causing diseases in susceptible host plant species. *Citrus exocortis viroid* (CEVd) causes an economically important citrus exocortis disease. Exocortis-infected 'Hamlin' orange [*Citrus sinensis* on Rangpur lime (*C. limonia*)] rootstock resisted *Phytophthora* infection, but there was no consistent differences between the effects of mild and severe strains on *Phytophthora* lesions (Rossetti et al. 1980). In a later investigation, citrus viroid-induced resistance to *Phytophthora* infection in citrus was assessed by the number of *Phytophthora* sporangia in 'Rio-Red' grapefruit (*C. paradisi*) bait tissue infected with citrus viroids, compared with non-inoculated controls. CEVd E9, a severe isolate of *Citrus exocortis viroid* significantly reduced the number of sporangia in leaves and bark. Citrus viroid mixtures significantly reduced the number of sporangia on bark, leaves and roots, compared with healthy plants. The results showed significantly reduced *Phytophthora* sporangia development resulting from a number of viroids in mixed infection. Efficient use of the right viroid isolate(s) might result in suppression of *Phytophthora* infection in citrus (Thomas et al. 2010).

7.2.3.2 Canker Disease

Cirus canker disease caused by *Xanthomonas citri* subsp. *citri* (*Xcc*) occurs all over the world. The bacterial pathogen infects all plant parts stem, leaves and fruits which bear cankerous lesions or spots. Application of organic compounds that induce systemic resistance to other crop diseases, has been assessed for its effectiveness for the suppression of citrus canker disease. Swingle citrumelo seedlings (*Citrus paradisi* x *Poncirus trifoliata*) were treated with imadacloprid, isonicotinic acid or acibenzolar-*S*-methyl (ASM) as soil drenches or with ASM as a foliar spray 1 week prior to inoculation of immature leaves with *Xcc*. Seedlings were re-inoculated four times over a 24-week period. Soil application of systemic acquired resistance (SAR) inducers reduced canker lesions up to 70 %, compared with the untreated inoculated plants. The populations of *Xcc* were also reduced due to soil treatment with imadacloprid and SAR inducers (Francis et al. 2009). In another investigation, imidacloprid, thiamethoxam and ASM were evaluated for the suppression of canker disease development on Ray Ruby grapefruit trees. All treatments significantly reduced foliar infection by *Xcc* on the combined spring-summer fall flushes. The standard chemicals copper hydroxide and streptomycin (11 sprays at 3-week interval) effectively reduced the canker incidence on shoot flushes produced throughout the season, compared with the untreated control. On the other hand, soil applied SAR-inducers reduced foliar infection, depending on the rate, frequency and timing of application. During tropical storm, SAR inducers were generally ineffective and the protective efficiency varied, depending on the concentration and timing of application (Graham and Myers 2011).

7.2.3.3 Citrus Tristeza Disease

Citrus tristeza virus (CTV) has been the primary cause for citrus decline and death of citrus propagated on sour orange rootstock, resulting in replanting of citrus, using decline-tolerant rootstocks. However, dispersal of severe isolates causing stunting and stem-pitting in grapefruit and sweet orange varieties irrespective of the rootstock used, necessitated to look for alternative strategies for containing the tristeza disease. Cross-protection provided by mild strain of CTV has been used extensively to reduce the losses due to the stem-pitting isolates of CTV in Australia (Broadbent et al. 1995). Some mild isolates of CTV from apparently healthy Marsh and Thomspson grape fruit trees in orchards declining with CTV stem-pitting isolate, when grafted into virus-free Marsh grapefruit, protected against stem-pitting isolates transmitted by the aphid vector *Toxoptera citricida*. The degree of protection varies with the CTV isolate and environmental conditions. Pre-immunization with a mild CTV isolate (PB61) protected Marsh grapefruit trees against stem-pitting isolate for about 35 years (Broadbent et al. 1995). In a later study, the efficacy of the mild strain PB61 to protect the seedlings against superinfection with a severe grapefruits stem-pitting isolate PB219 or two orange stem-pitting (OSP) isolates PB155 or PB235 was assessed in the glasshouse. Preimmunization with PB61 provided partial protection against superinfection, using aphid-inoculation and delayed superinfection, when challenge was by graft inoculation. Pre-immunization with mild isolate did not ameliorate expression of OSP symptoms, once superinfection with OSP-inducing isolates was initiated (Zhou et al. 2002).

Successful application of cross-protection as a disease management strategy for *Citrus tristeza virus* (CTV) depends on the precise identification and differentiation of isolates of CTV is usually dispersed to new areas by propagation of infected buds and then it is locally spread in a semi-persistent mode by different species of aphids. Identification of the seedling yellows (SY) and stem-pitting (SP) isolates being a major threat to citrus industry in many countries and the mild isolates (non-SY or non-SP) could be done by biological indexing on selected indicator plants. Several molecular and serological markers have been employed. But lack of knowledge on the actual genetic determinants of CTV pathogenicity, the presence of mixed populations in field trees and limited sensitivity of detection procedures have been formidable obstacles for their practical use to predict the pathogenic behavior of unknown isolates of CTV (Sieburth et al. 2005). A real-time RT-PCR amplification of 56 biologically characterized CTV isolates from 20 countries and melting curve analysis of the amplified DNA products was developed for identification of the CTV isolates. All the mild CTV isolates could be precisely identified and differentiated by this technique. The procedure developed in this study may be applied for quarantine and certification programs to avoid introduction or nursery propagation of potentially dangerous isolates in areas where only mild isolates may be prevalent (Ruiz-Ruiz et al. 2009).

7.2.4 Banana Diseases

7.2.4.1 Fusarium (Panama) Wilt Disease

Fusarium wilt disease of banana is incited by *F. oxysporum* f.sp. *cubense* race 4 (*Foc* R4) is considered as one of the most destructive diseases affecting banana crops in all countries around the world. The pathogen survives as chlamydospores that remain viable in the soil for a long time (decades). Hence, the pathogen introduced through suckers into the field, makes it almost impossible to have economic production within a few years. Preventive methods based on the use of pathogen-free plantlets produced through tissue culture technique and restriction of movement of people, vehicles and equipments may reduce the incidence of Fusarium wilt disease (Moore et al. 1999a, b). Cavandish varieties resistant to Foc R4 have not found market acceptability (Viljoen 2002).

Application of biocontrol agents as alternative strategy for the management of Fusarium wilt disease of banana has been examined. The microorganisms isolated from soils naturally suppressive to Fusarium wilt have been tested. Nonpathogenic strains of *Fusarium oxysporum* (Fo47) and *Pseudomonas fluorescens* were effective against Fusarium wilt diseases. The potential of nonpathogenic *Fusarium oxysporum* isolates from the rhizosphere soils of banana root to suppress the development of banana Fusarium wilt disease was assessed. Two nonpathogenic *F. oxysporum* isolates CAV 255 and CAV 241 reduced the Fusarium wilt disease incidence by 87.4 and 75.0 % respectively. The known Fo47 isolate did not suppress the disease significantly. *P. fluorescens* strain WCS417 reduced the disease incidence by 87.4 % under greenhouse conditions (Nel et al. 2006). The bacterial bioagents *Serratia marcescens* effectively suppressed the development of *F. oxysporum* f.sp. *cubense* race 4 (*Foc* R4). The bacterial culture was formulated using montmorillonite clay (carrier), nonfat skimmed milk (NFSM) and sucrose (enrichment materials). The materials used for formulation of *S. marcescens* improved the bacterial cell viability significantly. Bioformulation of *S. marcescens* with the above materials was found to be useful for both storage and field application (Ting et al. 2009).

The effectiveness of integration of compatible biological and cultural methods for the management of Fusarium wilt disease of banana was investigated. Application of nonpathogenic, endophytic *Fusarium oxysporum* (*Fo*) strains, *Trichoderma harzianum* EcoT®, silicon and mulching using macadamia alone or in combination were evaluated under greenhouse and field conditions. Nonpathogenic *Fo* strains were effective in reducing the disease incidence. Amendment with potassium silicate to plants exposed to *Fo* strains, improved plant health of cold-stressed banana plants reduced disease severity by more than 50 % and shoot yellowing and wilting by 80 %, compared to those treated with *Fo* strains alone. A field trial was conducted in a location in South Africa where the Fusarium wilt disease was a serious problem. Plants treated with combinations of nonpathogenic *Fo*, *T. harzianum* EcoT®, silicon and mulch had significantly better plant growth, compared with single application. Nonpathogenic strain N16 was superior to strain N7 in suppressing disease development and improving shoot height. The results showed that integration of nonpathogenic *Fo* strains, silicon and mulching treatments could provide an effective disease management option (Kidane and Laing 2010).

7.2.4.2 Root Rot Disease

Root rot disease caused by *Cylindrocladium spathiphylli* has assumed serious proportions, seriously affecting banana production. Adoption of cultural practices such as fallowing and chemical application has been attempted to reduce disease incidence with limited success. The usefulness of employing arbuscular mycorrhizal fungi (AMF) for the management of root rot disease of banana was assessed. The four AM fungi *Glomus* sp., *G. proliferatum*, *G. intraradices* and *G. versiforme* were tested. Preinoculation of banana plants with AMFs attenuated the detrimental effect of the pathogen *C. spathiphylli*. Lower disease severity and stimulation of plant growth were observed in plants inoculated with any of the AM fungi. The mean root rot intensity was 29 % in AMF colonized plants as against 57 % in the control plants. *Glomus* sp. and *G. proliferatum* stimulated the plant growth to the maximum extent and more effectively than the other AMFs tested. Postinoculation of infected banana plants with AMFs resulted in decreased intensity of AM fungal root colonization of the plants. The results indicated application of AM fungi prior to infection by *C. spathiphylli* could reduce the disease intensity, in addition to the stimulation of plant growth (Declerck et al. 2002). The potential of silicon (Si) in inducing resistance to banana toppling disease caused by *C. spathiphylli* was assessed. Dipping the root system of banana plants in conidial suspension of the pathogens and growing them on soils amended with 2 mM Si reduced the disease intensity. In addition, Si amendment exerted growth promotion effect, resulting in the alleviation of the negative effects of the pathogen. Application of Si appears to be an eco-friendly alternative to use of chemicals (Vermeire et al. 2011).

7.2.4.3 Leaf Spot Disease

The bacterial strain B106 isolated from the rhizosphere soil in banana field in China was identified as *Bacillus subtilis* based on its 16S rDNA sequence homology with related bacteria from GenBank as well as physiological and biochemical characteristics. The strain B106 was effective in suppressing the development of *Pseudocercospora musae* (teleomorph: *Mycosphaerella musicola*), causing banana leaf spot disease and *Colletotrichum musae*, causing postharvest anthracnose disease of banana fruits during storage. The cultural conditions for strain B106 were optimized for expressing higher antagonistic activity against *P. musae*. The antagonistic potential of the strain B106 was assessed under greenhouse and field conditions. Development of banana leaf spot was suppressed by 72.3 % in the greenhouse experiment at 10 days after pathogen inoculation. The efficacies of strain B106 (1×10^6 CFU/ml) for controlling both the banana leaf spot diseases in the field and the anthracnose disease at postharvest stage were 48.3 and 48.6 % respectively under optimized cultural conditions for the BCA to express its antagonistic potential. The results indicated that the BCA strain could be tested in different locations for further advancement (Fu et al. 2010).

7.2.5 Strawberry Diseases

7.2.5.1 Verticillium Wilt Disease

Verticillium dahliae, a soilborne pathogen, can infect a large number of plant species including strawberry. Cultural practices may reduce incidence and severity of soilborne diseases by different mechanisms. Crop rotation with broccoli reduced soil populations of *V. dahliae* and incidence of wilt disease. Inoculum densities of the pathogen are reduced to a greater extent, when broccoli or Brussels sprouts were grown as rotation crop, compared with lettuce. This corresponded to reduction in disease in the field and greater yields (Martin and Bull 2002). In another investigation, the effectiveness of rotations of strawberry with broccoli, Brussels sprouts and lettuce in reducing the soilborne inoculum of *V. dahliae* and *Pythium* spp. and severity of Verticillium wilt disease was assessed. Rotations did not alter the total population levels of *Pythium* spp. at two sites tested. On the other hand, *V. dahliae* microsclerotia were significantly reduced with broccoli and Brussels sprouts rotations compared with lettuce rotations at *V. dahliae*-infested site. Reduction in propagules of the pathogen was reflected by lower Verticillium wilt severity on strawberry plants in the broccoli and Brussels sprouts rotations, compared with lettuce rotations. Furthermore, strawberry plant vigor and fruit yield were significantly lower in lettuce-rotated plots. However, none of the rotation treatments were more effective than the standard chemical treatment with methyl bromide + chloropicrin for all variables quantified. Rotation with broccoli and Brussels sprouts could be an effective cultural practice for managing Verticillium wilt disease, when methyl bromide is to be phased out (Subbarao et al. 2007).

In a later investigation, the effects of broccoli and lettuce rotations on population densities of *Verticillium dahliae* and *Pythium* spp. in soil and on strawberry (*Fragaria* x *ananassa*) growth and yield were determined in conventional and organic production systems. Strawberry was planted after two successive crops of broccoli or lettuce. In the control treatment in the conventional field, the soil was fumigated with methyl bromide + chloropicrin prior to strawberry planting. No difference in the inoculum densities of *V. dahliae* was noted in organic and conventional plots following lettuce rotations. But in plots following lettuce rotations, decrease in the inoculum densities of *V. dahliae* in organic plots was greater than in conventional plots. Crop rotations did not have any consistent effect on the inoculum densities of *Pythium* spp. On the other hand, Verticillium wilt incidence in strawberry was lower by 12–24 % in fields rotated with broccoli, compared with fields rotated with lettuce. Severity of wilt was also reduced to a greater extent in fields rotated with broccoli, compared with that in fields rotated with lettuce. The usefulness of employing broccoli rotation coupled with postharvest incorporation of broccoli residue was indicated by this investigation (Njoroge et al. 2009).

An environmentally desirable disease management strategy was developed by employing a rhizobacterial species. *Serratia plymuthica* strain HRO-C48 isolated from the rhizosphere of oilseed rape was selected for its high chitinolytic activity, ability to produce indole-3-acetic acid and its relative harmlessness to human and environment. The biocontrol potential of the strain HRO-C48 was assessed for the suppression of the development of *Verticillium dahliae* and *Phytophthora cactorum*, causing Verticillium wilt and Phytophthora root rot diseases respectively. In the greenhouse assays, treatment with the strain HRO-C48 reduced Veriticillium wilt by 18.5 % and Phytophthora root rot by 33.4 %. In the field trials conducted in commercial strawberry farms, strawberry roots were dipped in the suspension of *S. plymuthica* prior to planting in naturally infested soils. Incidence of Verticillium wilt disease was reduced to different degrees ranging from 0 to 37.7 % with an average of 24.2 %, whereas the increase in yield ranged from 156 to 394 % with an average of 296 %. Reduction of Phytophthora root rot incidence ranged from 1.3 to 17.9 % with an average of 9.6 % in the BCA- treated plots. The yield increase due to BCA treatment was 60 %, compared with nontreated control. Under field conditions, *S. plymuthica* strain HRO-C48 could survive in the strawberry rhizosphere for 14 months, after root application. The results indicated that by selecting suitable strain of the rhizobacterial species, the success rate might be enhanced for the control of two important soilborne diseases of strawberry (Kurze et al. 2001).

7.2.5.2 Phytophthora Diseases

Phytophthora cactorum causes the destructive crown rot and wilt disease and leathery rot on strawberry plants and fruits respectively. *P. fragariae* var. *fragariae* is the causative agent of root rot disease. The thermal sensitivity of *Phytophthora* spp. was assessed to determine the feasibility of applying soil solarization as management strategy for Phytophthora diseases. Solarized plots reached a maximum temperature of 45 °C at the soil depth of 15 cm and *P. cactorum* was killed within 2 weeks at 15-cm depth, but withstood the adverse effects of soil solarization at depths of 30 and 45 cm (Juarez-Palacios et al. 1991). The effect of soil solarization on the disease incidence and severity of root rot disease was investigated. Increase in the soil temperature at 10 cm depth of the soil was between 7 and 17 °C over that prevailing in nonsolarized soils. Solarization significantly reduced root necroses ($P<0.05$) and increased root weight of plants. Infection of strawberry by *Pythium*, *Rhizoctonia* and *Cylindrocarpon* spp. was also reduced, in addition to infection by *P. fragariae* var. *faragariae*, due to soil solarization. In the second growing season, total number and number of healthy primocanes of cv. Qualicum raspberry plants were greater ($P<0.05$) in solarized plots, compared to nonsolarized controls. Combined application of solarization and the fungicide mefenoxam did not provide better protection than solarization alone which was more effective than mefenoxam alone. The beneficial effect of solarization in suppressing root rot disease in strawberry was obtained for two or more years after solarization. The results indicated that soil solarization could become a complementary component of integrated disease management system for strawberry and raspberry (Pinkerton et al. 2002).

The effectiveness of raised bed solarization (RBS) alone or in combination with chicken manure (CM) amendment, methyl bromide (MB), TeloDrip (1,3-dichloropropene + chloropicrin), short RBS combined with reduced doses of metham sodium (MS) and TeloDrip was assessed for the control of soilborne pathogens, *Phytophthora cactorum* and *Rhizoctonia solani* infecting strawberry. Field experiments were conducted in two cropping seasons between 2002 and 2004. In both seasons, raised bed solarization (for 7 weeks) alone or with CM amendment (10 t/ha), MS (50 ml/m^2) after 2-week RBS and MB (50 g/m^2)

significantly reduced both soilborne diseases. TeloDrip was less effective in reducing the disease incidence. All treatments effectively controlled four weed species, blue grass (*Poa annua*), common purslane (*Porutlaca oleracea*), red-root pigweed (*Amaranthus retroflexus*) and barnyard grass (*Echniochloa crus-galli*), commonly found in the strawberry fields. RBS treatment with or without CM and 2-week solarization + MS enhanced the total marketable yields to a level equal to that could be achieved by MB application in the first season. On the other hand, only RBS and CM amendment were found to be as effective as methyl bromide treatment in the second year. The results indicated that raised bed solarization procedure could be considered as an alternative to methyl bromide which faces imminent phasing out of field application (Benlioğlu et al. 2005). The effectiveness of soil solarization or *Trichoderma* spp. alone and in combination for suppressing the development of *Phytophthora cactorum*, causing crown rot and wilt disease of strawberry was assessed under field conditions for three consecutive years (2000–2003). Solarization was carried out during summer, using clear 50-μm low density polyethylene mulch. *Trichoderma* sp. was applied via drip and dip, adding to the soil at 7 days before planting. Strawberry roots were dipped in the suspensions of *Trichoderma* spp. prior to planting. Soil solarization reduced the soil population of *P. cactorum* by 100 % in the first year, by 47 % in the second year and by 55 % in the third year, relative to the untreated control. Application of *Trichoderma* spp. reduced soil populations of *P. cactorum* and also reduced leathery rot incidence on strawberry fruit by 76 % in the first year and by 33.8 % in the second year, compared with untreated control plots. The combination of soil solarization and *Trichoderma* spp. reduced the pathogen population to the maximum extent of 88.9, 97.6 and 99 % in the first, second and third years respectively. The results indicated that the effectiveness of the disease control could be markedly enhanced by the combination of soil solarization and the fungal BCA providing a promising alternative to chemical application (Porras et al. 2007).

Plant activators have been shown to promote the growth of plants and induce resistance to several diseases of plants as well. Acibenzolar-*S*-methyl (ASM) and chitosan were evaluated for their ability to induce resistance in strawberry against *Phytophthora cactorum*, causing rot and red stele disease. Application of ASM and chitosan reduced crown rot symptoms and the effect of amelioration of symptom intensity was enhanced when the time between treatment and challenge inoculation was increased from 2 to 20 days. No significant differences in the extent of suppression of disease symptoms could be seen due to increase in the concentrations of ASM was increased from 10 to 100 μg a.i./plant. Inoculation of alpine strawberry plants (*Fragaria vesca* var. *alpina* cv. Alexandria) with *Phytophthora fragarie* var. *fragariae* after treatment with ASM. Chitosan or fungicide fosetyl-Al showed that ASM was able to provide good protection to alpine strawberry. But chitosan had no effect at all. There was no significant difference between ASM and fosetyl-Al, when applied at the same time of inoculation with the pathogen. The results indicated the potential of ASM for use against the root-infecting fungal pathogens of strawberry (Eikeno et al. 2003).

7.2.5.3 Rhizoctonia Root Rot Disease

Strawberry root rot disease caused by *Rhizoctonia solani* (= *Macrophomina phaseolina*) causes high mortality of infected plants under favorable conditions. For solarization of the soil, polyethylene film (30 μm thick) containing different additives-ultraviolet (UV), UV + infrared (IR), UV + IR + antifog (AF) and anti-dust (AD) and used polyethylene film (260 μm thick) were employed. Soil solarization trials were conducted, using the above materials in commercial strawberry cv. Camarosa fields. The soil temperatures reached the peak respectively of 54 and 50.7 °C at a depth of 10 cm under the polyethylene sheet containing UV + IR + AF + AD in 2007 and 2008. Viability of the pathogen sclerotia was determined under various field conditions and temperatures. The sclerotia survived for more than 18 days at 45 °C and there was a sharp decline in the survival of sclerotia at 50 °C. Under field conditions, solarization did not reduce the viability of the pathogen at a soil depth of 10 or 20 cm. But significant reduction in pathogen survival (66 %) occurred at a soil depth of 5 cm. Several dicot and monocot weeds were entirely killed by soil solarization which, however, was not effective against the grass weed *Cyperus rotundus*. The marketable fruit yield was significantly increased in polyethylene sheet-covered plots (Yildiz et al. 2010).

7.2.5.4 Gray Mold Disease

Botrytis cinerea incitant of gray mold diseases, can infect several crops worldwide. Abundant sporulation of *B. cinerea* on dead and senescent plant tissues contributes significantly to the development and the maintenance of epidemic within the crop (Sosa-Alvarez et al. 1995). It may be expected that plant sanitation involving removal of dead and infected leaves and/or fruits may have some effect on the incidence of gray mold disease of strawberry. The effects of leaf and fruit sanitation on the severity of strawberry gray mold disease and leaf spot disease caused by *Mycosphaerella fragariae* were assessed. Neither leaf sanitation nor fruit sanitation had any beneficial effect on the incidence/severity of gray mold disease. However, the combination of the one-row-system, leaf- and fruit-sanitation reduced the gray mold disease by 50 % in the first crop year, compared with the two-row-system without leaf- and fruit-sanitation. Severity of gray mold disease correlated significantly and positively with plant biomass. On the other hand, leaf spot severity was significantly reduced by about 90 % due to leaf sanitation and by 50 % due to planting strawberry in one-row-system, instead of two-row-system. Based on the results, it was suggested that for the central humid central European conditions, adoption of a one-row-system combined with leaf sanitation in early spring and fruit sanitation during harvest might be effective in reducing the risk of damages caused by gray mold and leaf spot diseases in strawberry (Schmid et al. 2005).

The fungal biocontrol agent *Ulocladium atrum* (*Ua*) was found to be strong competitor in necrotic plant tissues. It could successfully compete with *Botrytis cine*rea, the gray mold pathogen. The effectiveness of a competitive BCA applied in an

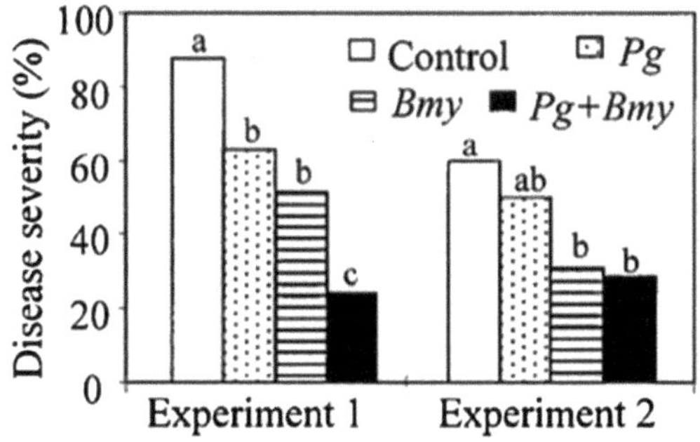

Fig. 7.10 Effect of root treatment with *Pichia guillermondii* (Pg), *Bacillus mycoides* (Bmy) or their mixtures on severity of gray mold disease caused by *Botrytis cinerea* in strawberry. *Bars with same letters* are not significantly different (P=0.05) as per LSD (Courtesy of Guetsky et al. 2002 and with kind permission of The American Phytopathological Society, MN, USA)

inundative release depends on the inoculum density, frequency of application, the application technique and microclimatic conditions around the crop canopy (Cook 1993). The biocontrol potential of *U. atrum* in suppressing the development of strawberry gray mold disease, using waiting-bed transplants, was assessed under field conditions. The efficacy of fungicide programs, *Ua*-spray programs and crop sanitation was compared with untreated controls. Under low disease pressure, *Ua*-spray programs effectively reduced gray mold at harvest in four of seven field trials. Sprays of *Ua* starting at transplanting resulted in better control of gray mold than sprays starting at the beginning of flowering in only one of the five experiments. Removal of necrotic leaves did not affect the level of gray mold which demonstrated that strawberry leaves were not a significant inoculum source for *B. cinerea* in the annual cropping system (Boff et al. 2002). The yeast *Pichia guillermondii* and the bacterial species *Bacillius mycoides* were evaluated, when applied alone or in combination for their efficacy in suppressing the development of the gray mold pathogen *Botrytis cinerea* infecting strawberry. Application of the biocontrol agents as a mixture to strawberry leaves suppressed the disease development more effectively than the individual BCAs. Further, the greater suppressive effect of the BCA mixture was observed under diverse conditions and also the inconsistency of disease control was significantly reduced. When the mixture of *P. guillermondii* and *B. mycodies* was applied as cell suspensions to the root zone of strawberry plants, the efficacy of suppression of *B. cinerea* development on inoculated leaves was at higher level than when they were applied separately in one experiment. In the other experiment, *B. mycoides* reduced the disease severity significantly, but *P. guillermondii* was not effective. Further, the presence of the yeast in the mixture did not improve the disease reduction that could be observed with *B. mycoides* applied alone (Fig. 7.10) (Guetsky et al. 2002).

Bacillus subtilis isolate S1-0210 was found to be antagonistic to *Botrytis cinerea* and a wettable powder formulation of the bacterial cells was prepared, using bean flour, rice flour, glucose $FeSO_4{\cdot}7H_2O$ and $MnCl_2{\cdot}4H_2O$. The formulation showed

85–89 % control efficacies of gray mold incidence on fruits of strawberry plants in the greenhouse. The formulated product was evaluated under field conditions. The fruits of all plants showed symptoms of infection in the untreated control plots. In contrast, the formulated product of *B. subtilis* effectively reduced the gray mold incidence with 70 % control efficacy. The bioproduct was as effective as the fungicide diethofencarb-carbendazim WP. The results indicated that the bioproduct containing *B. subtilis* isolate S1-0210 could be an effective alternative to the fungicide applied for gray mold disease in strawberries (Hang et al. 2005). A process for nutrient activation of the conidia of *Trichoderma harzianum* (*atroviride*) at 21 °C accelerated subsequent conidial germination of antagonist at temperatures from 9 to 21 °C. Both quiescent and preactivated conidia significantly reduced latent infection in greenhouse-grown strawberries at a mean temperature of 19 °C, whereas only preactivated conidia significantly reduced the gray mold disease in the field at a mean temperature of 14 °C on the day of application (Hjeljord et al. 2001). *Trichoderma atroviride* LU132 inhibited development of lesions induced by *B. cinerea* on detached senescing leaves and also reduced sporulation compared to control. The isolate LU132 was further evaluated under field conditions over a 2-year period, using three different strawberry cultivars. The BCA isolate significantly suppressed sporulation of *B. cinerea* on leaves and flower parts. The extent of disease control provided by *T. atroviride* LU132 was in equivalence to that of fungicide fenhexamid (Card et al. 2009).

7.2.5.5 Powdery Mildew Disease

Powdery mildew disease of strawberry has a worldwide distribution in various countries where strawberry is cultivated. Fungal and bacterial biocontrol agents have been evaluated for their ability to suppress the development of the powdery mildew disease. *Penicillium oxalicum* was effective in reducing disease development on different strawberry cultivars and genotypes in growth chamber and open-field nursery conditions. Treatment of strawberry cultivars and lines with *P. oxalicum* reduced powdery mildew spread on leaves (De Cal et al. 2008). In another investigation, the efficacy of *Ampelomyces quisqualis*, *Bacillus subtilis* and *Trichoderma harzianum* T-39 in suppressing the powdery mildew disease was assessed. The BCAs effectively controlled the powdery mildew, but none of them was more efficient than fungicide. When the BCAs were alternated with chemicals, the amount of chemical residues was significantly reduced, while good powdery mildew control was maintained. None of the biocontrol agents had any side effects either on the beneficial predatory mite *Amblyseius andersoni* population or any favorable effect on the pest *Tetranychus urticae*. Hence, the BCAs, tested in this investigation, can be regarded as safer for ecological balance for arthropods in strawberry greenhouses (Pertot et al. 2008).

7.2.5.6 Angular Leaf Spot Disease

Angular leaf spot disease (ALS) of strawberry caused by *Xanthomonas fragariae* (*Xf*), affects the seedlings in the nurseries. The infected seedlings are the principal mode of pathogen dispersal to the main field and elsewhere. Hence, disease management strategies are directed towards production of disease-free strawberry seedlings that are to be exported from USA to European countries. Populations of the bacterial pathogen *Xf* were examined for their thermosensitivity. The *Xf* cells exposed to 56 and 52 °C were killed completely after 15 and 60 min of exposure respectively. However, both treatments killed strawberry plants also. Strawberry seedlings of six cultivars were treated with hot water at 44 or 48 °C for 0, 60, 120, 180 and 240 min. Hot water treatments at 44 °C for 4 h (240 min) or 48 °C for 2 h (120 min) reduced pathogen population by 10^5 or 10^6 CFU/ml. These treatments minimally affected the vegetative growth of plants bagged dry or wet, but flowering was adversely affected. The strawberry cultivars varied in their sensitivity to heat. By selecting heat-tolerant strawberry cultivar, heat treatment for nursery stock might become feasible and it could be applied to supplement standard production practices for producing pathogen-free nursery stock (Turchek and Peres 2009).

7.2.6 Postharvest Diseases of Fruits

Various strategies, such as use of disease-free seed and planting materials, disease tolerant/resistant cultivars, application of soil solarization and/or antagonistic microorganisms and inducers of resistance to diseases have been employed under field conditions, to reduce the incidence and severity of diseases. These strategies have been demonstrated to reduce the pathogen load on the harvested fruits that are to be stored till they are shipped or marketed. The growing awareness and concern about the health hazards and environmental pollution due to chemical application have necessitated the acceleration of research and developmental efforts to find alternative nonchemical approaches for the effective management of postharvest diseases of fruits during storage. Biological control of diseases of fruits of involves the utilization of the biotic and/or abiotic agents that act through one or more similar mechanisms to reduce the potential of the pathogen directly or indirectly by activating the host defense systems to enhance the level of resistance of the treated fruits. Biocontrol strategies for postharvest diseases of fruits include use of antagonists, application naturally-derived compounds and induction of resistance of fruits by applying live microorganisms or their formulated products, physical or chemical agents. The postharvest environments provide some unique advantages and challenges for the use of biocontrol approaches.

7.2.6.1 Apple Diseases

Apples are infected by different fungal pathogens, of which *Penicillium expansum* causing blue mold disease is more widely distributed under storage conditions. Fungal and bacterial biocontrol agents have been applied alone or in combination with physical or chemical agents that could provide additive effects on the extent of

Fig. 7.11 A portable drencher developed for the biocontrol of *Penicillium expansum* infecting apples under storage (Courtesy of Janisiewicz et al. 2005 and with kind permission of The American Phytopathological Society, MN, USA)

disease control achieved. The yeast biocontrol agent *Metschnikowia pulcherrima* T5-A2 in combination with 1-methylcyclopropene (LMCP) was able to protect apple fruit against blue mold caused by *P. expansum* and also bitter rot disease caused by *Colletotrichum acutatum* on 'Golden Delicious' apples under controlled atmosphere (CA) conditions. Application of 1-MCP slowed down apple maturation and this might extend the action of natural mechanisms of disease resistance (Janisiewicz et al. 2003). A portable drencher, capable of drenching a single bin of fruit was built to simulate the commercial application of the test suspensions to harvested apples in small orchard operations. The drencher required only 125 l of treatment solutions and permitted various bin speeds. After placing the wounded apples at different positions in the bins, they were drenched with a suspension of *P. expansum*, followed by treatment with the yeast *Metschnikowia pulcherrima* alone or in combination with 2 % sodium bicarbonate (SB). Application of the yeast BCA reduced the decay on both cultivars and in combination with the yeast was the most effective treatment for protection of 'Golden Delicious' apple (Table 7.2). This portable drencher could be useful for evaluation of different treatments applied to apples after harvest at the commercial level (Fig. 7.11) (Janisiewicz et al. 2005).

The fungal pathogen *Penicillium expansum*, causing blue mold disease of apples, produces a mycotoxin patulin in the infected apple fruit tissues. The ability of the BCAs *Rhodotorula glutinis* LS11, *Cryptococcus laurentii* LS28 and *Aureobasidium pullulans* LD30 to grow in the presence of patulin and to decrease the mycotoxin concentration was assessed. The strain LS11 was more effective in reducing the decay due to infection by *Penicillium expansum*. In addition, accumulation of patulin was significantly lower in apples pretreated with the strain LS11. The results suggested that the yeast cells surviving in decaying apples could metabolize patulin and/or negatively affect its accumulation or synthesis (Castoria et al. 2005). In a later study, the compatibility of yeasts *Rhodosporidium kratochvilovae* LS11 and

Table 7.2 Efficacy of the yeast biocontrol agent alone or in combination with sodium bicarbonate in reducing blue mold disease incidence on two apple cultivars under cold storage at −2 °C for 3 months (Janisiewicz et al. 2005)

	Decay incidence (%)	
Treatment	Delicious[a]	Golden delicious
Control	66.0a	33.1a
Sodium bicarbonate (SB)	34.0b	2.7b
M. pulcherrima	26.0c	3.3b
M. pulcherrima + SB	8.8d	0.7b

Means with the same letter in a column are not significantly different according to Walter-Duncan multiple range test

[a]'Delicious' and 'Golden Delicious' apples were treated at 24 h after harvest

Cryptococcus laurentti LS28 with newly developed fungicides boscalid (BOSC), cyprodinil (CYPR) and fenhexamid (FENH) was examined to develop an integrated approach for the control of apple blue mold disease. Both BCA strains LS11 and LS28 were compatible with BOSC and CYPR, whereas they were strongly inhibited in in vitro assays. When applied alone at low dosage, LS11, LS28, BOSC and CYPR reduced blue mold by 35, 52, 67 and 72 % respectively. After 7 days of storage, only the integrated treatment based on BCA with BOSC or CYPR resulted in lower fungicide residues and patulin (PAT) contamination in apples. The results indicated that the effectiveness of the combination of yeast strains with a low dose of fungicide could be an efficient and safer strategy to control *P. expansum* and keep the fungicide residues as well as PAT contamination in treated apples low (Lima et al. 2011).

The effect of heat treatment combined with application of the bacterial BCA, *Pseudomonas syringae* on the development of blue mold pathogen, *Penicillium expansum* was determined. Apple fruits were first exposed to heat at 38 °C for 4 days and wound-inoculated with *P. expansum*, followed by treatment with *P. syringae*. Disease development was assessed, after storage for 7 days at 20 °C or 3 months at 1 °C. The heat treatment reduced the pathogen population on the apple surface, but provided little residual (transient) protection. Application of *P. syringae* further enhanced the level of disease control, because of the transient protection, indicating the synergistic action of heat and BCA. The combined of heat treatment and BCA resulted in substantial reduction of blue mold disease of apple caused by *P. expansum* (Leverentz et al. 2000). In another investigation, the eradicative activity of heat was integrated with the antagonistic activity of heat-tolerant yeasts. Prestorage hot air treatment was applied to apples at 12 h after inoculation with *P. expansum*, followed by application of yeasts. The biocontrol potential of yeasts was significantly improved by the heat treatment applied earlier (Leverentz et al. 2001). Pressure infiltration of $CaCl_2$ (0.27 M) followed by application of *Pseudomonas syringae* isolate ESC, (a component of BioSave 110) reduced blue mold decay on Golden Delicious apples inoculated with *Penicillium expansum*, incitant of blue mold decay to a greater extent, after cold storage for 6 months, compared to individual treatment with either $CaCl_2$ or BCA (Janisiewicz et al. 1998). Integration of $CaCl_2$ and BCA provided additional benefits, including reduction in the concentration of both components and alleviation of certain physiological disorders such as bitter pit, without loss of effectiveness of disease control (Janisiewicz and Korsten 2002a, b).

Gray mold of apples, another important postharvest disease is caused by *Botrytis cinerea*. Fungal and bacterial biocontrol agents have been shown to suppress the gray mold disease development to different degrees, when applied alone or in combination with other disease-suppressive agents. A biofumigant fungus *Muscodor albus* was able to inhibit and kill a wide range of storage pathogens including the gray mold and blue mold pathogens. Fumigation of apples for 7 days with the culture of *M. albus* grown on autoclaved rye grain provided complete control of *B. cinerea* and *Penicillium expansum*, causing gray and blue mold diseases of apples respectively. There was no need for direct contact between *M. albus* and the pathogenic fungi for its biocontrol activity. Hence, it coul be an attractive biological fumigant for controlling two important postharvest diseases of apples (Mercier and Jiménez 2004). The blue and gray molds of apples were more effectively controlled by the combined application of the yeast BCA *Candida saitoana* and glycochitosan (0.2 %). The effectiveness of disease reduction by the combined application was similar to that of the fungicide imazalil. A bioactive coating, consisting of *C. saitoana* and glycochitosan (0.2 %) controlled the decay of apples more effectively in many apple cultivars than either *C. saitoana* or glycochitosan alone. Furthermore, the level of disease control was comparable or even better than thiabendazole in reducing decay, depending on the apple cultivar tested (El-Ghaouth et al. 2000).

The effect of treatment of apples with the yeast antagonist *Cryptococcus laurentii* alone or in combination with indole-3-acetic acid (IAA) on the infection by *Botrytis cinerea*, causing gray mold disease during storage was assessed. The combined treatment with *C. laurentii* and IAA (20 μg/ml) was more effective in reducing gray mold disease incidence than the yeast alone. After 4 days of incubation, gray mold disease incidence in the combined treatment was about 18 % which was a 50 % reduction, compared with yeast treatment alone (Yu et al. 2008). *Cryptococcus laurentii* LS28 tolerant to high concentrations of thiabendazole (TBZ), when applied with a low dose of TBZ, was more effective than when it was applied alone in suppressing the development of gray mold disease on apple. The combined treatment provided prolonged control of fungal decay of apples. The integrated treatment was highly effective and durable showing high reduction of decay even after 18 days of storage (Lima et al. 2006). *Aureobasidium pullulans* isolate ApB was the most effective among the ten isolates tested, in reducing the gray mold disease development in apple. It was able to grow in a wide range of temperatures and it was resistant to TBZ, iprodione and imazalil commonly commercially applied fungicides for postharvest treatment of apples. The results indicated the adaptability of isolate ApB to temperature and feasibility of applying along with the fungicide application to provide more effective protection against gray mold disease during storage (Vero et al. 2009).

The possibility of enhancing the effectiveness of protection to apples against blue and gray mold diseases by combining the bacterial BCA with a compatible fungicide was examined. *Pseudomonas syringae* MA-4 and the fungicide cyprodinil were applied in combination on apples. The combined application provided greater than 90 % control against blue mold caused by *Penicillium expansum*. In contrast, gray mold caused by *Botrytis cinerea* could be controlled at a reduced concentration of cyprodinil (2.5 μg/ml) and a higher concentration of strain MA-4 (1×10^8 CFU/ml). As cypronidil does not have bactericidal activity, it could be safely combined with the bacterial antagonist *P. syringae* (Ting et al. 2002). Blue and gray mold diseases infect pears also in storage. The efficacy of *Cryptococcus laurentii* in combination with calcium chloride ($CaCl_2$) in suppressing these diseases was assessed. Combined treatment of pear fruit wounds with *C. laurentii* and $CaCl_2$ was more effective in inhibiting the growth of *Penicillium expansum* (blue mold) and *Botrytis cinerea* (gray mold) infections than the BCA or $CaCl_2$ alone. The chemical $CaCl_2$ elicited resistance responses against molds in fruits, when the time interval between $CaCl_2$-treatment and pathogen-inoculation was increased up to 24 h, being associated with an activation of the peroxidase activity of pear fruit. The results suggested that $CaCl_{2}$- reinforced biocontrol efficacy of *C. laurentii* might be due to induction of natural resistance in fruits (Yu et al. 2012).

The biocontrol potential of *Aureobasidium pullulans* strains L1 and L8 for suppressing the postharvest diseases gray mold (*Botrytis cinerea*), bitter rot (*Colletotrichum acutatum*) or blue mold (*Penicillium expansum*) was assessed. Washed cells of yeast strains controlled over 86 % of three types of decay. The cell concentrations of strains L1 and L8 were highly correlated with their efficacy (R^2 0.93 to 0.99). The highest concentration of L1 and L8 (10^8 CFU/ml) provided the best control of *B. cinerea*, *C. acutatum* and *P. expansum*. But for the control of gray mold, lower concentration (10^7 CFU/ml) was sufficient for effective suppression of disease development. Population dynamics of L1 strain in Gala apple increased almost by eight-fold, during the first 48 h, after treatment and remained at high level up to 7 days, indicating the good adaptation of L1 and L8 strains in wound environments. Dual culture assay revealed the production of volatile organic compounds (VOCs) by the yeast antagonists. The VOCs significantly inhibited the growth of all three pathogens tested, compared to the controls. When L1 strain was introduced into the wound at 12 h after pathogen inoculation, it exhibited curative effect by reducing blue mold and bitter rot by 38 and 50 % respectively, while the greatest suppression of gray mold occurred, when fruits were treated with the antagonist at 6 h after pathogen inoculation. The results indicated that *A. pullulans* strains L1 and L8 possessed antagonistic potential to the required level for the effective control of three important postharvest diseases of apple (Mari et al. 2012).

7.2.6.2 Citrus Diseases

Citrus spp. include several types of delicious fruits grown widely all over the world. Citrus fruits are infected by green mold disease caused by *Penicillium digitatum* and blue mold caused by *P. italicum* very commonly. Fungal and bacterial biocontrol agents either alone or in combination with physical or chemical agents have been applied to enhance the effectiveness of disease control. The combination of curing (at 37 °C for 72 h, 95 RH) of grapefruits inoculated with *P. digitatum* and after incubation for 1, 36 or 72 h at 25 °C and application of the yeast *Candida famata* (= *Torulopsis candida*) was evaluated for the control of decay. The combined treatments carried out within 36 h after inoculation with the pathogen, significantly reduced the decay percentage, both during storage and after a simulated marketing period of 1 week (D'hallewin et al. 1999). The yeast *Candida oleophila* was effective against the green mold disease infecting citrus. The commercial product Aspire, based on *C. oleophila* is recommended for application against citrus postharvest decay. It restricted the development of *P. digitatum* through multiple mechanisms of biocontrol activity like nutrient competition, site exclusion and direct mycoparasitism. Application of *C. oleophila* to surface wounds or intact 'Marsh Seedless' grapefruit, elicited systemic resistance against *P. digitatum*. Induction of pathogen resistance in fruit was pronounced 24 h after elicitation (Droby et al. 2002). Blue mold decay on oranges was completely suppressed by applying *Cryptococcus laurentii* prior to inoculation with *P. italicum* and incubation at 20 °C for 5 days at 4 °C for 30 days. Efficacy of *C. laurentii* was reduced, when inoculated oranges were treated with the BCA (Zhang et al. 2005).

The bacterial BCA *Pantoea agglomerans* CPA-2 was investigated for its potential to suppress the development of green and blue mold diseases caused respectively by *Penicillium digitatum* and *P. italicum* in citrus, when applied alone or in combination with sodium bicarbonate or sodium carbonate solutions under ambient (20 °C) and cold storage (3 °C) conditions. *P. agglomerans* controlled both the pathogens *P. digitatum* and *P. italicum* at a concentration of 2×10^8 CFU/ml. The efficacy of *P. agglomerans* was improved, when combined with sodium bicarbonate (SBC), resulting in complete and 97.6 % reduction of decay incidence at 3 and 20 °C respectively, when compared with untreated controls. Use of bicarbonate (2 %) treatment followed by application of the bacterial BCA could be an effective alternative to fungicide application for the management of postharvest disease of oranges (Teixidó et al. 2001). Commercial trials conducted with a formulated product based on the strains CPA-2 showed that natural decay due to *P. digitatum* and *P. italicum* was significantly reduced by the treatment of SBC (3 %) at 5 °C for 40 s. The effectiveness of the treatment was equal to that of chemical treatments in reducing both green and blue mold disease (Torres et al. 2007).

In another investigation, the effect of application of *Pantoea agglomerans* CPA-2 alone or in combination with a curing treatment at 33 °C for 65 h on the development of green mold (*P. digitatum*) on lemon stored at ambient and cold storage conditions was determined. Curing *P. agglomerans*-treated lemons at 33 °C for 65 h, entirely eliminated 24-h infections on artificially inoculated lemons stored at 20 °C for 14 days and on naturally infected lemons stored at 10 °C for 3 weeks plus seven additional days at 20 °C (Plaza et al. 2004). The effectiveness of the combination of *Pseudomonas* spp. with dipping in hot sodium bicarbonate solution for the control of green mold disease on oranges was also demonstrated, providing evidence for its potential to be considered as an alternative to the fungicides (Zamani et al. 2009).

The efficacy of *P. agglomerans* was evaluated for the suppression of *Penicillium digitatum*, causing green mold disease in Thomson Navel orange, when applied alone or in combination with dipping in 3 % sodium bicarbonate solution at 24 °C and 45 °C and stored at 20 °C (ambient) and 4 °C (cold storage). *P. agglomerans* alone reduced green mold incidence by 75 % at both temperatures. But the BCA alone was not as effective as the fungicide imazalil. *P. agglomerans* was completely tolerant to sodium bicarbonate (SBC) up to a concentration of 3 %. The efficacy of the bacterial BCA was enhanced at least by 5 and 11 %, when combined with 3 % sodium bicarbonate at 24 and 45 °C respectively. The combination of *P. agglomerans* and sodium bicarbonate provided protection to high levels consistently and has the potential for use as an alternative to chemical application for the control of postharvest decay in citrus (Zamani et al. 2009). The effects of pre- and postharvest application of sodium bicarbonate (SB), sodium carbonate (SC), sodium silicate (SS), potassium bicarbonate (PB), potassium carbonate (PC), potassium sorbate (PS), calcium chloride (CC) and calcium chelate (CCh) on the incidence of postharvest decay due to *Penicillium digitatum* and *P. italicum* on 'Commune' Clementine and 'Valencia Late' oranges were investigated. Aqueous salt solutions (2 % w/v, 20 hl/ha) were applied as preharvest sprays, or dipping after harvest and combination of sprays and dipping. Preharvest sprays and the combinations were more effective in suppressing decays than preharvest dipping. SC and PC completely suppressed decay development on Clementine and orange. Other salts reduced decays by 66–100 % and 78–100 % on oranges and Clementines respectively. Postharvest dipping was generally less effective, SC and PC being superior to other salts. The minimum inhibitory concentration (MIC) for both *P. digitatum* and *P. italicum* was 0.25 % for SB, SC, PB, PC, PS and SS. The results indicated that field application of salts could be a useful disease management strategy that might be included in integrated system for controlling postharvest diseases of citrus fruit (Youssef et al. 2012).

The biocontrol potential of *Bacillus* spp. for the control of green mold disease of citrus was evaluated. *Bacillus subtilis* strain GB03 was effective as the standard fungicide thiabendazole (TBZ) at 1,000 ppm concentration. The strain GB03 was most effective at 30 °C giving maximum reduction in disease incidence (Zhang and Dou 2002). The efficacy of *B. amyloliquefaciens* HF-01 in reducing the incidence of citrus green mold caused by *Penicillium digitatum* and blue mold caused by *P. italicum* was assessed, when applied alone or in combination with tea saponin (TS). The strain HF-01 significantly reduced the incidence of mold diseases, but it was less effective, compared with the fungicide treatment. The efficacy of HF-01 was significantly improved, when it was combined with TS (50 μg/ml) and the combined treatment was as effective as the fungicide treatment (>90 % control). *B. amyloliquefaciens* with a low dosage of TS significantly reduced postharvest decay without impairing any of the fruit quality parameters. The combination of HF-01 strain and TS could be useful as an effective alternative to synthetic fungicides for the control of citrus postharvest diseases (Hao et al. 2011). Wax coatings enriched with antifungals have been used to protect harvested citrus fruit against postharvest diseases. Essential oil (EO) of *Cinnamomum zeylandicum* was incorporated into commercial waxes shellac, carnauba, paraffin and polyethylene. The protective activity of these formulations against citrus blue and green mold diseases, as well as their viscosity and adherence to the orange fruit surface were assessed. Excellent disease suppression was provided by *C. zeylandicum* EO incorporated in shellac and/or carnauba wax, compared to other formulations tested. The effectiveness of the formulations depended on the volume that remained on fruit surface and retention of EO components on the fruit (Kouassi et al. 2012).

7.2.6.3 Grape Diseases

Postharvest decay of grapes is due to infection of berries by fungal pathogens *Botrytis cinerea*, *Alternaria* spp. and *Aspergillus niger*. The need for early detection of *Botrytis cinerea*, causing latent infections in the berry-pedicel attachment and stamens, as a tool to improve postharvest quality of table grapes was demonstrated. A quantitative real-time PCR (qPCR) method was developed, based on a probe designed on *B. cinerea* intergenic spacer (IGS) region. The method allowed reliable detection of pathogen in naturally infected asymptomatic tissues (Sanzani et al. 2012). The efficacy of application of the yeast *Metschnikowia fructicola* alone or in combination with ethanol or sodium bicarbonate (SBC) was assessed for the control of decay of table grapes. The vines were treated at 24 h before harvest and the extent of decay was assessed at different period of storage. All treatments significantly reduced the total number of decayed berries by the fungal pathogen after storage for 30 days at 1 °C, followed by 2 days at 20 °C. In three experiments, a mean gray mold incidence (*Botrytis cinerea*) of 34.2 infected berries/kg in untreated controls was reduced by *M. fructicola* (2×10^7 CFU/ml), ethanol (50 %) or SBC at 2 % to 12.9, 8.1 and 10.6 infected berries/kg respectively. The efficiency of *M. fructicola* was not improved, when combined with ethanol or SBC treatments (Fig. 7.12). Phytotoxic symptoms due to treatments with the yeast and SBC were visible on the rachis and berries. *M. fructicola* populations persisted on the berries during storage, when applied alone or after ethanol treatment, whereas SBC reduced the yeast population significantly (Karabulut et al. 2003).

The physical agents and naturally-derived organic compounds have been reported to suppress the development of postharvest diseases during storage. The effectiveness of chitosan treatment of table grapes alone or in combination with ultraviolet (UV-C) radiation to control gray mold caused by *Botrytis cinerea* was assessed. Clusters of cvs. Thompson Seedless, Autumn Black and Emperor were sprayed in the vineyard with chitosan (1 %) and then harvested daily for 5 days, followed by inoculation with *B. cinerea*. Decay incidence and disease severity were significantly reduced by chitosan which was most effective on berries harvested 1 or 2 days after treatment. In another experiment, grape berries were sprayed in the vineyard with chitosan, harvested 2 days later, irradiated for 5 min with UV-C and inoculated with *B. cinerea* 2 days later. Combined chitosan and UV-C treatments applied to cv. Autumn Black or selection B36-55 were synergistic in their action and reduced gray mold incidence and severity to a greater extent, compared with either treatment alone. The results indicated the usefulness of a

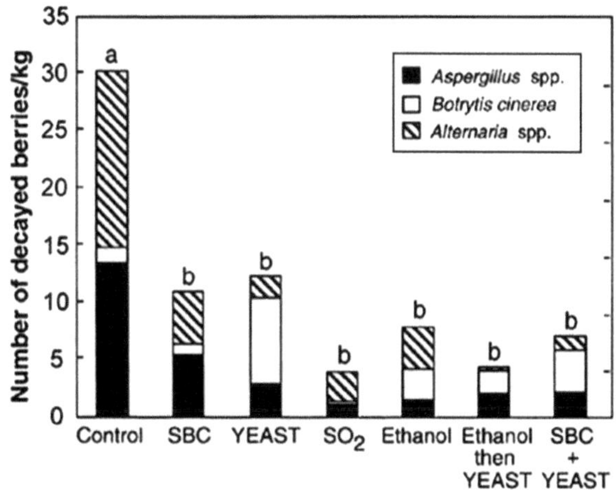

Fig. 7.12 Effect of application of sodium bicarbonate (SBC), Yeast *Metschnikowia fructicola* sulfur dioxide (SO_2) or ethanol alone or in combination on the postharvest decay of Thompson Seedless table grapes treated 24 h before harvest and stored for 6 days at 20 °C (Courtesy of Karabulut et al. 2003 and with kind permission of The American Phytopathological Society, MN, USA)

combination of preharvest and postharvest treatments with chitosan and UV-C irradiation in reducing gray mold disease incidence and severity effectively (Romanazzi et al. 2006).

7.2.6.4 Banana Diseases

The efficacy of sodium bicarbonate ($NaHCO_3$) in the suppressing the development of *Colletotrichum musae*, causing anthracnose disease was assessed. Addition of $NaHCO_3$ inhibited mycelial growth, spore production and germination and appressoria formation in vitro, by increasing pH (from 6.9 to 8.7) of potato dextrose (PD) broth. Postharvest dip of banana fruits in 300 mM $NaHCO_3$ for 10 min reduced the lesion area of anthracnose disease on artificially inoculated fruits. Natural infections of anthracnose, crown rot and blossom end rot were also reduced significantly in fruit following treatment with 300 mM $NaHCO_3$ for 10 min. The effect of combined treatment with $NaHCO_3$ and the bacterial antagonist *Burkholderia spinosa* on anthracnose disease development was investigated. Dipping banana fruit in 300 mM $NaHCo_3$ solution for 10 min, followed by dipping them in suspension of *B. spinosa* in nutrient broth (1 × 108 CFU/ml) effectively reduced anthracnose as well as crown rot and blossom-end rot diseases of banana cv. Kolikuttu. Increase in pH, total soluble solids and thickness of fruit peel due to treatment with $NaHCO_3$ might have an indirect or cumulative effect on reduction of postharvest disease incidence in banana (De Costa and Gunawardhana 2012).

The fungal pathogens *Lasiodiplodia theobromae*, *Thielaviopsis paradoxa*, *Colletotrichum musae* and *Fusarium veriticllioides* are associated with crown rot disease of banana fruit. *Trichoderma harzianum* strain DGA01 reduced the infection of banana fruit by crown rot pathogens. The effect of combination of *T. harzianum* with hot water treatment at 50 °C for 20 min on disease suppression was assessed, after storage at 22–25 °C and 90–95 % RH for 2 weeks. The bioefficacy of *T. harzianum* was enhanced by 11.4 % by combination with hot water treatment. The development of all crown rot pathogens was suppressed by the combined treatment. The strain DGA01 germinated on the fruit at 48 h after inoculation and parasitized the pathogens. Postharvest application of the BCA or hot water treatment separately was not effective, as the fungicide treatment. On the other hand, the combination of hot water treatment and DGA01 strain provided 93 % control of fruit decay which was in equivalence to the fungicide treatment (95 %). The quality of fruit was also markedly improved by hot water treatment in combination with the BCA to the level that could be achieved with fungicide treatment. The results indicated the usefulness of the combination of the fungal BCA and hot water treatment that could be an effective alternative to chemical treatment (Alvindia and Acada 2012).

7.3 Management of Diseases of Plantation Crops

7.3.1 Tea Diseases

7.3.1.1 Blister Blight Disease

Exobasidium vexans, the incitant of blister blight disease, causes extensive damage to leaf tissues, affecting tea leaf production. The efficacy of bioformulations based on the plant growth-promoting rhizobacterial species *Pseudomonas fluorescens* strain Pf1 in suppressing the development of blister blight disease was assessed under field conditions for two seasons. The bioproduct was applied as foliar spray at an interval of 7 days. Incidence of blister blight disease was consistently reduced by the biocontrol agent. The effectiveness of the strain Pf1 was almost equal to that of the fungicide. In addition, the tea yield was significantly enhanced by treatment with Pf1. The bacterial BCAwas considered to protect the tea plants by stimulating the natural defense systems, resulting in reduction in disease incidence and also plant growth (Saravanakumar et al. 2007).

7.3.1.2 Phomopsis Canker Disease

Phomopsis canker disease infecting tea is caused by *Phomopsis theae*, is responsible for extensive drying of twigs under high disease pressure conditions. The comparative efficacy of biological control agents and the fungicides was assessed under field conditions. Soil application and wound protection with biological control agents *Gliocladium* (= *Trichoderma*) *virens* and *Trichoderma harzianum* were more effective in protecting tea plants against *P. theae* than the fungicides. The systemic fungicides carbendazim applied as soil drench was more effective than copper oxychloride and mancozeb used for wound dressing. *G. virens* was more efficient in curing the cankers than *T. harzianum* and enhanced the yield of tea leaves as well. The initial carbohydrate level of 9.15 % increased to 11.89 %, when rejuvenated plants (pruned to healthy tissues) were treated with *G. virens* (Ponmurugan and Baby 2007).

7.3.1.3 Bird's Eye Spot Disease

Tea bird's eye spot disease caused by *Cercospora ocellata* (= *C. theae*) is responsible for considerable loss of foliage. The efficacy of plant products, bacterial and fungal biocontrol agents in reducing the disease incidence was assessed in comparison to fungicides under field conditions. Systemic fungicides were more efficient in controlling the disease. However, maximum green leaf yield was significantly increased by the biocontrol agents. *Streptomyces sannanensis* enhanced the leaf yield to the maximum extent, followed by *Trichoderma harzianum*. Two different carrier materials were used for formulating the BCAs. Vermicompost-based bioformulation was more effective than talc-based formulation. *S. sannanensis* protected the tea plants more effectively than *Pseudomonas fluorescens* and *T. harzianum* and also in enhancing the yield potential of tea plants. The parameters, plucking surface of the bush, number of plucking points/unit area, internodal length, leaf moisture and dry matter contents were greater in plots treated with biocontrol agents than those in plots treated with fungicides. In addition, the leaf quality parameters such as total liquor color, thearubigins, theaflavins, highly polymerized substance and caffeine were significantly improved by treatment with BCAs (Gnanambigai and Ponmurugan 2012).

7.3.1.4 Anthracnose Disease

Anthracnose disease caused by *Colletotrichum theae-sinensis* is another major disease affecting tea leaf production markedly. The antagonistic bacterial species *Bacillus subtilis* BD0310 effective against anthracnose disease was isolated from the phylloplane of tea trees. The suspension concentrate (SC) of the biofungicide was formulated. Cell viability and antifungal activity of *B. subtilis* were maintained in the formulation for more than 12 months at room temperature. The antagonist was sensitive only to copper sulfate among the four fungicides and seven insecticides tested. Under greenhouse conditions, the biofungicide controlled the disease more effectively in a protective mode than in a curative mode under field conditions. Alternated application of the biofungicide and fungicide were more effective in controlling the anthracnose disease than by the fungicide or biofungicide applied alone. The results showed that the anthracnose disease of tea could be effectively managed by alternate applications of *B. subtilis* and fungicide which could result in the restriction of chemical use (Kim et al. 2009).

7.3.2 *Coffee Rust Disease*

Coffee rust disease caused by *Hemileia vastatrix* has ruined coffee industry, because of its highly destructive nature. Development of alternative approaches to the conventional fungicide application has been considered as a priority for coffee production especially for organic coffee production. The endophytic bacterial species *Bacillus lentimorbus* and *B. cereus* were isolated from leaves and branches of *Coffea arabica* and *C. robusta. B. lentimorbus* isolate TG4-Ia and *B. cereus* isolate TF9-Ia were more effective in suppressing the development of leaf rust disease on leaf discs, detached leaves and coffee plants (Shiomi et al. 2006). Seven bacterial isolates obtained from coffee plants were evaluated for their efficacy in reducing the disease incidence. *Bacillus* sp. isolate B157 and *Pseudomonas* sp. isolate 286, selected because of their higher efficacy, were tested under field conditions, along with copper hydroxide and calcium silicate. None of the treatments was effective under high disease pressure (23.8 %) conditions. When the disease pressure was moderate (7.5 %), significant differences in disease incidence and severity between treatments could be observed. Both *Bacillus* sp. and *Pseudomonas* sp. were equally effective as copper hydroxide in reducing coffee rust disease incidence and severity. *Bacillus* sp. was more effective than *Pseudomonas* sp. The results indicated the potential of the bacterial BCAs as possible alternatives to application of fungicides (Haddad et al. 2009). In another investigation, the fungicides Bordeaux mixture and Bayleton and the bacterial BCAs *Bacillus subtilis* and *Pseudomonas fluorescens* were evaluated for their efficacy against coffee rust disease. The bacterial BCAs were less effective, compared with the fungicides in reducing the disease index. Bayleton was the most effective in reducing the disease intensity by 71.84 %, followed by Bordeaux mixture (53.37 %). On the other hand, *B. subtilis* could reduce the disease index by 42.98 %, whereas *P. fluorescens* reduced the disease index by 33.65 %. The combination of the biocontrol agents did not improve the effectiveness of rust disease control (Daivasikamani and Rajanaika 2009).

7.3.3 *Cocoa Diseases*

The efficacy of fungal biocontrol agents in suppressing the diseases of cocoa trees and pods has been assessed in in vitro assays and also under field conditions.

7.3.3.1 Witches Broom Disease

Witches broom disease caused by *Moniliophthora* (= *Crinipellis*) *perniciosa* is one of the most devastating diseases of cocoa, resulting in severe economical, social and environmental problems in certain countries like Brazil. Five native mycoparasite strains of *Clonostachys rosea* isolated from cocoa trees or basidiocarps of *M. perniciosa* were evaluated for their ability to suppress the symptoms of witches' broom disease. None of the strains of *C. rosea* was effective under field conditions. However, the combination of five strains (G-1–G-5) of *C. rosea* consistently showed high level of biocontrol activity against *M. perniciosa* (Krauss and Soberanis 2001). The antagonistic potential of *Trichoderma stromaticum* for suppressing the development of the pathogen *M. perniciosa* was investigated. Sixty three isolates of *T. stromaticum* were applied on brooms and placed under typical conditions of shaded cocoa plantations during two periods of 3 months each. The percentages of sporulation of the BCA and severity of disease were assessed to determine the effectiveness of treatment with the BCA. The results indicated a high phenotypic variation among isolates of *T. stromaticum*. The rates of sporulation of *T. stromaticum* were negatively correlated with the presence of *M. perniciosa* in the brooms. Many isolates of the BCA reduced the incidence of *M. perniciosa* more effectively than the reference isolate. Contrasting isolates with different efficiencies were assessed under field conditions. The results partially confirmed their biocontrol phenotypes, but also suggested isolate-specific responses to environmental variations. Inhibition of *M. perniciosa* basidiospore germination by total protein secreted in culture supernatants of *T. stromaticum* isolates correlated well with the field results. The procedure developed in this investigation may be useful for prescreening of large number of isolates for selection of efficient ones (Loguercio et al. 2009).

7.3.3.2 Cocoa Pod Diseases

Cocoa pods are affected by frosty pod rot disease caused by *Moniliophthora* (= *Crinipellis*) *roreri* and black pod disease caused by *Phytophthora palmivora*. The efficacy of two mycoparasitic fungi *Clonostachys byssicola* and *Trichoderma asperellum* field conditions. These BCAs were applied using motorized mist blowers (MMs) and hydraulic sprayers fitted with a narrow angle cone nozzle. Copper hydroxide, as prophylactic sprays at 1,500 g a.i./ha was found to be cost-effective and gave consistent protection against the black pod disease. The BCAs did not improve the yield of healthy pods (Bateman et al. 2005). Black pod rot disease affects the cocoa pods seriously in all countries inducing average losses above 30 %. The efficacy of endophytic fungus *Trichoderma martiale* strain ALF 247 in suppressing the development of *Phytophthora palmivora* was assessed. When the strain ALF247 was applied at concentrations ranging from 1×10^4 to 5×10^7 conidia/ml, the disease severity was reduced in proportion to the BCA propagules. The efficiency of the strain ALF247 was not affected by the addition of vegetable oil and/or sucrose in the formulation. The BCA was not sensitive to the fungicides copper hydroxide and fosetyl-AL, as reflected by the effect on conidial germination. The population of *T. martiale* was reduced progressively after application on cocoa pods after a period of 30–40 days, with concomitant increase in the severity of black pod disease. Rice grains as solid substrate supplemented with calcium carbonate favored rapid multiplication of *T. martiale*. The results revealed the potential of *T. martiale* for application against the black pod rot disease of cocoa pod (Hanada et al. 2009). Effectiveness of *Pseudomonas fluorescens* was applied as liquid formulation as soil

Fig. 7.13 Effect of application of *Pseudomonas fluorescens* on suppression of infection by Phytophthora palmivora under field conditions. Left: Untreated cocoa tree; Right: cocoa tree treated with *P. fluorescens* (Courtesy of Dr. P. Muthulakshmi, Tamil Nadu Agricultural University, Coimbatore, India)

drench and foliar spray for suppressing the infection of *Phytophthora palmivora*. The disease incidence was reduced significantly, in addition to increase in dry bean yield with higher cost-benefit ratio compared to other treatments (Fig. 7.13) (Dr. P. Muthulakshmi, Tamil Nadu Agricultural University, Coimbatore, India; Personal communication).

References

Abbasi PA, Al-Dahmani J, Sahin F, Hoitink HAJ, Miller SA (2002) Effect of compost amendments on disease severity and yield of tomato in organic and conventional production systems. Plant Dis 86:156–161

Abbasi PA, Conn KI, Lazarovits G (2004) Suppression of Rhizoctonia and Pythium damping-off of radish and cucumber seedlings by addition of fish emulsion to peat mix or soil. Can J Plant Pathol 26:177–187

Abo-Elyousr KAM, El-Hendawy HH (2008) Integration of *Pseudomonas fluorescens* and acibenzolar-*S*-methyl to control of bacterial spot disease of tomato. Crop Protect 27:1118–1124

Abo-Elyousr KAM, Ibrahim YE, Balabel NM (2012) Induction of disease defensive enzymes in response to treatment with acibenzolar-*S*-methyl (ASM) and *Pseudomonas fluorescens* Pf2 and inoculation with *Ralstonia solanacearum* race 3 biovar 2 (phylotype II). J Phytopathol 160:382–389

Addy HS, Askora A, Kawasaki T, Fujie M, Yamada T (2012) Utilization of filamentous phage ØRSM3 to control bacterial wilt caused by *Ralstonia solana*cearum. Plant Dis 96:1204–1209

Afek U, Orenstein J, Nutriel E (1999) Steam treatment to prevent carrot decay during storage. Crop Prot 18:639–642

Akgül DS, Mirik M (2008) Biocontrol of *Phytophthora capsici* on pepper plants by *Bacillus megaterium* strains. J Plant Pathol 90:29–34

Alaphilippe A, Elad Y, David DR, Derridj S, Gessler C (2008) Effects of a biocontrol agent of apple powdery mildew (*Podosphaera leucotricha*) on the host plant and on non-target organisms: an insect pest (*Cydia pomonella*) and a pathogen (*Venturia inaequalis*). Biocontrol Sci Technol 18:121–138

Algam SAE, Xie G, Li B, Yu S, Su T, Larsen J (2010) Effects of *Paenibacillus* strains and chitosan on plant growth promotion and control of Ralstonia wilt in tomato. J Plant Pathol 92:593–600

Al-Mughrabi KI (2010) Biological control of Fusarium dry rot and other potato tuber diseases using *Pseudomonas fluorescens* and *Enterobacter cloacae*. Biol Control 53:280–284

Altamiranda EAG, Andreu AVB, Daleo GR, Olivieri FP (2008) Effect of ß-aminobutyric acid (BABA) on protection against *Phytophthora infestans* throughout potato. Australas Plant Pathol 37:421–427

Alvindia DG, Acada MA (2012) An integrated approach with *Trichoderma harzianum* DGA01 and hot water treatment on control of crown rot disease and retention of overall quality in banana. Biocontrol Sci Technol 22:1021–1033

Ambadar VK, Sood AK (2010) Effect of solarization on tomato wilt incidence and population dynamics of *Pseudomonas fluorescens* and *Ralstonia solanacearum*. J Mycol Plant Pathol 40:120–123

Amith KN, Momol MT, Kloepper JW, Marois JW, Olson SM, Jones JB (2004) Efficacy of plant growth-promoting rhizobacteria, acibenzolar-*S*-methyl and soil amendment for integrated management of bacterial wilt on tomato. Plant Dis 88:669–673

Antoniou PP, Tjamos EC, Andreous MT, Panago-Poulos CG (1995) Effectiveness, modes of action and commercial application of soil solarization for control of *Clavibacter michiganensis* subsp. *michiganensis* of tomato. Acta Horticult 382:119–128

Askary H, Carriére Y, Bélanger R, Bordeur J (1998) Pathogenicity of the fungus *Verticillium lecanii* to aphids and powdery mildew. Biocontrol Sci Technol 8:23–32

Baker KF (1971) Fire blight of pome fruits: the genesis of the concept that bacteria can be pathogenic to plants. Hilgardia 40:603–633

Balogh B, Jones JB, Momol MT, Olson SM, Obradovic A, King P, Jackson L (2003) Improved efficacy of newly formulated bacteriophages for management of bacterial spot on tomato. Plant Dis 87:949–954

Basu A (2009) Employing eco-friendly potato disease management allows organic tropical Indian production systems to prosper. Asian J Food Agro-Indust 2(Spl Issue):S80–S87

Bateman RP, Hidalgo E, García J, Arroyo C, Ten Hoopen GM, Adonijah V, Krauss U (2005) Application of chemical and biological agents for the management of frosty pod rot (*Moniliophthora roreri*) in Costa Rican cocoa (*Theobroma cacao*). Ann Appl Biol 147:129–138

Baumgartner K, Rizzo DM (2001) Ecology of *Armillaria* spp. in mixed-hardwood forests of California. Plant Dis 85:947–951

Baumgartner K, Warnock AE (2006) A soil inoculant inhibits *Armillaria mellea* in vitro and improves productivity of grapevines with root disease. Plant Dis 90:439–444

Baysal Ö, Soylu EM, Soylu S (2003) Induction of defence-related enzymes and resistance by the plant activator acibenzolar-*S*-methyl in tomato seedlings against bacterial canker caused by *Clavibacter michiganensis* ssp. *michiganensis*. Plant Pathol 52:747–753

Baysal Ö, Gürsoy YZ, Örnek H, Duru A (2005) Induction of oxidants in tomato leaves treated with DL-ß-aminobutyric acid (BABA) and infected with *Clavibacter michiganensis* ssp. *michiganensis*. Eur J Plant Pathol 112:361–369

Beagle-Ristaino JE, Papavizas GC (1985) Biological control of Rhizoctonia stem canker and black scurf of potato. Phytopathology 75:560–564

Bélagner RR, Labbe C (2002) Control of powdery mildews: prophylactic and biological alternative for horticultural crops. In: Belanger RR, Bushnell WR, Dik AJ, Carver TLW (eds) The powdery mildews – a comprehensive treatise. APS Press, St. Paul, pp 256–267

Bélanger RR, Bowen PA, Ehret DL, Menzies JG (1995) Soluble silicon: its role in crop and disease management of greenhouse crops. Plant Dis 79:329–336
Bello GD, Mónaco C, Rollan MC, Lampugnani G, Arteta N, Abramoff C (2008) Biocontrol of postharvest grey mould on tomato by yeasts. J Phytopathol 156:257–263
Benlioğlu S, Boz Ö, Yildiz A, Kaşkavalci G, Benlioğlu K (2005) Alternative soil solarization treatments for the control of soilborne diseases and weeds of strawberry in the Western Anatolia of Turkey. J Phytopathol 153:423–430
Bi Y, Tian SP, Gao YR, Ge YH, Qin GZ (2006) Sodium silicate reduces postharvest decay on Hami melons: induced resistance and fungistatic effects. Plant Dis 90:279–283
Blok WJ, Lamers JG, Termorshuizen AJ, Bollen GJ (2000) Control of soilborne plant pathogens by incorporating fresh organic amendments followed by tarping. Phytopathology 90:253–259
Boff P, Köhl J, Jansen M, Horsten PJFM, Lombaers-van der Plas C, Gerlagh M (2002) Biological control of gray mold with *Ulocladium atrum* in annual strawberry crops. Plant Dis 86:220–224
Bonanomi G, Chiurazzi M, Caporaso S, Del Sorbo G, Moschetti G (2008) Soil solarization with biodegradable materials and its impact on soil microbial communities. Soil Biol Biochem 40:1989–1998
Boukaew S, Chuenchit S, Petcharat V (2011) Evaluation of *Streptomyces* spp. for biological control of Sclerotium root and stem rot and Ralstonia wilt of chilli pepper. BioControl 56:365–374
Bouws H, Finckh MR (2008) Effects of strip intercropping of potatoes with nonhosts on late blight severity and tuber yield in organic production. Plant Pathol 57:916–927
Bradley GG, Punja ZK (2010) Composts containing fluorescent pseudomonads suppress Fusarium root and stem rot development on greenhouse cucumber. Can J Microbiol 56:896–905
Brisset MN, Cesborn S, Pauling JP (2000) Acibenzolar-*S*-methyl induces the accumulation of defense-related enzymes in apple and protects from fire blight. Eur J Plant Pathol 106:529–536
Broadbent P, Dephoff CM, Franks N, Gillings M, Industo J (1995) Preimmunization of grapefruit with a mild protective isolate of *Citrus tristeza virus* in Australia. In: Proceedings of the 3rd international workshop CREC, Lake Alfred, pp 163–168
Budge SP, Whipps JM (2001) Potential for integrated control of *Sclerotinia sclerotiorum* in glasshouse lettuce using *Coniothyrium minitans* and reduced fungicide application. Phytopathology 91:221–227
Burkhead KD, Bothast RJ, Slininger PJ, Schister DA (2003) Postharvest biological control of potato sprouting by Fusarium dry rot suppressive bacteria. Biocontrol Sci Technol 13:477–494
Cañamás TP, Viñas I, Torres R, Usall J, Solsona C, Teixidó N (2011) Field applications of improved formulations of *Candida sake* CPA-1 for control of *Botrytis cinerea* in grapes. Biol Control 56:150–158
Card SD, Walter M, Jaspers MV, Sztejnberg A, Stewart A (2009) Targeted selection of antagonistic microorganisms for control of *Botrytis cinerea* of strawberry in New Zealand. Australas Plant Pathol 38:183–192
Carisse O, Rolland O (2004) Effect of timing of application of the biological control agent *Microsphaeropsis ochracea* on the production and ejection pattern of ascospores by *Venturia inaequalis*. Phytopathology 94:1305–1314
Carisse O, Philion V, Roland D, Bernier J (2000) Effect of fall application of fungal antagonists on spring ascospore production of the apple scab pathogen, *Venturia inaequalis*. Phytopathology 90:31–37
Carisse O, Rotland D, Tremblay D-M (2006) Effect of *Microsphaeropsis ochracea* on production of sclerotia-borne and airborne conidia of *Botrytis squamosa*. BioControl 51:107–126
Castoria R, Morena V, Caputo L, Panfili G, De Curtis F, De Cicco V (2005) Effect of the biocontrol yeast *Rhodotorula glutinis* LS11 on patulin accumulation in stored apples. Phytopathology 95:1271–1278
Chalutz E, Droby S (1998) Biological control of postharvest disease. In: Boland GJ, Kuykendrall LD (eds) Plant-microbe interactions and biological control. Marcel Dekker, New York, pp 157–170
Charles MT, Mercier J, Makhlouf J, Arul J (2008) Physiological basis of UV-C-induced resistance to *Botrytis cinerea*. I. Role of pre- and post-accumulation of phytoalexin- rishitin. Postharvest Biol Technol 47:10–20

Chellemi DO, Olson SM, Mitchell DJ (1994) Effects of soil solarization and fumigation on survival of soilborne pathogens of tomato in northern Florida. Plant Dis 78:1167–1172

Chen M-H, Nelson EB (2008) Seed-colonizing microbes from municipal biosolids compost suppress *Pythium ultimum* damping-off on different plant species. Phytopathology 98:1012–1018

Chen R, Guo YB, Wang JH, Li JY, Wang HM (2007) Biological control of grape crown gall by *Rahnella aqualis* HX2. Plant Dis 91:957–963

Chun WWC, Shetty KK (1994) Control of silver scurf disease of potatoes caused by *Helminthosporium solani* Dur. and Mont. with *Pseudomonas corrugata*. Phytopathology 84:1090 (Abst.)

Chung WC, Huang JW, Huang HC (2005) Formulation of a soil fungicide for control of damping-off of Chinese cabbage (*Brassica chinensis*) caused by *Rhizoctonia solani*. Biol Control 32:287–294

Clarkson JP, Payne T, Mead A, Whipps JM (2002) Selection of fungal biological control agents of *Sclerotium cepivorum* for control of white rot by sclerotial degradation in a UK soil. Plant Pathol 51:735–745

Clarkson JP, Mead A, Payne T, Whipps JM (2004) Effect of environmental factors and *Sclerotinia cepivorum* isolate on sclerotial degradation and biological control of white rot by *Trichoderma*. Plant Pathol 53:353–362

Clarkson JP, Scruby A, Mead A, Wright C, Smith B, Whipps JM (2006) Integrated control of Allium white rot with *Trichoderma viride*, tebuconazole and composted onion waste. Plant Pathol 55:375–386

Cook RJ (1993) Making greater use of introduced microorganisms for biological control of plant pathgoens. Annu Rev Phytopathol 31:53–80

Coşkuntuna A, Özer N (2008) Biological control of onion basal rot disease using *Trichoderma harzianum* and induction of antifungal compounds in onion set following seed treatment. Crop Prot 27:330–336

Coventry E, Noble R, Mead A, Whipps JM (2005) Suppression of Allium white rot (*Sclerotium cepivorum*) in different soils using vegetable wastes. Eur J Plant Pathol 111:101–112

Coventry E, Noble R, Mead A, Marin FR, Perez JA, Whipps JM (2006) Allium white rot suppression with composts and *Trichoderma viride* in relation to sclerotia viability. Phytopathology 96:1009–1020

D'hallewin G, Arras G, Dessi R, Dettori A, Schirra M (1999) Citrus green mould control in stored "Star Ruby" grapefruit by the use of biocontrol yeast under curing conditions. Acta Horticult 485:111–115

Daayf F, Adam L, Fernando WGD (2003) Comparative screening of bacteria for biological control of potato late blight (strain US-8) using in vitro, detached-leaves and whole-plant testing systems. Can J Plant Pathol 25:276–284

Daivasikamani S, Rajanaika (2009) Biological control of coffee leaf rust pathogen, *Hemileia vastatrix* Berkeley and Broome using *Bacillus subtilis* and *Pseudomonas fluorescens*. J Biopest 2:94–98

Davis JR, Huisman OC, Everson DO, Nolte P, Sorenson LH, Schneider AT (2010a) The suppression of Verticillium wilt of potato using corn as a green manure crop. Am J Potato Res 87:195–208

Davis JR, Huisman OC, Everson DO, Nolte P, Sorenson LH, Schneider AT (2010b) Ecological relationships of Verticillium wilt suppression of potato by green manures. Am J Potato Res 87:315–326

De Boer M, van der Sluis I, van Loon LC, Bakker PAHM (1999) Combining fluorescent *Pseudomonas* spp. strains to enhance suppression of Fusarium wilt of radish. Eur J Plant Pathol 105:201–210

De Boer M, Bom P, Kindt F, Keurentjes JJB, van der Sluis I, van Loon LC, Bakker PAHM (2003) Control of Fusarium wilt of radish by combining *Pseudomonas putida* strains that have different disease suppressive mechanisms. Phytopathology 93:626–632

De Cal A, Melgarejo P (2001) Repeated applications of *Penicillium oxalicum* prolongs biocontrol of Fusarium wilt of tomato plants. Eur J Plant Pathol 107:805–811

De Cal A, García-Lepe R, Melgarejo P (1999) Effects of timing and method of application of *Penicillium oxalicum* on efficacy and duration of control of Fusarium wilt of tomato. Plant Pathol 48:260–266

De Cal A, Redondo C, Sztejnberg A, Melgarejo P (2008) Biocontrol of powdery mildew by *Penicillium oxalicum* in open-field nurseries of strawberries. Biol Control 47:103–107

De Costa DM, Gunawardhana (2012) Effects of sodium bicarbonate on pathogenicity of *Colletotrichum musae* and potential for controlling postharvest diseases of banana. Postharvest Biol Technol 68:54–63

De Curtis F, Lima G, Vitullo D, De Cicco V (2010) Biocontrol of *Rhizoctonia solani* and *Sclerotium rolfsii* on tomato by delivering antagonistic bacteria through a drip irrigation system. Crop Protect 29:663–670

Deadman M, Al Hassani H, Al Sa'di A (2006) Solarization and biofumigation reduce *Pythium aphanidermatum*-induced damping-off and enhance vegetative growth of greenhouse cucumber in Oman. J Plant Pathol 88:335–337

Declerck S, Risede JM, Rufyikiri G, Delvaux B (2002) Effects of arbuscular mycorrhizal fungi on severity of root rot of bananas caused by *Cylindrocladium spathiphylli*. Plant Pathol 51:109–115

Ding J, Shi K, Zhou YH, Yu JQ (2009) Microbial community responses associated with the development of *Fusarium oxysporum* f.sp. *cucumerinum* after 24-epibrassinolide application to shoots and roots in cucumber. Eur J Plant Pathol 124:141–150

Dinz LP, Maffia LA, Dhingra OD, Casali VWD, Santos RHS, Mizubuti ESG (2006) Avaliaçao de produtos alternatives para controle da requeima do tomateiro. Fitopatol Bras 31:171–179

Domenech J, Reddy MS, Kloepper JW, Ramos B, Gutierrez-Mañero J (2006) Combined application of the biological product LS213 with *Bacillus*, *Pseudomonas* or *Chrysobacterium* for growth promotion and biological control of soilborne diseases in pepper and tomato. BioControl 51:245–258

Douglas SM (2010) Late blight of tomato and potato in Connecticut – 2010. The Connecticut Agricultural Experiment Station. www.ct.gov/caes

Droby S, Vinokur V, Weiss B, Cohen L, Daus A, Goldschmidt EE, Porat R (2002) Induction of resistance to *Penicillium digitatum* in grapefruit by the yeast biocontrol agent *Candida oleophila*. Phytopathology 92:393–399

Eikeno H, Stensvand A, Tronsmo AM (2003) Induced resistance as a possible means to control diseases of strawberry caused by *Phytophthora* spp. Plant Dis 87:345–350

Elad Y, Malathrakis NE, Dik AJ (1996) Biological control of *Botrytis*-incited diseases and powdery mildews in the greenhouse crops. Crop Protect 15:229–240

El-Ghaouth A, Smilanick JL, Brown GE, Ippolito A, Wisniewski M, Wilson CL (2000) Application of *Candida saitoana* and glycochitosan for the control of postharvest diseases of apples and citrus under semi-commercial conditions. Plant Dis 84:243–248

Elmer WH (2004) Combining nonpathogenic strains of *Fusarium oxysporum* with sodium chloride to suppress Fusarium crown rot of asparagus in replanted fields. Plant Pathol 53:751–758

Elmer PAG, Reglinski T, Wood PN, Hill RA, Marsden SM, Parry FJ, Taylor JT (2003) Suppression of *Botrytis* in grapes using a combination of elicitors and fungal antagonists. In: 8th international plant pathology congress, Christchurch, p 43

Elmer PAG, Hoyte SM, Vanneste JL, Reglinski T, Wood PN, Parry FJ (2005) Biological control of fruit pathogens. N Z Plant Protect 58:47–54

El-Mougy NS, Abdel-Kader MM (2008) Long-term activity of bio-priming seed treatment for biological control of faba bean root rot pathogens. Australas Plant Pathol 37:464–471

El-Tarabily KA (2004) Suppression of *Rhizoctonia solani* diseases of sugar beet by antagonistic plant growth-promoting yeasts. J Appl Microbiol 96:69–75

Eshel D, Regev R, Orenstein J, Droby S, Gan-Mor S (2009) Combining physical, chemical and biological methods for synergistic control of postharvest diseases: a case study of black root rot of carrot. Postharvest Biol Technol 54:48–52

Etebarian HR, Scott ES, Wicks TJ (2000) *Trichoderma harzianum* T-39 and *T.virens* DAR 74290 as potential biological control agents for *Phytophthora erythroseptica*. Eur J Plant Pathol 106:329–337

Evans JR, Evans RR, Regusi CL, Rademacher W (1999) Mode of action, metabolism and uptake of BAS 125W, prohexadione-calcium. Hortscience 34:1200–1201

Fallik E, Ilic Z, Alkalai-Tuvia S, Copel A, Polevaya Y (2002) A short hot water rinsing and brushing reduces chilling injury and enhances resistance against *Botrytis cinerea* in harvested tomato. Adv Hortic Sci 16:3–6

Fiedler Z, Sosnowska D (2007) Nematophagous fungus *Paecilomyces lilacinus* (Thom.) Samson is also a biological agent for control of greenhouse insects and mite pests. BioControl 52:547–558

Flaherty JE, Jones JB, Harbaugh BK, Somodi GC, Jackson LE (2000) Control of bacterial spot on tomato in the greenhouse and field with h-mutant bacteriophages. Hortscience 35:882–884

Fourie PH, Halleen F (2006) Chemical and biological protection of grapevine propagation material from trunk disease pathogens. Eur J Plant Pathol 116:255–265

Francis MI, Redondo A, Burns JK, Graham JH (2009) Soil application of imidacloprid and related SAR-inducing compounds produce effective and persistent control of citrus canker. Eur J Plant Pathol 124:283–292

Fravel DR, Deahl KL, Stormmel JR (2005) Compatibility of the biocontrol fungus *Fusarium oxysporum* strain CS-20 with selected fungicides. Biol Control 34:165–169

Freeman S, Zveibil A, Vintal H, Maymon M (2002) Isolation of nonpathogenic mutants of *Fusarium oxysporum* f.sp. *melonis* for biological control of Fusarium wilt in cucurbits. Phytopathology 92:164–168

French-Monar RD, Rodrigues FA, Korndörfer GH, Datnoff LE (2010) Silicon suppresses phytophthora blight development on bell pepper. J Phytopathol 158:554–560

Fu G, Huang S, Ye Y, Wu Y, Cen Z, Lin S (2010) Characterization of a bacterial biocontrol strain B106 and its efficacies on controlling banana leaf spot and postharvest anthracnose diseases. Biol Control 55:1–10

Gao Y, Reitz SR, Wang J, Tamez-Guerra P, Wang E, Xu X, Lei Z (2012a) Potential use of the fungus *Beauveria bassiana* against the western flower thrips *Frankliniella occidentalis* without reducing the effectiveness of its natural predator *Oris sauteri* (Hemiptera: Anthocoridae). Biocontrol Sci Technol 22:803–812

Gao Y, Reitz SR, Wang J, Xu X, Lei Z (2012b) Potential of a strain of the entomopathogenic fungus *Beauveria bassiana* (Hypocreales: Cordycipitaceae) as a biological control agent against western flower thrips, *Frankliniella occidentalis* (Thysanoptera: Thripidae). Biocontrol Sci Technol 22:491–495

Gent DH, Schwartz HF (2005) Management of Xanthomonas leaf blight of onion with a plant activator, biological control agents and copper bactericides. Plant Dis 89(6):631–639

Georgakopoulos DG, Fiddaman P, Leifert C, Malathrakis NE (2002) Biological control of cucumber and sugarbeet damping-off caused by *Pythium ultimum* with bacterial and fungal anatagonists. J Appl Microbiol 92:1078–1086

Ghorbani R, Wicokson SJ, Giostis C, Leifert C (2004) Potato late blight management in organic agriculture. Outlooks Pest Manag 15:176–180

Gilardi G, Garibaldi A, Gullino ML (2007) Effect of antagonistic *Fusarium* spp. and of different commercial biofungicide formulation on Fusarium wilt of lettuce. Phytoparasitica 35:457–465

Glass JR, Johnson KB, Powelson ML (2001) Assessment of barriers to prevent the development of potato tuber blight caused by *Phytophthora infestans*. Plant Dis 85:521–528

Gnanambigai BM, Ponmurugan P (2012) Evaluation of various fungicides and microbial based biocontrol agents against bird's eye spot disease of tea plants. Crop Protect 32:111–118

Graham JH, Myers ME (2011) Soil application of SAR inducers imidacloprid, thiamethoxam and acibenzolar-*S*-methyl for citrus canker control in young grapefruit trees. Plant Dis 95:725–728

Gravel V, Martinez C, Antoun H, Tweddell RJ (2005) Antagonistic microorganisms with the ability to control Pythium damping-off of tomato seeds in rockwool. BioControl 50:771–786

Grosch R, Faltin F, Lottmann J, Kofoet A, Berg G (2005) Effectiveness of 3 antagonistic bacterial isolates to control *Rhizoctonia solani* Kühn on lettuce and potato. Can J Microbiol 51:345–353

Guetsky R, Shtienberg D, Elad Y, Fischer E, Dinoor A (2002) Improving biological control by combining biocontrol agents each with several mechanisms of disease suppression. Phytopathology 92:976–985

Haddad F, Maffia LA, Mizubuti ESG, Teixeira H (2009) Biological control of coffee rust by antagonistic bacteria under field conditions in Brazil. Biol Control 49:114–119

Hanada RE, Pomella AWV, Soberanis W, Loguercio LL, Pereira JO (2009) Biocontrol potential of *Trichoderma martiale* against the black pod disease (*Phytophthora palmivora*) of cacao. Biol Control 50:143–149

Hang NTT, Oh S-O, Kim GH, Hur J-S, Koh YJ (2005) *Bacillus subtilis* S1-0210 as a biocontrol agent against *Botrytis cinerea* in strawberries. Plant Pathol J 21:59–63

Hao J, Subbarao KV, Koike ST (2003) Effects of broccoli rotation on lettuce drop caused by *Sclerotinia minor* and on population density of sclerotia in soil. Plant Dis 87:159–166

Hao W, Li H, Hu M, Yang L, Rizwan-ul-Haq M (2011) Integrated control of citrus green and blue mold and sour rot by *Bacillus amyloliquefaciens* in combination with tea saponin. Postharvest Biol Technol 59:316–323

Hassan MAE, Buchenaeur H (2007) Induction of resistance to fire blight in apple by acibenzolar-*S*-methyl and DL-3-aminobutyric acid. J Plant Dis Protect 114:151–158

Hassan MAE, Buchenauer H (2008) Enhanced control of bacterial canker of tomato by DL-3-aminobutyric acid and the fluorescent *Pseudomonas* isolate CW2. J Plant Dis Protect 115:199–207

Heller WE, Gessler C (1986) Induced systemic resistance in tomato plants against *Phytophthora infestans*. J Phytopathol 116:323–328

Herman MAB, Davidson JK, Smart CD (2008) Induction of plant defense gene expression by plant activators and *Pseudomonas syringae* pv. *tomato* in greenhouse-grown tomatoes. Phytopathology 98:1126–1232

Hjeljord LG, Stensvand A, Tronsmo A (2001) Antagonism of nutrient-activated conidia of *Trichoderma harzianum* (*atroviride*) P1 against *Botrytis cinerea*. Phytopathology 91:1172–1180

Hong JC, Momol MT, Ji P, Olson SM, Colee J, Jones JB (2011) Management of bacterial wilt in tomatoes with thymol and acibenzolar-*S*-methyl. Crop Protect 30:1340–1345

Hopkins DL (2005) Biological control of Pierce's disease in the vineyard with strains of *Xylella fastidiosa* benign to grapevine. Plant Dis 89:1348–1352

Houeto P, Bindoula G, Hoffman JR (1995) Ethylenebisdithiocarbamate and ethylenethiourea: possible human health hazards. Environ Health Perspect 103:568–573

Huang H-C, Erickson RS (2007) *Ulocladium atrum* as a biological control agent for white mold of bean caused by *Sclerotinia sclerotiorum*. Phytoparasitica 35:15–22

Isnaini M, Keane PJ (2007) Biocontrol and epidemiology of lettuce drop caused by *Sclerotinia minor* at Bacchus Marsh, Victoria. Australas Plant Pathol 36:295–304

Jager G, Velvis JH (1986) Biocontrol of *Rhizoctonia solani* on potatoes by antagonists: the effectiveness of three isolates of *Verticillium biguttatum* as inoculum for seed tubers and of a soil treatment with low dosage of pencycuron. Neth J Plant Pathol 92:231–238

Jaime MDLA, Hsiang T, McDonald MR (2008) Effects of *Glomus intraradices* and onion cultivar on Allium white rot development in organic soils in Ontario. Can J Plant Pathol 30:543–553

Janisiewicz WJ, Korsten L (2002a) Biological control of postharvest diseases of fruits. Annu Rev Phytopathol 40:411–441

Janisiewicz WJ, Korsten L (2002b) Microbial control of postharvest disease and spoilage. In: Bartz JA, Brecht JA (eds) Postharvest physiology and pathology of vegetables. Marcel Dekker, New York, pp 543–562

Janisiewicz WJ, Conway WS, Glenn DM, Sams CE (1998) Integrating biological control and calcium treatment for controlling postharvest decay of apples. Hortscience 33:105–109

Janisiewicz WJ, Leverentz B, Conway WS, Saftner RA, Reed AN, Camp MJ (2003) Control of bitter rot and blue mold of apples by integrating heat and antagonist treatments on 1-MCP-treated fruit stored under controlled atmospheric conditions. Postharvest Biol Technol 29:129–143

Janisiewicz WJ, Peterson DL, Yoder KS, Miller SS (2005) Experimental bin drenching system for testing biocontrol agents to control postharvest decay of apples. Plant Dis 89:487–490

Janousek CN, Lorber JD, Gubler WD (2009) Combination and rotation of bacterial antagonists to control powdery mildew on pumpkin. J Plant Dis Protect 116:260–262

Jayaraj J, Radhakrishnan NV, Kannan R, Sakthivel K, Suganya D, Venkatesan S, Velazhahan R (2005) Development of new formulations of *Bacillus subtilis* for management of tomato damping-off caused by *Pythium aphanidermatum*. Biocontrol Sci Technol 15:55–65

Jetiyanon K, Fowler WD, Kloepper JW (2003) Broad-spectrum protection against several pathogens by PGPR mixtures under field conditions in Thailand. Plant Dis 87:1390–1394

Ji P, Momol MT, Olson SM, Pradhanang PM, Jones JB (2005) Evaluation of thymol as biofumigant for control of bacterial wilt of tomato under field conditions. Plant Dis 89:497–500

Ji P, Campbell HL, Kloepper JW, Jones JB, Suslow TV, Wilson M (2006) Integrated biological control of bacterial speck and spot of tomato under field conditions using foliar biological control agents and plant growth-promoting rhizobacteria. Biol Control 36:358–367

Ji P, Momol MT, Rich JR, Olson SM, Jones JB (2007) Development of an integrated approach for managing bacterial wilt and root-knot on tomato under field conditions. Plant Dis 91:1321–1326

Ji P, Yin J, Kone D (2011) Application of acibenzolar-*S*-methyl (ASM) and standard fungicides for control of Phytophthora blight on squash. Crop Protect 30:1601–1605

Jiang Z-Q, Guo Y-H, Li S-M, Qi H-Y, Guo J-H (2006) Evaluation of biocontrol efficiency of different *Bacillus* preparations and field application methods against Phytophthora blight of bell pepper. Biol Control 36:216–223

Jindal KK, Singh H, Meeta M (1988) Biological control of *Phytophthora infestans* on potato. Indian J Plant Pathol 6:59–62

John S, Wicks TJ, Hunt JS, Lorimer MF, Oakey H, Scott ES (2005) Protection of grapevine pruning wounds from infection by *Eutypa lata* using *Trichoderma harzianum* and *Fusarium lateritium*. Australas Plant Pathol 34:569–575

Johnson TPC, Grout BWW, Bishop CFH, Perera A (2003) The use of radiant heat to reduce the inoculum level of silver scurf (*Helminthosporium solani*) on potato tubers (*Solanum tuberosum*) before storage. Acta Horticult 619:207–211

Jones AL (2001) Challenges of controlling fire blight. Comp Fruit Tree 34:86–91

Juarez-Palacios C, Felix-Gastelum R, Wakeman RJ, Paplomats EJ, De Vay JE (1991) Thermal sensitivity of three species of *Phytophthora* and the effect of soil solarization on their survival. Plant Dis 75:1160–1164

Karabulut OA, Smilanick JL, Gabler FM, Mansour M, Droby S (2003) Near harvest applications of *Metschnikowia fructicola*, ethanol and sodium bicarbonate to control postharvest diseases of grape in central California. Plant Dis 87:1384–1389

Kato M, Mizbuti ESG, Goodwin SB, Fry WE (1997) Sensitivity to protect fungicides and pathogenic fitness of clonal lineages of *Phytophthora infestans* in the United States. Phytopathology 87:973–978

Kawaguchi A, Inoue K (2012) New antagonistic strains of nonpathogenic *Agrobacterium vitis* to control grapevine crown gall. J Phytopathol 160:509–518

Kerr A (1980) Biological control of crown gall through production of agrocin 84. Plant Dis 64:28–30

Kidane EG, Laing MD (2010) Integrated control of Fusarium wilt of banana (*Musa* spp.). Acta Horticult 879:315–321

Kiewnick S, Jacobsen BJ, Eckhoff JLA B-KA, Bergman JW (2001) Integrated control of Rhizoctonia crown and root rot of sugar beet with fungicides and antagonistic bacteria. Plant Dis 85:718–722

Kim CH (1989) Phytophthora blight and other diseases of red pepper in Korea: disease and pest problems from continuous cropping. II. Soil-borne diseases. FFTC Ext Bull 302:10–17

Kim S-T, Yun S-C (2011) Biocontrol with *Myxococcus* sp. KYC 1126 against anthracnose in hot pepper. Plant Pathol J 27:156–163

Kim YC, Jung H, Kim KY, Park SK (2008) An effective biocontrol formulation against Phytophthora blight of pepper using growth mixtures of combined chitinolytic bacteria under different field conditions. Eur J Plant Pathol 120:373–382

Kim GH, Lim MT, Hur J-S, Yum K-J, Koh YJ (2009) Biological control of tea anthracnose using antagonistic bacterium of *Bacillus subtilis* isolated from tea leaves. Plant Pathol 25:99–102

Klein E, Katan J, Gamliel A (2011) Soil suppressiveness to Fusarium disease following organic amendments and solarization. Plant Dis 95:1116–1123

Konstantinidou-Doltsinis S, Markellou E, Kasselaki A-M, Siranidou E, Kalamarakis A, Tzembelikou K, Schmitt A, Koumakis C, Malathrakis N (2007) Control of powdery mildew of grape in Greece using Sporodex®L and Milsana®. J Plant Dis Protect 114:256–262

Kosaka Y, Fukunishi T (1997) Multiple inoculation with three attenuated viruses for the control of cucumber virus disease. Plant Dis 81:733–738

Kosaka Y, Ryang B-S, Kobori T, Shiomi H, Yasuhara H, Kataoka M (2006) Effectiveness of an attenuated *Zucchini yellow mosaic virus* isolate for cross-protecting cucumber. Plant Dis 90:67–72

Kouassi KHS, Bajji M, Jijakli H (2012) The control of postharvest blue and green molds of citrus in relation with essential oil-wax formulations, adherence and viscosity. Postharvest Biol Technol 71:122–128

Krauss U, Soberanis W (2001) Biocontrol of cocoa pod diseases with mycoparasite mixtures. Biol Control 22:149–158

Kurze S, Bahl H, Dahl R, Berg G (2001) Biological control of fungal strawberry diseases by *Serratia plymuthica* HRO-C48. Plant Dis 85:529–534

Lafontaine JP, Benhamou N (1996) Chitosan treatment: an emerging strategy for enhancing resistance of greenhouse tomato plants to infection by *Fusarium oxysporum* f.sp. *radicis-lycopersici*. Biocontrol Sci Technol 6:111–124

Lambert DH, Salas B (2001) Pink rot. In: Stevenson WR, Loria R, Franc GD, Weingartner DP (eds) Compendium of potato diseases. APS Press, St. Paul, pp 33–34

Larena I, Melgarejo P, De Cal A (2003) Production, survival and evaluation of solid-substrate inocula of *Penicillium oxalicum*, a biocontrol agent against Fusarium wilt of tomato. Phytopathology 92:863–869

Larkin RP, Fravel DR (1998) Efficacy of various fungal and bacterial biocontrol organisms for control of Fusarium wilt of tomato. Plant Dis 82:1022–1028

Larkin RP, Fravel DR (2002) Effects of varying environmental conditions on biological control of Fusarium wilt of tomato by nonpathogenic *Fusarium* spp. Phytopathology 92:1160–1166

Larkin RP, Honeycutt CW, Griffin TS, Olanya OM, Halloran JM, He Z (2011) Effects of different potato cropping system approaches and water management on soilborne diseases and soil microbial communities. Phytopathology 101:58–67

Leverentz B, Janisiewicz WJ, Conway WS, Saftner RA, Fuchs Y, Sams CE, Camp MJ (2000) Combining yeasts or a bacterial biocontrol agent and heat treatment to reduce postharvest decay of "Gala" apples. Poshtharvest Biol Technol 21:87–94

Leverentz B, Janisiewicz WJ, Conway WS, Saftner RA (2001) Effect of combining biocontrol heat treatment and MCP-treatment on the reduction of postharvest decay of 'Delicious' apples. Phytopathology 91:S55 (Abst.)

Liang YC, Sun WC, Si J, Römheld V (2005) Effects of foliar- and root-applied silicon on the enhancement of induced resistance to powdery mildew in *Cucumis sativus*. Plant Pathol 54:678–685

Liljeroth E, Bengtsson T, Wük L, Andreasson E (2010) Induced resistance on potato to *Phytophthora infestans*: effects of BABA in greenhouse and field tests with different potato varieties. Eur J Plant Pathol 127:171–183

Lima G, De Curtis F, Piedimonte D, Spina AM, De Cicco V (2006) Integration of biocontrol yeast and thiabendazole protects stored apples from fungicide sensitive and resistant isolates of *Botrytis cinerea*. Postharvest Biol Technol 40:301–307

Lima G, Castoria R, De Curtis F, Raiola A, Ritieni A, De Cicco V (2011) Integrated control of blue mold using new fungicides and biocontrol yeasts lowers levels of fungicide residues and patulin contamination in apples. Postharvest Biol Technol 60:164–172

Lin T-C, Ishizaka M, Ishii H (2009) Acibenzolar-*S*-methyl induced systemic resistance against anthracnose and powdery mildew diseases on cucumber plants without accumulation of phytoalexins. J Phytopathol 157:40–50

Ling N, Xue C, Huang Q, Yang X, Xu Y, Shen Q (2010) Development of a mode of application of bioorganic fertilizer for improving the biocontrol efficacy to Fusarium wilt. BioControl 55:673–683

Lobato MC, Olivieri FP, Altamiranda EAG, Wolski EA, Daleo GR, Caldiz DO, Andrew AB (2008) Phosphite compounds reduce disease severity in potato seed tubers and foliage. Eur J Plant Pathol 122:349–358

Loguercio LL, de Carvalho AC, Niella GR, De Souza JT, Pomella AWV (2009) Selection of *Trichoderma stromaticum* isolates for efficient biological control of witches' broom disease in cacao. Biol Control 51:130–139

Lopez-Mondejar R, Bernal-Vicente A, Ros M, Tittarelli F, Canali S, Intrigiolo F, Pascual JA (2010) Utilisation of citrus compost-based growing media amended with *Trichoderma harzianum* T-78 in *Cucumis melo* L. seedling production. Biores Technol 101:3718–3723

Louws FJ, Wilson M, Campbell HL, Cupples DA, Jones JB, Shoemaker PB, Sahin F, Miller SA (2001) Field control of bacterial spot and bacterial speck to tomato using a plant activator. Plant Dis 85:481–488

Malolepsza U (2006) Induction of disease resistance by acibenzolar-*S*-methyl and *O*-hydroxy-ethyolrutin against *Botrytis cinerea*. Crop Protect 25:956–962

Manikandan R, Saravanakumar D, Rajendran L, Raguchander T, Samiyappan R (2010) Standardization of liquid formulation of *Pseudomonas fluorescens* Pf1 for its efficacy against Fusarium wilt of tomato. Biol Control 54:83–89

Mari M, Martini C, Spadoni A, Rouissi W, Bertolini P (2012) Biocontrol of apple postharvest decay by *Aureobasidium pullulans*. Postharvest Biol Technol 73:56–62

Martin FN, Bull CT (2002) Biological approaches for control of root pathogens of strawberry. Phytopathology 92:1356–1362

Matheron ME, Porchas M (2002) Suppression of Phytophthora root and crown rot on pepper plants treated with acibenzolar-*S*-methyl. Plant Dis 86:292–297

Matheron ME, Porchas M (2010) Evaluation of soil solarization and flooding as management tools for Fusarium wilt of lettuce. Plant Dis 94:1323–1328

Maxson-Stein K, He S-Y, Hammerschmidt R, Jones AL (2002) Effects of treating apple trees with acibenzolar-*S*-methyl on fire blight and expression of pathogenesis-related protein genes. Plant Dis 68:785–790

Mazzola M (1998) Towards the development of sustainable alternatives for the control of apple replant disease in Washington. In: Proceedings of the annual international research conference on methyl bromide alternatives and emissions reductions, Fesno, pp 8.1–8.3

Mazzola M (1999) Transformation of soil microbial community structure and Rhizoctonia-suppressive potential in response to apple roots. Phytopathology 89:920–927

Mazzola M, Gu Y-H (2000) Impact of wheat cultivation on microbial communities from replant soils and apple growth in greenhouse trials. Phytopathology 90:114–119

Mazzola M, Gu Y-H (2002) Wheat genotype-specific induction of soil microbial communities suppressive to disease incited by *Rhizoctonia solani* anastomosis group (AG)-5 and AG-8. Phytopathology 92:1300–1307

McLean KL, Hunt JS, Stewart A, Wite D, Porter IJ, Villalta O (2012) Compatibility of a *Trichoderma atroviride* biocontrol agent with management practices of *Allium* crops. Crop Protect 33:94–100

Md. Hamiduzzaman M, Jakab G, Barnavon L, Neuhaus J-M, Mauch-Mani B (2005) ß-amin-obutyric acid-induced resistance against downy mildew in grapevine acts through the potentiation of callose formation and jasmonic acid signaling. Mol Plant-Microbe Interact 18:819–829

Melero-Vara JM, Prados-Ligero AM, Basallote-Ureba MJ (2000) Comparison of physical, chemical and biological methods of controlling garlic white rot. Eur J Plant Pathol 106:581–588

Mercier J, Jiménez J (2004) Control of fungal decay of apples and peaches by the biofumigant fungus *Muscodor albus*. Postharvest Biol Technol 31:1–8

Meshram SU (1984) Suppressive effect of *Azotobacter chroococcum* on *Rhizoctonia solani* infestation of potatoes. Nether J Plant Pathol 90:127–132

Minuto A, Spadaro D, Garibaldi A, Gullino ML (2006) Control of soilborne pathogens of tomato using a commercial formulation of *Streptomyces griseoviridis* and solarization. Crop Protect 25:468–475

Momol MT, Olson SM, Funderburk JE, Stavisky J, Marois JJ (2004) Integrated management of tomato wilt on field-grown tomatoes. Plant Dis 88:882–890
Moore NY, Pegg KG, Bentley S, Smith LJ (1999a) Fusarium wilt of banana: global problems and perspectives. In: Molina AB, Mabdek NHN, Liew KW (eds) Banana fusarium wilt management: towards sustainable cultivation. Proceedings of the international workshop on banana Fusarium wilt disease, Malaysia, pp 11–30
Moore NY, Pegg KG, Smith LJ, Langdon PW, Bentley S, Smith MK (1999b) Fusarium wilt of banana in Australia. In: Molina AB, Masdek NHN, Liew KW (eds) Banana fusarium wilt management: towards sustainable cultivation. Proceedings of the international workshop on banana Fusarium wilt disease, Malaysia, pp 64–75
Moss WP, Byrne JM, Campbell HL, Ji P, Bonas U, Jones JB, Wilson M (2007) Biological control of bacterial spot of tomato using *hrp* mutants of *Xanthomonas campestris* pv. *vesicatoria*. Biol Control 41:199–206
Murphy JF, Eubanks MD, Masiri J (2008) Reflective plastic mulch but not a resistance-inducing treatment reduced *Watermelon mosaic virus* incidence and yield losses in squash. Int J Veg Sci 15:3–12
Narisawa K, Ohki T, Hashiba T (2000) Suppression of clubroot and Verticillium yellows in Chinese cabbage in the field by the root endophytic fungus *Heteroconium chaetospira*. Plant Pathol 49:141–146
Narisawa K, Usuki F, Hashiba T (2004) Control of Verticillium yellows in Chinese cabbage by the dark septate endophytic fungus LtVB3. Phytopathology 94:412–418
Neher OT, Johnston MR, Zidack NK, Jacobsen BJ (2009) Evaluation of *Bacillus mycoides* isolate BmJ and *B. mojavensis* isolate 203-7 for the control of anthracnose of cucurbits caused by *Glomerella cingulata* var. *orbiculare*. Biol Control 48:140–146
Nel B, Steinberg C, Labuschagne N, Viljoen A (2006) The potential of nonpathogenic *Fusarium oxysporum* and other biological control organisms for suppressing Fusarium wilt of banana. Plant Pathol 55:217–223
Nielssen TH, Thrane C, Christophersen C, Anthoni U, Sørensen J (2000) Structure, production characteristics and fungal antagonism of tensin – a new antifungal cyclic lipopeptide from *Pseudomonas fluorescens* strain 96.578. J Appl Microbiol 89:992–1001
Njoroge SMC, Kabir Z, Martin FN, Koike ST, Subbarao KV (2009) Comparison of crop rotation for Verticillium wilt management and effect on *Pythium* species in conventional and organic strawberry production. Plant Dis 93:519–527
Norelli JL, Jones AL, Aldwinckle HS (2003) Fire blight management in the twenty-first century. Plant Dis 87:756–765
Ojaghian MR (2011) Potential of *Trichoderma* spp. and *Talaromyces flavus* for biological control of potato stem rot caused by *Sclerotinia sclerotiorum*. Phytoparasitica 39:185–193
Olivieri FP, Lobato MC, Altamiranda EG, Daleo GR, Huarte M, Guevara MG, Andreu AB (2009) BABA effect on the behavior of potato cultivars infected by *Phytophthora infestans* and *Fusarium solani*. Eur J Plant Pathol 123:47–56
Omar I, O'Neill TM, Rossall S (2006) Biological control of Fusarium crown and root rot of tomato with antagonistic bacteria and integrated control when combined with the fungicide carbendazim. Plant Pathol 55:92–99
Orober M, Siegrist J, Buchenauer H (2002) Mechanisms of phosphate-induced resistance in cucumber. Eur J Plant Pathol 108:345–353
Özaktan H, Bora T (2004) Biological control of fire blight in pear orchards with a formulation of *Pantoea agglomerans* strain Eh24. Braz J Microbiol 35:224–229
Özer N, Köycii ND (2006) The ability of plant compost leachates to control black mold (*Aspergillus niger*) and to induce the accumulation of antifungal compounds in onion following seed treatment. BioControl 51:229–243
Palma-Guerrero J, Jansson H-B, Salinas J, Lopez-Llorca LV (2008) Effect of chitosan on hyphal growth and spore germination of plant pathogenic and biocontrol fungi. J Appl Microbiol 104:541–553

Park K, Paul D, Kim E, Kloepper JW (2008) Hyaluronic acid of Streptococcus sp. as potent elicitor for induction of systemic resistance against plant diseases. World J Microbiol Biotechnol 24:1153–1158

Patricio FRA, Sinigaglia C, Barros BC, Freitas SS, Neto JT, Cantarella H, Ghini R (2006) Solarization and fungicides for the control of drop, bottom rot and weeds in lettuce. Crop Protect 25:31–38

Paulitz TC, Bélanger RR (2001) Biological control in greenhouse systems. Annu Rev Phytopathol 39:103–133

Percival GC, Noviss K, Haynes I (2009) Field evaluation of systemic resistance inducing chemicals at different growth stages for the control of apple (*Venturia inaequalis*) and pear (*Venturia pirina*) scab. Crop Protect 28:629–633

Pertot I, Zasso R, Amsalem L, Baldessari M, Angeli G, Elad Y (2008) Integrating biocontrol agents in powdery mildew control strategies in high tunnel growing systems. Crop Protect 27:622–631

Peters RD, Sturz AV, Carter MR, Sanderson B (2003) Developing disease-suppressive soils through crop rotation and tillage management practices. Soil Tillage Res 72:181–192

Peters RD, Sturz AV, Carter MR, Sanderson JB (2005) Crop rotation can confer resistance to potatoes from *Phytophthora erythroseptica*. Can J Plant Sci 85:523–528

Pinkerton JN, Ivors KL, Reeser PW, Bristow PR, Windom GE (2002) The use of soil solarization for the management of soilborne plant pathogens in strawberry and raspberry production. Plant Dis 86:645–651

Plaza P, Usall J, Smilanick JL, Lamarca N, Viñas I (2004) Combining *Pantoea agglomerans* (CPA-2) and curing treatments to control established infections of *Penicillium digitatum* on lemons. J Food Protect 67:781–786

Ponmurugan P, Baby U (2007) Evaluation of fungicides and biocontrol agents and microbial based biocontrol agents against Phomopsis canker of tea under field conditions. Australas Plant Pathol 31:68–72

Porras M, Barrau C, Arroyo FT, Santos B, Blanco C, Romero F (2007) Reduction of *Phytophthora cactorum* in strawberry fields by *Trichoderma* spp. and soil solarization. Plant Dis 91:142–146

Pradhanang PM, Ji P, Momol MT, Olson SM, Mayfield JL, Jones JB (2005) Application of acibenzolar-*S*-methyl enhances host resistance in tomato against *Ralstonia solanacearum*. Plant Dis 89:989–993

Press CM, Loper JE, Kloepper JW (2001) Role of iron in rhizosphere-mediated induced systemic resistance of cucumber. Phytopathology 91:593–598

Punja ZK, Rose S, Yip R (2002) Biological control of root diseases. In: Proceedings of the Canadian greenhouse conference, pp 1–5

Purcell AH (1990) Homopteran transmission of xylem-inhabiting bacteria. In: Harris KF (ed) Advances in disease vector research, vol 5. Springer, New York, pp 243–266

Pusey PL (2002) Biological control agents for fire blight of apple compared under conditions limiting natural dispersal. Plant Dis 86:639–644

Pusey PL, Stockwell VO, Rudell DR (2008) Antibiosis and acidification by *Pantoea agglomerans* strain E325 may contribute to suppression of *Erwinia amylovora*. Phytopathology 98:1136–1143

Raupach GS, Kloepper JW (2000) Biocontrol of cucumber diseases in the field by plant growth-promoting rhizobacteria with and without methyl bromide fumigation. Plant Dis 84:1073–1075

Reglinski T, Elmer PAG, Taylor JT, Parry EJ, Marsden R, Wood PN (2005) Suppression of Botrytis bunch rot in Chardonnay grapevines by induction of host resistance and fungal antagonism. Australas Plant Pathol 34:481–488

Reid TC, Hausbeck MK, Kizilkaya K (2002) Use of fungicides and biological controls in the suppression of Fusarium crown and root rot of asparagus under greenhouse and growth chamber conditions. Plant Dis 86:493–498

Rekanovic E, Milijasevic S, Todorovic B, Potocnik I (2007) Possibilities of biological and chemical control of Verticillium wilt in pepper. Phytoparasitica 35:436–441

Reuveni M, Zahavi T, Cohen Y (2001) Controlling downy mildew (*Plasmopara viticola*) in field-grown grapevine with ß-aminobutyric acid (BABA). Phytoparasitica 29:125–133

Riley DG, Pappu HR (2000) Evaluation of tactics for management of thrips-vectored *Tomato spotted wilt virus* in tomato. Plant Dis 84:847–852

Ristaino JB, Parra G, Campbell CL (1997) Suppression of Phytophthora blight in bell pepper by a no-till wheat cover crop. Phytopathology 97:242–249

Romanazzi G, Gabler FM, Smilanick JL (2006) Preharvest chitosan and postharvest UV irradiation treatments suppress gray mold of table grapes. Plant Dis 90:445–450

Romero D, de Vicente A, Zeriouh H, Cazorla FM, Fernández-Ortuño D, Tores JA, Pérez-García A (2007) Evaluation of biological control agents for managing cucurbit powdery mildew on greenhouse-grown melon. Plant Pathol 56:976–986

Rong RF, Feng SQ (2001) Effect of UV-C light irradiation on ripening and disease infection of postharvest tomato. J China Agric Univ 6:68–73

Rossetti V, Pompeu J, Rodriguez JO, Vechiato MH, da Veiga ML, Oliveira DA, Sobrinho JT (1980) Reaction of exocortis-infected and healthy trees to experimental *Phytophthora* inoculations. In: Proceedings of the 8th conference on IOCV, Riverside, pp 209–241

Ruiz-Ruiz S, Moreno P, Guerri J, Ambrós S (2009) Discrimination between mild and severe *Citrus tristeza virus* isolates with a rapid and highly specific real-time reverse transcription-polymerase chain reaction method using TaqMan LNA probes. Phytopathology 99:307–315

Sabuquillo P, De Cal A, Melgarejo P (2006) Biocontrol of tomato wilt by *Penicillium oxalicum* formulations in different crop conditions. Biol Control 37:256–265

Sabuquillo P, De Cal A, Melgarejo P (2010) Development of a dried *Penicillium oxalicum* conidial formulation for use as a biological agent against Fusarium wilt of tomato: selection of optimal additives and storage conditions for maintaining conidial viability. Biol Control 54:221–229

Sadeghi A, Hesan AR, Askari H, Qomi DN, Farsi M, Hervan EM (2009) Biocontrol of *Rhizoctonia solani*, damping-off of sugar beet with native *Streptomyces* strains under field conditions. Biocontrol Sci Technol 19:985–991

Sadfi N, Chérif M, Hajlaoui MR, Boudabbous A (2002) Biological control of the potato tubers dry rot caused by *Fusarium roseum* var. *sambucinum* under greenhouse, field and storage conditions using *Bacillus* spp. isolates. J Phytopathol 150:640–648

Sağir A, Eylen M, Pirinc V (2005) Effect of irrigation methods on pepper crown rot disease caused by *Phytophthora capsici*. Int J Agric Biol 7:804–806

Saito T, Sugiyama K (2005) Pathogenicity of three Japanese strains of entomopathogenic fungi against the silverleaf whitefly, *Bemisia tabaci*. Appl Entomol Zool 40:169–172

Sakuma F, Maeda M, Takahashi M, Kondo N (2011) Suppression of common scab of potato caused by *Streptomyces turgidiscabies* using lopsided oat green manure. Plant Dis 95:1124–1130

Salman M, Abuamsha R (2012) Potential for integrated biological and chemical control of damping-off disease caused by Pythium ultimum in tomato. BioControl 57:711–718

Sanzani SM, Schena L, De Sicco V, Ippolito A (2012) Early detection of *Botrytis cinerea* latent infections, as a tool to improve postharvest quality of table grapes. Postharvest Biol Technol 68:64–71

Saravanakumar D, Vijayakumar C, Kumar N, Samiyappan R (2007) PGPR-induced defense responses in the tea plant against blister blight disease. Crop Protect 26:556–565

Schisler DA, Slininger PJ, Kleinkopf G, Bothast RJ, Ostrowski RC (2000) Biological control of Fusarium dry rot of potato tubers under commercial storage conditions. Am J Potato Res 77:29–40

Schisler DA, Slininger PJ, Miller JS, Woodell LK, Clayson S, Olsen N (2009) Bacterial antagonists, zoospore inoculum retention time and potato cultivar influence pink rot disease development. Am J Potato Res 86:102–111

Schmid A, Daniel C, Weibel F (2005) Effect of cultural methods on leaf spot (*Mycosphaerella fragariae*) and gray mold (*Botrytis cinerea*) damage in strawberries. BioControl 50:179–194

Schuerger AC, Hammer W (2003) Suppression of powdery mildew on greenhouse-grown cucumber by addition of silicon to hydroponic nutrient solution is inhibited at high temperature. Plant Dis 87:177–185

Sengonca CM, Thungrabeab M, Blaeser P (2006) Potential of different isolates of entomopathogenic fungi from Thailand as biological control agent against western flower thrips *Frankliniellla occidentalis* (Pergande) (Thysanoptera-Thirpidae). J Plant Dis Protect 113:74–80

Seyb AM (2004) *Botrytis cinerea* inoculum sources in the vineyard system. Doctoral thesis, Lincoln University, Lincoln

Shiomi HF, Silva HAS, de Melo IS, Nunes FV, Bettiol W (2006) Bioprospecting endophytic bacteria for biological control of coffee leaf rust. Sci Agric Braz 63:32–39

Shorten PR, Elmer PAG, Soboleva TK (2003) A dynamic model to describe the interaction between *Botrytis cinerea* and the biological control agent, *Ulocladium oudemansii*. In: 8th international plant pathology congress, Christchurch, p 23

Shtienberg D, Elad Y, Bornstein Z, Ziv G, Grava A, Cohen S (2010) Polyethylene mulch modifies greenhouse microclimate and reduces infection of *Phytophthora infestans* in tomato and *Pseudoperonospora cubensis* in cucumber. Phytopathology 100:97–104

Sieburth PJ, Nolan KG, Hilf ME, Lee RF, Moreno P, Garnsey SM (2005) Discrimination of stem-pitting from other isolates of *Citrus tristeza virus*. In: Proceedings of the 16th conference, IOCV, Riverside, pp 1–10

Slininger PJ, Schisler DA, Ericsson LD, Brandt TL, Frazier MJ, Woodell LK, Olsen NL, Kleinkopf GE (2007) Biological control of postharvest late blight of potatoes. Biocontrol Sci Technol 17:647–663

Slininger PJ, Dunlap CA, Schisler DA (2010) Polysaccharide production benefits dry storage survival of the biocontrol agent *Pseudomonas fluorescens* S11:P:12 effective against several maladies of stored potatoes. Biocontrol Sci Technol 20:227–244

Slusarski C, Pietr SJ (2009) Combined application of dazomet and *Trichoderma asperellum* as an efficient alternative to methyl bromide in controlling the soilborne disease complex of bell pepper. Crop Protect 28:668–674

Someya N, Tsuchiya K, Yoshida T, Tsujimoto-Noguchi M, Sawada H (2007) Combined application of *Pseudomonas fluorescens* strain LRB3W1 with a low dosage of benomyl for control of cabbage yellows caused by *Fusarium oxysporum* f.sp. *conglutinans*. Biocontrol Sci Technol 17:21–31

Sosa-Alvarez M, Madden L, Ellis MA (1995) Effects of temperature and wetness duration on sporulation of *Botrytis cinerea* on strawberry leaf residues. Plant Dis 79:609–615

Soylu EM, Soylu S, Baysal Ö (2003) Induction of disease resistance and antioxidant enzymes by acibenzolar-*S*-methyl against bacterial canker (*Clavibacter michiganensis* subsp. *michiganensis* in tomato. J Plant Pathol 85:175–181

Soylu EM, Soylu S, Kurt S (2006) Antimicrobial activities of the essential oils of various plants against tomato late blight disease agent *Phytophthora infestans*. Mycopathologia 161:119–128

Sriram S, Roopa KP, Savitha MT (2011) Extended shelf-life of liquid fermentation derived talc formulations of *Trichoderma harzianum* with the addition of glycerol in the production medium. Crop Protect 30:1334–1339

Srivastava R, Khalid A, Singh US, Sharma AK (2010) Evaluation of arbuscular mycorrhizal fungus, fluorescent *Pseudomonas* and *Trichoderma harzianum* formulations against *Fusarium oxysporum* f.sp. *lycopersici* for the management of tomato wilt. Biol Control 53:24–31

Stavisky J, Funderburk J, Broadbeck BV, Olson SM, Anderson CP (2002) Population dynamics of *Frankliniella* spp. and tomato spotted wilt incidence as influenced by cultural management tactics in tomato. J Econ Entomol 95:1216–1221

Strashnov Y, Elad Y, Sivan A, Rudich Y, Chet I (1985) Control of Rhizoctonia solani fruit rot of tomatoes by *Trichoderma harzianum* Rifai. Crop Protect 4:359–364

Steddom KC, Menge JA (1998) Continuous application of the biocontrol bacterium, *Pseudomonas putida* 06909, increases soil populations over a single application. Phytopathology 88:S85 (Abst.)

Steddom KC, Menge JA (1999) Continuous application of the biocontrol bacterium, *Pseudomonas putida* 06909 improves biocontrol of *Phytophthora parasitica* on citrus. Phytopathology 89:S75 (Abst.)

Steddom KC, Menge JA (2001) Evaluation of continuous application technology for delivery of the biocontrol agent *Pseudomonas putida* 06909-rif/nal. Plant Dis 85:387–392

Steddom KC, Becker O, Menge JA (2002a) Repetitive applications of the biocontrol agent *Pseudomonas putida* 06909-rif/nal and effects on populations of *Phytophthora parasitica* in citrus orchards. Phytopathology 92:857–862

Steddom KC, Menge JA, Crowley D, Borneman J (2002b) Effect of repetitive applications of the biocontrol bacterium *Pseudomonas putida* 06909 rif/nal on citrus soil microbial communities. Phytopathology 92:857–862

Stevens C, Lui J, Khan VA, Lu JY, Wilson CL, Igwegbe ECK, Kabwe MK, Chalutz E, Droby S (1998) Application of hermetic UV-C for delayed ripening and reduction of Rhizopus soft rot in tomatoes: the effect of tomatine on storage rot development. J Phytopathol 146:211–221

Stockwell VO, Johnson KB, Sugar D, Loper JE (2002) Antibiosis contributes to biological control of fire blight by *Pantoea agglomerans* strain Eh252 in orchards. Phytopathology 92:1202–1209

Suárez-Estrella F, Vargas-Garcia C, Lopez MJ, Capel C, Moreno J (2007) Antagonistic activity of bacteria and fungi from horticultural compost against *Fusarium oxysporum* f.sp. *melonis*. Crop Protect 26:46–53

Subbarao KV, Koike ST, Hubbard JC (1996) Effects of deep plowing on the distribution and density of *Sclerotinia minor* sclerotia and lettuce drop incidence. Plant Dis 80:28–33

Subbarao KV, Hubbard JC, Schulbach KF (1997) Comparison of lettuce diseases and yield under subsurface drip and furrow irrigation. Phytopathology 87:877–883

Subbarao KV, Kabir Z, Martin FN, Koike ST (2007) Management of soilborne diseases in strawberry using vegetable rotations. Plant Dis 91:964–972

Summers CG, Mitchell JP, Stapleton JJ (2004) Management of aphid-borne viruses and *Bemisia argentifolii* (Homoptera: Aleyrodidae) in zucchini squash using UV reflective plastic and wheat straw mulches. Environ Entomol 33:1447–1457

Sundaramoorthy S, Raguchander T, Ragupathi N, Samiyappan R (2012) Combinatorial effect of endophytic and plant growth-promoting rhizobacteria against wilt disease of *Capsicum annuum* L. caused by *Fusarium solani*. Biol Control 60:59–67

Sundheim L, Amundsen T (1982) Fungicide tolerance in the hyperparasite *Ampelomyces quisqualis* and integrated control of cucumber powdery mildew. Acta Agric Scand 32:349–355

Sundheim L, Krekling T (1982) Host-parasite relationships of the hyperparasite *Ampelomyces quisqualis* and its powdery mildew host *Sphaerotheca fuliginea*. Phytopathol Z 104:202–210

Sundin GW, Werner NA, Yoder KS, Aldwinckle HS (2009) Field evaluation of biological control of fire blight in the eastern United States. Plant Dis 93:386–394

Szentiványi O, Kiss L (2003) Overwintering of *Ampelomyces* mycoparasites on apple trees and other plants infected with powdery mildews. Plant Pathol 52:737–746

Szezech M, Shoda M (2006) The effect of mode of application of *Bacillus subtilis* RB14-C on its efficacy as a biocontrol agent against *Rhizoctonia solani*. J Phytopathol 154:370–377

Tamietti G, Valentino D (2006) Soil solarization as an ecological method for the control of Fusarium wilt of melon in Italy. Crop Protect 25:389–397

Teixidó N, Usall J, Palou L, Asensio A, Nunes C, Viñas I (2001) Improving control of green and blue molds of oranges by combining *Pantoea agglomerans* (CPA-2) and sodium bicarbonate. Eur J Plant Pathol 107:685–694

Thirumalachar MJ, O'Brien MJ (1977) Suppression of charcoal rot in potato with a bacterial antagonist. Plant Dis Rep 61:543–546

Thomas TP, Kunta M, da Graça JV, Sétamou M, Skaria M (2010) Suppression of *Phytophthora* infection in citrus infected with viroids. Hortscience 45:1069–1072

Thuerig B, Binder A, Boller T, Guyer T, Jiménez S, Rentsch C, Tamm L (2006) An aqueous extract of the dry mycelium of *Penicillium chrysogenum* induces resistance in several crops under controlled and field conditions. Eur J Plant Pathol 114:185–197

Ting Z, Northover J, Schneider KE, Lu XW (2002) Interactions between *Pseudomonas syringae* MA-4 and cypridinil in the control of blue and gray mold of apples. Can J Plant Pathol 24:154–161

Ting ASY, Fang MT, Tee CS (2009) Assessment on the effect of formulative materials on the viability and efficacy of *Serratia marcescens* – a biocontrol agent against *Fusarium oxysporum* f.sp.*cubense* race 4. Am J Agric Biol Sci 4:283–288

Torres T, Nunes C, García JM, Abadias M, Viñas I, Manso T, Olmo M, Usall J (2007) Application of *Pantoea agglomerans* CPA-2, in combination with heated sodium bicarbonate solutions to control the major postharvest diseases affecting citrus fruit at several Mediterranean locations. Eur J Plant Pathol 118:73–83

Tränkner A (1992) Use of agricultural and municipal organic wastes to develop suppressiveness to plant pathogens. In: Tjamos EC, Papavizas GC, Cook RJ (eds) Biological control of plant diseases. Plenum Press, New York, pp 35–42

Turchek WW, Peres NA (2009) Heat treatment effects on strawberry plant survival and angular leaf spot caused by *Xanthomonas fragariae* in nursery production. Plant Dis 93:299–308

Turney JK, Yang CH, Cooksey DA, Menge JA (1994) Mechanisms of biocontrol of *Phytophthora citrophthora* by *Pseudomonas putida*. Phytopathology 84:1091 (Abst.)

Uppal AK, El Hadrami A, Adam LR, Tenuta M, Daayf F (2008) Biological control of potato Verticillium wilt under controlled and field conditions using selected bacterial antagonists and plant extracts. Biol Control 44:90–100

Utkhede RS, Koch C (2004) Biological treatments to control bacterial canker of greenhouse tomatoes. BioControl 49:305–313

Utkhede RS, Mathur S (2006) Preventive and curative biological treatments for control of *Botrytis cinerea* stem canker of greenhouse tomatoes. BioControl 51:363–373

van den Boogert PHJF, Luttikholt AJG (2004) Compatible biological and chemical control systems for *Rhizoctonia solani* in potato. Eur J Plant Pathol 110:111–118

Van der Schaaf DA, Ravenberg WJ, Malais M (1991) *Verticillium lecanii* as a microbial insecticide against whitefly. IOBC/WPRS Bull 14:120–123

Vanneste JL, Cornish DA, Yu J, Voyle MD (2002) P10c: a new biological control agent for control of fire blight which can be sprayed or distributed using honey bees. Acta Horticult 590:231–236

Velvis H, Jager G (1983) Biological control of *Rhizoctonia solani* on potatoes by antagonists: preliminary experiments with *Verticillium biguttatum*, a sclerotium-inhabiting fungus. Neth J Plant Pathol 89:113–123

Verhaar MA, Hijwegen T (1994) Biological control of cucumber powdery mildew by mycoparasites. In: Strink PC, Vredenberg WJ, Renkema JA, Parlevliet JE (eds) Plant production on the threshold of a new century. Kluwer Academic Publishers, Dordrecht, pp 373–374

Vermeire M-L, Kablan L, Dorel M, Delvaux B, Risede J-M, Legréve A (2011) Protective role of silicon in the banana-*Cylindirocladium spathiphylli* pathosystem. Eur J Plant Pathol 131:621–630

Vero S, Garmendia G, González B, Garat F, Wisniewski M (2009) *Aureobasidium pullulans* as a biocontrol agent of postharvest pathogens of apples in Uruguay. Biocontrol Sci Technol 19:1033–1049

Vestergaard S, Butt TM, Gillespie AT, Schreiter G, Eilenberg J (1995) Pathogenicity of the hyphomycete fungi *Verticillium lecanii* and *Metarhizium aniopliae* to the western flower thrips *Frankliniella occidentalis*. Biocontrol Sci Technol 5:185–192

Viljoen A (2002) The status of Fusarium wilt (Panama disease) of banana in South Africa. S Afr J Sci 38:341–344

Wang Y, Xu Z, Zhu P, Liu Y, Zhang Z, Mastuda Y, Toyoda H, Xu L (2010a) Postharvest biological control of melon pathogens using *Bacillus subtilis* EXWB1. J Plant Pathol 92:645–652

Wang Y, Yu T, Xia J, Yu D, Wang J, Zheng X (2010b) Biocontrol of postharvest gray mold of cherry tomatoes with marine yeast *Rhodosporidium paludigenum*. Biol Control 53:178–182

Wen A, Balogh B, Momol MT, Olson SM, Jones JB (2009) Management of bacterial spot of tomato with phosphorus acid salts. Crop Protect 28:859–863

Wiggins BE, Kinkel LI (2005) Green manures and crop sequences influence potato diseases and pathogen inhibitory activity of indigenous streptomycete. Phytopathology 95:178–185
Wilson M, Campbell HL, Ji P, Jones JB, Cuppels DA (2002) Biological control of bacterial speck of tomato under field conditions at several locations in North America. Phytopathology 92:1284–1292
Yan Z, Reddy MS, Ryu C-M, McInroy JA, Wilson M, Kloepper JW (2002) Induced systemic protection against tomato late blight elicited by plant growth-promoting bacteria. Phytopathology 92:1329–1333
Yang DM, Bi Y, Chen XR, Ge YH, Zho J (2006) Biological control of postharvest diseases with *Bacillus subtilis* (B1 strain) on musk melon (*Cucumis melo* cv. Yindi). Acta Horticult 712:735–740
Yildiz A, Benlioğlu S, Boz Ö, Benlioğlu K (2010) Use of different plastics for soil solarization in strawberry growth and time-temperature relationships for the control of *Macrophomina phaseolina* and weeds. Phytoparasitica 38:463–473
Yoder KS, Miller SS, Byers RE (1999) Suppression of fire blight in apple shoots by prohexadione-calcium following experimental and natural inoculation. HortScience 34:1202–1204
Youssef K, Ligorio A, Sanzani SM, Nigro F, Ippolito A (2012) Control of storage diseases of citrus by pre- and postharvest application of salts. Postharvest Biol Technol 72:57–63
Yu T, Zhang H, Li X, Zheng X (2008) Biocontrol of *Botrytis cinerea* in apple fruit by *Cryptococcus laurentii* and indole-3-acetic acid. Biol Control 46:171–177
Yu T, Yu C, Lu H, Zunum M, Chen F, Zhou T, Sheng K, Zheng X (2012) Effect of *Cryptoccus laurentii* and calcium chloride on control of *Penicillium expansum* and *Botrytis cinerea*. Biol Control 61:169–175
Zamani M, Tehrani AS, Ahmadzadeh M, Hosseininaveh V, Mostofy Y (2009) Control of *Penicillium digitatum* on orange fruit combining *Pantoea agglomerans* with hot sodium bicarbonate dipping. J Plant Pathol 91:437–442
Zhang J, Dou H (2002) Evaluation of *Bacillus subtilis* as potential biocontrol agent for postharvest green mold control on 'Valencia' orange. Proc Fla State Hortic Soc 115:60–64
Zhang H-Y, Zheng X-D, Xi Y-F (2005) Biological control of postharvest blue mold of oranges by *Cryptococcus laurentii* (Kufferath) Skinner. BioControl 50:331–342
Zhou XG, Everts KL (2004) Suppression of Fusarium wilt of watermelon by soil amendment with hairy vetch. Plant Dis 88:1357–1365
Zhou XG, Everts KL (2012) Anthracnose and gummy stem blight are reduced on watermelon grown on a no-till hairy vetch cover crop. Plant Dis 96:431–436
Zhou CY, Hailstones DL, Broadbent P, Connor R, Bowyer J (2002) Studies on mild strain cross-protection against stem-pitting *Citrus tristeza virus*. In: Fifteenth IOCV conference, pp 151–157

Additional References for Further Reading

Snowdon AL (1992) A color atlas of postharvest diseases and disorders of fruits and vegetables, vol 2, Vegetables. Wolfe Publishing, London
Wakil W, Ghazanfar MU, Kwon YJ, Shamas-ul-Islam AK (2012) Toxicity of *Paecilomyces lilacinus* blended with non-conventional agents to control cotton thrips (*Thirps tabaci* Lind.) (Insecta: Thysanoptera: Thripidae). Afr J Microbiol Res 6:526–533.

Index

P. Narayanasamy, *Biological Management of Diseases of Crops*,
Progress in Biological Control 16, DOI 10.1007/978-94-007-6377-7,

L

If you have any concerns about our products,
you can contact us on
ProductSafety@springernature.com

In case Publisher is established outside the EU,
the EU authorized representative is:
Springer Nature Customer Service Center GmbH
Europaplatz 3, 69115 Heidelberg, Germany

Printed by Libri Plureos GmbH
in Hamburg, Germany